Studies in Applied Philosophy, Epistemology and Rational Ethics

Volume 60

Studies in Applied Philosophy, Epistemology and Rational Ethics (SAPERE) publishes new developments and advances in all the fields of philosophy, epistemology, and ethics, bringing them together with a cluster of scientific disciplines and technological outcomes: ranging from computer science to life sciences, from economics, law, and education to engineering, logic, and mathematics, from medicine to physics, human sciences, and politics. The series aims at covering all the challenging philosophical and ethical themes of contemporary society, making them appropriately applicable to contemporary theoretical and practical problems, impasses, controversies, and conflicts. Our scientific and technological era has offered "new" topics to all areas of philosophy and ethics – for instance concerning scientific rationality, creativity, human and artificial intelligence, social and folk epistemology, ordinary reasoning, cognitive niches and cultural evolution, ecological crisis, ecologically situated rationality, consciousness, freedom and responsibility, human identity and uniqueness, cooperation, altruism, intersubjectivity and empathy, spirituality, violence. The impact of such topics has been mainly undermined by contemporary cultural settings, whereas they should increase the demand of interdisciplinary applied knowledge and fresh and original understanding. In turn, traditional philosophical and ethical themes have been profoundly affected and transformed as well: they should be further examined as embedded and applied within their scientific and technological environments so to update their received and often old-fashioned disciplinary treatment and appeal. Applying philosophy individuates therefore a new research commitment for the 21st century, focused on the main problems of recent methodological, logical, epistemological, and cognitive aspects of modeling activities employed both in intellectual and scientific discovery, and in technological innovation, including the computational tools intertwined with such practices, to understand them in a wide and integrated perspective. **Studies in Applied Philosophy, Epistemology and Rational Ethics** means to demonstrate the contemporary practical relevance of this novel philosophical approach and thus to provide a home for monographs, lecture notes, selected contributions from specialized conferences and workshops as well as selected Ph.D. theses. The series welcomes contributions from philosophers as well as from scientists, engineers, and intellectuals interested in showing how applying philosophy can increase knowledge about our current world. Initial proposals can be sent to the Editor-in-Chief, Prof. Lorenzo Magnani, lmagnani@unipv.it:

- A short synopsis of the work or the introduction chapter
- The proposed Table of Contents
- The CV of the lead author(s).

For more information, please contact the Editor-in-Chief at lmagnani@unipv.it.

Indexed by SCOPUS, zbMATH, SCImago, DBLP.

More information about this series at http://www.springer.com/series/10087

Joseph E. Brenner · Abir U. Igamberdiev

Philosophy in Reality

A New Book of Changes

Springer

Joseph E. Brenner
International Center for the Philosophy of Information
Xi'An Jiaotong (Social Sciences) University
Xi'An, China

Abir U. Igamberdiev
Department of Biology
Memorial University of Newfoundland
St. John's, NL, Canada

ISSN 2192-6255 ISSN 2192-6263 (electronic)
Studies in Applied Philosophy, Epistemology and Rational Ethics
ISBN 978-3-030-62756-0 ISBN 978-3-030-62757-7 (eBook)
https://doi.org/10.1007/978-3-030-62757-7

This Springer imprint is published by the registered company Springer Nature Switzerland AG
The registered company address is: Gewerbestrasse 11, 6330 Cham, Switzerland

Foreword by Lorenzo Magnani

First of all, I would like to say that the book illustrates the passion for philosophy, logic, and science of the authors and their ability to cause interest and admiration into whoever listens to their telling the beautiful philosophical and scientific narratives regarding stability and change, being and becoming. The book you are holding in your hands is the latest fruit of the authors' impeccable passion. The title they called it by, *Philosophy in Reality:* A *New Book of Changes*, tells a lot about their intent. The book aims at a new synthesis of science and philosophy covering all aspects of real change: (1) application of a new logic and dialectics to the areas of communication, information, and meaning, (2) application of the new logical-philosophical approach to the standard domains of philosophy, (3) the emphasis on the emerging new role for systems theory in philosophy, (4) the question of principles and their utility, (5) the dynamics of social change and the common good.

When you give someone a tool, or when you explain to someone that a thing is already a tool for some particular purpose, you are not just increasing her capabilities, but you are often hoping that she will find new amazement in her increased capabilities. By offering a toolbox to a child, your objective is not only that he will fix small things, and tear apart others, but also at the wonder he will experience in all that. As I will show in the next few paragraphs, Joseph E. Brenner and Abir U. Igamberdiev entrust the reader with a similar gift, consisting in a most comprehensive and accurate journey through plenty of philosophical, scientific, logical, and epistemological disciplines and traditions, ranging from an intelligent review of the intellectual milestones regarding change and stability to the heights of their own research, to which both experts and novices will be able to relate, and from which both can benefit.

In this book, a starting general philosophical point is soon revealed: In the first pages, the authors illustrate the centrality of the recovery of dialectics and semiotics from reductionist interpretations and promise their application in a new synthetic paradigm for science and philosophy, and at the same time admit, they derived further fresh inspiration from the ancient *Chinese Book of Changes* (*I Ching*), which contains a copious scientific-philosophical discussion of the so-called real change. The book consists of three parts, Part I Theory and Science; Part II Toward a New Natural Philosophy; and Part III The Philosophy of Structures and Systems. The approach also aims at addressing the question of the Unity of Knowledge in the ambitious perspective indicated in the sub-title: *A New Book of Changes*.

This text is both difficult and rewarding, affording a new perspective on philosophy *in* reality, interpreted, as I said, in terms of change and stability, being and becoming. The system of the authors is based on the logical philosophy of the twentieth Century Franco-Romanian thinker Stéphane Lupasco, and its precursor was a book entitled, *Logic in Reality*, published in 2008. The treatment represents a modality of philosophically reflecting upon the various topics involved which certainly avoids the typical traditional concentration on purely analytic arguments. The authors present a new description of the dynamic, interactive structures of real processes, physical and cognitive, and their implications for philosophy: They approach complicated issues—through a process they call of *de-fragmentation* of knowledge—related to ontology, metaphysics, epistemology, logic, semiotics, physics, biology, chemistry, systems theory, dialectics, social and human sciences, information and communication theories and ecology, by describing and criticizing the various ways of proposing the characterization of reality at the same time establishing and deepening new and unsuspected relationships between concepts and ideas.

Chapter 4 importantly illustrates a core target of the book: We need go beyond the mere abstract and formal aspects of logical analysis, and offer, also following the teaching of Stéphane Lupasco, a new architecture of logic that sees it as applied not only to the "reasoning processes," but also as directly concerned with various entities, events, and phenomena of a general kind. The new logic is "a logic extended to real systems and processes" that the authors fruitful expand and enhance, in the important Chap. 15, thanks to the analysis of recent developments of system theory. One of the aims of the book is to show and elaborate such a constructive way of thinking, that offers a new logical perspective on the structure of reality in general, together with the detection of how philosophy is implicated "in" reality, also taking advantage of the suggestions derived from several types of models provided by the most recent advance of natural and human sciences.

It is in the framework I have just described that the mainstream logical tradition can be emancipated, abandoning merely formal and abstract accounts, constrained by linguistic concepts, showing that by rendering logic a constructive conceptual tool, it can be seen not only as a logic "of" reality but also "in" that reality which is intrinsically governed by several of essential dualities (observer and observed, self and not-self, internal and external, etc.): Indeed the examination and the conceptual exploitation of "contradictions" play a dominant role in the book. It is in this perspective that Brenner and Igamberdiev clearly show that logic is also embedded in physical reality, as explained by physics. Logic is not only a conceptual tool about theories of reality such as the ones presented by as philosophy and metaphysics, but also a scientific concept, because its principles are a reflection of the underlying physical structure of the universe: "The opposing movements in every process from actual to potential, and potential to actual, are the origins, as operators, of structures and systems. For Lupasco, such a process is a *dialectogenesis*, whose logic is, also, a dialectology. His dialectomethodology [...] is operational in the sense that to (a) know and (b) know that one knows, one must act dialectically, operate (Lupasco's word) or actualize cognitively dialectical actualizations and potentializations and grasp their inherent antagonisms, their homogeneous and heterogeneous aspects and so on."

The whole structure of Brenner and Igamberdiev's book is about the exploitation of this new interpretative perspective of/in reality as a way of extracting cognitive suggestions from, and providing criticism to, the standard tradition of logics (Chaps. 3–5), as well as other disciplines: physics, mathematics, and chemistry (Chaps. 6–8), semiotics, (Chap. 9); metaphysics, metaphilosophy, and phenomenology (Chap. 10), information, and communication sciences (Chaps. 11 and 12), philosophy, natural philosophy and the so-called anti-philosophy (Chaps. 13–15), complex systems, living, social, and ecological systems (Chaps. 16–19). The treatment touches and deepens various topics that especially urge re-examinations and new clarifications, ranging from the pivotal role of axiomatics and the relevance of building new open and uninterpreted formal systems more able to account for real processes, entities, properties, and dynamics to the reinterpretation of formal ontology as a "process ontology," from the conceptual status of quantum theories to the role of "new energy ontologies", from the problems related to René Thom's catastrophe theory to metaphysics and the naturalization of phenomenology.

I have also to draw the attention of the reader that Brenner and Igamberdiev's book is strongly interdisciplinary: Indeed, as I have already said, their way of theorizing departs from the obsession with specific issues typical of the analytic tradition. Indeed, this tradition, beyond its obvious merits, often penalizes the primary philosophical ambition—instead central in this book—to produce new levels of intelligibility of the world, grounding them on the available most important achievements of current culture and science.

The authors deploy a gallery of intellectual subjects that must undergo a systematic work of compelling re-examination. Let us remember some of the major themes that dominate the text, which includes several cases of successful logical and philosophical modeling. Great parts of the book deal with the criticism of many non-classical logics (from the paraconsistent ones to the abductive and quantum logic), the epistemological character of abduction and of model-based reasoning, quantum and relativistic physics, ancient, Kantian and Hegelian philosophy, central aspects of authors such as Marx, Peirce, Husserl, Heidegger, Gödel, Whitehead, and Popper, and traditionally debated philosophical problems such as the analytic/synthetic distinction, determinism/indeterminism, and the concepts of "causality" and of "scientific explanation."

Brenner and Igamberdiev's text is really complex; an enormous quantity of information and knowledge about the widest range of issues relevant to their concern is present. This fact could present one of the major problems for the reader: Since their arguments are based on numerous criticisms of cognitive strategies belonging to various disciplines and areas of thought, the analysis of which feeds the architecture of the "philosophy in reality" project, the abundance of references present in the book is really remarkable. Of course, for the reader who is suitably grounded in the huge literature that Brenner and Igamberdiev afford, the reference to specific topics is surely not only profoundly illuminating because of the original interpretation provided, but also directly informative of the *status quaestionis* at hand. The reader who is less acquainted with the various areas expressed by the texts criticized must fruitfully rely on the cases she already knows, to get extremely informative and epistemologically rewarding insights.

Brenner and Igamberdiev open up an exceptionally innovative prospect for the analysis of the classic philosophical problem of *reality* by integrating its relational and dynamic features within the entire of recent culture, science, and philosophy, rather than by restricting the treatment to a one-sided perspective. Indeed, this brilliant and challenging book also teaches philosophers it is important to do more than abstractly analyze concepts, requiring that they become acquainted with the complexity of current knowledge assembling cognitive practices offered by special disciplines.

But now, enough said! When you offer someone a tool, you do not want to overly describe it, but rather let her get her hands on it. It is time to let the reader and Brenner and Igamberdiev begin their journey through "philosophy in reality" as a "tool for reason."

May 2020

Lorenzo Magnani
Computational Philosophy Lab
Department of Humanities, Philosophy Section
University of Pavia
Pavia, Italy

Preface

The origin of this book lies in prior work on a *logic* of and in reality by Joseph Brenner. Its subsequent development and expansion is a consequence of the collaboration established in 2019 between the authors, Joseph E. Brenner and Abir U. Igamberdiev. Both have backgrounds in science (chemistry and biology, respectively), as well as logic and philosophy. A joint article, "Philosophy in Reality: Scientific Discovery and Logical Recovery" was published in 2019 in the journal *Philosophies*. Its emphasis was on the recovery of dialectics and semiotics from reductionist interpretations and their application in a new synthetic paradigm to science and philosophy. It was in fact intended as a *Prolegomenon* to the current book. The latter consists of three parts, Part I Theory and Science; Part II Toward a New Natural Philosophy; and Part III The Philosophy of Structures and Systems. The further inspiration for this book was the ancient Chinese *Book of Changes (I Ching)*, and the scientific-philosophical discussion of real change that is its major theme. In Part I, the relation between our concepts of logic and change and those of the *I Ching* are explored, and change is a thread that runs throughout the various areas of science and philosophy discussed subsequently in relation to Logic in Reality. Many references are also made to the work of Aristotle and other classical Western authors. Our book is not intended as a history of Western philosophy, but it contains many references to philosophers relevant to our perspective: (1) Descartes, Spinoza and Leibniz; (2) Kant, Hegel, and Schelling; (3) the neo-Kantians, and then to Husserl, Heidegger, and Levinas in the mid-twentieth century. The logical philosophy of the Franco-Romanian Stéphane Lupasco, on which our theory is based, dates from this period. Our theory addresses also areas of current interest at the interface of science and philosophy including mathematics, information and communication. In these areas and others, we have made our presentation as non-technical as possible, but some familiarity of the reader with issues in the philosophy of science has been assumed. Several less familiar but in our view essential disciplines within philosophy are included such as the philosophy of chemistry, anti-philosophy, and the philosophy of complex systems.

We also address key questions for philosophy such as the Unity of Knowledge and, in the social sphere, the Common Good. The same framework is applied to other problems of society such as its origins of society, the transformation of reality by human subjects, and the emergence of a global, sustainable information society. These new perspectives and references support research by both philosophers and physical and social scientists concerned with the many facets of reality.

All of the above subjects are for us essential parts of what a 'Philosophy in Reality' should be about. In the final chapter, on the basis of the potential implications of our theses for philosophy, science, and society, we conclude that the book deserves the sub-title of *A New Book of Changes.*

Les Diablerets, Switzerland — Joseph E. Brenner
St. John's and Winnipeg, Canada — Abir U. Igamberdiev

Acknowledgements

This book is the result of a first encounter between the authors in Vienna, in the context of the 2015 International Summit Conference on Information in Vienna, Austria. Subsequent encounters in the *agora* of the journal *Philosophies* led to the emergence, as noted, of a joint article in that journal as a *prolegomenon* to this book in winter 2019. We each have people to whom we are most indebted to for its appearance today; they will be listed separately for each of us.

Joseph E. Brenner:

I should first like to thank, for their on-going interest and encouragement, the people who were directly involved in its predecessor, *Logic in Reality*, published in 2008 also by Springer. We recall first the name of Basarab Nicolescu, Professor of Theoretical Physics at the University of Paris and co-founder, with Stéphane Lupasco, of the International Center for Transdisciplinary Research. Nicolescu is our link to Lupasco, the originator of the logical philosophy that is at the heart of this book. Also of importance are the Professors of philosophy John Symons and Johanna Seibt, at the Universities of Kansas and Aarhus, respectively, whom I thank for their invaluable criticism and support. We are also fortunate in having the involvement and support of Lorenzo Magnani, Editor-in-Chief of the SAPERE Book Series. Magnani wrote the foreword to both *Logic in Reality*, and this book, providing a continuity over twelve years which, if not unique, is certainly most welcome.

Next, I wish to thank my friend, the mathematical physicist and computer industry leader and scientist Robert Bishop, for his support and suggestions to the current effort. Bishop heads the International Center for Earth Simulation in Geneva, Switzerland. Its transdisciplinary perspective on the physical and social science of climate change is awaiting its inclusion in the philosophy of the future and vice versa.

Several people will be referred to frequently in the manuscript who I consider not only experts but friends. I consider their work as part of the most solid scientific and philosophical foundations possible for a philosophy in and of reality. In alphabetical order, they are

Mark Burgin: Professor of Mathematics, University of California, Los Angeles; Editor of *Information.*

Rafael Capurro: Professor (em.) Dr. Rafael Capurro, Stuttgart Media University, Germany, who has been my mentor in philosophy.

Wolfgang Hofkirchner: Ao. Univ. Prof.i.R. Dr.phil. Wolfgang Hofkirchner; Director, GSIS. The Institute for a Global Sustainable Information Society, Vienna, Austria.

Loet Leydesdorff: Professor (em.) University of Amsterdam, Amsterdam School of Communication Research; Guest Professor Zhejiang University, Hangzhou; Visiting Professor, Institute of Scientific and Technical Information of China, Beijing.

Pedro Marijuan: Head, Bioinformation Group, Aragon Institute of Health Sciences, Spain, and Founder of the Foundations of Information Science initiative, another mentor, together with Hofkirchner and Wu, in the field of information science and philosophy.

Wu Kun: Professor, Director of the International Center for the Philosophy of Information, Xi'An Jiaotong (Social Sciences) University, Xi'An, China

I would also like to thank Marcin Schroeder and Gordana Dodig-Crnkovic, Editors of *Philosophies*, who helped a chemist like me gain some understanding of the public role of philosophy and the contribution one can make a reviewer as well as author.

Abir U. Igamberdiev:

I should first like to thank the people who supported me in my interest in natural science and natural philosophy from the early childhood. My first mentors were my grandparents who were teachers in the local school in Central Russia. My grandfather Fyodor Dashkov expressed a deep interest in natural history and philosophy. His ideas on the nature of physical universe, life, and social processes significantly impacted my development. Among those who influenced and shaped my scientific and philosophical views, I would like to mention the late Professors Sergei Meyen, Lev Beloussov, Efim Liberman, and Vladimir Lefebvre. Communication with them helped to broaden my views on different aspects of the phenomena of life, complexity, evolution and sociogenesis.

I acknowledge those who helped me in preparation of my previous books, "*Physics and Logic of Life*" and "*Time, Life, and Civilization*," published in 2012 and 2015 by Nova Science Publishers. In these books, I summarized my views on the nature of physical world, life, and consciousness that were earlier developed in my papers in *Semiotica*, *BioSystems* and other journals. I realized that the semiotic approach in its static interpretation is insufficient for the description of reality that appears as a dynamic process. Meeting with Joseph Brenner revived my interest to dialectics in its realistic interpretation. I realized that the formulation of the new version of dialectical logic by Stéphane Lupasco and Joseph Brenner became an essential basis for developing real foundations of scientific thought and natural philosophy.

For my involvement in the development of natural philosophy, an important place belongs to my exciting journey with the journal *BioSystems* that started in 1993 with publication of my work on the quantum basis of biological processes. As a member of the editorial board of the journal and then editor-in-chief, I had the privilege of communication with many brilliant scientists who are involved in theoretical and philosophical comprehension of physics, chemistry, biology, and social sciences.

I would like to mention several people, communication with whom I consider the most fruitful for development of the scientific and philosophical foundations of *Philosophy in Reality*: Koichiro Matsuno (Nagaoka University of Technology, Japan), Yukio-Pegio Gunji (Waseda University, Tokyo, Japan), Pedro C. Marijuán (Aragon Health Sciences Institute, Zaragoza, Spain), Richard Gordon (Gulf Specimen Marine Laboratory, Panacea, FL, USA), Plamen L. Simeonov (Dahlem Centre for Genome Research, Berlin, Germany), Nikita Shklovskiy-Kordi (National Research Center for Hematology, Moscow, Russia), Sergei Petoukhov (Institute of Machines Studies, Moscow, Russia), and many others.

Contents

Part II Toward a New Natural Philosophy

Part III The Philosophy of Structures and Systems

Part I
Theory and Science

Chapter 1
Introduction

1.1 The 'Science' of Philosophy

Human knowledge has been categorized in several major fields of which those of concern here are philosophy and science. Subfields that concentrate notions of the relations between the two, as well as iterations of one, are now part of the literature. Thus, the philosophy of science is a major field within philosophy concerned with the foundations, methods and results of science and their implications for society. The philosophy of philosophy has also attracted some attention, but much more limited; a recent definitive text is that of Williamson [29]. The field is overlaps that of metaphilosophy, which has its own journal and benefits from substantial formal discussion.

A *science of philosophy* does not appear to exist as a separate concept, let alone field of study. What are the experiments, foundations, methods and implications of such a putative field? Are there features of philosophy which are so regularly found that they might in principle constitute a science? We are not sure. All attempts to reference 'science of philosophy' on the Internet yield entries for the philosophy of science. However, this state-of-affairs led us to the conclusion that the concept of a science of philosophy, if taken together with the ideas underlying the other four fields—philosophy of science, philosophy of philosophy and metaphilosophy—defined a philosophy of and in reality as whole.

Hooker, the Editor of and contributor to the important compendium 2011 Philosophy of Complex Systems has said that he came to philosophy after science, and that this has of course coloured his approach [15]. The authors of this book have followed a similar path: Joseph Brenner in organic chemistry and Abir Igamberdiev in biophysics and biochemistry. In our discussion here, we have maintained links between science and philosophy to bring out the philosophical implications of critical aspects of the individual sciences without losing sight of their specificity. The consequence is, however, that not all readers will have a direct experience of or 'feeling' for certain scientific concepts such as the transition state in chemistry or recursion in communication theory. We have accordingly made an effort to make the

J. E. Brenner and A. U. Igamberdiev, *Philosophy in Reality*, Studies in Applied Philosophy, Epistemology and Rational Ethics 60, https://doi.org/10.1007/978-3-030-62757-7_1

operative statements and conclusions of the different Chapters as non-technical as possible. As we will propose in the next Section, the links within and between the major topics in this book are *logical*, based on a non-binary, non-propositional logic, grounded in physics and operative at all levels of reality.

1.2 Reality: Process, Logic, Philosophy and Change

The origin of this book lies in the research of Joseph Brenner, starting in 2005, to make the philosophical logic of the Franco-Romanian thinker Stéphane Lupasco (Bucharest 1900—Paris 1988), published in French, available to English-language readers. This theory was intended by Lupasco to have as its major subject "On Logical Becoming and Affectivity; Toward a New Theory of Knowledge" [21]. However, as we will see, application of this theory requires a revision of all standard philosophical concepts of the role and operation of the real physical processes underlying phenomena. The continuity of the Lupasco logical system with philosophy was suggested by the Memorial Volume on Lupasco [25] and as reflected in the title of the first paper by Brenner "Process in Reality" [1], following the major work of the British philosopher Alfred North Whitehead "Process and Reality". In a book that is the precursor to this one, *Logic in Reality* [2], the concepts of Lupasco were up-dated and expanded and relations established to current science, philosophy and logic. In particular, as will be discussed below, a formalization of the Lupasco system as a non-propositional, non-truth-functional logic of processes was established. This work was carried by one of us (Brenner) in close collaboration with and with the support and encouragement of Basarab Nicolescu, emeritus professor of Theoretical Physics at the University of Paris (VI). Nicolescu was a close friend of Lupasco and continuator of his work. Their concept of Transdisciplinarity, exemplified in their foundation of the International Center for Transdisciplinary Research (CIRET) and publications by Nicolescu will be discussed in connection with current systems philosophy.

Part of the core thesis of *Logic in Reality* is a categorial principle of Non-Separability for all complex processes, including cognitive ones. This principle applies, in our view, to philosophy and the logical system we have adopted, subsequently designated as Logic in Reality (LIR). In the last two years, the two authors, separately and together, have explored in much greater detail the application of LIR to philosophy as such, especially the to the definition of natural philosophy [4, 5, 16]. As this work developed, it became more and more clear that one underlying concept underlying was a new approach to *change*. The key next Chapter of this book is thus entitled and devoted to "Change in Reality", and a demonstration of its roots in the classical Chinese *Book of Changes*.

We avoid sterile arguments about what reality is 'exactly', because we claim that such a question is otiose. Our logic is one observable in real phenomena to which the term exact applies only to simple physical structures. One can perfectly well use a common-sense notion of reality without being concerned about 'catastrophic' counterfactuals, that is, ones with a probability close to zero. We do not presume to offer a definition of reality that would be either provable in some way or acceptable to all or most people. We refer the interested reader to the book of Colin McGinn [22] on the 'basic structures' of reality.

Comprehension of a dual operation of the principle of contradiction together with one of non-contradiction and the abrogation of the 2nd and 3rd Aristotelian axioms requires a new concept both of reality and of levels of reality. It must accommodate viewpoints of ontology, metaphysics and physical science, especially, fundamental physics, but also biological science and cognitive science where complex, partly contradictory principles are operative. This first requires a distinction between reality and the real. Our vision of the real should be understood as similar to, or rather starting from, that of d'Espagnat [9], namely, weak objectivity. An independent real exists, despite our inability to define it precisely (the veiled real, or veiled reality) or to give it a meaning that is independent of our intuition. This view is related to the so-called weak anthropic principle that states the obvious fact that the fundamental constants of the universe are such that observers such as ourselves can exist. The strong anthropic principle claims that our existence is the explanation of *why* the constants have the values they do, introducing teleology—that it was the purpose of the universe to produce us. Kauffman [18] appears to suggest this:

> If we are, in ways we do not see, natural expressions of matter and energy coupled together in non-equilibrium systems, if life in its abundance were bound to arise, not as an incalculably improbable accident, but as an *expected* (emphasis ours) fulfillment of the natural order, then we are truly at home in the universe.

This statement can be read as a statement equivalent to the self-evident weak anthropic principle. In any event, the latter is sufficient for our analysis, and the appearance of teleology can be reinterpreted as a phenomenon, not a fundamental principle. There is a deep insight by Luhn which should be noted in this connection [20]. He sees the universe as *operating* to increase the number of future states available to *it*. We are small, conscious elements of this grand process.

Other theories define *reality* as what is, or what constitutes, our experiences, representations, theories, logics, images and so on. The real is what *is*, but what absolutely resists our attempts to represent it or to see it clearly [26]. The real offers us the possibility, so to speak, of understanding it via a concept of levels of reality and the structure and properties of phenomenal reality that we can perceive (non-veiled reality). Rescher [28] cites Peirce to this effect: "Given that human intelligence is a resource developed over time by creatures that are themselves an evolved part of nature, our capacity to understand the world should not be seen as all that surprising." The most useful approach may be to maintain a description of reality as being both veiled *and* not veiled.

Non-veiled reality is divided into the familiar domains of *observables* and *non-observables*. When to say that an entity is a non-observable because not *directly* observable, and that hence its reality and existence independently of the human mind is open to question, is a major issue of scientific realism. The logic of/in reality is relevant to both the philosophy and physics applicable, as we will discuss in later chapters.

A book that purports to discuss reality also requires some minimum discussion of the metaphysical problem of what it is to be or exist, the nature of being, and well as the meaning of "coming into being" or becoming. Without some indication of what constitutes being, what it means for something to exist, and why something exists rather than nothing, any description of the inhabitants of existence, as pointed out by Jacquette [17], may be inconsequential unless it is in some way independently constructed *without* reference to our experience.

Perhaps the least incorrect thing one can say is that reality is something like what one thinks it is. We insist only on the existence of the dialectic between reality and appearance and the operation of the mind moving from one to the other according to the Principle of Dynamic Opposition (Chap. 3). In this approach, anti-realist positions that claim all reality is a social construct are unnecessary to avoid reductionism.

Our philosophy of/in reality is not and is not intended to be a 'philosophy of everything'. There is no intent to dismiss transcendental philosophy or philosophy based on beliefs about the origin or a possible purpose of the universe. What characterizes such positions in general is their reference to an inaccessible *un*-reality, expressed by the first and last letters of the Greek alphabet, alpha and omega in the metaphor chosen by the French philosophers Teilhard de Chardin and Xavier Sallantin.

We will limit the scope of our Philosophy in Reality to what is also the content of *natural* philosophy as it is currently being redefined by Schroeder and Dodig-Crnkovic [8] as well as by both authors [5]. The potentially useful criterion which Brenner [4] has proposed for delimiting natural philosophy from philosophy as a whole is the Logic in Reality itself which underlies this study and will be the specific subject of Chap. 3. For the prolegomenon to an integral paradigm of natural philosophy see Igamberdiev [16]. We conclude that the most useful description of reality is, therefore, that it is something like what one thinks it is neither more nor less.

Starting from our respective backgrounds in science, we thus present a new description of the dynamic, interactive structures of real processes, physical and cognitive, and their implications for philosophy. As outlined in a *prolegomenon* [5] published in the journal *Philosophies*, the purpose of our project could be described as the *de-fragmentation* of knowledge. We attempt to achieve this result through the joint application of new conceptions of logic, dialectics, semiotics, mathematics, and systems theory. These are recovered from reductionist interpretations and reunited in a new synthetic paradigm centered on information, meaning and its communication. The weak point of previous synthetic and process philosophies as contributions to knowledge, both before and after the emergence of quantum mechanics, has been their lack of grounding in science. Important examples are Hegel, Whitehead, Heidegger and Rescher. Further origins of our approach can be found in Heraclitus, the *De Anima* of Aristotle; in the more recent but much less well-known work of the twentieth Century Russian thinkers Bogdanov, Ilyenkov, and Voevodsky, as well as in the philosophical logic of the Franco-Romanian Stéphane Lupasco mentioned above.

Further major current contributors to our conception of what philosophy is, or should be, are Rafael Capurro on the fundamentals of being and becoming, Loet Leydesdorff in the area of communications, Gianfranco Minati on post-standard systems theory, and Michel Maffesoli on social dynamics and post-modernism. Without being complete—an impossibility—this book constitutes a demonstration of the *possibility* of the Unity of Knowledge, covering fields from physics to sociology [6].

We quote already from the major contribution of Gianfranco Minati and Eliano Pessa [23] to systems science and philosophy: "The ontological meaning of existence should here be considered as the properties of change. By adopting the sayings of *Heraclitus*, we can consider change as *coming first* as the quantum vacuum precedes matter, not being simply the lack of matter. Levels and states should be intended as *simplifications* at certain levels of descriptions." We will return later in detail to Minati's exemplary work to show its relation to our logic, dialectics and the structure of processes.

1.3 The Limitations of Propositional Logic

The enormous progress being made in science, especially, the physical sciences, justifies its application of principles derived from a propositional logic that is consistent and as complete and exact mathematically as possible. While some sciences, such as medicine and the 'social' sciences, remain in part outside this paradigm, the vast majority, from physics to technology do not. Documentation of both efforts and results is presented in a substantially linear fashion, moving in a more or less orderly sequence from premises to conclusions, from preliminary experimental evidence through to theory and successful prediction. Knowledge as a whole is organized into categories which follow rules of exclusivity and exhaustivity that mirror the logic from which they are derived.

Philosophical debates continue about aspects of these processes, for example whether or not the development of knowledge is evolutionary—continuous—or revolutionary—discontinuous as in the well-known description of Thomas Kuhn [24]. Other movements in thought at perhaps a more easily definable sociological level, such as that from logical positivism to a less dogmatic modernism and post-modernism, although often subject to violent controversy, still conform to a familiar pattern of discussion and dialogue, in part to satisfy the need to be integrated into the educational process.

One may still ask whether such a comforting paradigm, in which contradiction plays no functional role, adequately reflects the complexities of interactions at social, cognitive, biological and even physical levels of reality. As one leaves the domain of chemistry and goes farther down into quantum mechanics, gravity and cosmology, standard logics become inapplicable and a linearity of time is no longer a self-evident given. The problem has been well recognized by the philosopher of information Luciano Floridi [11]. In his 2019 book, as a counter to the strong naturalism that he sees as the consequence of the modern scientific revolution, he proposes a non-naturalist, realist constructivism as a better way of approaching our responsibilities. We will explore in depth below the crucial role that the philosophy of information, in the conceptions of the Chinese philosopher Wu [30] as well as of Floridi [11], is playing in the understanding of knowledge processes.

The additional complexities of human existence have also been well recognized in art, especially, the cinema The Slovene philosopher and deep critic of our time [32] has taken other examples from cinema to support his thesis that there is a strong current of irrationality in our society that has not been dealt with from a rigorous, that is scientific standpoint.

The reader should begin to suspect, from our approach as outlined briefly above, that the scientific basis of our thesis is substantially deeper than as it appears in recent concepts of philosophy. We will need to position it in relation to two major trends in philosophy in the last century, namely neo-Kantianism and the Critical Theory of the Frankfurt school. As we will show, it is possible to point to specific statements in modern philosophy that are compatible with the concepts of reality in our Philosophy in Reality. However, it is our logic that enables a more comprehensive synthesis, which includes the physical properties of real processes, to be made. In order for our approach to be understandable, we first need to present its dialectics and its grounding in science. In Chap. 15, then, we will see how the latter lead 'naturally' to our proposed new Natural Philosophy.

1.4 Logic and Philosophy in Reality: A New 'Book of Changes'

The paradigmatic-historical model of knowledge and the logical-semiotic model of thought tend to incorrectly emphasize the separation and differences between the respective domains vs. their overlap and interactions. We present Logic in Reality as a sublation of linguistic logics of objects and static forms by a dynamic logic of real physical-mental processes.

In the book of Brenner [2], Logic in Reality and its Principle of Dynamic Opposition were characterized as follows: "The metalogical properties of LIR are thus of an entirely different kind, since it is based on a view of nature that does not consider fundamental either the abstract entities of pure classical propositional or mathematical logic or the anthropomorphic ontological concepts of phenomenology. The most fundamental metalogical principle of LIR is that of opposition or antagonism, without which, in this view, nothing could exist. This is, therefore, at the same time the most fundamental metaphysical principle of LIR. Nothing exists independently of something else in the formal ontology of LIR."

We could have used the term Natural Logic for the Lupasco logic of complex natural processes. We have not done so because, as Brenner pointed out, the term

Natural Logic has been defined by Grize [12], following Piaget's *Logique Opératoire*, as a logic used spontaneously for customary reasoning performed by means of everyday language.

Wu Kun, in a 2018 article entitled "Natural Philosophy and Natural Logic" [31], defines the latter basically as the logic of nature, which is very close to Logic in Reality. A much earlier book, unfortunately published only in Chinese in 1990, has the title *Logic of Nature.* Wu's Natural Logic has the same concern for the natural ontological aspects of the logic of nature as Logic in Reality, and both are relevant to the construction of scientific rationality. As we will see later, Lupasco's approach was grounded in the concepts of quantum mechanics emerging in the 1920s, and its consequences for systems, while that of Wu emerged in his deep analysis, starting in 1980, of the properties and dynamics of information. Both information and systems will be discussed further in detail later.

But the theme underlying our entire discussion is *change*, a vision of the dialectical movement from actuality to potentiality, from identity to diversity that is characteristic of real processes. It is this processual view of nature in which the Franco-Romanian thinker Stéphane Lupasco found the basis of his non-standard logic that is at the heart of this book.

As noted, the inspiration for this book was the ancient Chinese *Book of Changes (I Ching).* The importance of the *I Ching*, in the still definitive version of Legge [19] was its attempt (or series of attempts) to find the underlying features common to the experience and knowledge of change. Two easily observable examples of *non*-change, the absence of change were available in antiquity: the so-called fixed stars, fixed in relation to the planets (wanderers) and the stable, identical sequence of human life of birth, life and death. In these cases, sameness and stability is exemplified in the middle of change, in one case the phenomenon itself, in the other the sequence abstracted from the phenomenon. It is perhaps not irrelevant to mention that these two phenomena have not changed, although the second has given rise to some rather hysterical talk about doing so. This is simply another way of saying that from a philosophical and psychological standpoint, there are no changes in human nature that accompany or are coupled to changes in technology. We call attention to this in line with our general strategy for this book of insuring that proper attention is paid to the opposite or contradiction of a phenomenon or position. We will refer to significant studies of Chinese thought in the West, including that of Fritjof Capra to illustrate the additional dimensions which our logic can bring to the 'East-West' relation.

The use of the *I Ching* in divination by the assignment of meaning to hexagrams is of primarily historical and psychological interest. Our theory constitutes a meta-thesis, composed of elements from science, logic and philosophy. It is a project within natural philosophy, which will also address aspects of the dynamics of the change and growth of civilizations and their implications for the information society. We will thus further emphasize the following:

1. Continuity of Philosophy in Reality and the rigorous, transdisciplinary approach of LIR to both philosophy and science; our approach is distinguished from speculative philosophies with only an anecdotal relation to science such as that of Peirce.
2. The change in perspective at the boundary of philosophy and science that is required, in which current conceptions of causality, space-time, consciousness and evolution are given new interpretations. Our basic methodology, following Lupasco, is to apply a dialectical-logical approach to these basic long-standing issues.
3. The relevance of our logical-dialectical approach to new developments in the areas of information, ecology and social dynamics

In an interesting, transdisciplinary parallel to our work, Farmer [10] suggested that neuroscience "promises to act as a catalyst in seeking unification of the fragmented social sciences (e.g., political science and economics) and social action subjects ... that concern governance. *Neuroscience can achieve this because it reveals that taken-for-granted concepts, and the language used to express them, should be challenged* (emphasis ours)." The author considers his approach as part of a "movement to repair increasing fragmentation of the social sciences". This statement applies to our approach to the fragmentation of knowledge in general. We will explore the field of metaphilosophy later. However, given a simple definition of metaphilosophy as a set of statements about the content of philosophy itself, we make one such statement in this introductory Chapter: to say, without qualification, that philosophy is determined by standard bivalent logic is not scientifically grounded and limits the domain of philosophy without justification.

We have benefited from many of the articles in The *Stanford Encyclopaedia of Philosophy*, published in the United States by Edward Zalta. They provide highly authoritative summary articles on the 'state-of-the-art', and sets of references which it would be difficult for any individual researcher to equal. Within the limited universe of discourse described in these references, the application of all the rules of standard, truth-functional logic and the historical positions taken since antiquity regarding what constitutes philosophy and logic are acceptable. Many challenges to the standard view have been made, but so far it remains the 'default' position against which newer concepts are judged. This is what happens in all research, but, as Brenner pointed out in regard to critiques of Logic in Reality as a logic, defining logic in a way that a priori excludes a logic of processes on the grounds that it is not linguistically based is circular and should be recognized as such. We cannot, like Alice in Wonderland, say that a book cannot be interesting and worthwhile because it has no 'pictures' in it—read—syllogisms.

1.5 The Nature and Role of Energy

The concept of energy in existence was and is central, underlying thought as philosophy and science, in both East and West, discussed by Brenner in relation to the Tao [3]. The way in which energy enters into our Logic in Reality (LIR) brings out what we believe are new aspects of the way it constitutes and evolves in reality. We describe LIR in detail in Chaps. 3 and 4 as a logic with its own axioms and categories. However, because the entire axiomatic structure of Logic in Reality and consequently of Philosophy in Reality depends on the nature and properties of energy, it is necessary to establish a consensus about these properties or characteristics in order to establish LIR as a formal system. We will not repeat here the construction by Brenner [2], for the purposes of this discussion we can agree on a number of descriptive statements about energy, first, that it comes in different forms. Its major property is expressed in the 1st Law of Thermodynamics, namely, that energy cannot be created or destroyed but only changed in form, e.g., electricity to light, chemical to mechanical, as in a muscle. Several dual, physical and metaphysical aspects of energy must be taken into account for a satisfactory description. We summarize these briefly here. For further details see Chaps. 3 and 6 and the book that is the precursor to this one [2].

1.5.1 Intensity and Extensity

It was recognized early that different forms of energy had properties that were in some way primarily either external or internal. The terms intensity and extensity refer to the quality of having intension and extension respectively, in a different sense from their use in classical logic [7]. The German electrochemist Ostwald [27] looked in detail at the extensive and intensive properties of various forms of energy. The 'measures' of intensive energies are vectors or tensors, whereas extensity always implies a measurement in terms of a number of identical, scalar units. Ostwald pointed out that an intensity and an extensity could be both actual and potential, but not at the same time. He thus provided the philosophical basis for both the alternation of actualization and potentialization and the relation in energy itself of intensity and extensity.

The smallest unit of energy, the Planck energy, is the product of a frequency and a constant, called the Planck constant. The Planck energy thus instantiates duality as it has aspects that are both continuous, since frequencies can take any value, and discontinuous (a constant). The photon has aspects of both a particle (discontinuity) and wave (continuity). According to the Heisenberg uncertainty principle, the energy (momentum) and position of a particle are both localized and non-localized in the sense that to the extent that one is actualized more precisely by some measurement, the value of the other becomes more imprecise.

The essential concept here is that energy is inherently dual and antagonistic, and the implication is that this duality will be present in all higher (more organized) forms of energy, although it will not necessarily always be the same duality. Our view is that the ability of fundamental microphysical elements to carry information to more complex levels resides in their irreducible, constitutive antagonistic dualism. Energy as Fire has been a topic of philosophical discourse since Heraclitus, and its properties are clearly relevant to our thesis: it is called not just fire but "ever-living fire—*pyr aeizoon*."; Heisenberg [13] considered Heraclitean 'fire' as energy in the terminology of modern physics. The phrase "ever-living" anticipates the 1st Law of Thermodynamics mentioned above. The opposition between substance (ontological) and number (epistemological) also appears in Heraclitus.

1.5.2 Identity and Diversity

The second most fundamental aspect of energy is expressed by the dialectic between its entropic and negentropic properties. Energy moves from diverse, heterogeneous high-level forms toward a single, homogeneous low-level form (heat), governed by the 2nd Law of Thermodynamics. Let us designate such a tendency toward a single undifferentiated state of affairs as exemplifying identity or, simply, identifying or 'homogenizing'—an 'identification'. At the same time, energy, as apparently indistinguishable electrons, shows a fundamental duality governed by the Pauli Principle of Exclusion. Energy and accordingly all existence thus also show a tendency toward or instantiate an opposing process of heterogeneity, or non-identity or diversity, a 'heterogenizing' process, a diversification. Combining this idea with the one in the previous paragraph, we suggest that homogeneity, exteriority and objectivity characterize the process of extensity, and heterogeneity, interiority and subjectivity that of intensity.

1.5.3 Actuality and Potentiality

The third introductory statement to be made is that for energy to manifest itself in change, with regard to an observer or not, it must go from a certain state of potentiality to a certain state of actuality. If everything were completely actualized or realized, everything would be definitively static; no event or change could take place. We have judged it essential for the comprehension of our Philosophy in Reality for the reader to have these principles in mind as a 'skeleton key' to understanding our illustrations.

There is one additional point to be made: one should remember that we as living systems, animal inhabitants of the world, are constituted by energy in the form of atoms and molecules, and the complex structures built up from them. Electricity is present as electrical nerve impulses and so on. It is accordingly both easy and difficult to describe energy, for when we describe energy, we are describing ourselves.

1.6 Outline of the Book

The book is organized into three parts, as follows, to each of which a brief sub-section will be devoted.

1.6.1 Part I Theory and Science

Following this Introduction in Part I, we proceed in Chap. 2 with the presentation of change as a critical theoretical topic in philosophy that has not been successfully discussed due to the lack of an appropriate logic. It introduces the aspects of the *I Ching* most relevant to our thesis that our book is a "New Book of Changes". We will present here the concept that establishes the link between Chinese thought and ours that of energy. Chapter 3 further summarizes the precursor book of Brenner, *Logic in Reality*, as the necessary basis of our current project and an approach to a new paradigm, based on our Logic of Energy. The concept of transdisciplinarity in the acceptation of Nicolescu will be introduced as a basis for later chapters on systems. Chapter 4 extends our concepts of both logic and change in parallel, and some other recent ideas of change are mentioned. In Chap. 5, we present an extensive argument for the recovery and rehabilitation of dialectics, supported by LIR, as a mode of inquiry. Chapter 6 is devoted to the physics of our external reality, in which our logic is grounded; included is a discussion of the implications of Quantum Mechanics and the domain to transition between quantum and classical mechanics. Chapter 7 presents our approach to mathematics, the concept of a new Mathematics in Reality (MIR), parallel to our concepts of Logic and Change in Reality, as a further basis for philosophy. Chapter 8 on Chemistry is a science in which the essential principle of our logic of potentiality finds concrete illustrations.

1.6.2 Part II Toward a New Natural Philosophy

As Part II of this book, we proceed from physics to philosophy and use our theory or theories to analyze the set of general *operative principles* in nature which are at the interface of philosophy and science, in particular that of meaning. Chapter 9 gives our view of the role of semiotics and semiosis in knowledge, and presents concepts for a directed discussion of meaning. Our dialectics and logic define a new set of working relations between the concepts of meaning and semiotics. Meaning, for which we will have provided a grounding in physics, is at the interface between the general principles we see operating in nature and their more specific applications in philosophy. Among other things, we will propose a new concept of ontolons as the dynamic 'units' of complex reality. Chapter 10 presents our new views of the principles underlying three important domains in philosophy—Metaphysics, Metaphilosophy and Phenomenology. In Chap. 11 we show that the preceding work is linked by the concept of information. We will discuss the essential role in knowledge played today by the logic and physics of information, including the concept of the contemporary Chinese philosopher Wu Kun that the philosophy of information is a metaphilosophy, and that our interpretation of the informational aspects of science and philosophy supports Wu's proposal that they are converging. In Chap. 12, we look at information again from the standpoint of its role in meaning and its communication, and communication further contrasting analytic and synthetic viewpoints. Chapter 13 is our attempt at a new synthesis, a new Natural Philosophy. It is based on very recent new perceptions of the content and role of contemporary natural philosophy, as well as our key logical and dialectical principles. To illustrate our own penchant for insuring that proper attention is paid to the contraries and oppositions to our preferred theses, Chapter 14 is devoted to *Antiphilosophy* as a valid subject for dialogue and critique. The concluding Chap. 15 in Part II is a directed search for links between our logic and recent developments in philosophy based followers of Kant and Kant himself. We look at our own Philosophy in Reality as a process, linked by our logic to science.

1.6.3 Part III the Philosophy of Structures and Systems

Chapter 16 of Part III discusses applications of our concepts to the science and philosophy of structures and systems. It begins with their presentation as dynamic processes, concepts found in the seminal work of Stéphane Lupasco. In addition to the logic and dialectics of change, an additional, closely-related theme running through this book is that of systems, as redefinition of the foundations of systems science and philosophy has also proven to be necessary, as outlined in Chap. 16. The concepts of Gianfranco Minati of the Dynamic Usage of Systems (DYSAM) to describe collective behavior, emergence and quasi-systems can be joined to the dynamic principles of Logic in Reality to generate a new systems approach that has a seamless relation to philosophy and other science. Chapter 17 is then a discussion of living systems; Chap. 18 of social systems and external reality and Chap. 19 of social systems in connection with the Global Sustainable Information Society proposed as a goal by the social scientist and information theorist Wolfgang Hofkirchner.

We accept from the outset that it is neither possible nor desirable to separate the scientific and philosophical aspects of systems and their changes, as it is impossible to do in the theory and praxis of information. We will see again that the evolution of processes in these fields, which are at different levels of reality, nevertheless follow the Logic in Reality, expressed in laws which are not identical at each level but isomorphic to one another. We make a statement of principle about the role of our project in the emerging information society, directed toward the common good. In future work, we plan to extend these ideas to address broader aspects of the dynamics of the growth of civilizations. Our concluding Chap. 20 returns, *da capo*, to the *I Ching*. We review what *we* have said about change and attempt to justify why, as suggested by its sub-title, our book may be considered “A New Book of Changes”.

As a glance at the Table of Contents of this book will show, the branches of knowledge whose philosophy is discussed range from quantum mechanics and metaphysics to human consciousness to sustainable development. In doing this, however, we wish to assure the reader that we are not seeking to promote some impossible, abstract conception of a Unity of Knowledge or Science. Recent developments relevant to the problematic of unity have been well summarized by Cat in his article on the Unity of Science referred to above [6]. These can be read as avoiding any principle of ‘unity for the sake of unity’ and this is exactly our position. Cat notes several positive, anti-reductionist trends in the philosophy of science, for example in moving from dependence on concepts primarily from standard, classical notions from physics. One is struck in fact by the diversity of senses which can be given to the definition of unity! These range from the possible single nature of the underlying ‘stuff’ of the universe, to the unification of the disciplines within science and, with regard to these, whether on is dealing with concepts or terms or the higher-level entity of theory and whether the relations are ones of reduction, explanation or logical inference. Cat suggests that the concept of unity can have value at the practical level in *science*, to justify approaches and goals, and in *philosophy* to help choose what philosophical questions to pursue and what target areas to explore. He suggests that “unities and unifications help us meet cognitive and practical demands upon our life ..., contribute to our self-image and can be seen as a source of aesthetic value, providing a grip on our intellectual imagination.”

Our logical-philosophical approach in this book supports but also goes beyond this view: the valence of standard approaches to the Unity of Knowledge and/or Science, self-referentially imply a unity of a sort that neglects, intentionally or not, the value of diversity. We believe it is necessary to point to it explicitly. Thus informational processes, as well as all other complex processes, in our theory and logic, are composites of unities and diversities. There is a kind of ‘unity of purpose’ in intellectual research that is best described by the historical philosophical term of ‘unity-in-diversity’. Hofkirchner, writing about a Unified Theory of Information (UTI) [14], avoids its reduction to a monolithic identity through his concept of the union of identity-diversity necessary to characterize information. Our Chap. 15 on philosophy as process refers to dialectical concepts we consider critical, such as ‘transcendence-in-immanence’, associated with the name of Emmanuel Levinas. Exploring Unity through Diversity is the title of a book series published by the Bertalanffy Center for the Study of Systems Science under the direction of Hofkirchner [14]. The explorations in this book are made in the same spirit.

References

1. Brenner, J.E.: Process in reality: a logical offering. Logic Logic. Philos. **14**, 165–202 (2005)
2. Brenner, J.E.: Logic in reality. Springer, Dordrecht (2008)
3. Brenner, J.E.: Linking the Tao, biomathics and information through the logic of energy. Prog. Biophys. Mol. Biol. **131**, 15–33 (2017)
4. Brenner, J.E.: The naturalization of natural philosophy. Philosophies **3**(4), 41 (2018)
5. Brenner, J.E., Igamberdiev, A.U.: Philosophy in reality: scientific discovery and logical recovery. Philosophies **4**, 22 (2019)
6. Cat, J.: The unity of science. In: Zalta, E.N. (ed.) The Stanford Encyclopaedia of Philosophy. Spring 2017 Edition (2017). https://plato.stanford.edu/archives/spr2017/entries/scientific-unity
7. Chalmers, D.: On sense and intension. In: Tomberlin, J. (ed.) Philosophical perspectives 16: Language and Mind. Blackwell Publishing, Oxford (2002)
8. Dodig-Crnkovic, G., Schroeder, M.: Contemporary natural philosophy and philosophies. Philosophies **3**, 2 (2018)
9. D'Espagnat, B.: A la recherché du réel. Gaulthier-Villars, Paris (1979)
10. Farmer, R.L.: Neuroscience and Social Work Practice—The Missing Link. SAGE Publications, Los Angeles (2008)
11. Floridi, L.: Semantic Conceptions of Information. Stanford Encyclopaedia of Philosophy. https://plato.standford.edu/archives/win2019/entries/information-semantic (2019)
12. Grize, J.B.: Logique naturelle et communications. Presses Universitaires de France, Paris (1996)
13. Heisenberg, W.: Physics and Philosophy: The Revolution in Modern Science. Harper Perennial Modern Classics (Reprint ed.) (1958). HarperCollins, New York (2007)
14. Hofkirchner, W.: Emergent Information. A Unified Theory of Information Framework. World Scientific, Singapore (2013)
15. Hooker, C. (Ed.): Philosophy of Complex Systems, vol. 10. Handbook of the Philosophy of Science. Elsevier, Oxford/Amsterdam (2011)
16. Igamberdiev, A.U.: Time and life in the relational universe: Prolegomena to an integral paradigm of natural philosophy. Philosophies **3**, 30 (2018)
17. Jacquette, D.: Ontology. McGill-Queen's University Press, Montreal (2002)
18. Kauffmann, S.: At home in the universe. Oxford University Press, New York (1995)
19. Legge, J. (ed.): 1963. The I Ching. The Book of Changes. Dover Publications, New York (1899)
20. Luhn, G.: The causal compositional concept of information. Part I: from decompositional physics to compositional information. Information **3**, 151–174 (2012)
21. Lupasco, S.: Du devenir logique et de l'affectivité; vol. 1: Le dualisme antagoniste. Essai d'une nouvelle théorie de la connaissance. Paris: J. Vrin. (Originally published in Paris: J. Vrin, 1935) (1973)
22. McGinn, C.: Basic Structures of Reality. Essays in Meta-Physics. Oxford University Press, Oxford (2011)
23. Minati, G., Pessa, E.: From Collective Beings to Quasi-Systems. Springer, New York NY (2018)
24. Nickles, T.: Scientific revolutions. In: Zalta, E.N. (ed.) The Stanford Encyclopaedia of Philosophy (Winter 2016 Edition); Stanford University: Stanford, CA, 2014 (2016); https://plato.stanford.edu/archives/win2016/entries/scientific-revolutions/. Accessed 17 Mar 2017
25. Nicolescu, B.: Le tiers inclus. De la physique quantique à l'ontologie. In: Lupasco, S., et l'œuvre, L., Badescu, H., Nicolescu, B. (eds.). Éditions du Rocher, Monaco (1999)
26. Nicolescu, B.: Nous, la particule et le monde. Éditions du Rocher, Paris. (Originally published in Paris : Éditions Le Mail, 1985) (2002)
27. Ostwald, W.: Vorlesungen über Naturphilosophie. Verlag von Veit & Comp, Leipzig (1902)
28. Rescher, N.: Process Metaphysics. State University of New York Press, Albany NY (1996)
29. Williamson, T.: The Philosophy of Philosophy. Blackwell Publishing, Malden (2007)

30. Wu, K.: The Basic Theory of the Philosophy of Information. In: Proceedings of the 4th International Conference on the Foundations of Information Science, Beijing, China, 21–24 Aug 2010
31. Wu, K., Wang, Z.: Natural Philosophy and Natural Logic. Philosophies **3**, 27 (2018)
32. Žižek, S.: Living in the End Times. Verso, London (2011)

Chapter 2
Change in Reality: The I Ching

2.1 Change in Reality

Change is ubiquitous in existence and experience. Western theories of change, however, have focused on making it mathematically, computationally and logically tractable, within the framework of standard logic. In fact, change is contradictory: it is regular *and* irregular; consistent *and* inconsistent; continuous *and* discontinuous. Since the only logics available have been propositional bivalent logics, incapable of accepting real contradictions, they have been incapable of describing change.

The number of kinds of quantitative changes is relatively small: there can be a change in the number of a group of things; a change in the relations between them; and a change in the measure of a property, size, shape, or other scalar quantity or direction, momentum or other vector quantity. In addition, however, there are changes in quality which cannot be directly expressed using standard categories. One would nevertheless like to be able to refer to such changes in a reproducible, scientific manner without reduction. In Chap. 9 below we introduce the notion of an 'ontolon' as a possible unit of dynamic change in reality.

A major barrier to the understanding of change is psychological. Human beings are constantly involved in change, in a sense they are nothing but change. However, change has two aspects: change is birth and growth but also decline and death, in progress at the same time. Change is welcome and unwelcome, desired and feared, in other words in contradiction. People do not like contradiction, for it recalls the negative side of change; they tend to focus on stability and identity and relegate change and diversity to a second-rate ontological level or, better, to the unconscious.

All approaches to change thus suffer from more or less explicit dependence on underlying logics, bivalent or multivalent, that are semantic and truth-functional, including their most recent modal, epistemic, doxastic and paracomplete (intuitionist) versions. These logics require absolute separation of, for example, continuity and discontinuity, space–time and matter, chance and necessity, etc. and lead to the paradoxes with which we are all familiar. Even paraconsistent logics, which accept contradiction, capture only the linguistic as opposed to the physical aspects of processes,

J. E. Brenner and A. U. Igamberdiev, *Philosophy in Reality*, Studies in Applied Philosophy, Epistemology and Rational Ethics 60, https://doi.org/10.1007/978-3-030-62757-7_2

although some real inconsistencies in simple change (Sorites problems) are accepted. Logics of epistemic change, frequently referred to as change in belief, are based on linguistic abstractions from which the physical processes of change have been largely eliminated. Under these conditions, it is not surprising that no generally applicable theory of change has been developed for the real extant domain of complex interactive processes.

Summarizing abusively, the most significant forms of knowledge in Western philosophy have been about things that do not change. This has changed only with the advent of quantum mechanics and uncertainty and of relativity and the loss of fixed frames of reference. It is only now, in the twenty first Century, with the advent of complex systems dynamics, that we have begun to define "a physics of change, a true science of contingency". Our reference is the major compendium of Hooker [13] on the philosophy of complex systems to which we will refer in detail in Chap. 17. Ancient Chinese cosmology presupposed change, not stability or constancy, as ontologically primary.

These contrasting viewpoints will be illustrated throughout this book. For now, we can say that classical Western medicine sought atemporal explanations where possible. The framework of classical Chinese medicine was a holistic view of *reality* as a system of complex interacting processes, a constantly changing universe. A consequence is that knowing for the Chinese was the ability to respond to changes in oneself in ways that corresponded to the rhythms of changes in nature, modelling the processes of living systems. Herfel goes as far as to use the phrase "the obsession with change in Chinese philosophy". This does not need to be read literally and negatively; the point is to insure that the emphasis, even today, on stability and identity is counterbalanced.

2.2 The Chinese Book of Changes (I Ching)

The *I Ching* is one of four major documents that have come down to the West from China. Its origins go back to the twelfth century BCE, and it has undergone innumerable changes and reconstructions. As noted, the most authoritative, usable version is that of Legge [17], which is an open access, on-line document, as well as in more recent translations and commentaries in English. The other three documents, which are the necessary starting points for any serious discussion of Eastern in relation to Western traditions are: (1) the *Tao Te Ching* (TTC), the *Book of the Way of Virtue or Harmony*, a 5000 word treatise attributed to the Chinese thinker Lao Tzü in the sixth century BCE, again the result of a complex process of re-edition and change [11], (2) the *Chuang Tzü*, attributed, incorrectly to a single author of the same name, ca. 370–319 BCE [9], (3) the *Analects of Confucius*, compiled after the thinker's death ca. 470 BCE. We will be concerned in this study primarily with the I Ching.

The *I Ching* is as inexhaustible as the human mind and spirit, for which Chinese uses the same symbol. On its first level, it consists of sets of diagrams composed of long and short lines arranged groups of three and six—trigrams and hexagrams which can be used in several ways: having meaning in themselves; as a tool for divining the future; suggesting an interpretation of the meaning of a physical or mental phenomenon; pointing to an underlying structure of change, through selective interpretations. Some of these commentaries and interpretations have become codified and combined with the first in what is no called the *I Ching*. For us, the importance of the *I Ching* was its attempt (or series of attempts) to find the underlying features common to the experience and knowledge of change and consequently of human life and the universe in their entirety. The *I Ching* can be understood on many levels, and it contains many forms of truth; one need not believe oneself in divination to know that it satisfies a human need, but we consider that this particular use is secondary.

As Brenner pointed out at a Conference on Change in 2009 [2], little progress has been made either in the understanding of change in general, nor in the reasons for that lack of understanding. In this book, we propose some recent developments in science, logic and philosophy as the basis for new insights into the nature of change and its meaning in today's world. Like the original Chinese, it offers not a recipe for thought, but some new/old ways of looking at the world and our position in it.

In Western antiquity, [3], logic and metaphysics also began as ways of reasoning about nature, or reality. However, relatively early, logic developed into a tool for determining the truth or falsity of propositions. Deductive reasoning per se was disconnected both from processes of scientific inference and from ordinary experience. The most common current definition of logic is that it is an analytical theory intended to formalize principles of valid reasoning as well as a theory of valid inference to provide insight into the foundations of mathematics. Logical relations, it is alleged, can obtain only between propositions, not between concrete entities, nor between abstract entities that are not propositional in nature. Examples of standard logics are classical term or syllogistic logics and their recent modifications, first and higher order predicate logics, modal logics, and ampliative adaptive logics. Further, most such propositional logics are based on the principles of bivalence, absolute non-contradiction and the law of the excluded middle. These principles are also often characterized as following a Boolean logic or algebra (see below).

In contrast to the abstractions that have constituted logic, real phenomena are a direct consequence of human observation and human ratiocination, but the introduction of logical considerations in the corresponding disciplines, phenomenology and epistemology, has proven extremely complicated. The gap between reality—real entities—and propositions about reality remains the source of a large number of metaphysical problems. Talking about a logic of/in reality might thus be considered either a contradiction or a vague metaphor, referring, but only informally, to some of the observed regularities in nature and science.

There is, however, no theoretical reason why this should be so. One thesis of this book is that the limitation of logic to linguistic concepts, which do not and in fact cannot apply to real entities, should be removed. The discussion can and perhaps should center on the non-abstract and often conflictual changes characterizing real entities or phenomena as objects of analysis, while retaining the advantages of the formal structure, rigor and generality of the logical approach. The new paradigm we are developing could be designated as a search for truth without form in the standard sense, that is, without a totally fixed set of premises and their tautologies. This picture of functional relations and interactions in the world that are *logical* corresponds to the concepts underlying the *Book of Changes.*

To see the relation between Chinese thought and the objective in this book of a formulation of a Philosophy in Reality, we should familiarize the reader with the concept of a 'form' of thought, one that is central to Eastern and in particular Chinese thought. Some of the similarities between Chinese thought and LIR were demonstrated by Brenner in (2017), drawing on the work of the French sinologist and philosopher François Jullien. Throughout this book, in general, we will strive to maintain a balance between form and content of thought, giving the proper ontological status to the former and explicating its dialectic relationship to the latter.

The characteristics of change, from its foundations in science to its interpretation in human thought, are an integral part not only of the *I Ching* but also other classical Chinese texts. Their value has been recognized in the West in the late nineteenth and twentieth Centuries by Needham, and other philosophers and scientists, of whom the best known is Capra [4]. The relation of Logic in Reality to this body of knowledge was discussed in detail in [1]. Some of the conclusions of this paper are relevant to the focus of this Book as they support the underlying concept of the scientific structure of a universal basis of knowledge that has *not* changed since the fifth Century B.C.E.

In an article by Igamberdiev [14] in the same Special Issue of *Progress in Biophysics and Molecular Biology* on thought in East and West, other contributions of Chinese thought to both science and philosophy were outlined. Since the Chinese texts are not accessible to us or to most Western readers, we summarize what for us are the most important points in the *I Ching*, followed by a discussion of the unique viewpoint of Jullien. The direct relation between Logic in Reality and modern Western concepts of change will be presented in Chap. 4, as they are in a sense, together with causation which is closely related to change, at the core of *all* philosophy, of East and West.

2.3 Main Principles in the I Ching

We will refer several times in this book to the pioneering work in the field of the Philosophy and Science of Information of Professor Wu Kun, of the Xi'an Jiaotong (Social Sciences) University, China. According to Wu, the *I Ching* is the most important of a group of six great books in ancient Chinese culture (the other five are: *The Book of Songs, The Book of Documents, The Books of Rites, The Book of Music, and The Spring and Autumn Annals.*). We give here just a very small sample of Wu's work in this field.

The basic theories of the *I Ching* can be summarized as three main principles: Bian Yi, Jian Yi and Bu Yi. Bian Yi means that everything in the world is in a process of continual change including inorganic matter, organic matter, human beings, human thoughts and the universe. It is similar to the concept of Wu Chang, in Buddhism—Variability.

Jian Yi means that even though everything in the world is changing, the world has principles and laws of existing and changing. With the necessary wisdom, these principles and laws are revealed and what seems complicated and unrecognizable becomes simple and normal. This is called "seeing phenomena without reason". Despite everything changing all the time, one thing stays constant, Bu Yi, which means that the origin of everything is itself invariant and eternally existing. Some ancient Chinese philosophers referred to it as 'Tai Chi' or 'Tao'. As noted also by Capra, it is similar to the 'noumenon' or 'thing-in-itself' in Western philosophy.

The theories in the *Book of Changes* are built on the dualism of Yin-Yang. The theory of Eight Trigrams derived from the structured coding of Yin-Yang dualism classifies everything in the world according to its 'character', evaluates the properties of specific things and phenomena, and predicts the tendency of their development.

Wu Kun studied *The Book of Changes* from the perspective of modern information theory, starting from the emphasis of ancient Chinese philosophy on the structuration (Giddens' term for structuring) of things and part-whole relationships as the basis of explanations. The eight trigrams theory in Ancient China was in this sense the first informational structuration theory, presented through philosophical logic, in human history. Its binary logical character is the same as in the development of modern information science (use of 1 and 0 for coding and computing).

We will not discuss here the Eight Trigrams theory, the lines generated by them, and their interpretations. As far as its form is concerned, the system of Eight Trigrams is an information coding method of symbol combination, reorganization and matching. As far as the interpretation and symbolization of explanation of trigrams and component lines are concerned, it is a formal way of information generalization and symbolic representation. These are metaphors, an 'image-numerology' of the changes in things, fate and life, misfortune or fortune. It is said that image-numerology theory "existed before the birth of Sky and Earth, so it is called "primordial graph" and its theory is called "primordial theory". Shao Yong believed that the origin of the universe is "Tai Chi", which is also called "Tao" or "Mind". The Tai Chi is eternal and other things grow and die, have beginnings and ends, changing periodically along a Primordial Graph, what would be called a *world-line* in the West. In spite of the subjective assumptions of such an image-numerology theory, its viewpoint of everything having "images" and "numbers" and the method of using symbols and graphs to explain those "images"and "numbers" are embedded in information thinking.

We can also relate to the complex evolutionary theory of the *I Ching*, which seeks some kind of orderly sequence in the emergence of more and more complex phenomena: "Everything appears after the appearance of the sky and the earth; man and the woman appear after the appearance of everything, then couples, then father and son, then monarch and subjects and so on. Interestingly, the *I Ching* avoids the problem of an infinite regress, introducing a 'Stop' trigram, because "things cannot move forever". As in LIR, processes do not go to idealized extremes: "things will develop in the opposite direction when they become extreme." There is no perpetual motion or perpetual motionless. The last trigram of sixty-four trigrams, the Wei Ji trigram, it is not a real 'last'. It is a step in an endless sequence that is moving, changing and transforming all the time.

Wu finds further examples of holism and a theory of processes in the work of Ruan Ji (A.D. 210–263), who presented an ontological view based on "nature". He formulated a view of 'hard' relativism that is not unlike that of Richard Rorty. Since "All things are united in one and have no differences between them, we should not distinguish right and wrong, good and evil, and should live a free and unfettered life, letting things take their own course." Ruan Ji wrote that the *I Ching* was the metaphysical truth of and from remote ages. "It comes from darkness and subtleness and spreads brightness and eminence."

2.4 Classical Chinese Logic

In classical Chinese thought, there is no single work or body of works directed toward logics alone, although various schools of logical thought did exist. Concepts which fit the standard Western description of logical were debated in period 400–200 BCE, that is, in parallel with some major commentaries on the *I Ching*. According to Jana Rošker (2015), Chinese interest in logical problems grew out of the methodology of disputes or disputations. It studied procedures for assessing the validity of conflicting assertions, but without using a unique term corresponding to the notion of semantic truth. We think the Chinese would have appreciated the dictum of Lupasco that "truth is the truth of reality, not of propositions".

It is the existence of concepts such as this which justify, in our view, the idea that Chinese logical thought was a 'precursor' to that of Lupasco. Rather than truth-functional contradiction it used paradoxes to express a relativism in reality in relation to human behavior and values. Reasoning and argumentation were not based on syllogisms involving premises and conclusions but a system of real analogies and recognition of structural patterns, modes of informal thought, rather than abstract formalisms.

Most importantly for showing the relation to the LIR Principle of Dynamic Opposition, in place of the Aristotelian Laws of Identity and Non-Contradiction that lead to the dichotomies of Western logic, "dual oppositions were seen as mutually defining and interdependent, guided by the underlying principle of complementarity or correlativity." As discussed below, the principle of complementarity in Lupasco is to be preferred to relatively static concept of the early Bohr. How the Chinese understood *their* principle of complementarity is not clear, but the intent does seem clear: Rošker quotes Jiang Xingyan who stated that "following ancient Chinese logic, the meaning of a word can be understood or clarified by looking at its opposite". This is not, yet, the dynamic opposition of LIR, but the further correlation between dual but complementary oppositions, based on relational rather a subject-object structure, represents a specific approach of ancient Chinese logic. The classical Chinese relationally structured world-view of 'things' interacting within continuous, dynamic patterns leads to the concept of a universe behaving as an organic entirety with the *parts reflecting the structure of the whole.* In his mereology, Lupasco used the same concept, adding to the received wisdom that the whole reflects or instantiates the structure or properties of the parts.

Rošker states that due to the absence of the linguistic and cognitive category of subject and the absence in Chinese of the verb 'to be', traditional Chinese philosophy never developed an explicit formal concept of ontology. But she further states that the Chinese worldview was based on a dynamic and changeable structure of being, that is, on change as the primary property of being. Thus classical Chinese philosophers dealt with ontological questions within the scope of an implicit ontology, in which this discipline was not clearly distinguished from epistemology and ethics. As we will see on several occasions later, the absence of such distinctions is not to be criticized but welcomed!

2.4.1 Classical Chinese Logic and Aristotle

A paper by Yuan (2005) in the *Journal of East-West Studies* gives a pertinent description of classical Chinese logic as an informal system of reasoning about the world. He shows how this logic should be related to the historical introduction, in China, of Aristotelian logic as part of an ethical–political system.

It is difficult to understand, except as an intuition, what Yuan means by referring to Aristotelian logic as having a third term, since the relevant axiom is generally considered to be the law of the *excluded* middle. In my view, and in my reading of Lu of Chinese logic as science, there is an included middle but it requires expansion of its qualitative aspects, that we consider Chinese Logic to exemplify.

A major difference between Chinese and Western thought is in the former's tradition of non-separability between science and philosophy. As Wing-Tsit Chan points out in his major *Source Book of Chinese Philosophy* [6], it is the scientific approach in the writings of masters such as Chu Hsi (1130–1200) that has led to an on-going scientific tradition in China that does not exclude philosophical insights. Thus, in the last Century, Chang Tung-Sun, in *his* theory of knowledge, maintained that in psychology, act and content cannot be separated: they are two aspects of the same thing, (but how they interact does not seem to have been discussed in terms we can use).

In contrast to the above, the pragmatic doctrines codified by Confucius and his followers seem almost a cartoon of primitive, ideological Western thought: emphasis on the ruler, the state and the family. One can find concepts in Confucius related to those in the *TTC* and Chuang Tzü, but they do not add anything fundamental to the understanding of 'life and mind' that is the main concern of this paper. This is patently unfair to a vast body of learning, but I mean no disrespect to it. "If you do not despise, there will not be despising … So they (the sages) discard 'that' and take up 'this' (*Tao Te Ching*, Sect. 72, Allegiance and Respect [17])."

We also note the controversies between Taoists and Confucians regarding the nature of and relation between being and non-being. One finds clear parallels to Heraclitus and Parmenides, but what is more significant for us is the existence of the two *primarily* 'antagonistic' viewpoints. A principle like that of dynamic opposition can be considered as present throughout the intellectual heritage of the species. Whatever current convergences between current Western and Eastern thought may be found, nothing would indicate that the classical interactions have been eliminated or rendered irrelevant, rather that they have been rediscovered in the conjunction/rejunction in progress.

This is the link that can be made between Logic in Reality and its Principle of Dynamic Opposition and Chinese philosophy and logic, the latter being a precursor of the former. With regard to the concepts of Yin and Yang, Brenner had found, in the doctrine of Chuang Tzü on the relation of what the latter called the 'void' or 'confusion', a similarity to the LIR conception of potentiality. Some Neo-Confucian scholars do not consider Chuang Tzü a 'sage' due to what they consider his excessive transcendence. We disagree, and feel that his 'going back and forth' between the void and ordinary, actual things corresponds to a description of the complex dialectical interactions between and the evolution of complex real processes, actual to potential and vice versa. His 'stories' are for us accurate descriptions of psychological states.

2.5 The Chinese Philosophy of Change and LIR

As this book was in preparation, an extraordinarily useful summary of the Chinese Philosophy of Change was published under that title by Tze-Ki Hon in the *Stanford Encyclopedia of Philosophy* [12]. We outline briefly here some of the key points in it to which our logical and dialectical approach can bring additional perspectives.

At this point we should again state clearly that this book does not aspire to achieve the ethical, religious and political heights and utility of the *I Ching*. There are no hexagrams, flying dragons, or equivalent devices which serve as 'attractors' for concentrating the mind on cosmic forces like *yin* and *yang* nor that correspond to the fear and anxiety that the authors of the major commentary, *Ten Wings*, sought to relate to the uncertainties of human life. However, as we will see below, our Logic in Reality assigns ontological value to uncertainty and other falsely labeled negative aspects of existence, so that LIR can be considered as having something of the same function as the *I Ching*.

Hon's article mentions several aspects of the 'philosophization' of change described in the *I Ching*: it suggests strategies for individuals to position themselves in an evolving universe, still a vital necessity today. Part can be considered a 'spatio-temporal grid' to enable people to reflect on their lives, look for alternative strategies and anticipate dangers. Anticipation, a subject of substantial current interest and discussed her in Chap. 12 on Communication, has perhaps its first reference here. The work of further commentators in the early CE can be divided into the following three areas: (1) the cosmology of change; (2) the ontology of change; and (3) the moral-metaphysics of change.

We will not attempt to discuss all of these objectives in this book, but focus on the point at which thinkers such as Wang Bi (206–249) started a turn toward the real world. According to Hon, Wang Bi saw the hexagrams as pointers toward "human creativity and ingenuity and that every aspect of human life is governed by the part-whole relationship". To cope with change, that is, live, one should find the part-whole relationship in each given situation. This statement is very similar to that of Lupasco that one should seek, in every complex situation, the elements in logical and physical opposition, as discussed in the next Chapter. Most importantly, rather than as a dichotomy, Wang Bi saw Being (*you*) and Non-being (*wu*) as co-dependent. Again, to be sure that the reader understands that we presenting an objective and not idolizing view of Lupasco, we note that the he adopted the former position, assigning ontological reality only to what he called affectivity or affect.

Not all commentators followed Wang Bi's movement from divination to philosophy, nor as a solution to all human problems. What is essential is the 'dialogue' between the reader and the text, to the extent that he/she is capable of one. The goal of the *I Ching*, again, is to "provoke thought, command attention and above all, make readers aware of the contingency of human existence". They are a potential source of hope in the face of the "fear and anxiety" associated with living with incessant change. However, because change takes place in an at least partly orderly manner, the *I Ching* and perhaps this book can help human beings find ways to understand its patterns, facing fear with humility and honesty.

2.6 Western Sources. The Tao of Physics

We will discuss some Western philosophies of change in Chapter 4. However, before that, we believe that the American sociologist Fritjof Capra, mentioned above, deserves the historical credit for presenting a link between fundamental classical Chinese thought and modern Western science. In *The Tao of Physics* (2010) Capra established the overlap, over a vast interval of time and space, between what can best be called the universal insights of the human species. We should not, however, forget that the sub-title of Capra's book is *An Exploration of the Parallels between Modern Physics and Eastern* Mysticism. This may not have been intended in a pejorative sense, but if Capra's objective was to give Eastern thought the value it deserves, than the label of mysticism without qualification seems at first glance to contradict it. Thus (p. 215) it is "Eastern mysticism" that emphasizes a dynamic between the 'Void' and the forms it creates as the apparently contradictory Western concepts of the continuum and the discontinuity of quanta. Suppose that Capra had written "Parallels between Modern Physics and Eastern Insights"? One could then *start* from the idea that these are the same in the sense of being part of the expression of a human capacity for (re)-cognition of their own place in a universe of change.

In his discussion of Patterns of Change (p. 275), Capra makes the 'classical' mistake of extension of the mathematician Geoffrey Chew's S-matrix theory of elementary particles to macroscopic phenomena: the sum of the probabilities of a reaction must be equal to one. A world view limited by 'unitarity' is clearly limited. But it is unjustified to accept, *literally*, that Eastern mystics "tell us again and again that all things and events we perceive are creations of the mind", based on the oneness of the totality of things." Capra cites as saying that the "innumerable things seen as multiplicity: the body, property, and above—all these I say, are nothing but mind." Charles Sanders Peirce, the prestigious American philosopher whose work we will critique later, echoed this is his famous dictum that "matter is nothing but effete mind".

Capra makes what we consider a correct reading of a part of the *I Ching*: "The eight trigrams are symbols standing for changing transitional states (*sic*); they are images that are constantly undergoing change. Attention centers not on their state of being, as is chiefly the case in the Occident, but upon their movements in change. The eight trigrams therefore are not representations of things as such but of their tendencies in movement."

What are missing here are the implications of such a world view based on Chew's 'bootstrap': it is not proven that mutual self-consistency is indeed a property of an indivisible universe, in which inconsistency is relegated to a lower ontological level. As Nicolescu has also stated [21], everything in the universe is connected to everything else and no part of it is fundamental, and no law governing it is more fundamental than any other.

Capra's vision is of a direct interaction between physics and Eastern mysticism, leading to a "completely new form of human intellectual endeavor", characterized by theories of natural phenomena "with ever-increasing accuracy". As we will see throughout this book, our claim is that it reflects what is important in Chinese and other Eastern thought by rejecting the complete application of such a Promethean concept. In his Epilogue, Capra partly answers the above critique by suggesting that science and mysticism are two complementary manifestations of the human mind, and that the attitudes of scientists nevertheless remain frozen in the abstract, In his further Afterword to the 4th Edition of his book, of 2010, used here, Capra points (finally) to the shift of the scientific paradigm from structure to process: "Energy, however is associated with activity, with processes…" "Process thinking is also a main characteristic of Eastern mystical (again) traditions." But what mysticism? Do mystics in deep meditation arrive at a point where the distinction between observer and observed breaks down *completely*, where subject and object *fuse*? Physicists and mystics allegedly deal with different aspects of reality, levels beyond ordinary sensory perception; the former levels of matter, the latter other levels of mind. We will try to show, throughout this book how intuitions like those of Capra are *partly* correct and *partly* incorrect, and that for any socially effective operation of a new scientific—philosophical paradigm, at least one major additional dynamic property of the world, real contradiction or counteraction, must be taken into account.

To illustrate the minimum continuity of this book, with due humility, and the *I Ching*, we quote from its Appendix III, The Great Appendix, Sect. I, Chap. 12; 76, 78:

"The sages made their emblematic symbols to set forth fully their ideas; appointed (all) the diagrams to show fully the truth and falsehood (of things); appended their explanations to give the full expression of their words; and changed (the various lines) and made general the method of doing so, to exhibit fully what was advantageous." "Transformation and shaping is what we call change; carrying this out and operating with it is what we call (generalizing) the method; taking the result and setting it forth for all the people under heaven is, we say, (securing the success of) the business of life."

As for the fundamentality of opposition as described in Chap. 3 of *Logic in Reality* (Brenner 2008), of the many similar formulations made in the *I Ching*, we will cite just one here, also in the Great Appendix, Sect. II, Chap. 1; 5: "Good fortune and ill are continually prevailing against each other by an exact rule."

2.7 Jullien's "Way of Thought"

Following the French Academician and sinologist Jullien [15], we suggest that what Chinese Logic captures is an additional level of mental processing, consciously or not, in its references to what is vague, left unsaid or is absent in art, especially, painting. Even more than duality, the concept of the 'binomial' as an integral part of the Chinese language expresses two aspects of an evolving process linked dynamically, for example, wind-sentiment for the atmosphere associated with a particular individual. As Chinese readers will know better than we, the concept of energy as an 'energy-breath', is another fundamental discovery of classical Chinese thought. As noted above, Logic in Reality was in fact called the Logic of Energy by Lupasco [19].

The concept goes farther in Chinese with the pair, also associated with wind, 'effective-evanescent'. Jullien suggests that this conveys a non-ontological status to the wind, since for him an ontology depends on an assignment in the two senses of attribution and determination. While we agree with this author's characterization of the non-objectifiability of the wind, it is not necessary to separate them in the dynamic ontology of Logic in Reality: wind_as_process is both non-ontological *and* ontological, and one focuses on one or the other in the manner previously described. As the link between classical Chinese views of change and those of contemporary philosophers, we have thus benefitted from the perspective of Jullien. For anyone interested in the logic, dialectics and philosophy outlined in this Book, Jullien's work is unique in pointing clearly to aspects of Chinese thought (the translations of the French text are by JEB) should become, in our view, central to any new Philosophy in Reality. One commentary about Jullien is indicative of this, and brings him close to the 'revolution in philosophy' discussed below in relation to the work of Wu Kun: "He (Jullien) brings Chinese thought into play like a theoretical operator that shakes to their foundations the choices made by philosophy and open it to question itself in new ways."

A key part of Jullien's work is his approach to the 'dealgorithmization' of knowledge and society. We note that *three* commonly used notions in philosophy and elsewhere have something on common—all bad in the sense of their exclusion, in a reductionist paradigm, of the dynamics of change: algorithms, Peircean signs and models. These all have a role and epistemological function but cannot provide an ontological basis for change and should not be used, for example, as a basis for morality. This was, of course, one of the major functions of Chinese thought, more or less successful as is still discussed.

2.8 East–West Studies

The domain of East-West Studies possesses what one might call a built-in benign dualism—comparisons of and contrasts in the operation of cognitive processes in two regions of the world. We claim that the logic most applicable to them is the non-propositional, non-truth-functional system designated by us as Logic in Reality (LIR). LIR is based on the fundamental dualities of physics and their dynamic interactive evolution between potentiality and actuality, leading under favorable circumstances to the emergence of new entities at higher levels of reality or complexity. LIR is a *logic* of change that can accept and permit inferences from real contradictions–better, counteractions—in real processes, which include dialectical relations broadly such as those between theories and the subjects of those theories. LIR is a formal system for justifying claims of, for example, complex mutual entailment of Eastern and Western perspectives on art and life.

The concept of (partial) non-separability, applicable to the macroscopic as well as to the quantum level of reality, any 'Western' phenomenon or person is also always partly 'Eastern', the same and different, 'the same and not the same' (cf. Chap. 8). LIR thus provides a non-transcendental basis for the fact, to which many others have called attention, that I have a selfish interest in altruism and morality because the other is always partly I.

The domains of Eastern and Western logics span an enormous range of formal and informal concepts, but nevertheless share and utilize a number of common metalogical and metatheoretical principles. We have suggested that LIR provides a framework for these principles by showing their common origin in the microphysical properties of matter-energy as they operate at the macrophysical level. The intuition that the Tao, as a paradigmatic example of Eastern 'logic' in the broad sense, expressed in art as well as philosophy [18], has something to offer to Western thought, that this perspective receives rigorous new support from LIR.

We conclude that the differences between Eastern and Western logic and the thought patterns underlying them are as important as the similarities but for a different reason than is usually adduced. In its partial rejection of some of the critical categorizations by the former of the latter, a dynamics of human knowledge is established in which the two approaches complement one another. This is perhaps the essential original Eastern insight, which needs to be further integrated into Western thought, of the Tao being both the whole and a part of itself.

The LIR approach to the hermeneutic relations outlined in this paper might be considered as a kind of Relational Quantum Mechanics (RQM; [16]) of thought. In standard RQM, the relations between entities are as important as the entities themselves, the relata that are the 'end-points' of the relations. In my opinion, the relations between world-views in East and West, can be expressed as logical processes following Logic in Reality, and are as important as, and inseparable from, those views or positions. The unique dualism of information has *ipso facto* major implications for science and philosophy as a new form of cognitive object that forces their mutual integration in East and West. In the conception of Wu Kun, the perspective of the 'scientification of philosophy' and the 'philosophization of science' is not intended to eliminate the specificity of either discipline nor their individual development at theoretical or practical levels, but requires the acceptance of the non-separability of certain kinds of science and philosophy. The consequence may be an improved understanding the ethical and social level of an integrated, more logical approach, in the sense of relation to reality, to eventual resolution of on-going conflicts in the information society. The monograph by the European philosopher of ethics and information, Rafael Capurro [5] is of particular interest in this context because it relates these two subjects to the Tao (Dao as usually written in German documents).

2.9 A World Without Change

It is important to recognize that to all intents and purposes, the quantum level of reality is not characterized by change in the humanly accessible, thermodynamic sense. Many thinkers have referred to the 'dance' of virtual particles in and out of existence, and mathematical theories exist describing these purported phenomena. This is a world with energy that may indeed be related to the emergence of 'something' from 'nothing', but it cannot be discussed in the context of a logic of and in reality of which humans can have some experience, even through the intermediary of instruments. As we will see in Chap. 3 on the grounding of Logic in Reality and also in Chap. 6, macroscopic processes have selected properties that are isomorphic to quantum properties at a different level of reality, but they are not the *same* processes nor, and this is essential given the principles of our logic in other cases, the same *and* not the same. If there is a philosophy that can be attributed to the inhabitants of the quantum vacuum, we will be glad to read it as a study in imagination, but it will not be included in our Philosophy in Reality.

2.10 Chuang Tzü and the Logic of Energy

The major classical Chinese work, designated by the name of Chuang Tzü, (ca. 399–295 BCE), is most relevant to this study. In the form in which his thought has been preserved, that of some 30 Chapters [17], 'Chuang Tzü' sets out a doctrine as a series of teaching stories or fables. Rather than categorizing them reductively as 'mystical' doctrines, we consider them as elliptical (shorthand) ways of referring to the extremely complex real relations obtaining in life and mental phenomena. From the perspective of Logic in Reality, these relations are indeed real, involving the transfer or change in form of energy, and are therefore within the purview of science as well as philosophy, hence of logic, hence a part of Chinese Logic.

Capra's dichotomy mentioned above—Taoism is intuitive 'mysticism' versus rational science as two complementary manifestations of the human mind—is thus open to the critique of being too 'Western'. In the Lupasco approach, unification of knowledge, which Capra also calls for, should not be taken to involve the loss or conflation of the specificity of two opposing parts of processes or doctrines. This is a typical weakness of 'Western' thought; in LIR, (the expressions of) Yin and Yang are not complementary in the weak sense of Bohr but linked dynamically hence logically. A related energetic concept appears in another Chinese classic, Sun Tzu's *The Art of War* [20]. Chapter 5 has the title 'Potential Energy', an acceptable translation of *shi*[4], which also means power and momentum. There is a clear indication by this author of his understanding of a dynamic complementarity of opposites, direct and indirect, action and non-action, how they are interlocked, merge and transform into one another. The comment is also made that the "entire premise of the *Book of Changes* (*I Ching*) is that it is possible to see into potential changes before they occur".

In Graham's discussion of the Chuang Tzü [9], a Chapter entitled *Rejection of Logic* helps to position Logic in Reality (LIR) as an alternative to standard logic and philosophy, compatible with Chinese thought. The Chuang Tzü throws out claims that reason can give us certainty, and LIR denies any absolute character to certainty (or uncertainty). LIR states, like the Chuang Tzü not only that "neither side of a dichotomy can be wholly true" but also the cognitive movement from one 'side' to the other and the possibility of emergence of a unification or resolution. Chuang Tzü rejected the simplistic, binary disputations that passed for logic in his time. Graham summarizes Chuang Tzü's Taoism as a whole philosophy of life derived from "a single imperative to deal with things as they objectively are, not as one would like them to be". This message remains rather pertinent for the West in 2020.

In the West, a complex energetic structure of reality had been worked out by Aristotle. Movement or change is instantiated in a triad of concepts: (1) *dynamis*—power; (2) *energeia*—the operation of power and (3) *entelechia*—the product or outcome of that operation (see also the discussion of Aristotle in Chap. 5). The middle term, the 'energy' links rather than separates the other two. The contribution of Lupasco was to see the recursive overlap between the first and third states in complex ontological processes, for example at the social level of reality. These essential details have, unfortunately, been occulted in the West in the oversimplifications of linguistic bivalent and multivalent logic and category theory.

The well-known Peircean idealized classifications of Firstness, Secondness and Thirdness fail to function as adequate descriptions of reality due to the absence of an ontological commitment to a physical energy underlying it. The Aristotelian conception of power is recovered, however, in that of energy as the capacity to do work in the onto-biology of Terrence Deacon's *Incomplete Nature* [7].

There are many relations between Western and Eastern thought and culture to which the principles of a logic of energy (LIR) do and do not have to apply. We suggest, however, that Western doctrines which have the capacity of 'making the connection' to the Tao must possess a degree of complexity that is something like Ashby's Law of Requisite Variety. One formulation of this is as follows: "During regulatory or control processes involving feedback, in the face of perturbations with a variety of possibilities for action, the relevant regulative mechanisms must be able to produce at least as many types of counteractions as there are disturbances. The regulator should thus have a maximum potential of internal variety or diversity."

The contemporary Chinese thinker Wayne Wang [23] has stated that Aristotle and Lao Tzü can not be compared since the latter has been viewed as totally anti-logical. This because traditional Taoism distorted or ignored the logic in the Tao Te Ching, while Wang's non-formal logic suggests that the logical structures of Lao Tzü can be discussed in terms of Aristotle's Square of Opposition. In terms of standard logical principles, Lao Tzü and Aristotle are thus completely consistent, and on this basis the inclusion of logic in the Tao Te Ching would make it a universal system. While we are sympathetic to Wang's approach, the non-formal logic described in his "The True Meaning of Tao Philosophy" [24] remains an epistemic system, as in standard bivalent Western logic. As we have seen, the logic of the Tao is dynamic in an ontological way and is best described by LIR.

2.11 Further Aspects of Chinese Philosophy and Logic

2.11.1 Harmony

The concept of the harmony throughout nature is present implicitly or explicitly all of the Chinese classics. It is clearly one of homeostasis, an integrated dynamic equilibrium with the potential for the emergence of new phenomena. The complete title of Hansen's version of the Tao Te Ching includes the critical 'On the Art of Harmony'. Throughout Jullien's treatise on art one finds references to the harmonic dimension in music and elsewhere, harmonic coherence, images with a regulatory effect that are generators of order. Harmony is that of image and phenomenon, also, for which Chinese uses the same word. Harmony appears also in the work of the Western ecologist Ulanowicz [22]. In his semi-classical conception, under some non-equilibrium circumstances, the production of order becomes an inevitable feature of increasing entropy. In particular, perpetual harmonies can emerge from the collapse of non-equilibrium configurations. Thus in Ulanowicz' view, "an appropriate metaphor for the course of nature seems to be less the headlong drive towards an endpoint than a more nuanced interplay between agonistic tendencies. Here the Hegelian dialectic or the Eastern dance between Yin and Yang spring to mind."

This is, in another context, a further expression of the operation of the Lupascian Principle of Dynamic Opposition at the social level of reality. A society which is respectful of the environment is in a relation to and appreciation of it which is also a harmony. This is the ground of ecology which one can and should be happy to see is a rapidly expanding field in the West, as discussed further in Sect. 2.8.

2.11.2 The Logic of Tao Philosophy

In a recent monograph entitled *The Logic of Tao Philosophy*, Wang [23] goes beyond Bohr's notion of a complementarity of Yin and Yang toward an understanding of the dynamics of their interaction. Wang's approach is thus compatible with the concept of the unfolding of actuality and potentiality in real phenomena proposed by Logic in Reality.

Logic in Reality establishes an ontology of complementarity based on the Principle of Dynamic Opposition as a "meaning principle". The necessary concepts are the non-separability of individuality and non-individuality; part and whole; subjectivity and objectivity in relation to the experiment-experimenter pair; and of object and meta-levels of theories. The difference, which is not absolute between reciprocity and complementarity is only that the former is more quantitative and the latter more qualitative. Deacon [7, 8] discusses complementarities of symmetry and shape as the source of self-amplifying non-linear processes and also refers to the "reciprocal complementarity of these self-organizing processes."

As Brenner has done elsewhere, Wang establishes the equivalence of his concept of process with that of Whitehead. (Whitehead for us is a 'guarantor' of the authenticity of one Western philosophical tradition.) When Wang says that Tao philosophy has a 'complete logical structure', he clearly does not refer to a logic of propositions but a logic of the fundamental nature of reality and of phenomena themselves as described in the *TTC*.

Among later Chinese scholars cited by Chan, we note Chang Tsai (1020–1077) to the effect that 'completion' is not possible unless there is mutual influence between similarity and difference (change) and between being and non-being (becoming in a sense of not-yet-being). We see this as a basis for the emergence of new states. Few Western philosophers talk about the existence of a *dynamic* relation between similarity and difference, but this is what we consider the contribution of Lupasco. In his Great Synthesis, Chu Hsi said that while seemingly dualistic, principle and material force are never separate. While one can argue endlessly about what 'principle' is, for me there is a clear intuition of a potentiality inhering in real systems. Continued in the materialism of Wang Fu-Chih (1619–1692), principles are in any case only found in concrete objects and systems, as in the approach of Lupasco.

Chan summarizes the debates on metaphysical concepts by the group of philosophers known as the logicians. This group included Hui Shih, a contemporary of Chuang Tzü, and their references to binary aspects of language and life and are no better or worse than corresponding ideas in the West. Also, in the last Century, the attempt by Fung Yu-Lan to convert neo-Confucian ideas into logical concepts and a philosophy of transcendence has not added much to the original insights of the 'sages'. Hansen has stated that the Chinese had a semantic theory but not a logic, because of the absence of syllogisms. However, if any reader is aware of an explicitly Chinese logic, ancient or contemporary that is a logic in and of real phenomena we would be most interested in learning about it.

2.12 Other Chinese and Eastern Traditions

The apparent differences between Eastern and Western thought are often referred to as a 'divide', consistent with the standard tendency in the West to seek differences rather than similarities. This problem in the tradition of thought has been addressed directly in a recent important article by Gu and Guo (GG) [10]: "How can we cross the intellectual divide between East and West?" This article is a commentary on another, by Fan Meijun and Wang Zhihe which calls for concerted efforts to remove obstacles to genuine dialogue. It is accepted that, by and large, the Western tradition is founded on a disjunction between nature and culture, subject and object, transcendence and immanence, the Eastern is based on a continuum between the human and natural worlds. GG point to other classical authors, in particular Chuang Tzü (that is, his original work plus the documents collated under this designation), whose work provides possible bridges, To anticipate, 'Chuang Tzü' conceives the world as a web of interrelated processes of which human beings are integral parts. This is, in other words, Heidegger's core idea of *Dasein*, that of a being entangled in the world, neither and both subject and object, a unity of Being-in-the-world. For GG, an informal and to a large extent unconscious cultural logic in both East and West constitutes the fundamental barrier to the East–West dialogue and any consequent potential integration that would preserve the specificity of both.

As a body of knowledge, in both East and West, logic originally referred to a way of reasoning about reality, nature and its changes. Relatively early in the West and in India, however, it developed into a tool for assessing the truth or falsity of propositions or their mathematical equivalents. Comparison of Eastern and Western logics has tended to focus on the degree to which Eastern logics meet this criterion, in the absence of any other. Formal logical relations can hold only between propositions, not between concrete entities or processes, etc. From this standpoint, standard Eastern and Western logics are the same. Quantum logic and paraconsistent logics (see Chap. 3) have challenged part of the Aristotelian paradigm, but the latter remain linguistic, truth-functional systems.

In our view, GG's correct description of the dichotomy between Eastern and Western thought, in an oppositional paradigm, can be seen as an artifact of Western thought in the first place, as both metaphysics and logic. The authors cite Whitehead's view of the difference as one between emphasis on process and fact, or process and substance [25]. The disjunction can also be described as being between subject and object, interconnection and separation, transcendence *and* immanence *versus* transcendence *or* immanence. As is obvious, the Western worldview reflects and is reflected in classical and neo-classical, bivalent, linguistic, truth-functional logic. All these adjectives refer to slightly different aspects of separability as a fundamental categorial feature. What, however, is the Eastern counterpart to Western logics, the specific 'divide' that may and perhaps should also be bridged?

We argue that the comparison between the relevant systems of thought has in fact not been properly made. First of all, the proper geographical division should be between a 'West' constituted by India and the Middle East as well as Europe—and an 'East' by China and Japan. As discussed further below, formal linguistic logics were studied and developed in India as early as 500 C.E. and before. In contrast to the propositional logic of classical India, in this East, an informal logic also developed inhering in biological and cognitive processes and art.

This 'Eastern' logic has remained uncodified. The non-standard, non-propositional Logic in Reality (LIR), based on pioneering work in the twentieth Century by Stéphane Lupasco, as an extension of logic to real energetic processes (Brenner 2008). Grounded in quantum mechanics, LIR permits inferences about the emergence and evolution of such processes. It is a 'logic of energy' which can be related to traditional dualistic views of energy in, especially, Chinese thought. This development is taking place contemporaneously with advances in China in other domains to which LIR can be shown to apply, e.g., Informational Philosophy and Ecology. A rigorous revalorization of traditional insights in both West and East is thus possible on this new logical basis. In view of its central role in this book, we offer an overview of LIR in the next chapter.

References

1. Brenner, J.E.: Linking the Tao, biomathics and information through the logic of energy. Prog. Biophys. Mol. Biol. **131**, 15–33 (2017)
2. Brenner, J.E.: The logic of change in reality. Paper presented at the Conference of the Italian Society for Logic, Bergamo, Oct 2009
3. Brenner, J.E.: Logic in Reality. Springer, Dordrecht (2008)
4. Capra, F.: The Tao of Physics, 35th Anniversary Edition. Shambala, Boston (2010)
5. Capurro, R.: 2010. The Dao of the Information Society in China and the Task of Intercultural Information Ethics, Copyright Rafael Capurro. https://www.capurro.de/home-cn.html.
6. Chan, W.T.: A Source Book in Chinese Philosophy. Princeton University Press, Princeton (1963)
7. Deacon, T.W.: Incomplete Nature. How Mind Evolved from Matter. W. W. Norton & Co, New York (2012)

8. Deacon, T.W.: What is missing from theories of information? In: Davies, P., Gregersen, N.H. (eds.) Information and the Nature of Reality: from Physics to Metaphysics. Cambridge University Press, Cambridge (2010)
9. Graham, A.C.: Chuang-Tzü. The Inner Chapters. Hackett Publishing, Indianapolis (2001)
10. Gu, M.D., Guo, J.: How can we cross the intellectual divide between East and West? Reflections on reading "toward a complementary consciousness and mutual flourishing of Chinese and Western cultures: the contributions of process philosophers." Philos. East West **65**(1), 298–315 (2015)
11. Hansen, C. (ed.): Laozi. Tao Te Ching; on the Art of Harmony. Metro Books, New York (2009)
12. Hon, T.-Z.: Chinese philosophy of change (Yijing). In: Zalta, E.N. (ed.) The Stanford Encyclopedia of Philosophy (Summer 2019 Edition) (2019). https://plato.stanford.edu/archives/sum2019/entries/chinese-change/
13. Hooker, C. (ed.): Philosophy of Complex Systems, Handbook of the Philosophy of Science, vol. 10. Elsevier, Amsterdam (2011)
14. Igamberdiev, A.U.: Evolutionary transition from biological to social systems via generation of reflexive models of externality. Prog. Biophys. Mol. Biol. **131**, 336–47 (2017)
15. Jullien, F.: La grande image n'a pas de forme. Éditions du Seuil, Paris (2003)
16. Laudisa, F., Rovelli, C. : Relational quantum mechanics. In: Zalta, E.N.: The Stanford Encyclopedia of Philosophy (Spring 2002 Edition). https://plato.stanford.edu/archives/spr2002/entries/qm-relational/
17. Legge, J. (ed.): The I Ching. The Book of Changes. Dover Publications, New York (1899)
18. Lu, Z.: Scientific Holism: China Meets West. Journal of East-West Studies **2**(3), 34–49 (2012)
19. Lupasco, S.: Le principe d'antagonisme et la logique de l'énergie. Éditions du Rocher, Paris (Originally published 1951, Éditions Hermann, Paris) (1987)
20. Minford, J. (ed.): Sun Tzu. The Art of War. Penguin Books, London (2002)
21. Nicolescu, B.: Relativité et Physique Quantique. Éditions Albin Michel, Paris (1998)
22. Ulanowicz, R.E.: Increasing entropy. Heat death or perpetual harmonies? Int. J. Des. Nat. Ecodyn. **4**(2), 83–96 (2009)
23. Wang, W.L.: The Logic of Tao Philosophy. The Principle of Oneness. Helena Island Publishers, Darien (2013)
24. Wang, W. L.: The True Meaning of Tao Philosophy: Its Logic and Applications. Createspace Independent Publishers (Amazon), Scotts Valley CA (in Chinese) (2016)
25. Whitehead, A.: Process and Reality. Corrected Edition. Free Press, New York, NY (Originally published 1929)(1998).

Chapter 3
Logic in Reality

3.1 Introduction. Logic and Logical Philosophy

The title of this Section is the name of an important *Journal* which raises the issue, relevant for this book, of the relation between the two disciplines. The *Journal* states that "it is devoted to philosophical logic and the application of logical rules to philosophical problems". The Wikipedia entry for "Philosophy of Logic" [53] gives four definitions, as follows:

- Philosophy of logic is the area of philosophy devoted to examining the scope and nature of logic.
- Philosophy of logic is the investigation, critical analysis and intellectual reflection on issues arising in logic. The field is considered to be distinct from philosophical logic.
- Philosophical logic is the branch of study that concerns questions about reference, predication, identity, truth, quantification, existence, entailment, modality and necessity.
- Philosophical logic is the application of formal logical techniques to philosophical problems.

Two critical assumptions are made in these definitions, in (1) and (2) that there is a clear separation between philosophy and logic and in (3) that logic is limited to a binary linguistic domain limited to the familiar issues of truth-functionality, quantification and modal conceptions of the structure of the world.

The further comment is made that the demarcation between the philosophy of logic and philosophical logic is a recent one and "is not always entirely clear". To cut short this part of the discussion, we suggest simply that this is a metalogical question which is itself a consequence of a conception of logic limited to a linguistic domain of propositions and their mathematical equivalents. The solution will emerge from the extension of logic discussed in this Chapter and its application to a new, more adequate definition of Natural Philosophy in Chap. 13.

J. E. Brenner and A. U. Igamberdiev, *Philosophy in Reality*, Studies in Applied Philosophy, Epistemology and Rational Ethics 60, https://doi.org/10.1007/978-3-030-62757-7_3

3.2 A Logic of Change

In the preceding Chapter, change has been discussed from essentially a phenomenological standpoint, which has allowed for continuity between classical and modern conceptions. In this book about change, about changing and real-world processes, we are interested in phenomenological and ontological change. Little has been written about the former from a formal logical standpoint. As to the latter, real change is ubiquitous in existence and experience, but theories of change have focused on making it mathematically, computationally and logically tractable, within the framework of standard logic. For example, differential equations provide an excellent description of continuous change, but what if the change in question is partly discontinuous, recursive and/or random, or, and above all, qualitative? In fact, change is contradictory: the most familiar thing about change is that it never occurs in isolation from stability. Change is regular *and* irregular; consistent *and* inconsistent; continuous *and* discontinuous. Since the only logics available have been propositional bivalent logics, incapable of accepting the real contradictions present or implied in the description of real phenomena, they have been incapable of describing change.

As noted in Chap. 1, in the mid-twentieth Century, the Franco-Romanian philosopher Stéphane Lupasco (Bucharest, 1900—Paris, 1988) saw, in the emerging particle-field view of the universe, the basis for a new kind of logic that extends its domain to real phenomena, relating them to an underlying particle/field view of the universe [37]. Its axioms and rules provide a framework for explaining the evolution of real world entities and processes at all levels of reality and complexity, responding in part to the need perceived by von Bertalanffy in his grounding of systems, without going outside the laws of physics. Brenner has written several articles in the last decade which describe the non-standard, non-linguistic logic that we see instantiated in changing processes [8]. The extension of Lupasco's system made by Brenner is called Logic in Reality (LIR), and it has been applied most recently to the fields of information [10] and the philosophies of information and ecology, as we will see.

Despite its publication in some fifteen books in his lifetime and its continuation by his associate Basarab Nicolescu [44], Lupasco's system of thought has remained unknown outside France, where it had been rejected by the academic community. With a few notable exceptions, Brenner's publications have suffered the same fate. The reason here is less obvious, but in our opinion it has to do with the fact that acceptance of such a logic of and in reality requires the acceptance of a new, scientifically grounded concept of the dynamics of change in all complex, interactive phenomena at biological, cognitive and social levels of reality.

The comparison of the Lupasco logic with standard logics is rendered difficult due to the limitation of the latter, noted above, to the linguistic domain. These semantic logics, bivalent or multivalent and their most recent epistemic, paraconsistent and paracomplete versions still require absolute separation of, for example, continuity and discontinuity, space–time and matter, chance and necessity, etc. and lead in linguistics to the paradoxes with which we are all familiar. Paraconsistent logics, which accept contradiction, capture only the linguistic as opposed to the physical aspects of processes, although some real inconsistencies in simple change (Sorites problems) are accepted. Logics of epistemic change are based on linguistic abstractions. No logical characteristics are ascribed to the physical processes of change. It is thus not surprising that no generally applicable theory of change has been developed for the extant domain of macroscopic, complex and interactive processes.

3.3 Logic in Reality: Actuality, Potentiality and Emergence

The major contribution by Lupasco to a restructuring of logic that now grounds it in physics was a view of phenomena in terms of processes instantiating dynamically different degrees of actuality and potentiality. Its fundamental postulate is that (1) every real complex process is accompanied, logically and functionally, by its opposite or contradiction—Principle of Dynamic Opposition (PDO). We reproduce from [9] what is still the best and simplest expression of this Principle: "The antagonistic dualities of our world can be formalized as a structural, logical, and metaphysical principle of opposition or contradiction instantiated in complex higher-level phenomena (Principle of Dynamic Opposition—PDO). The fundamental postulate of LIR is that for all energetic phenomena (all phenomena) alternate between degrees of actualization and of potentialization of themselves and their opposites or 'contradictions' but without either going to the absolute limits of 0% or 100%. The point traversed at which a logical element and its opposite are equally actualized and potentialized is one of maximum interaction from which new entities can emerge. It is designated by Lupasco and Basarab Nicolescu, the physicist colleague and major continuator of Lupasco [44], as a 'T'-state, T for included middle or third (*Tiers-inclus*). It is thus the opposite itself of the principle of the *excluded* middle in standard bivalent propositional logic. A relatively simple example of a physical T-state is the transition state in a chemical reaction. This is the point at which the number of molecules of reactants moving toward more thermodynamically favored products and the number moving in the reverse direction is the same. We use the concept of T-states to evaluate both philosophical and scientific theories, including patterns of human individual and social behavior. A dynamic systems view can be used to focus on the feedback or recursion present in all natural processes."

The absence of debate of these concepts of the dynamic general properties of natural phenomena has given them the appearance of a statism when the opposite was intended. The choice of the domains of application of the PDO and LIR in Sects. 3.2 and 3.3 of this book was in part based on the difficulties of explaining them through the use of standard doctrines, as well as the desire to have a clear demonstration of the utility of the PDO as an explanatory methodology.

3.3.1 The Grounding in Physics

Classical precedents to Lupasco's concept of a logic of and in reality go back to Heraclitus. They were based on the observation that the macroscopic world evolves in a manner that is both regular and irregular, in other words that change itself changes. This is an oversimplified formulation of the dynamics of reality, of real processes in nature which express the dynamic properties of energy, in its various forms, above what can be called the thermodynamic threshold at which to conscious observers, the phenomenon designated as 'time' appears. It is a Natural Philosophy although Lupasco did not designate it as such, given the lack of any grounding to the concept in the mid-twentieth Century. Our view is that in fact it is the Lupasco logic that grounds a natural philosophy and differentiates from philosophy as a whole, that is, one which includes its idealistic or transcendental components.

It was recognized early in the development of quantum mechanics that the principles of standard logic did not apply to the equations describing quantum phenomena: the terms of the latter do not commute or distribute. The relevant logic is thus non-Boolean, but it is noteworthy that this term is scarcely used in the literature; it is always 'Boolean' logic that is accepted as the default logic, without consideration of an obvious alternative (see below, Sect. 3.4). In addition, the simplistic property of 'time' does not appear in these equations. The 'laws' of non-contradiction and the excluded middle (the second and third axioms of Aristotle) do not apply. It was the contribution of Lupasco to restate these Axioms positively in way which recognizes the real contradictions, better counteractions in the real world and the existence of emergent entities from an 'included' middle state of maximum interaction or counteraction.

The 1st Law of Thermodynamics, in our thermodynamic world, states that energy is neither created nor destroyed but only changed in form. This is also true, of course, of catastrophic quantum events such as the mutual annihilation of particles and their anti-particles, but this phenomenon has no consequence for our daily existence, to which quantum mechanics and its logic do not apply. (We will see later to what extent catastrophe theory is and is not relevant to our thesis.)

Lupasco thus postulated that there is a logic of energy [37] that can describe energetic phenomena in this thermodynamic world, that is, all phenomena, and their evolution. This conception of logic, however, runs immediately into conflict with standard notions of logic as constructed from propositions, which can be true, false or neither, about real states of affairs. Propositional or truth-functional logics, and their mathematical equivalents, remain the basis for the overwhelming majority of scientific and philosophical discourse, at least, until the emergence of quantum mechanics in the early twentieth Century. We ask the reader to at least suspend a view of logic that *requires* it to refer to propositions and accept, as a basis for discussion, the existence of another form of logic as both a scientific and metalogical principle.

In the book which precedes this one [9], and in subsequent papers Brenner updated and extended the Lupasco logical system, now designated as Logic in Reality (LIR). The six axioms of Logic in Reality in this formulation are the following: the

Table 3.1 Directions of change

Level	Major development toward
Quantum	Contradiction; T-state (superposition)
Physical (inorganic)	Non-contradiction of identity
Biological	Non-contradiction of diversity
Cognitive	Contradiction; T-state (emergence)

first three are modifications of the three classical axioms and the second three those necessary for the extension to real systems.

LIR1: *(Physical) Non-Identity*: No A at a given time is identical to A at another time.
LIR2: *Conditional Contradiction*: A and non-A both exist at the same time, but only in the sense that when A is actual, non-A is potential, reciprocally and alternatively.
LIR3: *Included (Emergent) Middle*: An included or additional third element or T-state emerges from the point of maximum contradiction at which A and non-A are equally actualized and potentialized, but at a higher level of reality or complexity, at which the contradiction is resolved.
LIR4: *Logical Elements*: The elements of the logic are all representations of real physical and non-physical entities.
LIR5: *Functional Association*: Every real logical element e-objects, processes, events—is always associated, structurally and functionally, with its anti-element or contradiction, non-e, in physics terms, they are conjugate variables. This Axiom applies to the classical pairs of dualities, e.g., identity and diversity.
LIR6: *Asymptoticity*: No process of actualization or potentialization of any element goes to 100% completeness.

3.3.1.1 The Properties and Directions of Change and Interaction

In sequence, these axioms express the *notion*[1] of change; the *mechanism* of change; the *product* of change, that is, emergence of a new entity or phenomenon; the *locus* of change in the elements of reality; the *structure* of reality and a *property* of change. The nature of these real-world elements can be assumed to be what are commonly termed 'facts' or extra-linguistic entities or processes.

The logic in and of reality (LIR) is thus a logic applying to interactive processes, in a process-ontological view of reality, to trends and tendencies, rather than to 'objects' or the steps in a state-transition picture of change. Processes are described formally as transfinite chains of chains of chains, etc. of alternating actualizations and potentializations. Table 3.1 summarizes the directions of change at the major levels of reality in Lupasco's conception.

[1]This notion of change is in a sense anticipated by the predicate reading of the classical axiom of identity.

The *primary* directions of change are thus either (1) toward stable macrophysical objects and simple situations, the result of processes of processes, etc. going in the direction of a "non-contradictory" identity or diversity: or (2) toward a state of maximum contradiction (T-state for included third term) from which new entities can emerge. The Lupasco logic is, therefore, a logic of emergence, a new non-propositional, non-truth-functional logic of change.

As noted by Igamberdiev, Aristotle outlines his fundamental concept of potentiality and actuality in the "*De Anima*", ("*On the Soul*"). In this already metaphysical context, he writes that potentiality is matter and actuality is form. We need to know what matter and form are, but we can assume they are different. Why, to use a more modern term, potentiality should be instantiated in matter and actuality in form is not clear. What is more important for this study, however, is the relation that *Aristotle* sees between potentiality and actuality. In *De Anima* there is no direct indication that one can be transformed or transform itself into the other, but this is implied by his concept of life as an internal transformation of the body. If it is implied by Aristotle's view of the movement of energy, there is certainly no indication of the possibility of movement from actuality to potentiality.

As discussed by Brenner in [9], Lupasco's definition of the two terms in terms of energy has place for this contrary forward movement. It is implied by the fact, observable in part, that no complex process goes to the absolute limits of 100% potentiality or 100% actuality, except in trivial cases or those in which there is no real interaction. The entire literature around Schrödinger cat fails as 'alive' and 'dead' are not interacting states. Given this interpretation, why is not more attention paid to it in philosophy when it fits modern physics?

The major distinguishing component of Logic in Reality is thus its foundation in the physical and metaphysical dualities in nature. LIR is based on the quantum mechanics of Planck, Pauli and Heisenberg, and subsequent developments of twentieth century quantum field theory. LIR states that the characteristics of energy—extensive and intensive; continuous and discontinuous; entropic (tendency toward identity or homogeneity—2nd Law of Thermodynamics) and negentropic (tendency toward diversity or heterogeneity—Pauli Exclusion Principle)—can be formalized as a structural logical principle of dynamic opposition, an antagonistic duality inherent in the nature of energy (or its effective quantum field equivalent) and accordingly of all real physical and non-physical phenomena. The overall theory is thus a metaphysics of energy and LIR is the formal, logical part of that metaphysical theory.

The main function of the Principle of Dynamic Opposition (PDO) is thus to describe the structure of *interactions*, emphasizing their character as *actions* involving mutual change of the aspects of the processes involved. The critical change, and this should be kept in mind for all the examples discussed in this book, is from actual to potential and from potential to actual.

3.3.2 The Res Potentiae of Heisenberg

Further insight into the operation of the PDO can be obtained by looking at the parts of the quantum world which a provide models for macroscopic processes. Thus, descriptions of the everyday world in terms of a Cartesian ontology of *res extensa* and *res cogitans* are still common. An example is the analytical theory of human communication of Loet Leydesdorff which we will discuss in detail in Chap. 12. In his development of quantum mechanics however, Werner Heisenberg found it necessary to introduce the concept of *res potentiae*, potential things to account for the behavior of quantum entities. Attempts to extend the quantum mechanics to the macroscopic world, e.g., to explain consciousness by imputing localized quantum processes in the human brain, as well as the effect proposed by Conrad of subquantum fluctuations in biological processes have so far not been demonstrated satisfactorily. In their recent article Kastner et al. (KKE; [33]) reinterpret Heisenberg's ontological duality of a *res extensa* and a *res potentia* as mutually implicative ontological extants that serve to explain some of the hard problems of physics.

The authors state that quantum states instantiate a form of quantum *res potentiae* in a non-substance dualism as suggested by Heisenberg, the other term being a more or less standard *res extensa.* These quantum potentiae, however, are non-actuals and as such not space–time objects "and they do not obey the (axioms of the) Law of the Excluded Middle (LEM) or the Principle of Non-Contradiction (PNC). … *Res extensae*, which constitute "structured elements of space–time", are actuals that do obey LEM and PNC."

In an article with Robert Bishop [6], Brenner showed that the concept of real macroscopic processes being constituted by potential as well as actual parts was consistent with the KKE interpretation of Heisenberg. As discussed by KKE, Heisenberg considered that the probability wave of quantum mechanics was a potentia, a new kind of physical reality "halfway between the massive reality of matter and the intellectual reality of the idea", and the reduction of the probability wave during measurement was a movement from potential to actual. His intuition that the choice of the term potentia implies is intriguing, as is the 'halfway' that reminds one of the Lupasco T-state. The work of KKE is thus essential to counter the on-going lack, in the literature, of ontological commitment as to the role of potentiality, going back to Russell's statement that logical constructions (in the standard sense) are to be substituted for inferred entities. No reference that we have uncovered suggests that non-localized potentialities not only have real existence, fitting the logical foundations of LIR, but also are functionally related to actualities, an Axiom in LIR. If one follows this line, one arrives at potentiality as an inferred physical entity. The philosopher and logician Aerts [1] stated, supporting the LIR picture, that "Change is described by potential properties becoming actual and actual properties becoming potential (emphasis ours)." The reciprocity of the interaction between actual and potential remains is central to LIR. What drives *change* or 'becoming' is the overall energy gradient of the universe, but what relates the two is the dynamic opposition inherent in energy and all of its manifestations.

We will return to the discussion of the role of potentiality in our Chapter 6 on Quantum Mechanics, but there is one additional quantum property that may underlie the structure of energy and hence of all reality: self-duality. The key underlying property of both matter/energy and the gravitational field may be that of self-duality, in its many aspects (including anti-self-duality). As suggested by Majid [40], the ultimate laws of physics may be no more than rules of looking at the universe in a self-dual way. At this fundamental level of reality, outside thermodynamics and temporal change, the classical conception of identity does not hold and can be replaced by the notion of the application of two sets of equations "simultaneously". For systems of two or more entities, self-duality is replaced by a "mirror" duality as noted above. Majid states that a fundamental theory of physics should focus on self-dual structures in a representation-theoretic sense, where objects and their representations can reverse their roles. In the metaphysics of LIR, this principle is extended to the description of change according to which two opposing entities are predominantly actualized or potentialized, alternately and reciprocally, with change driven by the overall energy gradient of the universe. I see the LIR principle of dynamic opposition as one of those "rules" for physics that may be critical for understanding higher levels of reality, going from "abstractification" towards a model of direct perception.

3.3.3 Probability

At several points in this book, we make references to probability in general terms which correspond to the common understanding of probability as the likelihood of the occurrence of some future event. Stated simply in this way, probability appears as an epistemological concept, one in any event that does not refer to a real physical state. It is an absence, as in the concept of information of Deacon to be discussed in Chapter 11. Much of the vast literature on probability is essentially devoted to examining two major types of probability as so defined, namely, logical and statistical properties and logic. For a very up-to-date discussion of this subject and its implications for computation, artificial intelligence and other issues, we refer the reader to the recent article by Chenguang Lu [36].

In the context of this paper, which focuses on the properties and evolution of real processes, the discussion of probability presents a problem. Probability or better probability function (parameter) is a measure function whose values are numbers between 0 and 1 or percentages between zero and 100 percent. The sources of the data are statistical analyses of sets of events, such that the numbers reflect the simple property of recurrence or frequency. Once a probability established, it is an epistemic object without its own dynamics. It undergoes no further change. Thus in a simple case (the appearance of one of the numbers on a die), is said to be more or less probable.

Bayesian and Bayesian inference and probability are also defined as the result of a relation between beliefs, hence of an essentially epistemic character. The complex processes of interest to our philosophy in and of reality are changing, processes in progress. We have characterized them as the resultant of two opposing forces, say attraction and repulsion such that at any point in time the process has a certain degree of potentiality and actuality. It has been actualized to a greater or lesser extent. These degrees of actuality and potentiality are also probability-like measures, but always in the sense that they are changing measures, not static entities or values. This means that if an event starts out with or later acquires a degree of actuality A and of potentiality P, and it is, for example, becoming more actual (or more potential) in reality, there is a certain probability that it will continue to do so. In this picture of the relation between probability and potentiality, probability is always a posteriori, an epistemic judgment after an ontic observation.

Lu [36] presents a very complete and authoritative overview of current views of probability and proposes a "probability framework" capable of unifying the subjective and objective aspects of probability, the relation and the applicability of statistical and logical probability at the same time." In a manner congenial to the interactionist aspects of Logic in Reality, Lu emphasizes the complementarity of the various interpretations of probability, referring to Hájek [30]. In particular the two kinds of probabilities - Shannon's for statistical and Kolmogorov's for logical are linked by Zadeh's membership functions as truth functions are connected by an extended Bayes theorem with which one can convert a statistical function (likelihood) and a logical truth function from one to the other. The LIR view, in contrast, starts from the complex explanatory system that is put into place by Lu, based on truth-functional logics, and complements it by extending the domain of application to the non-linguistic properties of real systems. We examine here the situation that obtains in real physical as well as cognitive processes such as learning or remembering. At any point in time one will have achieved a state or degree in my acquisition or re-acquisition of knowledge. This a dynamic state in that it is always moving in one direction or the opposite one: as one's knowledge becomes more actual or actualized, one's ignorance becomes more potential or potentialized and vice versa, alternately and reciprocally, without going to an idealized limit of one or the other. If one becomes 'tired', that is without the energy to pursue a learning process, it will stop, with a certain probability, of course, that it will start again. This means being in a range of discourse that amounts to speaking about two levels of reality at once or that something is and is not but neither totally. These are necessary expressions of the alternation between actuality and potentiality that we have proposed in this book as fundamental to all knowledge. What is of interest and we believe partly, but only partly, accessible is the extent to which a given process will continue on the same path, stop or reverse itself. This probability measure can be related, at first informally, to the potential state in which the system finds itself: the closer we are to finishing this book, the more likely it is that we will complete it. We will find the energy, literally, to overcome internal and external obstacles most of the time. It is the essence of this theory that an irreducible part of non-predictability is always present in any real process. Our assessment of a probability therefore is and must remain incomplete.

3.4 Implication and Deduction in LIR

3.4.1 Implication as Process. Dialectics

The key to the understanding of Logic in Reality from a *formal* standpoint point is the Lupasco concept of the relation of implication between logical elements as a *process*. As defined in Lupasco's Table of Deductions [37], ortho- and para-deductive chains of implication are an integral part of logic. A *disjunctive* dialectical oscillation is required between the first three implications of implications, and between the three implications of implications of implications controlled by the former, and so on, of which the following is the first sequence (A = actual; P = potential; T = T-state):

$$\left[(\supset A) \supset (\bar{\supset} P)\right] \vee \left[(\bar{\supset} A) \supset (\supset P)\right] \vee \left[(\supset T) \supset (\bar{\supset} T)\right]$$

This implication formula defines the meaning of disjunction as the mechanics of dialectics: no dialectic without disjunction and vice versa. It is disjunction that is implied by the fundamental postulate that permits the dialectic, and the dialectic implied by the same postulate, as principle of antagonism that permits and requires the disjunction, the connective 'or' (2000). In this paper, Aerts describes the failure of the classical 'or' but offers no adequate replacement.

The second critical concept is that the Principle of Dynamic Opposition, the fundamental structural logical principle of LIR, implies a *dialectics* at all levels of reality between the two terms of any interactive duality. In other words, the existence of the dialectical characteristics of energy, mentioned in Chap. 1—actual and potential, continuous and discontinuous; entropic and negentropic, identifying or homogenizing and diversifying or heterogenizing—describe a dialectical, dynamic structure for all interactive phenomena, physical and mental, including information, propositions and judgments. In Chap. 5, we will look more closely at the subject of dialectics in general.

3.4.2 Ortho-deductions and Para-deductions

The three "normal" chains of implication move asymptotically toward three absolute, unreachable limits, two of non-contradiction and one of contradiction. There are, however, six other "abnormal" chains of implications, each of which also subdivides into three further ones, which develop transfinitely without any particular orientation. Lupasco identifies these with the multiple combinations of linkages and ruptures (negative links) that build up the scaffolding of the various complex deductive configurations of experience and "weave the tapestry of a large number of existential forms."

The para-deductive chains are thus an integral part of logic. As can be derived from the Table of Deductions, a *disjunctive* dialectical oscillation is obligatory between the first three implications of implications, and between the three implications of implications of implications controlled by the former, and so on:

$$\left[(\supset A) \supset (\bar{\supset} P)\right] \vee \left[(\bar{\supset} A) \supset (\supset P)\right] \vee \left[(\supset T) \supset (\bar{\supset} T)\right]$$

One sees here the real meaning of disjunction: Lupasco considers it to be the mechanics of dialectic: no dialectic without disjunction and vice versa. It is the disjunction that is implied by the fundamental postulate that permits the dialectic, and the dialectic implied by the same postulate, as principle of antagonism that permits and requires the disjunction, the logical "or".

But disjunction itself, as discussed in the next Section, cannot itself be absolute and rigorous. Para-deductions, will always accompany, to some extent, ortho-deductions and vice versa. As long as the logical world exists, there will always by chains of deductions or causalities that have a sense, an orientation, and those that will have, rather, a non-sense, in the sense of Sect. 9.1, a negative sense of non-polar limits or divergence. Lupasco assigns, in fact, greater existentiality to para-deductions in that they do *not* go toward something impossible.

In other words, if the elements of the sets or classes of processes, and processes of processes follow an implicational process of ortho-deduction as indicated, and if ortho-deduction is a consequence of the operation of the principle of energetic antagonism in reality, then the concatenation of symbols developing in a transfinite manner according to their own internal dynamics represents *deductive necessity*. This is the metaphysical basis of logical necessity in *all* logics, both classical and LIR. This is another statement of my view that logic does not found metaphysics, as Kripke has been quoted as saying, but that metaphysics founds logic (see Chap. 10).

3.4.3 On Truth and the Absence of Proof in Logic in Reality

The calculus of standard logics has its major role in the construction of proofs of theorems. One of the major differences between Logic in Reality and standard semantic logic is that since its theorems do not refer to the truth of propositions, there are no proofs of a propositional kind. For a truth to be grounded is not for a binary relation to obtain between a proposition and a truth-maker, A grounding relation consists in the existence of sub-sentential thought/world relations and the fact that the object instantiates the property in question [22]. As pointed out by Kauffman, proofs are not primarily a matter of true and false, but of coherent indication, and this concept is compatible with the arguments of LIR.

There are no proofs in LIR, and also no syllogisms nor standard principle of logical consequence. In a sense, our work is always a search for a dynamic logical connectedness than for logical consequence, if one is seeking to explain the relation between A and B,

The above formulation of processes as deductions serves as a reminder that the complex transitions of reality can be modeled by a relatively limited number of logical generalizations that reveal their inner workings. The approach of Jacquette [31], looks at the combinatorial possibilities of the objects and properties of a pure classical logic, but makes no commitment as to forces at work in individual exemplars. In contrast, the 'combination' in LIR of real logical elements, the reality values referred to above, is not the result of random conjunction or adjunction of terms, as in category theory, but of a form of necessity that is both physical (causal) and logical in our extended sense.

3.4.4 *First (Predicate)- and Second-Order Logics*

To clarify further the innovation represented by Logic in Reality, we may compare it with the relatively recent concept and properties of first-order or predicate logic. This is essentially a system for quantifying (assigning numerical values to) variables in a domain of discourse rather than truth-values to propositions. These quantified variables can be further defined by a set of axioms which provide the basis for belief in those axioms and their applications in mathematics, computer science and so on. From the LIR standpoint, however, the domain of discourse remains an epistemological domain, not distinctly different in this respect from that of propositional logic. One is not dealing with real processes and the changes which they undergo. Second-order logic simply extends first-order logic by quantifying over relations, especially, the theoretical 'objects' of sets and functions collectively designated as well-formed formulas. There is a vast but readily accessible literature on the variants, models and proof systems that have been proposed for these logics which the interested reader may consult. For us, the term sometimes used of a "first-order logic in nature" refers only to what we also obviously accept, namely the existence of certain regularities of cause and effect. We can thus restate here, as it is perhaps necessary, that our logic of processes in dynamic opposition reduces to classical propositional or predicate logics in simple systems, without interactive dynamics, for whose description only a binary truth-functional logic is required. To characterize Logic in Reality in one phrase, with all the caveats required, we might say that it is the logic of qualitative change, where qualification has equivalent ontological purport as quantification.

3.5 Logic in Reality 'In Operation'

If we turn now to the causal evolution of processes in nature, that is, of change, one of the most general concepts is that of the entities responsible for effecting change, that of operators. The most general and complete discussion of the kinds of operators and their properties, of which we are aware, was given by Burgin and Brenner [14]. Operators, as such conceptual and physical entities, are found throughout the world as systems or subsystems in nature, the human mind, and the manmade world.

Operators, and what they operate *on*, that is, their substrates, targets, or operands, have a wide variety of forms, functions, and properties.

As was suggested by Burgin and Brenner, operators have explicit philosophical significance. On the one hand, they represent important ontological issues of reality. On the other hand, epistemological operators form the basic mechanism of cognition. At the same time, there has been no unified theory of the nature and functions of operators; they range from the most abstract formal structures and symbols in mathematics and logic to real entities, human and machine, and are responsible for effecting changes at both the individual and collective human levels. In our view, operators raise important ontological issues of reality reflecting their dynamical nature when some objects act on other objects. Our entire universe exists in interaction of its components and elements, and this interaction can be decomposed of separate actions involving the structure and functioning of an operator. In addition, epistemological operators can be seen as forming the basic structures and mechanisms ofcognition. Many cognitive processes, such as search, selection or recognition, are performed by physical operators, which can be natural or artificial and are represented by abstract operators decomposable into simpler operators. In this context, any scientist, measuring device or computer is a physical operator.

3.5.1 Classification and Properties of Operators

Our overall approach establishes that operators are characterized by their single or multiple classificatory definitions and their properties, as outlined by Burgin. For example, human beings are *material*, *natural*, *naturalized* and *socialized* operators. The indicated intrinsic, relational and pragmatic properties and the related disciplinary methodologies further characterize operators. Most significantly, the concept of a natural operator fits the process view of reality in LIR. Following itself the Principle of Dynamic Opposition, the discussion of operators moves from the primarily theoretical domain toward the operational characteristics of the human mind.

The applicability of logic and its symbolic operators and operations has been limited to language and mathematics. This is, however, based on the same discretionary metalogical principle introduced by the scholastic followers of Aristotle and maintained by the linguistic turn of the nineteenth–twentieth centuries to which we have called attention above.

The *function* and *process* information operators in Burgin's General Theory of Information [15] provide the basis for a more formal characterization of the calculus developed by Lupasco and outlined in [9]. The connectives, that is, what are usually defined as the symbolic logical operators of implication, conjunction and disjunction, all correspond in LIR to real operators on real elements in the evolution of real dynamic processes. These operators are also subject to being actualized, potentialized or in a T-state. They operate not on theoretical states-of-affairs or propositions, considered as the abstract meaning of statements, but on events, processes and properties, where properties also have the character of processes.

The key concept is that the LIR operators themselves must be considered as processes, subject to the same logical rules, fundamental postulates and formalisms as other real and hence, natural processes. This answers a potential objection that the operations themselves would imply or lead to rigorous non-contradiction. Real processes are, accordingly, seen as constituted by series of series of series, etc., of alternating actualizations and potentializations. These series are not finite, however, in reality, processes do stop, and they are thus not infinite. Following Lupasco, we use the termtransfinite for these series or chains, which are called ortho- or para-dialectics. The reader is referred to [9] for details of this applicable non-standard calculus.

3.5.2 Operators in Language and Causality

The concept of operators in language enables a junction to be made between the formal (non-LIR) and non-formal or qualitative properties of language. The issues and relations addressed are much more complex than those addressed by standard linguistic operators. Ghils [26] has shown, for example, that the spatio-temporal dialectics in the linguistic theory of Roman Jakobson [32] is best described by the movement between actual and potential, using the corresponding operators as expressed by the Lupasco (LIR) calculus.

These *natural* operators of Logic in Reality are extremely complex, being both *symbolic*, *material* and *mental*, but also in part *symbolized, naturalized* and *social*, since implication, conjunction and disjunction obviously also function within social systems as well as between propositions. The same picture applied to conjunction and disjunction as opposites provides the basis for a non-classical set theory, in which there is no absolute separation between sets and their members. According to de Morgan duality in classical logic, conjunction and disjunction are not *independent*, in the sense that a complementation operator takes any proposition to a similar one with the negative and operation inversed. This duality, however, still refers to a relation between abstract entities.

The discussion of the nature of operators allows us to explicate their relations to the notion of causality or cause and effect. It is easy to accept that all operators are efficient causes. However, are all causes operators? Further, what is the causal nature of the operator-operand interaction? By localizing the origin of action in the *potential aspects* of phenomenal processes, which in the LIR view are intrinsic properties of all operators, our approach cuts through the debate on whether causation by omission, absence and prevention are cases of causation or not. They are. This line of argument also applies to the artificial distinction between natural and causal change, as well as internal and external change. LIR thus supports and explicates Fair's transference theory according to which, as discussed by Dowe [23] causation is a transfer of energy and/or momentum although causation by absence does not include any transfer of energy or momentum. Besides, in the context of the general theory of information, causation necessarily is a transfer of information [11].

We have anticipated here—a totally linear sequence of subjects is impossible by our own theory—the discussion of information in Chap. 11. We simply remark here that the theory we are building will ultimately be grounded in new interpretations of what might otherwise be considered established concepts, namely, of systems, operators and processes.

3.6 Precursors of Lupasco

The seminal work of Lupasco in logic and philosophy has been summarized in Brenner [12], in the journal *Logic and Logical Philosophy.* The major difference in the Sections on precursors which follow is that Lupasco had studied Kant and Hegel in translation, but was unaware of contemporary work in German, Husserl and Heidegger or English, Peirce, Mead and others. French-language contemporaries of Lupasco, discussed below, were Gaston Bachelard, Ferdinand Gonseth and Emmanuel Levinas.

3.6.1 Kant and Hegel

Lupasco devoted a significant portion of his 1935 French State Thesis [38] to a critique of some major predecessors—Kant, Hegel, Schopenhauer and Bergson. Lupasco was able to show that each philosopher was able to maintain the coherency of his system only by ignoring or relegating to an inferior ontological status one or the other of the two essential aspects of existence, identity or diversity. The diametrically opposed approaches to time and space by Kant and Bergson could be read as an example of the way in which dialectics 'plays out' at the level of individual psychologies. Lupasco thus proposed the contradictory relation of their theories as a consequence of contradictory aspects of their logical assumptions. This alone could provide a basis for consideration of his work as a major contribution to twentieth Century thought.

Lupasco suggested that Kant, due to a tendency to actualize identity, was able to see only the form and the framework of the 'one', rejecting the diversity of experience into the mystery of the noumenon. For Kant, the logical act of understanding involves synthetic extensive and a priori judgments. Finding the connection or link between a concept A and an external predicate B, "two heterogeneous entities", constitutes the process of knowing. Non-identity was a fact, something static, a-rational and external. Analytic judgments were possible, but far from creating "real" heterogeneity, could only go over the ground of the synthesis, backwards.

Lupasco's basic dialectical concept is that, as in Kant, the connection between A and B is a phenomenon, but he added that so is the *non*-connection as well:

> A phenomenon is something which contains in itself, coming from nowhere else, its life and its death at the same time, its affirmation and negation of itself, without one being able to annihilate the other, because their existence is a function of their coexistence. This belief

> is behind all the examples we have discussed previously of the relations of contradictory terms. No thing has an absolute value. Nothing is not the absence of something but a logical quantity a positive absence of link, a contradiction of a thing by itself, a non-identity, but existential and constitutive of an analytical factor having the same value as the synthetic factor responsible for the link.

For Lupasco, the only solution is to focus on the relation, and see that when one "goes" from one concept to the other, one has in hand a basis for their difference, the potentiality of a link or a non-link, that enables one to carry out the synthesis. The synthetic approach takes place on top of the differentiating approach that evolves *between* A and B, and does not exist *in* A or B. Thus, both the synthesis and the discovery of difference, the true analysis, only exist in the relation between A and B taken as the field of investigation. Differentiation, in the case of real phenomena, like identity, is to be found only in the passage from A to B, and both have, and should be stated as having, the same existential value.

Both Hegel and Lupasco started from a vision of the contradictorial or antagonistic nature of reality; developed elaborate logical systems that dealt with contradiction and went far beyond formal propositional logic; and applied these notions to the individual and society, consciousness, art, history, ethics, and politics. Parallels to Hegel's dialectics, logic and ontology may thus suggest themselves to the reader, and so we first note that Lupasco considered that his system included and extended that of Hegel. However, one cannot consider Lupasco a Hegelian or neo-Hegelian without specifying the fundamental difference between Hegel's idealism and Lupasco's realism.

3.6.2 Husserl and Heidegger

As Brenner has pointed out in [11] , Husserl's bracketing of external existence and acceptance of only the phenomena of consciousness as reality establishes a permanent disjunction between them. As Capurro has shown [16], Kant brackets the issue of what things are in themselves (*Ding-an-sich*), independently of the knower of them. However, he is open to reality through the forms of subjectivity: experience as fundamental in the sense of the relation of the subject to objects he did not create. "So Kant creates a transcendental logic in reality, while Husserl's 'logic' appears to concern the constitution of phenomena only as correlates to consciousness. Heidegger takes Husserl's idea of intentionality but relates it to the world as the openness in which human beings meet each other and other beings, but this openness is neither a Kantian object nor the space of consciousness but what incorporates both, his *in-der-Welt-sein.* ... Our theories are then explicit or 'logical' interpretations of our pre-logical understanding of the world.

The domain of operation of the Lupasco logic of processes is the world given by this pre-understanding. In this, Lupasco is 'Kantian' in the sense that Kantian phenomena are primordial as they 'address' (call into question the existence of) the finite subject. We will return to some of the issues raised here in Chap. 10 on

Phenomenology and Chap. 15 on philosophy in reality as a philosophy of process (see also next Section). Wu Kun has made a detailed study of the relevance of Husserl to the philosophy of information [54]. Our conclusion is, however, that aspects of Husserl's conception of philosophy as rigorous science can be interpreted in a way that is consistent with the nature of physical reality in our approach. Our position will be outlined further in our concluding Chap. 20.

3.6.3 Philosophers of Process

To our knowledge, there have been no recent philosophers, even if they sympathize with the *panta rhei*, who have designated themselves explicitly as philosophers of change. We take the hopefully not too radical position that all real processes involve change and consider the group so designated in the compendium of Browning and Myers (1988). The group has nine members, and we make in this Chapter the following few comments about eight of them. With the exceptions of Nietzsche, Günther and Bergson, all are English or American.

One further preliminary remark: this book could be seen as a sequel to the *Process Metaphysics* of Nicholas Rescher [50], but the contours of the term 'process' are by no means clear. The general context is a rather reified conception of process, opposed to our tendency to try to see things in terms of change, and changes of change. We 'excuse' the thinkers mentioned in view of their contributions to thought in general, without becoming acolytes of any of them. We have taken from their essays those things which support our thesis. Our critique of Peirce is much broader, however, than that presented here, and we refer the reader to Chap. 9 on semiosis for our reasons. In many ways, this Section will be a 'preview' of the further philosophical and scientific consequences of adopting the principles of Logic in Reality. Further discussion of the relation between LIR and other philosophy and of philosophy *as* process will be made in Chap. 15.

3.6.3.1 Charles S. Peirce

It is difficult for us to associate the name of Peirce with notions of real change and process. He combines some extremely deep insights with apodictic statements for which he makes, on closer inspection, no ontological commitment. Peirce states that his logical system includes a doctrine of thoroughgoing evolutionism, that is, change, and objective idealism, which for us could be acceptable in principle. However, he grounds the latter in principles of absolute continuity and chance or indeterminism that are incompatible with reality as we see it and are excluded in Logic in Reality (LIR).

Peirce states that mental 'law' follows the forms of this logic. However, this logic is grounded in the above absolute principles. He can understand, for example, the real opposition between the personalities of two human beings, and the means by which one can recognize the other, but he cannot cast the interaction so that it describes "intercommunication between two minds" as anything but some form of direct hypothetical transfer. It is Peirce's concepts of meaning and semiosis that have received the most attention, and we will return to these in Chap. 9.

3.6.3.2 William James

In the James' Chapter on determinisms, the authors of the Compendium argue cogently against absolute determinism. Change is not completely determined, and deterministic views leave not room for or explanation of morality. LIR solves James' dilemma for us, but we would suspect that he would not accept our argument.

3.6.3.3 Friedrich Nietzsche

Nietzsche is congenial to our theses in his rejection of a priori received wisdom and dogma. "No one is responsible for the fact that he (Man) exists at all, that he is constituted as he is, and that he happens to be in certain circumstances and in a particular environment. The fatality of his being cannot be divorced from the fatality of all that which has been and will be." In other words, the universe is 'free', even if we are not. That such an idea is unacceptable to many people is also part of our reality.

3.6.3.4 Samuel Alexander

The British philosopher, considered a member of the 'emergentist' school, captures the essence of what must be at the ground of any natural philosophy of change and process: "Life is therefore resoluble without remainder into physico-chemical processes, but it cannot be treated as *merely* (italics his) physico-chemical." For this statement to hold, however, modifications must be made in the understanding both of process and of 'physico-chemical'.

3.6.3.5 Henri Bergson

Bergson is one of the few important philosophers for whom we possess a critique by Lupasco. As described in [9]: "time in the empirical philosophy of Bergson is a heterogeneous duration, psychological, biological, vital, etc. However, it was defined as being outside logic, involving another classical *distinguo* that was simply opposite to that of Kant—what is heterogeneous in Bergson is logical, what is homogeneous is not.

Bergson's discussion of the possible and the real, of appearance and reality, however, has language that favors something like our view of change and process:

"What we perceive as being a succession of states is conceived by our intellect, once the fog has settled, as a system of relations. The real becomes once more the eternal, with this single difference, that it is the eternity of the laws in which the phenomena are resolved instead of being the eternity of Ideas which serve them as models. …Time is immediately given. That is sufficient for us and until its inexistence or perversity is proved to us we shall merely register that there is effectively a flow of unforeseeable novelty." Exactly, unforeseeable, epistemological 'chance' accompanies ontological determinism. Bergson's closing sentence is very strong: "In this speculation on the relation between the possible and the real, let us guard against seeing a simple *game* (italics ours). It can be a preparation for the art of living."

3.6.3.6 John Dewey

The Chapter in the Dewey Section of most interest to us has the title "Qualitative Thought". It is therefore most appropriate to our valorization of the qualitative aspect of phenomena, and Dewey clearly states that to be universal, logic should be able to deal with more than propositions. Logic should have 'something to do' with forms of thought characteristic of qualitative objects.

Dewey complexifies the simplistic notion of an independent time. "Human individuality requires a notion of time as a fundamental reality, but for physical individuals time is not simply a measure of predetermined changes in mutual positions, but is something that enters into their being."

His idea of potentiality as an operational principle is close to that of Lupasco. Potentiality is not *only* a process of unfolding what was previously implicit or latent. "Positively it is implied that potentiality is a category of existence, for development cannot occur unless an individual has powers or capacities that are not actualized at a given time. But is also means that these powers are not unfolded from within, but are called out through interaction with other things." We only change this phrase 'to not *totally* from within'.

3.6.3.7 Alfred N. Whitehead

Whitehead's *Process and Reality* is one of our major precursor documents, despite the fact that his categories appear at first sight to be relatively idealistic ones. As discussed in Chap. 13, the movements between forms can be 'naturalized', placed into some correspondence with the evolution of real mental processes; we agree that his 'prehensions' refer to occasions of experience as an activity, "analyzable into modes of functioning which jointly constitute its process of becoming". Whitehead's conception of space–time has too little science in it, and the scientific materialism of his day was certainly a barrier to its philosophical understanding. Without going into further detail here, let us simply agree with Whitehead that "the elucidation of meaning involved in the phrase "all things flow" (*panta rhei*) is one chief task of metaphysics."

3.6.3.8 George H. Mead

Mead's article is entitled "The Present as the Locus of Reality" and deserves mention in our book for that reason, but other reasons are his clear dialectical statements: "There is little purpose or profit in setting up antinomies and overthrowing the one by the other, or in relegating permanence to a subsistent timeless world while the event, in which there is nothing but passage, is made the substantial element in existent things. *The permanent character that we are interested in is one that abides in existence and over against which change exists as well* (italics ours)." "The pasts that we are involved in are both irrevocable and revocable. ... The irrevocable past and the occurring change are the two factors to which we tie up all our speculations in regard to the future. Probability is found in the character of the process which is going on in experience."

In closing this overview, we invite the reader to position other authors in a similar manner with respect to their ideas of change and process. The reference by Mead to metaphysics as a concept that is possible to use in some way that is independent of physics is contrary to our entire approach, as we will see in Chap. 10. We now turn to a thinker who better fits the description of a direct precursor to Lupasco and the principle of Logic in Reality.

3.6.4 Gotthard Günther: Transcendental Logic and Trans-classical Rationality

Gotthard Günther was a contemporary of Stéphane Lupasco (*ob.* 1988), but it is safe to assume that they never met. Günther emigrated from his native Germany to the United States in 1940, and Lupasco was unfamiliar with both German and English and most of the literature in those languages. His paper [27] is of philosophical interest in that it records the irrational aspects of the defence of classical bivalent logic ("fanatical consensus"; "dogma"; "sacrosanct tradition". We have noted that many-valued logics are just as limited as standard bivalent logic when it is based on the same requirement of absolute truth-functionality.

Günther's solution to the problem was to relax the dichotomy of form and matter, and the disjunction of truth and falsity for many valued systems. His logic still remains within a linguistic framework—a calculus, *with one major and critical exception*: the datum of conscious self-reflection is formally allowed in his trans-classical logic. He shows that the laws of self-reflection govern an opposition between Subject and Object, involving a "tri-partite division between an individual subject, individual object and 'general subjectivity'." The conjunction of Subject and Object is what confronts us in the world that surrounds us. General subjectivity reflects the capacity of human beings to process a 'distributed' form of subjectivity or what Husserl called intersubjectivity.

His critique of Brouwer and intuitionist logic—the failure of *tertium non datur* is identical to that of Lupasco: intuitionism does not go far enough. Günther's principle of distribution invalidates not only *tertium non datur* but simple identity and contradiction, in other words, all of the cornerstones of Aristotelian logic. He concludes in the necessity of a higher order logic (>2) to properly represent human rationality, since each human being is "subject-center".

The differences between this approach and Lupasco's are clear: there is no grounding of the former in energy and the properties of real phenomena. This is however partly compensated by the inclusion of the obviously non-binary properties of human beings as subject-objects. Günther's work was a step in breaking the stranglehold of classical logic that Vasiliev called for, and that we consider vital. However, his attempt to "provide Time with an ontological locus of its own" fails due to Günther's reliance on new symbolic representations for the loci of thought and being called 'kenograms' [28]. This term is derived from kenoma which means ultimate metaphysical emptiness in Gnostic philosophy. The diagrams are an attempt to capture complex ontologies essentially with geometric diagrams—the "four ontological loci of the subject's entanglement with reality—Thought, Life, World and Time." We cite this article not because we agree that that it correctly mirrors reality, but because we feel it should not be lost. Its value is human.

3.6.5 Gaston Bachelard

Readers familiar with twentieth Century French philosophy may be reminded of that of Gaston Bachelard, a contemporary of Lupasco, but much better known for his work on the philosophy of science (he was a chemist) and poetics. Bachelard knew Lupasco's work, and, as in the case of the Swiss philosopher Ferdinand Gonseth, saw its originality but did not accept its central theses. In his book *Philosophy of (the) Not* [3], Bachelard provides a lengthy discussion of potentially needed "dialectical" changes in the axioms of logic, based on his view of the Heisenberg principle and other discoveries in science, that Bachelard said might lead to a form of ternary system. "One must agglomerate a ternary system around each dialectic." However, his view of logic as a simple distribution of true and false shows that he missed seeing the relation between logic and dialectics, and the dynamic aspects of both.

Lupasco said, according to Bachelard, one must dialectize everything as a method in the service of knowledge "sole judge and conqueror." However, Bachelard's own dialectics were not based on true contradictions with a basis in reality, but on semantic oppositions, "a sort of benevolent rupture, a phenomenon of intellectual mutation." Nothing could demonstrate more clearly Bachelard's idealized picture of existence, change and dialectics itself. Bachelard did not accept (our translation) the integration of a principle of contradiction in the "intimacy of knowledge". Bachelard's dialectical dualism of the mind was limited to putting in place a logical kaleidoscope that changed the relationships in knowledge, but kept their forms. His "surrationalism" is constituted by rational systems that are simply juxtaposed. His dialectics "serves only to change from one system to another".

3.7 Recent Developments in Logic

We recall that the term "Logic in Reality" (LIR) is intended to imply both (1) that the principle of change according to which reality operates is a *logical* principle embedded in it, *the* logic in reality; and (2) that what logic really *is* involves this same real physical-metaphysical but also logical principle. The LIR approach can be seen as a non-arbitrary method for including in theories or models contradictory elements whose acceptance would otherwise be considered as invalidating them entirely. It is a way to "manage" contradiction, a task that has also been undertaken by paraconsistent, inconsistency-adaptive and ampliative-adaptive logics only in relation to linguistic propositions or their mathematical equivalents. In order to see the 'revolution' in philosophy that is implied by the Lupasco logic, it is necessary to analyze some other recent developments in logic and establish the relation of LIR to them.

The major logical developments of the twentieth Century involved the demonstration, by da Costa and others, that rigorous logics could be constructed by rejecting the universal application of the second and third classical axioms: logics in which the axiom of non-contradiction does not apply but that of the excluded middle does are called paraconsistent; in the opposite case they are paracomplete. It is in this sense that Béziau has described them as 'duals'.

3.7.1 Paraconsistent Logic

Paraconsistent logic (PCL) is a *linguistic, propositional logic* albeit with a non-truth-functional semantics. Paraconsistent logic is *defined* such that contradiction does not entail triviality or explosion. In some paraconsistent logics, such as those of Priest [38], an ontological commitment is made and real contradictions are allowed. In others, such as the logics of formal inconsistency of Carnielli and Marcos [17], they are not. These authors have shown the relation in his logics of formal inconsistency that the indicated equation is valid:

$$\text{Contradiction} + \text{Consistency} = \text{Triviality}$$

In LIR, Triviality is equal to zero: nothing real is 'trivial'. Accordingly, Contradiction is equal to Inconsistency. In all cases, however, I believe that there is a relation to Gödel's theorems which found the existence of inconsistency in paradoxes and proof procedures. Gödel, while sticking to first-order predicate logic, saw the need for a more powerful logic, and Hintikka suggests it should be second-order or in any case with "stronger and stronger" formal rules of logical inference and better logical proof methods. In contrast, LIR relates inference to real change.

Paraconsistent and paracomplete logics do mirror some of the contradictory aspects of real phenomena. Priest has shown this in his work on inconsistency in the material sciences [48]. However LIR captures in addition the contradictorial structure of the dynamics involved in real processes, in particular of thought. Dialetheism and the Logic of Paradox (LP) proposed by Priest provide a solid basis for understanding set-theoretic and semantic contradictions in terms of the truth or falsity of certain sentences or formulas, and a 'nexus' (nodal point of transition) which is both true *and* false. It is thus an appropriate logic for aspects of language and mathematics in isolation, with single-level or two-dimensional self-reference. Priest has analyzed some real changes in these terms, such as simple motion (to which the Eleatic paradoxes apply, cf. Chap. 5 and the apparent passage of time, in which the contradictions that appear are considered to be of this type. However, the concepts of truth and falsity as used by Priest do not differ fundamentally from those used by classical logicians. The Principle of Dynamic Opposition focuses on the dynamics of the *processes* by which changes come about in reality, in terms of their alternating actualization (A), potentialization (P) and the nexus here—the T-state—which is an included third term at another level of reality or complexity.

Thus we not only agree with the paraconsistent logician Greg Restall that there is "paraconsistency everywhere", as discussed in the paper with the same name, but we add that LIR gives the ontological basis for distinguishing assertion of a negation and denial at the lower level of standard classical and non-classical propositional and predicate logics. We consider that our logic is semantically incomplete, but a genuine logic need not be semantically complete. Incompleteness means only that a Turing machine cannot exhaustively enumerate the valid patterns of inference. This is a limitation of a Turing machine, not a refutation of logic in our extended sense [18].

Thus despite some early references to real phenomena by Priest [48], an extension to real processes was never made, by Priest or by others. The result is to drastically limit the application of PCL to the real world, e.g., to the qualitative dualistic aspects of information. In one exception, Krause [34] applied PCL to quantum particles and showed that they both *are* and *are not* individuals, or both parts *and* wholes. The semantics of LIR are non-truth-functional in the different sense that their elements are not propositions at all, and the concept of truth-functionality (defined as valuations based on homeomorphisms (mappings) between formulas and an algebra of truth functions defined on a given set of values, (0 or 1 in binary logic, several values in many-valued logics) should not be applied.

Readers who are interested in more detailed discussion of why semantic dialetheic logics cannot handle true contradictions should consult Martin [41]. Our basic principle is perhaps hinted at in his statement that, but there is no move toward a non-semantic process solution of the kind we propose. The Liar Paradox, is, unfortunately, still considered a problem for a logic of truth and falsity. Other paraconsistent approaches refer to inconsistencies and 'actions', but are wholly in the realm of epistemology [5]. The presentation by Gödel of indeterminism refers to an epistemological indeterminism, it is a consequence of our inability to measure all the forces impacting as process, such as rolling dice.

3.7.1.1 Qualitative Phenomena

It is unfortunate that the qualitative aspects of the world have received little formal attention. One exception is the work of Forbus on qualitative modeling [25]. In the related concept of qualitative reasoning, group qualitative representations provide formalisms for conceptual knowledge. Qualitative mathematics expresses partial knowledge and supports causal reasoning. Qualitative reasoning captures important aspects of human reasoning. It can use partial, incomplete information and includes a qualitative process theory of change. 'Influences' provide a qualitative mathematics that is compositional and captures causal intuitions, resulting in qualitative relationships between parameters. The corresponding, qualitative mathematics provides a natural level of detail; allows for partial knowledge and expresses an intuition of causality. Qualitative direct Influences as defined above provide information about relative rates of processes that can disambiguate conflicts. Despite the abstract nature of qualitative properties, there is no loss of generality in expressing ordinary differential equations qualitatively.

The seminal article of Kuipers [35], Qualitative Simulation, provides a paradigmatic example of the limitations of qualitative approaches to knowledge. Qualitative structure and descriptions of behavior are defined as "abstractions of differential equations and continuously differentiable functions". Qualitative simulation is an inference process, but it is one for which algorithms can be written. It is even admitted that a qualitative simulation algorithm may produce spurious qualitative behavior which cannot be derived from the given constraints.

As we have argued elsewhere, such strategies invert the nomological sequence we believe correctly reflects reality and change. With a stroke of the pen, so to speak, qualitative behavior that cannot be simulated by differentiable functions is not considered worthy of further analysis. Qualitative simulation is proposed as furthering a causal explanation of phenomena, but the value of the method is reduced by limitation of the phenomena which meet the criteria for analysis. An area of inquiry in which qualitative structures are given more ontological value is that of medical diagnosis. Dynamic models are abducted from pathophysiological systems and used to create new hypotheses. Computer processing allows the integration of knowledge representation frameworks and the indicated models to yield a sound framework for diagnostic reasoning.

What is missing from the still, in comparison to other fields, limited discussion in the literature of qualitative processes is a sense of their philosophical purport. We have chosen to focus on this domain and its relations to our principles, logic and dialectics in this preliminary attempt to support an alternative, more balanced vision of the importance of quality versus quantity.

3.7.2 Paracomplete Logic

The most significant example of a paracomplete logic is intuitionist logic. Brouwer [13] and Heyting were prompted to develop intuitionist logic as a basis for mathematical reasoning about infinite sets. Brouwer claimed that the law of the excluded middle cannot apply in mathematics, "once it has been recognized to be an autonomous interior constructional activity which, although it can be applied to an exterior world, neither in its origin nor in its methods depends on an exterior world". He thus rejected the application to mathematics of a classical binary logic of 'truth' and 'falsehood', and of the concept of truth as a relationship between language and an extra-linguistic reality. However, there is no indication in this work of a basis (or need) for applying such principles outside mathematics. He did not, apparently, formulate or show the necessity of a law of the *included* middle. His formulations contain idealized distinctions and processes that apply only within mathematics. As the law of non-contradiction is maintained, however, as shown by Priest and others, intuitionist logic remains closely related (congruent) to classical logic.

3.7.3 Quantum Logics

Quantum physics, including the Pauli Exclusion Principle; the uncertainty principle of Heisenberg; and the Gödel principle of the reciprocity of completeness and consistency in mathematics has led to the development of quantum logics embodying these principles. Quantum logics offer a basis for discussing the dual nature of phenomena such as information in our interpretation. Their elements are similar to non-standard probabilities in that the laws of commutation or distribution are not followed. Aerts' quantum formalism [2] can be applied to complex macroscopic phenomena, including the emergence of biological form and human cognition. Sklar wrote in his *Philosophy of* [52] that quantum logics were inadequate to resolve quantum paradoxes related to the superposition of states. LIR provides a natural interpretation for quantum superposition in terms of semi-actual and semi-potential states, eliminating the philosophical problems associated with the modified Copenhagen and many-worlds interpretations of quantum mechanics.

The relation between quantum logics and Logic in Reality was suggested by Nicolescu: the superposition of quantum states is isomorphic to the T-state in LIR at macroscopic levels of reality. Together with uncertainty, a physical basis is provided for the Lupasco physical-logical Principle of Dynamic Opposition. LIR can thus be usefully compared with quantum logics, but further conceptual changes are necessary to make quantum-type logics applicable to macroscopic change. Further relations between LIR and Quantum Mechanics will be established in Chap. 6.

3.7.4 Abduction and Abductive Logic

Abduction, defined as reasoning to the best explanation and abductive logic, like paraconsistent logics are closer "in spirit" to LIR, although they are essentially restricted to a place on a linguistic logical map [25] and follow the rules of propositional, truth-functional logics. LIR is also capable of capturing degrees of complexity that are not captured by abductive logic, and some of the facts about real-life abduction that in most theories are suppressed, ignored or idealized can be described by LIR.

The above notwithstanding, development by Lorenzo Magnani and Nancy Nersessian and their colleagues of extending the concept of abduction to reasoning, especially, model-based reasoning goes well beyond standard conceptions of sentential abductive logic. The complexity of the interaction between abduction, considered as equivalent to abductive reasoning, models and reasoning as such is illustrated in Lorenzo Magnani's important paper "Inconsistencies and Creative Abduction in Science" [39]. Abductions of many kinds are model-based if they involve a reasoned problem solving process. The further definition of "model-based reasoning", due to Magnani and Nersessian, is "the construction and manipulation of various kinds of non-verbal representations, not necessarily sentential or formal". We attribute a process character to models, the primarily non-physical cognitive models that are most easily seen as processes, whereas most formal models have the mathematical structure that Batterman has called "applied mathematics" [4].

Formal models have inherent limitations in accounting for inconsistencies inherent in reasoning, and accordingly, in sentential abduction itself. There is a problem, however, with the expression "models of abductive reasoning", which are, presumably, cognitive models. Standard logical accounts of abduction fail to capture, in Magnani's words, "much of what is important in abductive reasoning", *i.e.* abduction, including, and we here exit a threatening circularity, the existence of model-based abductions, in which empirical, as well as theoretical inconsistencies should not be eliminated arbitrarily.

Magnani calls attention to the need, in creative scientific reasoning, to allow two rival theories not only to coexist but to compete, as epistemological and "non-logical" inconsistencies. We claim that it is possible to describe all of these concepts of interaction in a more rigorous manner by extending the domain of logic to real phenomena, including theories and their "competition". As we will see, this means redefining the meaning of the phrase "simultaneous satisfaction of a set of positive and negative constraints, without the need for question-begging term of "spontaneous" in the connectionist model.

In their "Advice on Abductive Logic", Gabbay and Woods (2006) outlined a schema of abduction in which its formal epistemological properties are clearly outlined. We agree in part with their differentiation between deduction, induction and abduction. Their proposition is that whereas deduction is truth-preserving and induction is probability-enhancing, abduction is ignorance preserving in the sense that "if you knew the answer, or the hypothesis you are making does not lack some degree of epistemic virtue, it wouldn't be abduction any more." A simplification introduced is to restrict the abductive inferences considered to those that close a cognitive agenda, that is, a drive to execute a reasoning process. We would be reluctant to make this restriction; in the LIR conception of dynamics, processes go substantially, but not completely to "completion", potentialities remain for new actualizations, for new sequences of sequences, etc. Their open-ended existence is at least as ontologically important as closure. It is fascinating to note that even in their formal model of abduction, Gabbay and Woods find it necessary to introduce a technical rule that provides for backward propagation of the abductive process "against" the flow of deduction, in fact following the lines of non-formal natural reasoning described by LIR.[2]

The difficulties for applying the Gabbay-Woods approach to model-based reasoning lie in its essential restriction of abduction to a place on a linguistic logical map, where inquiry involves premises searches and inference with premises projections, both following the rules of propositional, truth-functional logics. Both the degrees of complexity that the authors admit are not captured by their abductive logic, and some of the facts about real-life abduction that in most theories are suppressed, ignored or idealized are exactly those described by LIR, namely, the dialectical, *non-linguistic properties* of the abductive process.

Brenner [8] suggested that a logic extended to real systems and processes was the preferred logic for addressing issues in many areas in philosophy, epistemology and science, and had the potential for explicating outstanding problems in model-based reasoning. To summarize our current thesis, a logic is indeed the theory of complex reasoning, but it is not logic as it is generally understood. It is Logic in Reality, and it in part supports and explicates the Magnani approach. It also supports a anti-perspectivist thesis like that of Chakravartty [19]. Perspectivism states essentially that there is no reality to things in themselves that is free of individual bias. Chakravartty shows that it is *inconsistent* models that undermine this kind of perspectivism.

3.7.5 Dialogical Logic

We referred above to logical systems such as paraconsistent, paracomplete, free, modal and others. All of these, however, are limited to a truth-functional conception of logic as referring only to linguistic entities. If a logic, does not meet the criteria of these logics, that is not a—logical—reason for denying it the status of a logic of a different kind.

Dialogical logic [43] is among the examples of apparently non-classical or non-standard logic which the reader might wish to see compared to Logic in Reality. The term was correctly used in antiquity to describe the dialogue or argument between two

[2] In LIR, it is the existence of two parallel chains of causality in complex, e.g. cognitive processes that provides the ontological basis for backward and forward movement at the same time in reality, cf. Brenner [9].

people. Modern dialogical logic uses concepts from game theory to design dialogue games that provide a semantics for a wide variety of logical systems.

In his extensive and highly popular publications, Edgar Morin defines a dialogical principle, namely, that two mutually exclusive notions may be inseparable in reality. However, Morin's system has no grounding in science and simply restates the existence of an order/disorder/organization triple and the Heisenberg uncertainty principle as significant, without discussion of how such principles could apply to living beings. He allows contradiction, but as an example cites the familiar semantic Liar Paradox. Morin was associated with Lupasco, but his failure to refer to his more serious contemporary borders on intellectual dishonesty.

Rahman [49] a distinguished logician, has developed a dialogical concept of what was called 'Free Enrichment' by Ricanati as a tool for the dialogical constitution of meaning in an approach to contextualism. In the reference paper, free enrichment is defined as a method for a complete evaluation of the truth conditions for sentences by adding back necessary unarticulated aspects of context. Again, we remain firmly within the linguistic or semantic domain. The contextual meaning including unarticulated context has in principle more of a 'psychological reality'. This approach would appear to be desirable in some way, but seems to us to have little reality; the concept of meaning, also, is strictly limited to the meaning of sentences, whereas the LIR theory of meaning also refers to the meaning *of* reality. Our dynamic conception of meaning will be discussed further in Chap. 9.

We will not pursue this critique further, except to note the following: in the definition of modern dialectical logic, an important property of the ancient form has been lost, namely the real logical dialectic between people arguing or exchanging views. This is a process in which a real movement occurs between the positions of the participants which dominate alternately and reciprocally. The existence of the dynamic opposition is the basis for the emergence or new ideas or compromise. In Chaps. 19 and 20, we will try to disentangle the relation between communication and social systems which are, in a sense, nothing but communications!

3.8 Toward a Non-Boolean Logic

Since George Boole published his *Laws of Thought* [7], Boolean logic has been the canonical logic of science and philosophy. Boole demonstrated that classical, bivalent propositional logic could receive an algebraic formulation and proposed a general symbolic method for logical inference. His algebra contains terms for both quality and quantity and provides the basis for standard probability theory. However, his terms for *quality* are strictly limited to formalizable, binary properties of phenomena. Non-Boolean logics and algebras have been shown, relatively recently, to be really necessary only in the area of quantum mechanics, with a few interesting but constructed exceptions in the work of Aerts and D'Hooghe [2], Conte et al. [21].

However, as should be more widely appreciated, Boole himself was aware of the limitations of his own system and was completely open to others, sometimes of a striking modernity: "we sometimes find more just conceptions of the unity, the vital connection and the subordination to a moral purpose, of the different parts of truth, among those who acknowledge nothing higher than *the changing aspect of collective humanity* (italics ours), than among those who profess an intellectual allegiance to the Father of Lights". Further in a key Appendix, Boole writes that the central role of mathematics as derived from his Laws of Thought, it is not a sufficient basis either of knowledge or of discipline, "As truly, therefore, as the cultivation of the mathematical or deductive faculty is a part of intellectual discipline, so truly it is only a part."

Niestegge [45] proposes a non-Boolean extension of the classical probability model that reproduces typical quantum phenomena. The proposed model is more general and more abstract, but easier to interpret than the quantum–mechanical Hilbert space formalism and exhibits a particular phenomenon (state-independent conditional probabilities) that may provide new opportunities for an understanding of the quantum measurement process.

The problems with the assumption that there is an objective absolute physical reality behind quantum measurement now become problems of a non-Boolean extension of probability theory, or of a non-Boolean logic. State-independent conditional probabilities depend only on the underlying logico-algebraic structure of the events/propositions and may therefore be regarded more as a logical than a stochastic phenomenon. These probabilities themselves have an objective character and thus differ from classical probabilities the origin of which always lies in the observer's subjective lack of information. This approach does not apply directly to non-Boolean phenomena in the macroscopic world. It does, however, contain the essential point of Logic in Reality that its probabilities have an *objective* character.

Balance is necessary in any view of the operation of the human mind: "I would especially direct attention to that view of the constitution of the intellect which represents it as subject to laws determinate in their character, but not operating by the power of necessity; which exhibit it as redeemed from the dominion of fate, without being abandoned to the lawlessness of chance," Boole's logical Laws of Thought can manifest their presence "otherwise than by merely prescribing the conditions of formal inference." The distinctions between true and false, between correct and incorrect, are cornerstones for his and all other standard logics. But this distinction "exists in the processes of the intellect, not in the region of physical necessity".

Boole is honest in admitting not to have found a *construens* to accompany his *destruens*, but several of his remarks suggest that some aspects of Logic in Reality would have been congenial to him. One was to the effect that his Laws of Thought (logic) were capable of precise scientific expression, but were invested with a lower 'authority' than the laws of nature in general. "Were the correspondence between the forms of thought and the actual constitution of Nature proved to exist, whatsoever connexion or relation it might be supposed to establish between the two systems, it would in no degree affect the question of their mutual independence." "Wherever the phenomena of life are manifested, the dominion of rigid law in some degree yields to that mysterious principle of activity", a teleology accomplished "not, apparently, by the fateful power of external circumstances, but by the calling forth of an energy from within." We quote Boole *in extenso*, because we have seen no references to these passages elsewhere. It has been left to Lupasco to continue and talk about the 'laws of energy' and their deployment and to show that some real macroscopic processes follow a logic whose terms do not commute or distribute and are accordingly non-Boolean. It is interesting to note that while the term 'Boolean' is routinely used to qualify logic in descriptions of it, 'non-Boolean' is almost never used despite the fact that it would seem to designate a perfectly valid domain of knowledge. This is of course just a further reflection that the extension of logic proposed by Lupasco has not been widely accepted in the literature.

To those who continue to think that a major reconceptualization of logic is not needed, we call their attention to this again little quoted passage from Boole, pp. 241–242: "To what final conclusions are we then led respecting the nature and extent of the scholastic (Aristotelian) logic? I think to the following: that it is not a science, but a collection of scientific truths, too incomplete to form a system of themselves and not sufficiently *fundamental* (italics ours) to serve as the foundation upon which a perfect system may rest." (We do not need to parse this use of 'perfect'; it was standard for the nineteenth Century and earlier.) Boole then talks of the historical, functional way in which this logic has become a valuable 'organ' of thought. But he still asks whether the rules of conversion and syllogism, the 'mnemonic forms', possess any real utility or the mind should better be left to its 'own' resources are questions "it might be profitable to examine". In other words, what is fundamentally *non-natural* about binary logic is a barrier to progress. Although, as noted above, Boole sees no farther here than a "more perfect theory of deductive reasoning", the same considerations apply *a forteriori* to other forms of—Boolean—reasoning.

3.9 Logical Realism versus Natural Realism and Natural Philosophy

There are several hermeneutic stages through which it is necessary to pass between (1) a logic—a logic of/in reality and accordingly about nature, and (2) natural philosophy, which will be the subject of a later Chapter. The combination of logic and reality in the single phrase used to characterize our logical system might accordingly lead to a possible misunderstanding which it would be well to dispose of promptly. LIR, as the logic of and in reality, is a logic of real entities, and the latter implies a realist metaphysics. On the other hand, as discussed in [9], the term logical realism is usually taken to mean that the terms of *standard classical or neo-classical logics* are endowed with real characteristics. On this view, logical forms can provide a basis for logical necessities, and their connections could amount to logical 'facts'. This would yield a kind of logical realism as these 'facts' would be the ontological grounds for logical truth and inference.

Such a philosophical doctrine of logical realism, that there could be any facts or matters of *logic* that obtain independently of our holding them to be true has been criticized by Resnik [51]. His argument is that logical (and mathematical) truths are obtained through deductive proofs and are insulated from experience, even although not immune to empirical revision. We conclude that LIR has nothing to do with this doctrine of logical realism, despite or better due to the fact that the PDO, the fundamental postulate of LIR is physical—as well as metaphysical. *It* is independent of human psychology, etc. in the same way in which other objects of our theories, but on other ontological grounds, since the logic that is referred to in logical realism is a classical ideal and abstract logic.

Despite its name, Cocchiarella's concept [20], is another limited system of predication about natural properties and relations. A mode of being is ascribed to it that is analogous to that of other predicable concepts: they are part of human cognitive capacities to identify, characterize and refer to real world objects. LIR supports this view, but notes that its implications have not been explored: the fact that these capacities and concepts do not exist independently of human thought or language does not mean they are not "objective" as assumed in logical realism. On the contrary, I consider these entities to be *more* realist and hence more objective in the usual sense of the term, better, objective and subjective at the same time.

In conceptualism, being and concrete/actual existence are *only* formal, 'logical' concepts and not properties, or attributes, which things might or might not have. The nature of natural properties or relations, on the other hand, does not consist in its being a characteristic of some object at some time or other, but rather the causal possibility of its being *in re*—that is, having a mode of being and *becoming* within the causal structure of the world. This is in other terms the metaphysics which is supported by LIR.

3.10 The Scope of Logic in Reality: Structural Principles and Attitudes

The thesis of this book can be stated in the following summary fashion: we see the basis for both a science and philosophy of the world in the concomitant operation of four scientific, structural principles;

- The Gödel incompleteness theorems for mathematics, which describe the reciprocity of completeness and consistency; define the structure of reasoning.
- The Heisenberg Principle of uncertainty of position and momentum: defines the structure of quantum entities.
- The Pauli Principle of Exclusion for electrons: determines the possibility of macroscopic physical structures.

To these three well-known principles, we add a fourth, proposed by Stéphane Lupasco in the same mid-twentieth Century period:

- The Lupasco Principle of Dynamic Opposition (PDO): defines the dynamic structure of the evolution of real processes, of *change*.

The Gödel theorems and logic—as written—do not apply to physical or mental emergent phenomena, but we view the principle involved, the duality of consistency and completeness, axiomatically, as an instantiation of the fundamental duality of the universe. The logical and ontological development undertaken in this book provides a bridge between the PDO in the real world and Gödelian dualism. The relation between consistency (absence of internal contradiction) and completeness, in logic, language and mathematics, is between two abstract entities. For any application in physics or other science, what must be recognized is that an isomorphic relation of opposition or dynamic interaction can exist in the physical domain between real elements, processes, etc., with emergence of new phenomena as a consequence in certain cases.

What is novel in our proposal is that one should not only recognize the contradictions in science and knowledge, better, knowing, especially as regards time, space and change, but also as the basis for a new understanding, avoid the bias in the assumptions one makes and in the methodology one uses inherent in conceptual systems based on standard bivalent propositional logic. In the remainder of this book we will see the various ways in which these principles can be used to restructure philosophy so that it becomes—to a greater extent—a "Philosophy in Reality".

At this stage in our inquiry, we can only state that the adoption of the principles of Logic in Reality requires major changes in *all* major philosophical concepts, including of course change itself. As indicated in the Introduction, the necessity of moving away from a binary logic of propositions to something more dynamic has been perceived in a general way by workers in the area of systems, East-West studies and of course the life sciences. It is nevertheless the scientific aspects of the Lupasco logic which in our view provide the concepts necessary to bridge the numerous explanatory gaps in the current literature.

An additional set of changes is required in the general attitudes of philosophy to what is most important in reality. Logic in Reality is intended to give equal ontological status to what is incomplete, inconsistent or contradictory in the dynamic, process sense we give to this term as well as their positive counterparts. We therefore practice the 'New Skepticism' which was proposed in Brenner [9], of which the following is a slightly modified citation: "The epistemological definition of skepticism states that there may be inevitable errors in the knowledge of our own experiences that make it impossible to have adequate knowledge of the world. By pointing to the interactive relations between phenomena at and between all levels, our logic of/in reality reformulates this skepticism. We can know adequately, but we cannot know completely or absolutely, as is being recognized by the increasing acceptance of the implications of Gödel's work. This is not a novel position. To repeat, what is novel in LIR is the proposal that one should recognize not only the contradictions in science and knowledge, especially as regards time, space and change, but also another kind of more or less unconscious contradictory bias in the assumptions one makes, in the methodology one uses and in the public positions one takes. Any explanation, for example, always implies its potentialized contrary explanation, even if it can be readily disposed of. Thus the theoretical conclusions of LIR itself should always be understood conditionally: counterexamples and borderline cases can always be found. The core thesis of LIR is that such cases do not invalidate the theory. They are not pathological in the negative sense, any more than certain mathematical curves that display fractal properties are 'pathological'. They are a logical and hence philosophical aspect of the universe in which we find ourselves.

We have applied such skepticism to new developments in science and technology that helps in recognition of their reductionist aspects that exemplify primarily identity or homogeneity on the one hand, or excessive diversity on the other. Artificial intelligence work claims a possible elimination of any fundamental difference between living and non-living supports. Navigation in virtual reality, on the Internet and elsewhere, offers unusual possibilities of new forms of perception and human association, but it entails risks to fragile individuals. The point here is not to detract from the value of highly creative discoveries and inventions in these fields. It is to counter the ideological, absolutist components that may contaminate them by rehabilitating, 'within reason', by giving a logical status to, the related non-computable, imprecise and intuitive aspects of our 'classical' biological existence.

The application of this principle of skepticism applies also to theories and fields of thought. For example, the last words of the conclusion of Petitot's *Physique du sens* (Physics of Meaning) [46] are that the humanities *are* natural sciences ("*die Geisteswissenschaften sind Naturwissenschaften*"). This is a desirable, even a necessary conclusion, but it is not complete and even partly misleading in the sense that it obscures the—necessary—differences between them. We will refer several times to the phrase, the title of an important book by Hoffmann [29] that two disciplines, or entities in general, "both are and are not the same, by seeing the different ways in which the same underlying logical principles apply to and characterize the disciplines and those who practice them".

The systems scientist and philosopher Gianfranco Minati [42], whose work we will have occasion to discuss in detail in Chap. 16, has pointed to the treatment and the contrast between the concepts of completeness and incompleteness as knowledge to manage the complex knowledge society. An analogous transition is made from a probabilistic logic of certainty to one of uncertainty [24]. "The conceptual framework of completeness is not appropriate for dealing with complexity. Conversely, the conceptual framework of incompleteness is consistent and appropriate with interdisciplinary complexity."

Other characteristics of the Lupasco logical system and Logic in Reality will emerge in relation to the scientific and philosophical domains in subsequent Chapters. The reader may wish, however, to already have at this stage one or two incontrovertible examples of the operation of LIR in the sense of Sect. 6.5. One then has the '*embarras du choix*' of selecting from what amounts to the extant domain! Two examples which deserve and will receive further discussion below are from the areas of cognition and politics: (1) if one concentrates on (actualizes) knowledge as knowing something, the knowledge or knowing of knowing is potentialized and, as always, vice versa. (2) in the competition between two political factions or parties, provided it is made in good faith, that is, acceptance of the existence of the other, there is a good probability that a compromise (T-state or included third term) will emerge.

LIR presents a new approach to the inevitable problems resulting from the classical philosophical dichotomies, appearance and reality, as well as the concepts of space, time and causality as categories with separable categorial features, including, *e.g.*, final and effective cause. Non-separability underlies the other metaphysical and phenomenal dualities of reality, such as determinism and indeterminism, subject and object, continuity and discontinuity, etc. Non-separability at the macroscopic level, like that being explored at the quantum level, provides a principle of organization or structure in macroscopic phenomena that has been neglected in 'classic' science and philosophy.

As presented in the outline, but perhaps can now be seen more clearly, a statement needs to be made of the relation of LIR to domains in philosophy as it is commonly understood—especially metaphysics, metaphilosophy and natural philosophy which will be the subjects of Chap. 10. Our first application of LIR will be, however, to the further discussion of the phenomenon of change itself, amplifying and extending the discussion the previous chapter.

3.10.1 World Logic Day

Readers may be interested to know that while this book was in preparation, the first World Logic Day was held on January 14, 2020, under the auspices of UNESCO. (We thank Dr. Yu Liu of the Jules Verne University of Picardie, France for calling this event to our attention.) In her summary statement, Ms. Audrey Azoulay, the Director General of UNESCO quoted the phrase of the French author André Gide in his novel, *Nouvelles nourritures* that "the fear of stumbling makes the mind hold tightly to the handrail of logic". Azoulay suggests that the discipline of logic is becoming ever more important to current thought, "a contemporary universal", referring several times to its applications in Artificial Intelligence, computer science and their roots in algorithmic reasoning. UNESCO will soon be establishing a framework for studying the ethical principles of AI. While we applaud this initiative of giving proper value to logic, we can only regret that there is no reference to possible broader basis for broader applications to natural intelligence and the range of thought, which we have pointed to in this book, of processes and systems for which no algorithms can be written, but whose logic is included in our concept of Logic in Reality.

References

1. Aerts, D., D'Hondt, E., Gabora, L.: Why the logical disjunction in quantum logic is not classical. Found. Phys. **30**, 1473–1480 (2000)
2. Aerts, D., D'Hooghe, B.: Potentiality states: quantum versus classical emergence. arXiv **2012** (2012). arXiv:1212.0104v1.
3. Bachelard, G.: La philosophie du non. Paris: Presses Universitaires de France, 4th ed., Paris: Quadrige, PUF, 1994 (1940)

4. Batterman, R.W.: Idealization and modeling. Synthese **169**, 427–446 (2009)
5. Białek, L., Dunin-Ke, B., Szałas, A.: A paraconsistent approach to actions in informationally complex environments. Ann. Math. Artif. Intell. Published on line 31 May 2019
6. Bishop, R., Brenner, J.E.: Potentiality, actuality and non-separability in quantum and classical physics: Heisenberg's Res Potentiae in the Macroscopic World 2018. arXiv:1801.01471
7. Boole, G.: The Laws of Thought. On Which Are Founded the Mathematical Theories of Logic and Probabilities. Dover Publications, Inc., Mineola. Originally Published by Macmillan, New York, 1854 (1958)
8. Brenner, J.E.: Process in reality: a logical offering. Logic Logic. Philos. **14**, 165–202 (2005)
9. Brenner, J.E.: Logic in Reality. Springer, Dordrecht (2008)
10. Brenner, J.E.: Information in reality: logic and metaphysics. Triple-C **9**, 332–341 (2011)
11. Brenner, J.E.: The naturalization of natural philosophy. Philosophies **3**, 41 (2018)
12. Brenner, J.: The philosophical logic of Stéphane Lupasco. Logic and Logical Philosophy **19,** 243-285 (2010)
13. Brouwer, L.E.J., 1951. Cambridge lectures on intuitionism. https://home.mira.net/~gaffcam/phil/brouwer.htm
14. Burgin, M., Brenner, J.E.: Operators in nature, science, technology and society: mathematical, logical and philosophical issues. Philosophies **2**, 21 (2017)
15. Burgin, M.: Theory of Information. Fundamentality, Diversity and Unification. World Scientific, Singapore (2010)
16. Capurro, R.: Brenner. Personal communication, 15 Feb 2019
17. Carnielli, W.: Logics of formal inconsistency. State University of Campinas, Brazil, CLE e-Prints, vol 5(1) (2005)
18. Chaitin, G.: The Limits of Mathematics. Springer, Singapore (1998)
19. Chakravartty, A.: Perspectivism, inconsistent models, and contrastive explanation. Stud. History Philos. Sci. Part A **41**(4), 405–412 (2010)
20. Cocchiarella, N.:. Conceptual realism as a formal ontology. In: Poli, R., Simon, P. (eds.) Formal Ontology. Kluwer Academic Press, Dordrecht (1996)
21. Conte, E., Khrennikov, A., Zbilut, J.P.: The transition from ontic potentiality to actualization of states in a quantum mechanical approach to reality. arXiv:[quant-ph] (2006)
22. Dodd, J.: Negative truths and truth-maker principles. Synthese **156,** 383-401 (2007)
23. Dowe, P.: Causal processes. In: Zalta, E.N. (ed) The Stanford Encyclopedia of Philosophy (Fall 2008 Edition) (2008)
24. De Finetti, B.: Theory of Probability, vol. 1. Wiley, New York (1974)
25. Forbus, K.D.: An introduction to qualitative modelling. In: Gabbay, D.M., Woods, J. (eds.) Logic Journal of the IGPL, vol. 14(2), pp. 189–219 (2006)
26. Ghils, P.: Les tensions du langage. Peter Lang, Berne/Berlin (1994)
27. Günther, G.: The Tradition of Logic and the Concept of a Trans-Classical Rationality. Allgemeen Nederlands Tijdschrift voor Wijsbegeerte en Psychologie 54, 194-200, https://www.thinkartlab.com/pkl/archive/GUNTHER-BOOK/TRADIT1.html (Accessed 26 January 2009) (1962)
28. Günther, G.: Time, timeless logic and self-referential systems. Ann. N.Y. Acad. Sci. **138,** 397-406 (1967)
29. Hoffmann, R.: The Same and Not the Same. Columbia University Press, New York NY (1995)
30. Hájek, A.: Interpretations of Probability. Stanford Encyclopedia of Philosophy, Fall 2019 Edition, https://plato.stanford.edu/archives/fall2019/entries/probability-interpret/ (2019)
31. Jacquette, D.: Ontology. McGill-Queen's University Press, Montreal (2002)
32. Jakobson, R.: Essais de linguistique générale. Éditions de Minuit, Paris (1963)
33. Kastner, R.E., Kauffman, S., Epperson, M.: Taking Heisenberg's potentia seriously. arXiv: 1709.05395v1[quant-ph] 5 Oct 2017
34. Krause, D.: Separable non-individuals. Representaciones **1**, 21–36 (2005)
35. Kuipers, B.: Qualitative simulation. Artif. Intell. **29**, 289–338 (1986)
36. Lu, C.: The P-T probability framework for semantic communication, falsification, confirmation and Bayesian reasoning. Philosophies **5,** 25 (2020).

37. Lupasco, S.: Le principe d'antagonisme et la logique de l'énergie. Éditions du Rocher, Paris (Originally published in Paris: Éditions Hermann, 1951) (1987)
38. Lupasco, S.: Du devenir logique et de l'affectivité; Vol. 1: Le dualisme antagoniste. Essai d'une nouvelle théorie de la connaissance. J. Vrin, Paris (Originally published in Paris: J. Vrin, 1935) (1973)
39. Magnani, L.: Inconsistencies and creative abduction in science. In: AI and Scientific Creativity. Proceedings of the AISB99 Symposium on Scientific Creativity. University of Edinburgh (1999)
40. Majid, S.: Algebraic approach to quantum gravity I: relative realism (2007). Preprint, https://philsci-archive.pitt.edu/archive00003345/
41. Martin, B.: In defence of Dialetheism. a reply to Béziau and Tkaczyk. Logic Logic. Philos. **27**, 205–233 (2018)
42. Minati, G.: Knowledge to manage the knowledge society: the concept of theoretical incompleteness. Systems **4**, 26 (2016)
43. Morin, E.: On complexity. Hampton Press, Cresskill (2008)
44. Nicolescu, B.: Nous, la particule et le monde. Paris: Éditions du Rocher. (Originally published in Paris: Éditions Le Mail, 1985) (2002)
45. Niestegge, G.: An approach to quantum mechanics via conditional probabilities. Found. Phys. **38**, 241–256 (2008)
46. Petitot-Cocorda, J.: Physique du Sens. CNRS Éditions, Paris (1992)
47. Priest, G.: In contradiction. Martinus Nijhoff, Dordrecht (1987)
48. Priest, G.: Inconsistency in the empirical sciences. In: Meheus, J. (ed.) Inconsistency in Science. Kluwer Academic, Dordrecht (2002)
49. Rahman, S.: What the weatherman said enrichment, CTT and the dialogical approach to moderate contextualism 2019. https://hal.archives-ouvertes.fr/hal-02380627/
50. Rescher, N.: Process Metaphysics. State University of New York Press, Albany NY (1996)
51. Resnik, M.D.: Against logical realism. History Philos. Logic **20**, 181–194 (1999)
52. Sklar, L.: Philosophy of physics. Westview Press, Boulder and San Francisco (1992)
53. Wikipedia: en.wikipedia.org>wiki>Philosophy_of_logic, 1.20.2020 (2020)
54. Wu, K.: The complex characteristics of multidimensional emergence: critique of Husserl's phenomenological reduction. Paper at the International Symposium on Cutting Edge Issues of the Contemporary Philosophy of Science and Mind, Guangzhou—Macau, 13–17 Dec 2010.

Chapter 4
Change and Logic

4.1 Change in Modern Western Science and Philosophy

The general conception of change in Western thought is that of difference or non-identity of both quantity and quality. Simple differences of scalar (extensive) quantities, such as length or temperature, accompany more complex changes, involving vector changes and energy (intensive quantities).

Mortensen [15] notes that the problem of the instant of change is too general to admit a single solution, it requires specification of further metaphysical principles envisaged as constraints on possible *kinds* of solution. However, he separates the concept of change from the cognate concepts of cause, time and motion, suggesting that "the operation of a cause on a thing is neither necessary nor sufficient for change in that thing". In addition, he maintains the distinction between the monadic or internal or intrinsic properties of a thing and its relations or external or extrinsic features, with "real, metaphysical change" being change in the former. It is exactly this reification, these distinctions and separations, plus the perennial notion of 'thing' or substance that in their LIR interpretations are the source of the problems in understanding change.

The classical argument against the existence of change, in which a changing thing has different and mutually incompatible properties, the classical argument is that it implies a contradiction (is inconsistent). The heart of our argument is that if change is allegedly inconsistent because people and billiard balls are supposed to have self-identity across time and space, something has been missed in the picture of real particulars. That is, of course, that all real entities *do* change, (billiard balls do not, to all intents and purposes, over a short time span in the absence of a fire). But people clearly change and do not change, and in our approach this process is both *logical* and physical.

The interpretation of the unity of opposites in Logic in Reality expands and founds the intuitions of Heraclitus [10] that it is the *struggle* between opposites which drives change. In the LIR view, it *is* change. To say that this is a version of a Marxist dynamic of dialectical materialism (see Chap. 5) is to invert the causal relation:

J. E. Brenner and A. U. Igamberdiev, *Philosophy in Reality*, Studies in Applied Philosophy, Epistemology and Rational Ethics 60, https://doi.org/10.1007/978-3-030-62757-7_4

Marxist dialectics is a reductionist version of the real dialectics of change which, according to LIR need no further argument to accept inconsistency and contradiction. In particular, the contradictions (or better, counter-actions) are not the theoretical constructions of our minds, although they *appear* to be. They appear to be when looked at as expressions of propositions that require truth-functionality.

In another paper (2004), Mortensen [14] contrasts the accounts of change given by the Buddhist logician Dharmakîrti (7th Century CE) and the contemporary paraconsistent logician and analytical philosopher Graham Priest, mentioned above. We recall that Priest denies the universal applicability of the Law of Non-contradiction, as does Logic in Reality, while Dharmakîrti endorses it. Mortensen concludes that Dharmakîti's position is the more correct, being based on a theory of the momentariness of existence, construed to allow universals. A suitable physical field theory accomplishes this, allowing for the non-identity of atomic particulars. Logic in Reality is thus a theory that gives a basis for an atomic view of physical reality, while also rejecting the Law of Non-contradiction in its extension to real phenomena. Change is not a sentential property.

4.2 Chance and Change

4.2.1 Contingency Versus Determinism

In Aristotle's original intuition [1], 'underlying things' make statements true, and Aristotle's implication that these 'things' are logically structured is acceptable in LIR. However, Aristotle also defined facts and things as *contingent*, by opposition with the necessary truth of propositions, thus prejudging an indeterminist view of the real world. As we will see, the relationship of chance and necessity is also dialectical and interactive, governed by the LIR Principle of Dynamic Opposition (PDO).

For Aristotle also, the only modality of change in the universe was the possible, capable of evolving toward the necessary or contingent. In LIR, each logical value of a process or process element is a *probability* that is more or less necessary and more or less contingent. In contrast to possibilities, which admit only binary, yes–no outcomes, probabilities are related to the elements of real processes in progress. In addition to the two inverse probable processes in the PDO of evolution toward non-contradiction (identity and diversity) or logical transcendence, there is a third probable process that evolves towards contradiction or immanence, the symmetrical reciprocal inhibition of chance and necessity. If we look back at this point at some of the entities in the category of T-states (Chap. 3), things that have been characterized as emergent included middles, ideas, works of art, innovations of all kinds, all seem to have components of both chance and necessity. To our knowledge, no one has provided a place in *logic* for such events, as logical values. Accordingly, in LIR terms, one could perhaps best say that the universe overall is a-deterministic, an included middle T-state with local domains of determinism and indeterminism.

In practice, it would seem to be impossible to decide, for systems showing unpredictability, non-computability or randomness, between a model of the system being governed by underlying genuinely statistical, indeterminate laws of nature or by deterministic ones resulting in chaotic behavior. In the first case, apparent randomness is real randomness, in the second it isn't. As discussed below in relation to realism, one aspect of the world is the existence of reality and appearance, and we suggest a dialectical relation between chance and necessity, as between other dualities.

In Bohm's theory of quantum mechanics, all particles have at all times a definite position and velocity, whether or not one is able to determine them. The Schrödinger wave equation that describes the evolution of a physical system is taken to be perfectly deterministic. Bohm reinforced this by a guidance equation that determines, on the basis of the particles' wave function plus the positions and velocities, what their future states will be. The result is a fully deterministic theory that confirms the Copenhagen interpretation of quantum mechanics, that is, that the particle and wave descriptions of quanta are complementary. However, and the simplistic Copenhagen view has been largely superseded. Hoefer [11] stated the resulting dilemma as follows: if there is ever a "Final Theory" of the quantum structure of the world, it will not only be difficult to decide whether it is deterministic or not, but there seem to be today equivalent deterministic and indeterministic theories.

One way out of the dilemma is to apply the PDO and assume that quantum phenomena are and are not deterministic, sometimes primarily one and sometimes primarily the other. Both theories apply in reality, and the states and relations involved in individual quantum processes are always partly determined and partly non-determined. This view is consistent with the relational version of quantum mechanics discussed below.

If we return to the non-quantum macroscopic world, as Brenner has noted, there are no examples of pure ontological indeterminism, despite extensive epistemological non-determinability. As a consequence, we conclude that in the analysis of *change*, pure chance can be ignored in science and philosophy as fundamental property of nature.

4.2.2 The Weak Pinsker Conjecture

Despite extensive study, arriving at a satisfactory description of chance and change remains one of the most difficult problems in philosophy and mathematics. Pinsker [2] stated that every dynamical system is a 'blend' of chance and determinism, constituted by points or elements that move in a purely random or completely deterministic fashion in a continuous space of some kind. This conjecture was proven false, and a weaker one proposed in which deterministic was replaced by 'almost' deterministic, and the requirement for continuity was dropped. From an LIR standpoint, the change from an absolute to less absolute formulation is already of interest as necessary but it is not sufficient for our purposes.

The discussion both by Pinsker and his critics refers to systems of points moving in simplistic ways in a configurational space of low dimension, whereas the systems interest to us are macroscopic real systems of complex relations, as outlined in our discussion of Relational Quantum Mechanics. It is no problem in LIR, then, to see a concomitant operation of epistemologically random processes and strict ontological determinism. The original Lupasco theory provides a basis for macroscopic change being a mix of ordered and non-ordered elements, the latter being random epistemologically to all intents and purposes. From our point of view, it is in only in this sense that is all such phenomena, even Brownian motion are random. The movement of any given molecule or larger particle is the resultant of the impacts, "one after the other", of the other molecules which impact it. The so-called random walk is ontologically determinate, isolatable conceptually in a phase space of sufficient dimensionality.

The above notwithstanding, many people, for what we qualify without demeaning them are psychological reasons, will wish to retain a simple, phenomenological understanding of chance. This is equivalent to stating that the reality of chance defined as truly random events is determined by the subjective decision of a human agent! In the face of such anti-realist positions, the best 'philosophical' response is probably not to argue them, since, as in the case of others we will point to, there is no dialectic interaction with our realism at the level of the specific point, the existence of chance as such. Any dialectic takes place at a higher level—that of wisdom as in the hierarchy of Zhong [23]—the acceptance on both sides of the inevitability of the conflict.

4.3 Continuity and Discontinuity

The picture of time suggested above has direct implications for the on-going scientific research and metaphysical debate on the nature and role in the universe of continuity and discontinuity. One gets the impression that the existence of a connection between the two exists and is accepted, but no explicit discussion of such a connection is made. 'Space' and 'time' are alleged to share the property of continuity, which is the basis for space, time and space–time continua, all composed of infinitely many dimensionless points. However, whether there are such continua and whether they are composed of such points in reality was "a legitimate question in both physics and philosophy [12]" and remains so. As Penrose has pointed out [16], and as discussed in detail in Chap. 7, both Einstein's theories and standard quantum mechanics depend on the assumption of real number space-time continuity, but there is serious doubt as to whether its mathematics is appropriate for describing the ultimate constituents of nature [16].

The concept of continuity is problematic, but so is the basis for the emergence of discontinuity. In biological aspects of catastrophe theory, one goes from factors involving continuous thermodynamic change to mechanisms of genetic regulation that involve the discontinuous intervention of the biochemical structures needed for hereditary control. There is here some kind of immanent operation of discontinuity

[18] on the continuous substrates of biological or higher-level systems. We thus have a series of explanations of continuity, but they seem questionable. We have a concept of discontinuity, but no explanation.

4.3.1 A Summary Statement of the LIR View

To reduce the debate to a manageable size, we propose the following (1) physical and (2) metaphysical arguments in favor of an interactive, contradictorial relation between continuity and discontinuity:

(1) if there is continuity and discontinuity of real entities at the quantum level, that is, both discrete quanta and continuous frequency of wave phenomena, and they are intuitively and mathematically opposite, by the LIR theory, they instantiate the key axioms of Conditional Contradiction, Functional Association and Asymptoticity. Accordingly, continuity cannot exist without discontinuity (or discreteness) throughout nature, and continuity actualized implies discontinuity potentialized. Asymptoticity has another consequence: no real element can be an infinitely small point of space or time since in reality, a lower bound is determined by the Planck constant, 6.62×10^{-34} J-seconds. Infinitesimal quantities of space and time of differential calculus cannot exist in reality.
For us, continuity and discontinuity is a pair of ontological predicates, where the former is inherent to or related to homogeneous extensity and the latter to changes in levels of energy in phenomena. The differences in level between which energy as heterogeneous intensity falls are themselves extensities. It is the discontinuous *passage* from one level to another that represents the intensive quantity, the movement of transformation; higher and lower forms (e.g., chemical energy and heat) are actualized extensities, with greater or lesser potential for further transformation.

(2) The contradictorial relation between continuity and discontinuity, the impossibility of their simultaneous co-instantiation at the quantum level is mirrored in the processes of logic and thought. From the point of view of logic, the dynamisms, as processes, of affirmation and negation (better, affirming and negating) do not show any obvious or conceivable discontinuity. In other words when we affirm or negate something, we do not do so in steps. But logical thought itself, insofar as it is the potentiality of these two contradictory, antagonistic actions and both coexist, as inverse possibilities, demonstrates the existence of an immanent, constitutive discontinuity in reality at this as well as at the quantum level.

We review again here the two concepts essential for a discussion of continuity and discontinuity: the continuum hypothesis and the foundations of differential calculus.

4.3.2 The Continuum Hypothesis

The continuum hypothesis refers to a conception of the universe founded on geometry, the Cantor–Dedekind view, as discussed by Longo [13], which sees not only in mathematics, but everywhere, continuity as ontologically preceding the discrete: "The latter is merely an accident coming out of the continuum background." Points are derived concepts, even if 'non-dimensional'. In this view, geometry (statism, cf. the next Section on statism and dynamism) is in some deep sense more fundamental than dynamics, that is, energy in the standard view. This hypothesis has the advantage of corresponding to our intuition and experience, integrated into and confirmed by mathematics, of continuity in our perception of 'time' and linear movement.

Penrose, on the other hand, had the strong intuition that "physics and space–time structure should be based, at root, on discreteness (emphasis his), rather than continuity". This discreteness is evidenced in quantum mechanical spin, combined, however, with a fundamental notion of expressing phenomena in terms of a relation between objects, rather than between an object and some background space (Penrose 2005).

In the construction of the mathematical continuum, objective realities are not found in the mathematical entities involved, but in the process of constituting these entities as conceptions. There can be an interaction, dynamic in LIR terms, between the applicable mathematical logic and intuitions about continuity. Stating this somewhat more strongly, since the establishment by Gödel of the reciprocal relation between consistency and completeness in formal mathematical systems, the situation is no longer absolute. One should not be forced to choose between geometry and discontinuous objects and their relations. This opens the door to a different foundational principle, using the principles of LIR, in which both continuity and discontinuity are fundamental and are dialectically related.

A clear division between that we call 'matter' or 'substance' and what we call 'empty space'—supposedly the voids entirely free of matter of any kind—cannot be made. *Matter and space are not totally separate types of entity. Actual substance need not be clearly localized in space* (italics ours). This is basically another statement, at the most fundamental possible level, of the Principle of Dynamic Opposition. At our macroscopic level, governed by the laws of thermodynamics, we of course do not perceive this duality with the sense organs we possess. This leads to the familiar arguments about the non-reality of anything. Our counter is our logic, whose principles reflect or are isomorphous to the oppositions at the quantum level, including that of the quantum vacuum, 'space' and matter, energy [17].

It is possible to describe the vacuum by mathematical entities that are neither continuous nor discrete. One could even say, like Weingard [22], that "everything is made out of nothing", but at the quantum level, time and change do not exist, but the problem then becomes of how best to describe macroscopic change.

4.3.3 The Further Problem of Differential Calculus

In principle, the usual notion of differential calculus captures the apparently simultaneously continuous and discrete nature of change—changing phenomena. This position only displaces the philosophical and metaphysical problem. Change at an instant is what differential calculus presents in formal terms. It is well recognized, however, that this implies an inconsistency—continuity and discontinuity at the same time. It begs the question of whether reality is composed of 'points' and 'instants' in the sense used in the theory. If it is not, then differential calculus, like classical logic, is not capturing the essential property of real processes and systems, since it assumes that such points exist. Only in the most recent work on general relativity is the concept of a 'point-event', first codified in the Buddhist logic of the sixth century AD, receiving an adequate interpretation (see Sect. 7.6).

In *Logic in Reality*, it was suggested that a mathematical language for formalizing the contradictorial view of continuity and discontinuity is that of Smooth Infinitesimal Analysis (SIA), developed by Bell [3]. Bell quotes Weyl to the effect that "we are employing the principle of gaining knowledge of the external world from the behavior of its infinitesimal parts." However, Weyl did not prove that Bell's infinitesimals (or any others in standard calculus) and those of the external world, (if such exist), are the same.

The LIR metaphysical approach also looks at the implications of the logical reasoning process for continuity and discontinuity: the contradiction between continuity and discontinuity, the impossibility of their simultaneous co-instantiation at the quantum level is mirrored in the processes of logic and thought. From the point of view of logic, the dynamisms, as processes, of affirmation and negation (better, affirming and negating) do not show any obvious or conceivable discontinuity. In other words when we affirm or negate something, we do not do so in steps. But logical thought itself, insofar as it is the potentiality of these two contradictory, antagonistic actions and both coexist, as inverse possibilities, demonstrates the existence of an immanent, constitutive discontinuity in reality at this level as well as at the quantum level.

We conclude that Bell's SIA is a theory of abstractions, unsuitable for an ontology that purports to deal with the world of real change. As Bell himself concludes, SIA is a theory of infinitesimal geometric objects, designed to provide an intrinsic formulation of the concept of differentiability and perhaps not more than that. The real world is not differentiable as a whole, although a continuum of states exists between (almost) fully actual and (almost) fully potential. The infinitesimal units of which Bell's objects are constructed are, in LIR, pure intensity, and thus cannot exist, any more than can any idealized, abstract constructs. Despite their interesting properties, to assign them any role in real phenomena, with the exception of description of simple physical processes totally dominated by the 2nd Law of Thermodynamics, seems abusive. Applications to systems such as computer science, artificial intelligence and data processing are included in this group, but all of these require no more than a binary logical system, sufficient when there is no exchange of energy qua the elements or terms of analysis. They thus clearly belong in the sub-category of Separability in the LIR ontology. Everything else, life, growth and mind, as well as photons and the vacuum, requires a dynamic, ternary logic capable of handling the fundamental antagonism inherent in energy, and hence throughout nature, and the inconsistencies and contradictions that derive from it.

In the LIR approach, the heterogeneity of intensity is not a series of independent elements or extensive stages, it is an attempt to differentiate (movement of differentiation of) something that wants to stay the same, the extensity of which resists and opposes this change. In this movement, there is a continuity that is not measurable by extensive values. As these values are potentialized, it looks as if extensity contains discontinuity and intensity is a continuous dynamics. Lupasco saw the continuity in extensity, despite its divisibility and capacity for adding new entities, as for example new premises are added in defeasible deductive logic, in its aspect of identity extending from one thing to another. Thus, intensity is a continuous non-identity with respect to itself; extensity is a continuous identity with respect to the other. Intensity and extensity are continuous as dynamisms, considered as independent of one another, and from this point of view accessible to the techniques of differential calculus. But, discontinuity is inherent in their existentiality, since neither can exist without the other, without operating on the other: intensity and extensity reciprocally 'discontinuate' each other. The differences of energy level that result in 'something happening' are not due to intensity or extensity alone but to their interaction. Analysis and synthesis are continuous dynamisms, heterogenizing and homogenizing respectively, but and their necessary discontinuity constitutes their existentiality. This is our philosophical 'position' that will be reflected in the discussion of all the major further issues in this book.

4.4 General Theories of Change

The best available General Theories of Change (GTC) might be said to address the following, partially mutually dependent principles and issues:

- Difficulty of resolving the apparent continuity of General Relativity and the discontinuity of Quantum Mechanics;
- A substance and/or object metaphysics;
- Independent background time or space–time;
- Categorial separability of intrinsic and extrinsic accounts;
- Resistance to if not rejection of inconsistency (real contradictions);
- Dependence on the Leibniz Continuity Condition (LCC): what holds up to a limit holds at the limit.

In contrast, a GTC that is supported by the grounding of Logic in Reality in current cosmology has the following principles:

- Construction of continuity and discontinuity as physically coexistent and co-determining for both gravity and particles.
- A process metaphysics;
- No background space–time in which 'objects' evolve; process entities develop their own space–time;
- Categorial non-separability of intrinsic and extrinsic *properties*;
- Acceptance and explication of the inconsistency of change;
- Irrelevance of the Leibniz Continuity Condition; no real, complex, interactive processes—all the adjectives are necessary—go to a Kolmogorovian limit of 0 or 1.

As far as the LCC is concerned, the determinism of LIR argues against Humean rejection of the lack of determination of current states by past states. The burden of proof should thus be on proponents of the standard solution to show where the LIR conception is not valid. As an interim conclusion, since a "final" one would seem to be excluded by our own theory, we suggest that some readers might ask whether we believe that LIR has resolved all the problems of change and there is nothing further to explain. Our answer, as one might expect, is yes and no, again, with no intention of trivializing the question. LIR proposes a methodology for avoiding *all* problems whose source is only in the classical logical refusal of contradiction. But it also considers it probable that some people will *always* refuse contradiction and its accompanying interactive view of processes and relations. Further, although the combination of self-duality at the quantum field level and duality in the thermodynamic world seems both necessary and sufficient to describe existence, that is change, how the two fit together in detail requires much more discussion than has yet been provided. We are convinced, however, that LIR can 'operate' to remove barriers to consideration of other new and creative approaches to the characterization of continuity and its partner.

In epistemic change, non-punctuate intervals in which propositions and their negations are "dense in each other" can be described as intervals of change. An extrinsic view of change is as a matter of a relation between states at nearby instants of time. Taking the usual view of time, with us since classical India, as a continuously distributed collection of point-instants "… it is the nature of change, even change as a 'point', that it is relational in that it requires comparison (epistemically) or interaction (physically) with nearby points, hence the demand for an exclusively intrinsic conception of change is a mistake."

Even the identity condition for times is "smeared out". It is a theorem of LIR that not-(t = t) holds for each t by the Axiom of Non-Identity but only the sense of the Axiom of Conditional Contradiction that it holds *and* not-holds, excluding the simple opposite. In LIR, our slogan supports a degree of LCC causality, while retaining an essential discontinuity of a process that "non-locality is inconsistent locality", applies to dynamic change in general and to discontinuous change viewed as a temporary non-contradiction. Both are causal.

4.4.1 Epistemic Change and Epistemic Logic

In the comparison of Logic in Reality (LIR) with paraconsistent logic (PCL) in the previous Chapter, we noted their key points of similarity and difference: both allow for the existence of real contradictions that do not invalidate an argument, but PCL remains an epistemic or semantic, truth-functional system without direct ontological purport. Changes in such systems, especially what are called belief changes involve abstractions from reality to such an extent that they can be described by standard bivalent or multivalent logics. In the logic of epistemic change, either the agent has unlimited ability (read energy) to effect a change independently of the resistance to it, or, unrealistically, there is no barrier to change, internal or external. The principle of Guillaume de L'Hôpital (L'Hôpital's rule), the 17th Century mathematician who codified infinitesimal calculus, can be formulated rigorously. This principle states that it cannot be said of any two quantities separated by an infinitesimal whether they are the same or different. The continuity in an interval on the (real or hyperreal) number line is to be replaced by a paracontinuity. This concept also defines a paradiscontinuity, and that paracontinuity and paradiscontinuity are in fact the same. The principle also holds for relational entities, in our view, in the real world.

A volume of the journal *SYNTHESE* entitled "Logic of Change, Change of Logic" [7] seems relevant, but the change whose logic is discussed is completely limited to epistemic change! Must we conclude that other change is beyond the scope of any logic such as LIR? The concept of change used has a drastically reduced ontological scope; change of belief, for example means going (how?) from believing that A is true to believing that not-A or B is true. The focus here is, as usual, on the end-points of change. Any General Theory of Change, however, must have some dynamic logical aspects, and our answer is no.

Floridi [9] proposed a Logic of Being Informed or Information Logic (IL) which formalizes the relation of 'being informed' as different from 'having knowledge' or 'having a belief', described by epistemic (EL) logics and doxastic (DL) logics respectively. This proposal seems perfectly justified. The term "logic of being informed" recognizes something static and abstract about standard formulations of EL and DL, and opens the door to a more dynamic view of information as a process involving real change. Floridi proposes a non-doxastic informational approach to the acquisition of knowledge (change in epistemic state) that does not depend on the tripartite notion of knowledge as justified true belief.

However, the epistemology of LIR is only superficially similar: in the logic corresponding to *it*, (1) the concept of knowledge as justified true belief ($Kp \rightarrow Bp$) is to begin with not itself being grounded in science; (2) the classical triad of knower, knowing and known is given a physical, non-semantic interpretation; and (3) the LIR focus is on the actual mental movements and their origins in the oppositional conscious and unconscious processes involved.

4.4.2 The Boundaries of Epistemic Change

The nature and boundaries of epistemic change are at the center of the debate on change and meaning. Epistemic change is definable as change in the knowledge, opinion, beliefs or perceived truth values of a conscious human being, but the meaning and implications of this definition are extremely complex. The differences between the three kinds of epistemic form are equally vague. In our approach, as outlined in this book, most of the difficulties arise from neglect of the real, physical, that is, ontological processes taking place in connection or together with epistemic change. For example, if a conscious person changes his/her opinion, and this is a consequence of some fact, internal but unconscious or external, in the process of the change, the person is not operating as a some kind of mechanical transducer but is involved as a part of the change, inseparable from it.

The discussion of dynamic epistemic change seems to neglect this dimension. Action models of dynamic epistemic change "are simple relational structures that can be used to describe a variety of informational actions from public announcements to more subtle communications". We understand the authors' objective in limiting the domain of discussion, but only suggest, here are elsewhere, that it is the changing structures of the actions that are as interesting as their models.

It is a fair question, also, to ask where the boundaries are between epistemic and the cases of ontological change to which we have referred in Chap. 2 and above. If we invert the perspective, for a moment, we find the clearest examples of epistemic change in the emergent science and philosophy of communications. We are indebted here to Loet Leydesdorff, the sociologist and theoretician of communications, for his personal reflections to us of his detailed analytical approach to the structure and evolutionary processes involved in communications. In this work, both the roles and relations between information, meaning and communications emerge in a manner in which the differences between kinds of information and meaning emerge clearly. His theory, and its relation to our more ontological and synthetic orientation, will be discussed in Chap. 12. For the time being, we will simply state that changes in the way in which communications are construed a posteriori are, certainly, predominantly epistemic. We will argue, however, that to the extent they are processed by a human mind, they have an irreducible ontological component.

To make defining the interface between epistemology and ontology more difficult, if possible, 'ontology' is often used to refer to classificatory schemes, even if preceded by the adjective 'dynamic' which implies the presence of change. One definition of dynamic ontology refers to the following three branches of study:

1. Data Mining
2. Expert Feedback
3. Machine Reasoning.

These are all valid domains of research, but they are far from exhausting the dynamics of ontological change in reality.

4.4.3 Emergence and Emergent Change. Life and Autopoesis

The standard epistemic logic of change, referring only to change of belief or static knowledge (epistemic state), cannot capture complex physical emergence. However, the emergence in our universe of observers capable of knowledge of emergence and of organizing and interpreting that knowledge implies an epistemological dimension that must be included in any general theory of change.

For the authors of this book, emergence is an instance of change *par excellence.* One has entities A and B and then, somehow, they are joined or replaced by entity C, a changed situation. Nevertheless, emergence itself, as a process, is a subject of debate, as noted, some people going so far as to deny its existence as a meaningful concept. Life as a process of constant change is accepted in principle, but concepts of how it emerges from non-living matter as a consequence of some kind of self-organization separates living systems from their physical ground.

Maturana and Varela's concept of autopoiesis starts from the observation of the unique mechanisms developed by living systems to insure both their necessary emergent evolutionary adaptation to survive in a generally hostile environment while maintaining their identity. These authors propose that the capability to do this is a unique characteristic of such organisms to change the rules. However, the origin of this capability is not explained. Varela used linguistic and visual paradoxes as models for the 'benign circularity' he saw as the basis for biological phenomena. However, such paradoxes are abstract artifacts of language without intrinsic dynamics. Although Varela [21] admits the non-separability of subjective and objective, he uses it to suggest the "*groundlessness* (our emphasis) of experience". In LIR, logic and experience are not identical, but grounded in physics, not an "endless metamorphosis of interpretations".

To explain the origin of both life and change, some authors fall back on a radical indeterminacy, spontaneity or "chance" (see Sect. 4.3 above and [20]), a single conceptual chain random of random events ultimately leading to a point at which organization becomes self-sustaining, and emergence possible. The resistance to a contradictorial logical paradigm for change probably has a solid basis in our genetic makeup: we know that change is necessary, that total stasis is equivalent to death, but that the maintenance of some sameness or identity is also necessary. The only way to "make sense" of such a contradictory universe, and especially of the part of it we constitute as human beings, is to develop some principled understanding not only of change, but of *non*-change and the relation between them that naturalizes contradiction. This is what the system we describe as Logic in Reality has as one of its objectives.

All change is to some extent emergent, but we use the term emergent change to refer to the substantially irreversible emergence of new entities or processes, with the understanding that complex living entities are, also, processes. The multiple aspects of emergence will be discussed later in this book in a systems perspective.

4.5 Analyticity. The Unchanging Character of Analytical Philosophy

At several points in this book, we refer to approaches to knowledge that are analytical or involve analytical methods or points of view. These are usually presented, negatively, as in direct opposition to our own, synthetic point of view. We therefore owe the reader some explanation of our general position so that s/he may better understand, if not necessarily agree with, our preferred approach which focuses on the real physical aspects of cognition and its 'movement'.

In this Chapter, for example, we provide support for a concept that our logic is a logic of change and our philosophy is a philosophy of change, both being related to real, physical process. We are therefore concerned with the nature of change itself, rather than its myriad instantiations. Other philosophical attempts to understand the modern world have led to a number of general but still binary distinctions with rather vague contours, such as 'analytical' versus 'continental' philosophy. These currents can be discerned in the work of almost all the major philosophers active in the last 150 years, Quine in particular. Discussion of the history of the distinctions between logical positivism, logical empiricism and their developments would take us too far from our own logical approach, in which these distinctions are essentially erased in our emphasis on thought processes as instances of the transfer of energy as in other real processes.

One distinction we have adopted for study here and elsewhere is that between natural and non-natural philosophy ([4] and Chaps. 10 and 13). Non-natural philosophy deals with abstract concepts and relations, and can also be characterized as philosophy *tout court*. The most obvious property of the latter and analytical philosophy is that they are modes (ways) of thought designed to make the clearest possible distinctions between complex epistemic entities. The subsidiary, rarely stated property is that these distinctions are stable and can be expected to be valid for a substantial period. Analyticity and analytical philosophy can be considered legitimate approaches to deciding if there are parts of science that (1) do not rest on experience or (2) rest also on something else. It is true that as described by Logic in Reality, one does not directly 'observe' or experience potentiality, the extent to which a process is actual or potential, but one does experience their logical consequences in our extended sense of logic.

For sceptics like us, such an analytical program has the advantage of being almost surely successful, as the domain under study it not *in and of itself* undergoing real change. It thus provides certainty, and we know that for many if not most people certainty is considered a highly desirable consequence of cognitive activity. As discussed by Creath [6], the standard views of logic, mathematics, and mathematical geometry had traditionally seemed to have a unique status of self-confirmation. Leibniz had called them truths of reason. Hume said that they represented relations of ideas, and Kant had held that the truths in these areas were *a priori.* "Kant had two criteria of analyticity, apparently thinking them equivalent. First, in subject-predicate sentences, an analytic sentence is one in which the concept of the predicate is contained in that of the subject. Second, an analytic sentence is one whose denial is self-contradictory. This seems to include not only the sentences whose surface logical form would be of the required sort but also those that can be got from such logical truths by making substitutions that were conceptually equivalent. The more modern rough analog of this is to say that the analytic sentences are those that are true in virtue of logic and definition." Frege developed logic beyond that which was available to Kant, but he did not think of himself as changing its analytic status. "Logic is after all the only avenue we have for giving meaning to the notion of (logical) contradiction."

We quote this passage, also, *in extenso* in order to make as clear as possible the disparity between the indicated analytic conception of logic and ours. In LIR, one not only gives meaning to the standard notion of logical contradiction but to real contradictions, better oppositions, in nature and science. Thus, while analytical philosophy has been forced to take science into consideration (the monolithic designation of 'science' is characteristic of such approaches), its ultimate objective is truth or a set of truth values which do not depend on meaning alone. Here, meaning is, also a secondary, solely epistemic property of cognition, and does not inhere in phenomena in the phenomena themselves. For us, on the contrary, all existence has meaning—*is* meaning, even if it is only for us. The understanding of Logic in Reality and its dialectics, and their consequences for philosophy in reality, thus requires focus on ontology and the ontological, where a conscious observer is *of course* required, but is always is ontologically secondary to the natural environment in which s/he is embedded. Let us now explore some of the characteristics of that environment.

4.6 Change in Our Time. The Economics and Politics of Change

The characterization and management of change are critical issues and problems of a society that is struggling to maintain even a minimum level of decency in human relations. Growth as the only positive change possible has been and still is a key tenet of neo-capitalist ideology and the philosophies of exclusion that accompany and support it. In Chaps. 18 and 19, we will provide a detailed discussion of the application of our principles to social systems, again, with substantial reference to the foundational systems concepts of Minati.

In the meantime, we can say that big data, data mining, artificial intelligence and digital innovation are some of the key words and concepts that characterize the current scientific and technological landscape. We learn almost daily about new capacities for even more rapid data processing. Change for some people means the rapid changes in business models and machines (*Future Automation*; *Changes to Lives and to Businesses* [5]. We thus consider it essential, for the balance in thought that is a primary objective of this book, to pay attention to both familiar and less familiar aspects of non-change [19].

Advances in science, in particular that of neural systems, obviously offer new dimensions in the understanding of how human beings think. But such advances have not led to broad changes in individual or collective behavior or in what people think about. The 'digital world', which Capurro has studied extensively, is in fact one of increasing polarization of society, on the one hand toward a computer-driven standardization of ideas and needs and on the other toward increased attention to the traditional human values of diversity, in which the computer is a tool to be used but not a god to be sacrificed to.

At the socio-economic level, only in the last few years have the unsustainable aspects of growth begun to be properly recognized. Movements are emerging calling for zero or negative growth of economies and populations, the sustainable use of resources, and the frantic resistances such changes is a good measure of the opposing forces at work in the society.

4.7 Interim Conclusions: Change in Reality

In this Chapter, we have discussed change further in terms of our logic of real processes as the one applicable to it. Our thesis is that the behavior and evolutions of real systems can be described in logical terms, and change itself can be described in the same terms. In reality, change *is* that behavior and those evolutions. LIR is sufficient as a part of a description of change, but also necessary since without it the physical driving forces for change cannot be properly associated with conceptual epistemological descriptions of the relevant complex processes that are observer-dependent, while the change itself is not.

What we call Change in Reality occurs at simple macrophysical, biological, and both simple epistemic and complex cognitive and social levels. The differences in the way both LIR and the concept of change apply are closely related. Logic in Reality is a logic of inferences about changes *with meaning*, involving primarily *non*-Shannon information (see Chap. 12), that of biological and complex cognitive processes. LIR reduces to standard logic for processes in which no change occurs *to all intents in purposes* that is, where the real changes, for example at the impact of two billiard balls, the melting of ice, or even phase changes involving quantum decoherence and recoherence have no intrinsic meaning. The LIR methodology avoids the semantic paradoxes of change by incorporating real contradiction, recognizing that some people will *always* refuse contradiction and its accompanying interactive view of processes and relations.

4.7.1 Self-duality

Since the grounding by Lupasco of his logical system in quantum physics, the new principle of self-duality at the quantum field level adds further physical grounding of duality from a philosophical standpoint. Although the combination of this self-duality and duality in the thermodynamic world seems to us both necessary and sufficient to describe existence, that is, change, how the two fit together in detail requires the further discussion in Chap. 6. To repeat, Logic in Reality is not psychology, physics or sociology and does not claim to describe in detail how or why a change occurs or is made. LIR does establish that there are similar dynamic patterns throughout all change involving real interactions within or between the elements of real processes. As it deals with real physical (including mental) phenomena, we consider LIR the proper logic for a potential General Theory of Change. As it deals with real physical emergent phenomena, we consider LIR the proper logic for, and part of, a General Theory of Emergence, especially when embedded in a new dynamic theory of systems. Together, these theories form the core of a new Philosophy in Reality. With this background on the logic and philosophy of change let us look at how the role of our underlying Principle of Dynamic Opposition (PDO) appears in dialectics in the next Chap. 5 and then at the level of cognitive and other phenomena in Part II.

References

1. Aristotle: On the soul. In: Barnes, J. (Ed.); Smith, J.A., (Trans.) The Complete Works of Aristotle, pp. 641–692. Princeton University Press, Princeton NJ (1995).
2. Austin, T.: Measure Concentration and the Weak Pinsker Property (2017). arXiv: 1705.0032[mathDS].
3. Bell, J.L.: A Primer of Infinitesimal Analysis. Cambridge University Press, Cambridge UK (1998)
4. Brenner, J.E.: The naturalization of natural philosophy. Philosophies **3**, 41 (2018)
5. Carone, T.E.: Future automation; changes to lives and businesses. World Scientific, Singapore (2018)
6. Creath, R.: Logical Empiricism. In: Zalta, E.N. (Ed.). The Stanford Encyclopaedia of Philosophy. Summer 2020 Edition (2020). https://plato.stanford.edu/archives/sum2020/entries/logical-empiricism
7. Van Ditmarsch, H., Hill, B., Majer, O.: Logic of change, change of logic. Synthese **171**(2), 207–234 (2009)
8. Engels, F.: Dialectics of Nature. Progress Publishers, Moscow, Russia (Written 1883, Originally published 1925) (1976)
9. Floridi, L.: The logic of being informed. Logique Et Analyse **49**(196), 433–460 (2006)
10. Heraclitus: Fragments, T.M. Robinson (Trans.). University of Toronto Press, Toronto (1987)
11. Hoefer, C.: Causal determinism. In: Zalta, E.N. (Ed.) The Stanford Encyclopedia of Philosophy (Summer 2005 Edition). https://plato.stanford.edu/archives/sum2005/entires/determinism-causal/
12. Van Inwagen, P.: Metaphysics. Westview Press, Boulder (2002)
13. Longo, G.: The mathematical continuum. In: J. Petitot et al. (Eds.) Naturalizing Phenomenology. Issues in Contemporary Phenomenology and Cognitive Science. Stanford University Press, Stanford (1999)
14. Mortensen, C.: Dharmakîrti and Priest on change. Philosophy East & West **54**(1), 20–28 (2004)
15. Mortensen, C.: Change and Inconsistency. Stanford Encyclopedia of Philosophy, Winter 2016 Edition (2016). https://plato.stanford.edu/archives/win2016/entries/change
16. Penrose, R.: The Road to Reality. Knopf, New York (2004)

17. Penrose, R.: The mass of the classical vacuum. In: Saunders, S., Brown, H.R. (Eds.). The Philosophy of Vacuum. Clarendon Press, Oxford UK (1991)
18. Petitot-Cocorda, J.: Morphogenèse du Sens. Presses Universitaires de France, Paris (1985)
19. Priest, G.: Beyond the Limits of Thought. Oxford University Press, New York (2002)
20. Ulanowicz, R.E.: A Third Window. Natural Life Beyond Newton and Darwin. Templeton Foundation Press, West Conshohocken PA (2009)
21. Varela, F.: The creative circle: sketches on the natural history of circularity. In: P. Watzlavick (Ed.) The Invented Reality. Norton Publishing, New York (1984)
22. Weingard, R.: Making everything out of nothing. In: S. Saunders, H.R. Brown (Eds.) The Philosophy of Vacuum. Clarendon Press, Oxford (1991)
23. Zhong, Y.: Ecological methodology and mechanism approach—new perspective to intelligence science. Paper presented at the 4th Summit Conference of the International Society for Studies of Information, Gothenburg, Sweden (2017)

Chapter 5
Dialectics in Reality

5.1 Introduction. We and the Universe

We have referred in previous Chapters to a world of change and process, in which a principle of dynamic opposition (PDO) can be seen to be operating. It describes all complex processes as moving from a states in which opposing elements are more or less actual or actualized to ones in which they are more or less potentialized or potential. The movement is alternative and reciprocal, with no element ever going to an abstract limit of 0 or 100%. We have further stated that the PDO is 'grounded in physics', but in what aspect of physics and what is the world that physics purports to describe? We are thus driven, for a philosophically valid answer to this question to a cosmology, a fundamental structure of a universe that instantiates the change to be explained.

In *Logic in Reality*, Brenner showed among other things how the PDO was necessary and sufficient to ground the existence of systems. In fact, in a dualistic universe constituted by energy, the PDO is necessary and sufficient to ground its existence and evolution. The most fundamental dualism is that between the universe and our embedding in it as observers. We are and are not part of the universe, the same and not the same as it. This has been said many times in the past as philosophy, but our position is that this scientifically and logically correct. For us, change from an unconscious state to a conscious one has already taken place once, and the opposite movement from consciousness to unconsciousness will take place in the future. This is part of our personal reality of existence as process. The insight of Lupasco was to see *how* the Principle of Dynamic Opposition applied to this reality in practice and could describe its evolution.

J. E. Brenner and A. U. Igamberdiev, *Philosophy in Reality*, Studies in Applied Philosophy, Epistemology and Rational Ethics 60, https://doi.org/10.1007/978-3-030-62757-7_5

5.1.1 The Dialectical Methodology of Logic in Reality

As originally formulated by Lupasco [18, 17], the visible physical oppositions in nature, starting at the level of basic physics, define how all natural physical, biological and cognitive structures emerge and operate. The process of understanding nature in these terms, scientifically and philosophically, should proceed in the following dialectical manner:

> ...in the presence of every element or event, of every system and every structure, as of every system of systems and structure of structures, one must extract their internal dialectics and seek the contradictory antagonistic dialectics of every antagonistic and contradictory element or event, of every system and every structure, of every system of systems and structure of structures. It is this method of cognitive investigation of phenomena, whatever they may be, that I call the *Dialectomethodology* of knowledge." "...to discern, behind every operation of thought, a contradictory duality, a duality inherent in the thought itself, to show, at the same time, that a qualitative logical becoming (*devenir*), essentially existential and phenomenal, develops according to this transcendence of thought by itself, establishing a monism which signifies nothing, in reality, but the actualization of one of the terms of this duality on top of the potentialization of the contradictory term.

5.2 The Dialectics of Energy

As indicated in Chap. 3, we consider that energy is inherently dual and antagonistic. The implication is that this duality will be present in all higher (more organized) forms of energy, although it will not necessarily always be the *same* duality. Our view is that the ability of fundamental microphysical elements to carry information to more complex levels resides in their irreducible, constitutive antagonistic dualism. Wave-particle duality has been observed with large molecules, and there is no reason why the *basis* for such dualism should disappear as one goes to still larger more complex systems. As one does, however, the dualism is no longer analyzable in terms of fundamental physics. The form of the duality of the microsystems has been replaced by the duality of the macro-system—the antagonisms of macroscopic living and cognizing systems which are certainly as recognizable and potentially measurable, if not more so.

The LIR view implies a type of *dialectics* at all levels of reality between the two terms of whatever duality is being considered. As an extension of logic, it at the same time one takes into account, and gives proper metaphysical weight to, some of its characteristics that have tended to be neglected. These include the concepts, present also since antiquity, of dialectics—conflict as well as change and alternation between the different but closely related, interactive elements of a phenomenon. Dialectics can be considered neither more, nor less, than the generalization and mental expression of conflicts in nature and civilization, and their resolution, that man has observed from time immemorial. "Beings and things seem to exist and are able to exist only in function of their successive and contradictory conflicts." [17] (For Heraclitus, conflict did not mean the splitting or destruction of the unity of reality, but its constitution. The *logos*, the only "abiding thing", the orderly principle according to which all change takes place, is a 'binding-together'. Conflict (*polemos*) and *logos* are the same.)

5.2.1 Catastrophe Theory

As developed by René Thom and Jean Petitot, the mathematical description of phenomena in 'terms of catastrophes', specific states in which the image of a process can be seen as having a form involving one or more singularities. Graphs of these images can be made and the reader is referred to their basic books [21, 30].

From a philosophical standpoint, Petitot's major project was to reformulate the Husserlian phenomenological reduction to lived experience in terms of critical reduction to phenomena by reinterpreting the original giving intuitions in terms of transcendental esthetics and schemes. For this, a mathematical formalization of categorial intuitions was essential that resulted in a new conception of the relationship between mathematics and reality (see Chap. 7). The development of mathematics realizes an authentic dialectic of concepts which transcends them and this dialectic insures a schematizing function which makes the Kantian a priori historical and diversifies it in regional ontologies. If catastrophe theory can pretend to be implicated in a constituting fashion in experience, it is because it is based on mathematical theories which instantiate and re-solve problematic dialectic concepts (Hegelian syntheses suggested by Lautman as "Ideas"). The impression that one retains from these conclusions of Petitot is that of arriving back at real phenomena after a lengthy detour. If phenomena themselves instantiate dialectics, then they (the dialectical relations) can be considered as ontologically constitutive, and the dynamic view of dialectics in LIR provides physical explications of them.

5.2.2 The Emergent Materialism of Hofkirchner

Another classical philosophical problem, with which both we and the contemporary sociologist Wolfgang Hofkirchner are concerned, is the relation between the 'one' and the 'many'; identity and diversity; and identity and difference. This is a key concept in Lupasco's dynamic logic of/in reality. Movement from primarily diversity or heterogeneity to unity or homogeneity is as fundamental as that between actuality and potentiality in the basic structure of the universe. Hofkirchner proposes the relation between identity and difference as describing four ways of thinking: reductionism, projectivism, disjunctivism and integrativism. The first two yield unity without diversity, the third diversity without unity. Only the fourth, as also expressed by the contemporary philosopher Edgar Morin, yields the necessary unity-in-diversity and diversity-in-unity.

Hofkirchner applies these concepts to information in a doctrine of 'Emergent Materialism (EM) that goes beyond materialistic and idealistic monism and (even) interactive dualism. The dialectics of EM recognizes in particular the identity and difference of matter and information. EM is a philosophy of mind, "overarching all manifestations of information and not only mind."

Logic in Reality is compatible with part but not all of this approach. LIR provides a realistic interpretation of the physical, dialectical relation, a grounded "interactive dualism" between identity and diversity, a unity-in-diversity and a diversity-in-unity as well as between the terms of other critical physical and philosophical dualities. To keep this distinction between the two terms better in mind, the language of asymptotic reasoning developed by Batterman [2] is useful. The limits of identity and diversity (primacy of non-contradiction: contradictional) toward which phenomena tend asymptotically are instances of *regularity* at which the tenets of reduction and those of classical logic apply. The 'movement' toward an equal actualization and potentialization of two opposing elements results in a state of maximum contradiction. This limit, at which emergence is possible, is *singular*(contradictorial: T-state). As stated above LIR, in contrast to standard bivalent or multivalent propositional logics, provides the basis for an ethics as the finality for the intellectual process, a principle which also pervades the work of Hofkirchner.

5.2.3 *Life and Immanence*

In looking for dialectic pairs to study, we may choose Life and Immanence, where Life is simply a more affirmative and better-specified concept than Immanence. Life rather than Immanence opposes Transcendence not only as a general concept, but as the form in which (or by which) Transcendence is specified, namely dialectics itself! Term-to-term oppositions remain an essential part of philosophy—Negation-Transcendence *vs.* Affirmation—Immanence, and Dialectics, placed in opposition to Life, in opposition to an integral 'immanentism'. Transcendence is thus specified by Dialectics, but its relation to Immanence is not Dialectics. These are not dynamic, ontological relations in the LIR sense, in which the relations between non-absolute elements constitute the 'dialectics', and there is no difference between opposition and dialectics. We will return in more detail to the transcendence-immanence dualism later in this book.

5.3 The Scope of Dialectics in Lupasco

In a small monograph, published towards the end of his life (1979), Lupasco [17] presented the key physical, biological and cognitive concepts in his entire work in terms of dialectics. His focus was on the latter, indicated by the title *L'univers psychique (The Psychic Universe).* The opposing movements in every process from actual to potential, and potential to actual, are the origins, as operators, of structures and systems. For Lupasco, such a process is a *dialectogenesis*, whose logic is, also, a dialectology. His dialectomethodology (see above) is operational in the sense that to (a) know and (b) know that one knows, one must act dialectically, operate (Lupasco's word) or actualize cognitively dialectical actualizations and potentializations and grasp their inherent antagonisms, their homogeneous and heterogeneous aspects and so on.

Lupasco proposed the application of his scheme to the psychological problems of human beings, basically seeing different pathologies of as expressions of excesses of one form of thought over another. Such disequilibria were to be treated by a *dialectotherapy*, and death and affectivity or affect were also capable of being studied using this methodology. We will not go into these subjects here. As we have noted elsewhere [4], Lupasco's concept of affectivity as a cognitive domain outside that of logic, energy and contradiction amounted to a denial of the universality of his fundamental principles. His suggestion that affectivity 'accompanied' other mental processes in some way is incompatible with our best current views of mind. There is still one point in this theory of forty years ago which sounds quite modern (translation ours): "Cognitive and therapeutic dialectomethodology should be applied to the serious problem of *psychic poverty*, that is, contradictorial and emotional, of people, mentally ill historically, to try to ground a politics of the soul". Let us not take this passage literally but use it as a bridge to the past and to dialectics in Ancient Greece.

5.4 Dialectics in Ancient Greece

5.4.1 Parmenides, Plato and Anaxagoras: The Dialectical Discourse of Being in Its Existing Multiplicity

Dialectics appeared in Ancient Greece as a discourse between two or more people holding different points of view about a subject, but wishing to establish the truth through reasoned arguments. Formal logic was developed later as a referral to subjects concerned with the most general and universal laws of truth. This means that *dialectics refers to the world consisting of multiple acting subjects*, while formal logic refers to a *unified sub-set of the laws common to all observers*. A discourse between several people providing reasoned arguments about the reality common to all of them follows from the multiplicity of the world containing interacting observers. From the logical point of Parmenides that Being is undivided and unmoved, a position which was claimed to be substantiated by the Zeno paradoxes, a search for the explanation of movement in the perceived world led to (1) substantiation of the smallest existing spatio-temporal "quanta" of movement called "atoms", referring to physical reality and (2) the formulation of cognitive pure ideal forms called *eidoi*. This separation into physical and cognitive reality came from the necessity of attributing the property of existence to Being, basically something accessible only to conscious (human) beings. This attribution is analyzed in detail in Plato's dialogue "Parmenides", which is considered as the most outstanding example of antique dialectics and in which Plato, to analyze the notion of existing Being, abandons narrowly interpreted principles of his philosophy of ideas for the goal of explaining the phenomenon of existence.

The thesis substantiated in "Parmenides" is formulated in the following way: "The existing One appears as Many". Thus the existing world is present as a multiplicity that corresponds to the principle of multiplicity of primary elements called "seeds" in the philosophy of Anaxagoras. These were later defined by Aristotle as "*homoeomeria*". They include both potential and actual constituents and are harmonized in the omnium of their relations by the principle called "*Nous*" by Anaxagoras. A similar philosophical concept was formulated in modern times by Leibniz, who called these elements "monads". Below, we will discuss monad in relation to our definition of an "ontolon" as the element of real existence.

The definition of the principle uniting *homoeomeria* as "*Nous*" assumes that a non-physical 'discourse' reasoning between them takes place that results in their harmonization in the Universe. Such harmonization can be either based on the ideal pre-established principles defined by Leibniz as a "pre-established harmony", seen in antiquity in the concept of the Demiurges of Plato's dialogue "Timaeus", or it can be achieved through a kind of a principle of natural selection that was formulated explicitly in the poem "*De rerum natura*" of the Roman poet Lucretius, who followed the atomistic philosophy of Epicurus.

5.4.2 Heraclitus: Energy, Dynamic Oppositions, and Self-Growing Logos

Heraclitus was the first thinker in human history who considered energy as the ground of the real world and formulated a principle of dynamic opposition. Reality to him was the substance of fire (*pyr*) which determines that everything flows (*panta rhei*, "*rhe*" is the Greek root for "flow"). For Cosmos, to be 'ever-living fire' (Fragment 30), there should be a certain Logos expressed as a set of physical laws. Fire determines the tendency of a system to "flow" to another state, which permits the system to complexify or evolve. Heraclitus defined this property as "*logos heauton auxon*" (self-growing Logos). This property belongs to a soul, i.e. to an individual observer, while the whole objective world means a kind of a "common account" (Fragment 2). The unity of the world comes from the dynamic opposition: "Couples are things whole and things not whole, what is drawn together and what is drawn asunder, the harmonious and the discordant. The one is made up of all things, and all things issue from the one." (Heraclitus: Fragment 10). For Heraclitus, conflict of opposites does not mean the splitting or destruction of the unity of reality, but its constitution. The logos is the only "abiding thing", the orderly principle according to which all change takes place as a 'binding-together'. Conflict (*polemos*) is the core of logos, both appear rather as the parameters of 'fixed measures' according to which the Heraclitean fire (*pyr*) is kindling and extinguishing (Heraclitus: Fragment 30). According to Heraclitus, the opposites are the same not in matter, but in Logos:

Listening not to me but to the Logos it is wise to agree that all things are one (Fragment 50).

The assumption of "common account" is very important and not trivial at all. There are different types of objectivity: one (ideal, Platonic) is related to the objectivity of mathematical world ("thinking is common to all", Heraclitus, fragment 113) and it is timeless, another (real, which we can define as the Democritean) is the objectivity of observed physical world and it occurs in time. Really, Diogenes was not fully wrong in 'solving' the kinematic paradox by walking: the physical (but not logical) impossibility of the infinite division of space in a given universe fixes the paradox and creates a set of physical laws via introduction of a minimum unit (quantum in our lexicon) of action. The unity of dynamic oppositions in the Logos defines the main ethical statement of Heraclitean philosophy: "For God all things are fair and good and just, but men suppose that some are unjust and others just" (Heraclitus: Fragment 102).

In his philosophical paradigm, that represents a first *Philosophy in Reality*, Heraclitus also formulates the category of potential reality. Although not many fragments of his book have survived, this is clearly seen in what we have. Potential reality appears as "unapparent": "An unapparent harmony is stronger than an apparent one" (Fragment 54). The unity of opposites remains in the movement in which "beginning and end are common" (Fragment 103) and the cyclic movement keeps the structure solid and resistant to disintergration: "also *kykeon* (a kind of beverage or posset) disintegrates if it is not stirred (*kinoumenos*)" (Fragment 125). Reality unfolds as a perpetual game in which the potential and the actual are united: "Eternity is a child at play, playing draughts: the kingdom is a child's" (Fragment 52). The self-growing reality in the paradigm of Heraclitus represents its basic law (Logos) and is evolved in the totality omnium of individual souls (observers) that sense it via their perception and cognition.

5.4.3 Dialectical Discourse in Atomism

The philosophy of atomism of Leucippus and Democritus initially appeared as a response to Parmenides and Zeno as an attempt to explain the existence of movement in the world. At first glance, it avoided dialectics by claiming atoms as real existing elements, between which was emptiness (the vacuum). In the early versions of atomism, atoms were not considered as ideal essences that could be involved in a kind of discourse, although there is some evidence that Democritus sometimes called atoms "ideas". The strict determinism of the Democritean universe does not leave room for the possibility of a 'discourse'. Later, Epicurus postulated an unpredictable movement (swerve; swerving) of atoms called by Lucretius "*clinamen*" which introduces the concept of potentiality in the description of the universe of atoms and enables it be more diverse and variable. These deviations initially appear arbitrary, but they can be controlled resulting in higher levels of organization. In this picture, finally, human consciousness appears which controls them further. The atoms of Epicurus can be viewed as quanta in the process of actualization and the actualized Being is Multiplicity due to the *clinamen*. Without *clinamina* that enlarge the field of potentialities, a strictly deterministic universe would not be able to evolve and generate a multiplicity of events, phenomena and realizations. However, the principle defined by the term *clinamen* cannot be developed without the more advanced concept of potentiality and actuality introduced later in Greek philosophy by Aristotle, despite the fact that Epicurus lived after Aristotle. Aristotle developed his concept of potentiality without reference to atomism, and Epicurus was not influenced by Aristotle's understanding of potentiality.

The most important point here is that atoms possessing *clinamina* escape from the strict determinism of the Democritean world. They can enter into a process that reduces the field of potentialities defined by their *clinamen* and form a kind of discourse or 'communication'. This discourse is not strictly physical as it escapes physical determinism being based on persistency during interaction. The result of such interaction is a consistent history that is produced in it1. The complexification of interaction is associated with generation of predictability of the movement defined as *clinamen* and the possibility of its control. This corresponds to the development of cognition in systems of interaction of atoms possessing *clinamina*. Having the property of the *clinamen*, Epicurean atoms are no longer mechanistic structures but rather *ontolons* in our interpretation of the unity of reality outlined in Chap. 9. The constraints imposed on *clinamina* in the course of interaction between atoms become as a prerequisite of the formation of complex structures developed in the Universe.

By the introduction of *clinamina* as the properties of atoms, there is a basis for an interaction of atoms not only via application of deterministic forces but also via the influence of the potential fields of their possible unpredictable deviations, recursively modifying their capacity for such unpredictable movement. Moreover, if atoms unite in organized structures, these capacities become restricted to a certain extent. The interaction via the effects of *clinamina* on each other can be described as a mutual measurement process, and when atoms become united into organized structures this measurement appears as an internal measurement within the structure. This kind of measurement can lead to further complexification of the system via reduction of the potential field and self-referential 'memorization' of this reduction.

5.4.4 Aristotle and the Dialectics of Potentiality and Actuality

Aristotle's development of a discourse that involves potential reality makes him the greatest figure in 'Philosopher in Reality' and also in formal dialectics. The common view is that Aristotle mainly developed formal logic, which is the content of his Categories (part of his Organon). In fact, he also introduced a kind of logic *in reality* which can be considered as a version of the logic of/in reality of Lupasco and Brenner. This includes the transition from the potential to the actual, although the opposite transition was not analysed in detail. The great achievement of Aristotle is the inclusion of *potentia* in his dialectical logic in reality. This is seen in his

Metaphysics and in more detail in *Περὶ Ψυχῆς* – *Peri Psychēs* or *De Anima* in Latin (On the Soul). Heredity can be understood in Aristotelian terms as the transfer of information, defined as "entelechy as knowledge" in a seed as compared to the realized "entelechy as the actual exercise of knowledge" of the developed organism, *cf*. G. Stent's Molecular Genetics [28]. Matter in Aristotle's concept is a pure potentiality which is a prerequisite for two types of actuality: one is information ("entelechy as possession of knowledge") and the other is actual realization/presence ("entelechy as the actual exercise of knowledge"): "Now the word actuality has two senses corresponding respectively to the possession of knowledge and the actual exercise of knowledge" [1]. In the latter ("actual exercise"), a selection from two or more possible attempts have been made to make a 'consistent histories' approach to quantum phenomena. We consider these as tautological and ultimately reductionist: the basic Lupascian approach assigns ontological value to inconsistency.

Realizations takes place according to the first type ("possession of knowledge"), and the discourse becomes incorporated in the total reality instead of being a result of interaction of reasoned arguments of conscious subjects. Being in the philosophy of Aristotle includes the unity of potential reality and its realizations, a concept which was developed by him in detail but which had arisen in the concept of the primary substance as *apeiron* outlined more than two centuries earlier by Anaximander [24]. Another important aspect is the understanding of a soul as a capacity, inseparable from the body, which we later find in the concept of substance in the philosophy of Spinoza. According to Aristotle, "... the soul neither exists without a body nor is a body of some sort. For it is not a body, but it belongs to a body, and for this reason is present in a body, and in a body of such-and-such a sort" [1]. On the other hand, different bodies can be animated by the same set of capacities, by the same kind of soul, so soul can be used in a singular, not plural, sense. This was earlier substantiated in Plato's dialogue "Parmenides" formally, while Aristotle presents this as a foundation of a kind of natural science that was later defined as psychology. The discourse of different potential realizations becomes incorporated into the reality that evolves from the inanimate potential matter to the actual realization, bearing form and being capable of an information transfer through the discourse of the substance with itself and in which it appears as a multiplicity of "seeds" sharing the same "soul". In fact, Aristotle was the first who developed a something like a Philosophy in Reality, and despite certain shortcomings of its scientific basis, e.g. the absence of real reference frame in his physics, this philosophy in/of reality was the most successful realization of this paradigm over the centuries. Certainly, Plato was a predecessor of Aristotelian philosophy, but Aristotle successfully developed his own paradigm. The attempts by Marx to position the philosophy of Hegel in reality and by Heidegger that of Husserl did not solve the problem of the *res potentia* in reality. This was inherent in Aristotle and expressed by Lupasco.

5.5 Dialectics in European Philosophy

In modern times, the logic of discourse was incorporated into philosophical thought as a main feature of the new European philosophy, formulated by Rene Descartes as the distinction between a *res cogitans* and a *res extensa*. Being initially represented as two separate essences, they were unified in different philosophical systems in different ways, and dialectics appeared as the discourse for establishment of such unification. Thus, we can see the development of dialectical principles in the systems of Spinoza, Leibniz and Kant. After Kant, dialectics became associated with Hegelian idealism on the one side and Marxian materialism on the other. In this Section, we will discuss these advances, their limitations and possible future developments.

5.5.1 Spinoza, Leibniz and Kant

An important aspect here is that if substance is a *causa sui*, it cannot be fully cognizable, since only mechanisms can be fully cognizable in the framework of mechanics. That is why Spinoza ascribed to substance an infinite number of attributes, among which only cognition and spatio-temporality are perceivable by us. In relational terms, only these two attributes can be involved in relations, while other attributes cannot be cognized although they can be involved in some sense in the shaping of the *res potentia*. The statement of the infinite number of attributes by Spinoza has similar meaning to the statement of Leibniz that "monads have no windows" (there is no window through which to see these attributes) and to the statement of Kant about the non-cognizability of the *Ding-an-sich*. The latter also appears as a monad and should not be confounded with the physical object, as it cannot be reduced to its spatio-temporal structure or to its ideal form. Later in this book, we will call this primary unit "ontolon". The "non-cognizable" attributes shape the potential of monad/*Ding-an-sich*/ontolon. These attributes correspond to the Epicurean *clinamen* which, as noted above, is beyond shape and form.

The great progress in philosophical thought made by Immanuel Kant, which he called the Copernican revolution in philosophy, was the revelation [15] that mind is not the basis of Being but only the instrument of human cognition. Real Being is beyond mind and can be defined as a *Ding-an-sich* (thing-in-itself). The later development of German philosophy (Fichte and Hegel, but not Schelling who is often misinterpreted) abandoned this basic statement and returned to a concept of Mind in its totality that generates the material world as an *Anderssein (Andersheit)* via its dialectical discourse. On the contrary, Arthur Schopenhauer associated the *Ding-an-sich* with the primary energy called *Wille* (will) whose permanent goal is actualization via generation of representation (*Vorstellung*).

5.5.2 *The Process Philosophy of Schelling and His Followers*

Schelling (1775–1855) was an important philosopher who is often neglected. We consider a major contribution his development of a dialectical version of natural philosophy [26]. His revival as an important thinker in natural philosophy is, in particular, due to Gare [8, 9] who emphasizes the significance of the philosophy of Schelling to the development of theoretical biology in particular. Schelling argued that it is necessary to identify a "primordial realm" from which subjects and objects emerge simultaneously. Schelling equated the 'pure subject-object' with nature, through which pass 'unconscious' stages before finally consciousness finally emerges. Schelling attributed the self-constraining or limiting activity to nature, simultaneously including 'productivity' (or process) and 'products'. Nature as productivity is subject and as product it is object. According to Schelling, productivity consists in *opposed activities* limiting each other, *i.e.* of dynamic oppositions. His great statement is the following: "Give me a nature made up of opposed activities … and from that I will bring forth for you intelligence, and with it the whole system of its presentations" [8]. From opposed activities, Schelling derived force and matter, space and time, chemical systems and living organisms.

Schelling stated that a whole maintains itself by reciprocal causation, contrary to being a derivative form of causation, as Kant considered. This reciprocal causation should be taken as the primary form of causation, while the mechanistic cause-effect relations appear as its derivative. Schelling characterized living beings as inherently unbalanced (non-equilibrium systems in modern terms) and thereby actively engaged with their environments. Life involves creative activity which forms and reforms living bodies as products. Higher developments of life are associated with greater imbalances, assuming greater awareness of living organisms on the world around them. With the appearance of humans, greater levels of intersubjectivity and reflexivity arise. Gare [9] sees in Schelling's philosophy a program for a theoretical biology based on the dynamic opposition of process and product. This is a reciprocal causation which resembles the "closure to efficient causation" in the concept of Robert Rosen.

In fact, Schelling's program resolved the paradox of describing the origin of logic by logical means. In this view, since coming-into-being is not an entity but a becoming and always contingent, existence and movement cannot be a logical category. For the purpose of describing the process of actualization, we need an instrument that can in some way overcome the rigid framework of a purely logical description of the process of coming into being, Dialectics represents such a substitution for descriptions using classical logic. Gare concludes that Schelling developed a theory of emergence and a new concept of life relevant to current theoretical biology, as noted above. This theory is grasped in Logic in Reality in the Principle of Dynamic Opposition (PDO) operative at the most fundamental physical level as well as all higher levels of reality. The PDO and contradiction or counteraction, and a robust notion of potentiality, is required for understanding the relation between substances, events and processes.

The concept of emergence, introduced by Schelling, was further developed by Whitehead [35]. Following Whitehead, it is only events that are the actual entities of the physical world and need to exist in order to have a physics. This leads to a new idea of physical substances, that of a distribution of potentialities, "powers" or propensities, that are "reasonable consequences of a theory of processes" as Thompson [31] suggested.

Schelling also influenced Bogdanov [3] who developed the concept of organizational evolution of matter called tektology (see also Arran [7]. In this concept, Bogdanov put process and organization at the center of perpetually evolving reality seen as the (self)-organizational process of the Universe. Science itself, according to Bogdanov, is comprehensible as a development within and of nature. Tektology claims that processes are the primary reality rather than things or substances and their attributes. This assumes that a non-formal language is more appropriate than mathematics for characterising the basic characteristics of nature; mathematics should be seen as having a derivative status [33]. For understanding the phenomenon of complexification as a dialectical process, Schelling, Whitehead, Bogdanov and Lupasco represent an essential alternative to the positivist interpretation of reality that dominated science for many years.

5.5.3 Hegel's Dialectics

Hegel started from a vision of the contradictorial or antagonistic nature of reality; developed elaborate logical systems that dealt with contradiction and went far beyond formal propositional logic. Hegel incorporated contradiction in logic and rejected the idea of a classical 'formal' logic that claimed to be a study of the form of thought in abstraction. Hegel proposed three axioms to describe reality that differ from our first reformulation of the classic axioms: A is A; A is non-A; non- A is A after all, or else they are all present together. As written, they imply a primarily diachronic sequence of A, non-A, and A as thesis, anti-thesis, and synthesis.

Hegel's contradictions had their origin in the manifestations of Spirit as Idea or Concept. Governed by Absolute Necessity as their Internal Teleology, they struggle to return to it in an ascending dialectic via the vehicle of human- consciousness-in-history being finally in a position to understand the process. At first sight, Hegel seems to have accepted contradiction as fundamental, until one realizes that, although the most ontologically significant relation is one of opposition between two things that mutually define each other, what is essential is their inner identity. In fact, if an element is in contradiction with itself as its negation, it disappears. This argument suits only Hegel's ontological conclusion that finite things disappear or die because they are failed attempts to 'embody the infinite' and makes it clear that Hegel lacked a physical/metaphysical basis for life, form and diversity of equal ontological value. This absence of grounding was also noted by Brenner [4], referring to Taylor [29].

Hegel's logic is integrated into a metaphysical dialectics, in which the contradictory duality he introduced was continually abolished by successively purer and broader syntheses of antithetical terms, finally reaching the *Aufhebung*. Contradiction is inherent even in the supreme identity of absolute spirit (*Geist*), since it is both embodied and opposed to its embodiment. One may, rather, take Hegel's idea that every phenomenon is a 'fragment' of *Geist* that reflects the latter's properties to foreshadow the contradictorial, dynamic view of energy, provided the difference in their role and behaviour is not overlooked.

This is nothing more than the philosophical expression of macrophysical becoming, governed by the Second Law of Thermodynamics. The subsequent dialectics of Marx and Engels simply transpose, to the social level of reality, the same Hegelian drive toward a synthesis involving the suppression of, in contrast to Hegel, all contradiction. As pointed out by Taylor, Hegel's thesis depends on a premise of ontological necessity that in turn depends on the contradiction of the finite. Hegel established or expounded his ontological structure at 'high' levels, but his project required demonstration of his ontology at the lowest level of simply determinate beings, and his attempted proof of contradiction failed. The Hegelian picture of the world has ongoing relevance as the basis of a relevant philosophical vision of "embodied subjectivity, of thought and freedom emerging from the stream of life, finding expression in the forms of social existence, and self-discovery of humans in relation to nature and history." Hegel's system is not metaphysically and physically grounded at the "lowest level of simply determinate beings" that is, microphysical entities.

Some comments about dialectical logic may be appropriate here. As discussed also by Priest [22], distinguished between dialectics and formal logic—which was for him the Aristotelian logic of his day. The law of non-contradiction holds in formal logic, but it is applicable without modification only in the limited domain of the static and changeless. In what is generally understood as a dialectical logic, the law of non-contradiction fails. The subsequent developments of formal logic, starting with Frege and Russell, have forced Hegel's conception of contradiction to be rejected or interpreted non-literally. Neo-Hegelians have attempted to conserve this principle of contradiction by emphasizing the factor of time: A is not identical to A, because time has passed in which changes have occurred; contradictions take place one after the other, etc. Articles purporting to describe dialectical logics still appear. In one example, a relation is proposed with non-linear dynamics in which dialectical logic is enhanced by mathematical logic. These and other moves, however, do not address, any more than Hegel did, the question of what drives the change from thesis to antithesis to synthesis, that is, how any term cannot 'stand on its own' but 'goes over' into its opposite or contradiction. Russell demonstrated that Hegel's logic could be deconstructed because it still presupposed traditional Aristotelian logic, but not for this more important reason.

Piaget, also, did not go beyond the standard Hegelian form of Marxist dialectical materialism. This correctly accords a central role to conflict and contradiction in the transformation of social realities. However as noted in (Priest 1989), Marxist dialectics fail to give an adequate account of the true contradictions involved in society: an inconsistent or paraconsistent logic is necessary for such an account, albeit not sufficient. The aim of dialectics is to characterize the emergence of new structures from real contradictions, in other words, it should operate to reflect the properties of generic phenomena *in reality.*

5.5.4 Marxist Dialectics

Another trend, which overturned Hegelian idealism, resulted in Marxism in which dialectical discourse is located in an inanimate reality called matter. However, Marxism did not develop the concept of substance in the sense of Spinoza or other earlier philosophers. Dialectical discourse appeared, in particular, in Engels' interpretation, as a vaguely formulated set of "laws of dialectics", which emerge without proper substantiation.

Engels [6], in his unfinished work *Dialectics of Nature*, formulated "three laws of dialectics". He elucidated these laws as the immanent properties of dynamics of material substance, although his concept is, however, not elaborated well as compared, e.g., to Spinoza. These "laws" are the following: (1) the law of the unity and conflict of opposites (which present in Heraclitus); (2) the law of the passage of quantitative changes into qualitative changes (generally based on the Sorites paradox of the heap); and (3) the law of the negation of the negation (which may be considered as the invention of Hegel). In fact, in comparison to Hegel's philosophy where dialectics appears as an internal discourse of the Absolute Idea, in Engels' interpretation dialectics is positioned far away from any discourse and represents a set of vaguely, rather reductively defined formal laws. These nevertheless became the basis of "dialectical materialism".

To instantiate discourse, an element of the world should represent a monad that performs some kind of internal 'computation', which in a more materialistic view can be equated to an "ontolon" (see below the discussion on ontolons in Chap. 6). In this sense it is more logical to discuss "dialectical organicism" as suggested by Joseph Needham [20] than "dialectical materialism". In the organicism interpretation, the dialectical discourse of simple monads, ontolons in our terms, generates the complex structure of space–time, and, as in Spinoza's philosophy, this discourse describes a *causa sui* principle. We should state here again that a principle of "something being the cause of itself is a 'necessary' one only in standard philosophy, and makes no sense in science or the logic and philosophy of reality. The formulation of "dialectical materialism" without clear definition of the concept of substance as matter that would justify the deductive necessity of the "laws of dialectics" resulted in the difficulties of development of this concept by the next generations of philosophers. We will show later that the concept of an ontolon can resolve the difficulties in the ontological interpretation of dialectics in nature.

5.6 Post-Marxist Dialectics

We will turn now to the "post-Marxist" philosophers Mamardashvili [19] and Ilyenkov [14] who performed the most important re-evaluation and development of the Marxian dialectical concept of consciousness. Since the developed concept of substance is lacking in Marx' theory, Merab Mamardashvili refers to Descartes and Kant in its description, while Ilyenkov is grounded mostly in Spinoza. Both Mamardashvili and Ilyenkov refer rather to the logic of *Das Kapital* of Marx than to the *Dialectics of Nature* of Engels. To what extent the newly formulated principles really arise in Marx or are the result of the major reformulation performed by Mamardashvili and Ilyenkov is not so important, but, in our view, their interpretation goes far beyond the basic formulation of Marx.

We will outline first the approach to dialectics as it was formulated by Mamardashvili and then turn to Ilyenkov.

5.6.1 Mamardashvili: Dialectics of Social Consciousness

According to Mamardashvili (1930–1990), who significantly contributed to the rationalist theory of perception whose origin was in Descartes and Kant, the relation of subjective consciousness to objective reality is mediated by the potential set of the whole system of, in particular social and political relations organized hierarchically. This means that Marx, according to Mamardashvili, 'discovered' the phenomenological nature of consciousness via its quasi-objective nature by introducing an abstraction that allows the analysis consciousness as the objective transformation of external objects into quasi-objective patterns, without direct involvement of the processes taking place in internal subjective reality. This means that the nature of consciousness is placed beyond the phenomena that serve for the maintenance of a social system of communication. "Being–consciousness" is unified, so that Being and consciousness appear as the different aspects of one continuum in which the object and the subject, the reality and its representation, the real and the imaginary are not strictly separated, while remaining relatively differentiated and non-identical to one another. They are connected in the continuum via a relational operator of transformation expressed in the course of social processes. This is a representation of the "dialectical" nature of consciousness; the application of the principles of Logic in Reality allows one to cut through the endless discussion of organicism *vs.* realism *vs.* holism *vs.* reductionism. Complex dynamic part-whole relationships are possible, as in the first concept, without the system being 'alive', but which are not reducible to their energetic substrates.

The discourse between conscious subjects, mediated by the incorporation of actualized material reality into the discourse, generates the conditions for social dynamics and progressive social evolution. The actualized material reality appears primarily as the result of previous human activity that has formed the signified social memory that represents the basis on which the current social structure is built. This "basis" is not only material but also cultural, and it shapes social structure by providing already existing forms or models. In other words, it "geometrizes" the society in the same way as Spinoza's substance geometrizes the world in the course of its self-actualization. These concepts form part of our dialectical approach to systems in society outlined in Chaps. 18 and 19.

5.6.2 Dialectics in Ilyenkov's Conception and Beyond

The concept of dialectics as a metalogic based on reasoning and discourse was developed by Ilyenkov (1924–1979); he incorporated its principles into basic substance understood in the sense of Spinoza and having the basic property of a *causa sui*. Despite his close association with Marxist dialectic materialism and apparent rethinking of Hegelian idealism, Ilyenkov suffered from the attacks of both orthodox Marxists and anti-Marxists, which resulted in his premature death. According to Ilyenkov, substance perpetually generates objective forms of subjective activity which follow a logic external to a material body [14]. In this regard, the process of cognition which includes discourse and reasoning is not transcendent to the being, but immanent init. Before Ilyenkov, such a point of view was formulated in the Marxist psychology of Lev Vygotsky and Leontiev in the 1930s, later substituted by a reflexology grounded in the works of Ivan Pavlov. The main idea on which Ilyenkov's concept of dialectics is based is the unity of cognition and space–time, of a *res cogitans* and a *res extensa*, which are linked via a *res potentia* (cf. our discussion of Heisenberg's *res potentiae* in Chap. 6). The unity of *res cogitans* and *res extensa*, which is the central point of Spinoza's philosophy, is attributed to substance from the lowest levels of its organization and becomes highly expressed at the highest levels such as human civilization. The basic property of cognition, following Spinoza, is a capacity of a body to build a trajectory of its movement across other bodies according to a logic of arrangement of these bodies in the space external to that body. The idea of a thing in this regard becomes fully coincident with the way of its being, which is its generation based on this idea. In other words, the idea as eidos turns to be the idea as *technos* as defined by Mamardashvili [19], and these two aspects of idea (*eidos/technos*) are inseparable in the generating process. As an example, a geometric shape is ideal because it represents a way of conceptual formation/generation of all material bodies possessing this shape.

The basic function of intellect, according to Spinoza and Ilyenkov, is to move and arrange external objects, *i.e.* an anticipatory operation on and with externality. Humans perceive the world only because they actively move and arrange external bodies. This activity is not based on mechanical causality but on the causality which immanent to the primary substance which was a *causa sui*. Mechanical causality is always external to the body, while substance possesses causality in itself realizing self-movement via establishing relation to external bodies by its abstracting capacity which appears to generate pure (ideal) forms of reality. In this activity, the *ordo et connexio idearum* coincides with the *ordo et connexio rerum*, in other words, ideal activity results in valid practical implementations. At the level of social organization, intellect is involved in establishing relations not only to external material objects but to other intellectual beings, which becomes the basis of morality and successful social communication.

The grounding of Ilyenkov's dialectics in Spinoza's concept of substance represents a major advance, but it has certain limitations due to the apparent disregard of relational, that is, energetic, principles in the operation of cognitive activity. In fact, Spinoza's substance is a manifestation of the 'One', while Plato, in the dialogue "Parmenides", had already substantiated that the existing One appears as Many. In the universe of the forms of existence or "ontolons", the spatio-temporal order appears as a relation between objects established in their interaction. The intrinsic limits of computation shape the spatio-temporal order and also pose limitations on cognitive activity. They represent the principles that are inherent to our world and may result from the transcendent action of the establishment of the actualized physical world from the 'pure' logical principles that are manifestly insufficient for its appearance.

To what extent this transcendent action is really mirrors our cognitive activity remains open, but in this book we present what amounts to an 'immanent' alternative in the work of Lupasco. Hume [13] indicated a possibility of such similarity. If the basic property of consciousness, according to Spinoza and Ilyenkov, is the establishment of space–time, then the introduction of fundamental constants at the birth of the Universe is a 'conscious' act that sets the limits of (and for) actualization of the *res extensa* from the *res potentia*. The assumption of a *res cogitans* here seems apparent, otherwise we need to introduce the principle of natural selection between universes, as in the multiverse of Smolin [27] which assumes the unsubstantiated actual pre-existence of all of them. The anthropomorphic 'bootstrapping' by a basic 'intelligence' that determines the limiting conditions of the physical world is, however, beyond scientific reasoning and cannot be discussed in this book.

5.6.3 Ilyenkov and Dubrovsky. Information and Consciousness

In the development of dialectics as metalogic, Ilyenkov did not consider the psychophysiological problem as important for understanding the nature of consciousness. He emphasized consciousness as a primarily social phenomenon. His approach was criticized by David Dubrovsky (b. 1929), who claimed that the neurophysiological grounds of consciousness are essential for the study of conscious events [5]. At the present time, this dispute has lost its actuality. The dynamic opposites of neurophysiological and social realities, in fact, generate a "transaction" that leads to the appearance of consciousness. Dubrovsky can be considered as an important philosopher who developed philosophical comprehension of information as the basis of consciousness in its biological and social operation. Dubrovsky contributed significantly to understanding the information as a fundamental philosophical category having its own ontological status and to introduction of the problem of consciousness into natural philosophy.

5.6.4 The Dialectical Concepts of Wu Kun

Wu Kun defines information as a philosophical category in addition to those of to matter and energy. Wu Kun considers the philosophy of information as a metaphilosophy, and it will be analyzed in the corresponding Chap. 10. But it is also a dialectical concept representing an important aspect of development in dialectical thought. In the informational stance, the "opposites" in information are not captured by the classical concept of a classical, static "unity of opposites", but by the dialectical interaction of opposites, classified by Wu based on his general philosophy of natural ontological levels that captures the essence of the properties of information. The resulting doctrine of objective information, subjective information and human information in society constitutes Wu's information theory and establishes it as a unified philosophical foundation for information science grounded in dialectics. We discuss Wu's philosophy of information as a metaphilosophy also in Chap. 11 Information.

5.6.5 Other Trends in Post-Marxist Dialectics

It is worthwhile also mentioning the development of dialectical logic in the works of Zinoviev (1922–2006). He belonged to the same generation as Ilyenkov and Mamardashvili. He became a political writer and an opponent of the Soviet regime. He emigrated to the West but after returning to Russia in the 1990s, developed the most conservative and anti-Western views on social dynamics. In his earlier years, he outlined a "substantive multi-valued logic" limiting the applicability of Marxist dialectics to a special kind of objects (historical or social), which are an "organic whole" with a complex functional structure. According to his views, dialectics is a method for studying complex systems of empirical relationships, a concept that was introduced in his monograph "*Philosophical Problems of Many-valued Logic*" (1963) [34]. This logic explicitly introduced the ontological connection of dialectics to the description of the process of formation and development of complex systems in reality.

One of important concepts in the current philosophy is the dialectical concept of Nicholas Rescher (see also Chap. 3). According to Rescher, although classical logic has a similarity to the apparent object-property structure of reality and classical ontology reflects this structure, the logic of object and predicate does not prevail in nature. The foundational work of Rescher on process metaphysics and process semantics can be viewed as part of the creative development of dialectics. Rescher's mission for process philosophy is "enabling us to characterize, describe, clarify and explain the most general features of the real." Further, he relates his view of the processual structure of reality to energy, the entities of quantum mechanics entering into more and more complex arrangements. Rescher sees the development of a process semantics, as opposed to logic, as the conceptually most versatile and philosophically most fundamental tool for understanding reality. It is at the basis of his ideas of process philosophy and process metaphysics. What he called the mainstream logical theory of the West, which takes an approach to truth that is committed to its static fixity, was and is unable to meet this challenge.

Pragma-dialectics, the pragma-dialectical theory of van Eemeren and Grootendorst [35, 36], was an attempt to return dialectics to its original meaning in the Ancient Greece. This theory does not concern the natural science but is centred on argumentation as a phenomenon of communicative and interactional discourse. The dialectical nature of the concept originates from "critical rationalism" and formal dialectics, having the pragmatic dimension via inclusion of speech act theory, language study and discourse analysis. In this concept, socialization is viewed as a dialectical discourse describing the nature of interaction.

5.7 Conclusion

While formal semantic logic refers to a unified sub-set of the laws common to all observers, in the relational world that includes multiple non-equivalent observers, the standard logical approach needs to be expanded to the dialectical one, providing for the inclusion in logic of the grounding in science, proposed by Logic in Reality. Dialectics refers to the world consisting of multiple acting subjects. In this expansion, logic as extended to reality appears as dialectics. A conflict as well as change and alternation between the different but closely related, interactive elements of a phenomenon represent the basic nature of dialectics. Dialectics can be considered neither more, nor less, than the generalization and mental expression of conflicts in nature and civilization, and their resolution, that man has observed from time immemorial. For Heraclitus, conflict did not mean the splitting or destruction of the unity of reality, but its constitution [12]. The logos, the only "abiding thing", is the ordering principle according to which all change takes place as a 'binding-together'. To understand reality as a dialectical process, at noted in Chap. 1, we need to return to the original view of Heraclitus about energy (fire) as the natural substance that perpetually generates the dynamic oppositions in reality constituting the reality itself. Dialectics initially appeared as a discourse between two or more people holding different points of view about a subject, but wishing to establish the truth through reasoned arguments. While formal logic has a substantially epistemic nature and refers to a unified sub-set of the laws common to all observers, dialectics refers to the world consisting of multiple acting ontological units acting as observers. These units appear as conscious subjects only at the highest level of cognitive organization. Generally, these ontological units are the forms of existence defined as "ontolons" and representing elementary entities of the relational polyphonic Universe. To use the classical term, dialectics represents the discourse of ontolons.

References

1. Aristotle: De Anima (On the Soul). In Barnes, J. (Ed) The Complete Works of Aristotle, pp. 641–692, transl. by Smith, J.A. Princeton University Press, Princeton NJ (1995)
2. Batterman, R.: The Devil in the Details. Asymptotic Reasoning in Explanation, Reduction and Emergence. Oxford University Press, Oxford (2002)
3. Bogdanov, A.: Essays in Tektology: The General Science of Organization, 2nd edn., p. 1984. Intersystems Publications, Seaside (1921)
4. Brenner, J.E.: Logic in Reality. Springer, Dordrecht (2008)
5. Dubrovsky, D.I.: The hard problem of consciousness. Theoretical solution of its main questions. AIMS Neurosci. **6**, 85–103 (2019)
6. Engels, F.: Dialectics of Nature. Progress Publishers, Moscow (Written 1883, Originally published 1925) (1976)
7. Gare, A.: Aleksandr Bogdanov and systems theory. Democracy & Nat. **6**, 341–359 (2000)
8. Gare, A.: Approaches to the question 'What is life?': reconciling theoretical biology with philosophical biology. Cosmos Hist.: J. Nat. Soc. Philos. **4**, 53–77 (2008)
9. Gare, A.: From Kant to Schelling to process metaphysics: on the way to ecological civilization. Cosmos and Hist. J. Nat. Soc. Philos. **7**, 26–69 (2011)
10. Günther, G.: Time, timeless logic and self-referential systems. Ann. N Y Acad. Sci. **138**, 397–406 (1967)
11. Günther, G.: The tradition of logic and the concept of a trans-classical rationality. Allgemeen Nederlands Tijdschrift voor Wijsbegeerte en Psychologie **54**, 194–200 (1962). www.thinkartlab.com/pkl/archive/GUNTHER-BOOK/TRADIT1.html.
12. Heraclitus: Fragments, trans. by T.M. Robinson. University of Toronto Press, Toronto (1987)
13. Hume, D.: 1779. Dialogues Concerning Natural Religion. Cambridge University Press, Cambridge (2007)
14. Ilyenkov, E.V.: Dialectical Logic, Essays on Its History and Theory, p. 2008. Aakar Books, Delhi (1977)
15. Kant, I.: 1781. Critique of Pure Reason. Cambridge University Press, Cambridge (1998)
16. Kauffman, S., Logan, R.K., Este, R., Goebel, R., Hobill, D., Shmulevich, I.: Propagating organization: an enquiry. Biol. Philos. **23**, 27–45 (2008)
17. Lupasco, S.: L'univers Psychique. Denoël-Gonthier, Paris (1979)
18. Lupasco, S.: L'énergie et la matière vivante. Éditions du Rocher, Monaco , 1986 (Originally published in Paris: Julliard) (1962)
19. Mamardashvili, M.K.: Cartesian Meditations. Progress, Moscow (1993).((In Russian))
20. Needham, J.: Moulds of Understanding. George Allen & Unwin, Crows Nest, Australia (1976)
21. Petitot, J.: Logos et Théorie des Catastrophes. Editions Patiño, Geneva (1988)
22. Priest, G.: In Contradiction. A Study of the Transconsistent. Martinus Nijhoff, Dordrecht (1987).
23. Rescher, N.: Process Metaphysics. State University of New York Press, Albany (1996)
24. Rovelli, C.: The First Scientist. Anaximander and his Legacy. Westholme Publishing, Yardley (2011)
25. Rošker, J.S.: Classical Chinese logic. Philos. Compass 1–9 (2015)
26. Schelling, K.F.A., ed.: Friedrich Wilhelm Joseph Schelling's Sämmtliche Werke. I Abtheilung Vols. 1–10, II Abtheilung Vols. 1–4. Cotta, Stuttgart (1856–1861)
27. Smolin, L.: The Life of the Cosmos. Oxford University Press, Oxford (1997)
28. Stent, G.S.: Molecular Genetics; an Introductory Narrative. W.H. Freeman, San Francisco (1971)

29. Taylor, C.: Hegel. Cambridge University Press, Cambridge (1975)
30. Thom, R.: Esquisse d'une Sémiophysique. InterÉditions, Paris (1988)
31. Thompson, I.J.: Process theory and the concept of substance. Gener. Sci. (1990). https://www.generativescience.org/ph-papers/subst5c.html
32. Whitehead, A.N.: Process and reality. An essay in cosmology. Gifford lectures delivered in the University of Edinburgh During the Session 1927–1928. Macmillan, New York (1929)
33. Zelený, M.: Tektology. Int. J. Gen. Syst. **14**, 331–342 (1988)
34. Zinoviev, A.A.: Philosophical Problems of the Many-Valued Logic. D. Reidel Pub. Co., Dordrecht (1963)
35. van Eemeren, F.H. (ed.): Advances in Pragma-Dialectics. SicSat/Newport News, VA, Vale Press, Amsterdam (2002)
36. van Eemeren, F.H., Grootendorst, R.: Speech Acts in Argumentative Discussions: A Theoretical Model for the Analysis of Discussions Directed Towards Solving Conflicts of Opinion. Floris Publications, Dordrecht (1984)

Chapter 6
Physics: External Reality—Time, Space and the Observer

6.1 Introduction. Quantum and Classical Mechanics: Non-linearity, Downward Causation, and General Operative Principles in Nature

This book is about the world, of which we and how we look at the world are parts. The book is not about the abstractions which the human mind is capable of making about the contents of the world. However, it is definitely about the capacities of human beings to make those abstractions and create imaginary worlds, while keeping them in their place, so to speak. From a philosophical standpoint, therefore, our position is a realist one, based on a categorial principle of *non-separability*, derived from our Logic in Reality, corresponding to the most fundamental observed properties of real phenomena. It thus allows for a firm relation to science, without reduction and loss of rigor, but also without loss of first-person perspective and affect. We have presented above a theory of real processes based on the operation of the real, physical interactions between opposing elements of those processes. We have demonstrated, as far as possible, a sense that there is an isomorphism between this opposition at microscopic and macroscopic levels of reality. Examples of elements in which this opposition is clear are positive and negative charges, antibodies and antigens, people, teams or countries, or philosophical theories. This suggests a general principle designated as the Principle of Dynamic Opposition (PDO), and its operation is the basis for what we have called the Logic in Reality (LIR) (Chap. 3).

Although no other *logics* have been defined on the same or a similar basis, there are other operative principles in nature that have been so defined by both scientists and philosophers. They describe phenomena in terms, some of which are compatible with the LIR approach and supported by it. All of these principles are ways of looking at *change* in a non-reductionist perspective: its complexity and incompleteness are accepted not as artifacts to be eliminated or 'captured' by standard theories, mathematical or other, but as essential to the *being* of the process in the systems sense. We have made several references to physics in Chaps. 1 and 3 which demonstrate the necessity of physics for philosophy. Our theory brings discussion of physical

J. E. Brenner and A. U. Igamberdiev, *Philosophy in Reality*, Studies in Applied Philosophy, Epistemology and Rational Ethics 60, https://doi.org/10.1007/978-3-030-62757-7_6

aspects of reality, logical in our understanding of logic, as dialectics, even into the domain of theory itself. When we turn to the physical world, a major question we face is whether the world either digital or analogue or both. The attempts to resolve this question reveal a deep conjunction of physics and philosophy, which turns us to the foundations of dynamic oppositions in the world that we discussed in Chap. 5 on dialectics.

Some changes in nature are obviously the result of a simple energy gradient operating on inanimate substrates, essentially at the physico-chemical and cosmological levels of reality, reflecting the overall movement of energy in the universe from higher to lower levels. All other changes involving reciprocal interaction have meaning, in the sense of meaning for a conscious entity as discussed in Chap. 9. Our position can be summarized by saying that these principles are at the same time philosophical and scientific. At the level of basic physics, LIR allows for physical, dialectic interactions between entities, in principle eliminating the need for a background field substrate of the universe. Quantum particle and field descriptions are contradictorially related, in the LIR interpretation, describable in both particle and field terms. A metaphysical question remains, however, even within our system, as to whether entities at the quantum level *themselves* participate in change and can therefore be included in the category of real processes. This would imply a logico-philosophical distinction between a proton, that theoretically can decay, and an electron, that cannot. This distinction does not, however, affect the discussion here.

Our conclusion is that the PDO is an additional *necessary* condition for existence, life and evolution, but it is not *sufficient*, or rather that we do not know if it is sufficient or not, and if not what categories any additional principle might involve. We claim that there is something 'true' and potentially open and fecund about this ignorance. This is similar, albeit formally so, to the anti-realist position that propositions about reality are either true or false but we cannot tell which. Nevertheless, we *have* added one more explanatory step between us and the universe, consisting of a model of reality and a set of its categories that capture some essential aspects of living and higher level complex systems. To analyze the relation of observer to external reality, we will start from the analysis of the most fundamental phenomena, namely, space, time and movement.

6.2 Time

The emergence of quantum mechanics allowed the theoretical separation of the properties of physical reality from our observation of them. The three most important *explananda*, which have been with us since antiquity, are a space and time in which we appear to exist; and the capacity for movement from one position in space to another. We will not review here the familiar attempts to explain these phenomena, or, in some cases, prove their non-existence.

The Aristotelian idea of an external time was later relationally elaborated in the special and the general theory of relativity, while the concept of internal time, appearing as a reduction of potentialities, became the basis of quantum mechanics (see Igamberdiev [23] for details). This represented a fundamental duality in our picture of the physical world, now reflected in modern physics as a discrepancy between the foundations of the general relativity and quantum mechanics. In fact, it arises due to the incongruity of external and internal time in the Aristotelian sense. Temporality as a bridge between the ideal mathematical world and the real physical world can be seen as the central point of the philosophy of Aristotle. The logic of movement, both internal and external, according to Aristotle, involves a deep structural component in which the nature of temporal transformations is referred to as actualization of a potentiality, perhaps the earliest instance of what has become the basis of our Logic in Reality.

While Plato analyzed the forms he saw as outside of or beyond time, Aristotle placed them in the temporal world and developed a complex conceptual apparatus that describes the phenomenon of actualization. The space–time of Aristotelian physics is fundamentally relational; time appears as the measure of movement which, when viewed as external, can be the subject of measurement itself [27]. As noted in the previous Chapter, Aristotle developed the concept of a *res potentia* that has relevance to Anaximander's *apeiron* and through which the worlds of *res cogitans* and *res extensa* are interconnected and interacting. He called it an *entelechy* which can be either in the form of knowledge (as referred to the noumenal world of *res cogitans*) or as *energeia* (as linked to the phenomenal world of *res extensa*). The theory of actualization is represented in Aristotle's philosophy in great detail, but later philosophers often misinterpreted or ignored it. The most radical view (in some positivist theories) claims that no *res potentia* exists "in reality". As discussed above, it was not until some years after the advent of quantum mechanics that Heisenberg [2] revived the *res potentia*, which we have related to our concepts of actuality and potentiality (Chap. 3).

A unification of space and time was attempted in the 4-dimensional space–time of special relativity, but this is in contradiction to their phenomenological separation. Our approach to a coherent theory runs as follows: what are primary are energetic events and they, logically and physically imply dynamic interactions. In our theory, time, space and space–time, as these terms are commonly understood are *not* categories themselves. Rather, they are dynamisms, patterns of energy being deployed (literally, unfolded, *dépliés*), that follow the rules of the Principle of Dynamic Opposition in evolution (change). As discussed in Brenner [7], the actualization and potentialization of dynamic events are not a function of time, but time is a function of the dynamics of actualization and potentialization. If the actualization of an element or its opposite is rigorous and absolute, there is no more time; the logical element is fixed and immutable, an identity. The notion of movement in time comes from the concepts of wear, change and transformation, all of which require modification of identity. Complete actualization or potentialization would be equivalent to an end of time, in the heat death of the universe, for example, the absolute homogenization of energy at the lowest level. Time is only possible due to the existence of *contradictory dualities* whose energetic antagonism is both the source and necessary condition of partial, non-infinite actualizations and potentializations. Time is, like the dynamisms that generate it, neither finite nor infinite but transfinite. Reversing Kant, time is not a condition of phenomena, but conditioned by phenomena, due to their logical dynamic structure. To repeat, time, like space, as will be discussed, results from the development of a *process that actualizes itself*, the necessary consequence of the dynamic structure of an energetic world.

It is impossible to overestimate the consequence of this view for philosophy. Some of the deepest insights of the philosophy of our era, those of Heidegger, nevertheless depend on a notion of an independent time in which, or against the background of which, human drama is acted. We will study some of the consequences of this, view in Chap. 15 on Philosophy in Reality as Process. For us process is term that applies to change, although some people think of process only epistemically, without relation to ontological change.

6.3 Space

In our view, as soon as the illusion that we are 'swimming' in something called time is dispelled, at least temporarily (*sic*), certain features of space and its relation to time become accessible. Thus, the moment two elements exist as such at the same time, they imply, simply because they do not coalesce, a space, a location with a distance between them. The link between space and time is achieved as follows [32]:

> In every actualization, and precisely because there exists an antagonistic actualization relative to a contradictory potentialization and never the possibility of an absolute actualization, a contradictional conjunction (of some sort) essentially and irreducibly accompanies the development of the process of actualization and potentialization. In other words, there is always a simultaneity of an element or event that is more or less actualized and of an element or event more or less potentialized.

This contradictional conjunction is present during any actualization such that the logical temporality that is generated by energy, 'time', is always tied to a logical spatiality that is generated similarly. The space of the expanding universe is the space of a dynamism that actualizes and potentializes itself alternately and reciprocally. This concept is thus based on the structure of energy itself and its logical aspects. There are both spaces and times, which are *proper to* individual phenomena, functions of their actualization.

Thus, as described for time, objects and events do not exist or take place in space, but are the sources of their own space and time. Objects are not in space, but space is in objects; objects are not localized, but localize, create localizations. It is in this admittedly informal way that such a space has the same characteristics as a configuration space, that is, it is a function of the number of its elements and of their degrees of freedom; it is what links the elements, their relations, that permits their co-existence in a system and their simultaneity. There is no spatial location outside of what is inside it. Logical space and logical time constitute a space–time proper to each system, a configuration space-time. Time cannot be separated from space, and only space–time exists in reality, despite appearances. At this point, we ask readers to accept this position as a working hypothesis, 'for the time being'.

6.4 Simultaneity and Succession. Movement

Simultaneity is characterized in the Lupasco approach as a contradictional *conjunction* that requires the presence of elements that are both identical and diverse; otherwise, the simultaneity would tend to disappear. Simultaneity is spatializing in that it defines and is, accordingly, an energetic operation, not a static given. An energetic simultaneity is thus a 'simultaneization'. It *implies* a passage from a degree of potentialization to a degree of actualization, of a certain quantity of potential energy to a certain quantity of actualized energy, therefore movement, succession and time: "No space without passage from succession to simultaneity and thus without time, and no time without passage from simultaneity to succession, no time without space." [33].

Thus, succession is considered as a contradictional *disjunction* or *exclusion*, opposed to the simultaneity or conjunction. It also involves identity and diversity of elements at the same time, but succession involves an actualization of a series of identities and potentialization of a series of diversities, something like a choice between contradictory elements or events. Succession in time is a dialectical series of metaphysical 'choices' by a system. Succession is also, in this picture, the passage from a certain potential state to an actual state, and inversely. There could be no succession possible, in energetic phenomena, if there were no potential of succession, of dynamic 'successionizations' and an actualization of this potential. There is actualization of a succession—which is essentially actualization itself—because there is a potentialization of the contradictory succession. Every succession implies this dialectic.

The conclusion is the following: logical space, in the sense of simultaneity or conjunction is dynamically opposed (in a contradictional relation) to logical time as succession or disjunction. The simultaneity of elements in space is based on their succession of time, and vice versa. Neither of the contradictional relations being processes ever going to absolute completion, each will always have an irreducible residue of the other; there will always be some space in time, some time in space. This picture is sufficiently novel to warrant a further formulation of the basic points.

For succession to exist, that is, temporality, there must also be simultaneity, spatiality, in which and by means of which succession can operate and develop. Inversely, for there to be space and contradictional conjunction, that which what constitutes disjunction and entails succession and temporality and coexists with space must be potentialized. Since these processes never go to completion, there is always some space in time, some time in space. Spaces and times develop dialectically, moving toward, but not reaching, the ideal non-contradictory limits of identity or diversity, or, alternatively, toward a limit of maximum contradiction. The dialectic of the three energetic times and three energetic spaces defined above and their interaction by contradiction and antagonism constitute space-times or time-spaces. To the spatialization or homogenizing simultaneization of positive space is linked a temporalization or heterogenizing successionization, contradictory and antagonist, a negative time, potentialized by the actualization of that positive space, such that it tends to disappear asymptotically.

The relativization of the neo-classical concept of simultaneity (or of a plane of simultaneity) to a frame of reference results in paradoxes in assigning times, or rather, temporal sequences, positive, negative or zero for different observers, a concept whose metaphysical significance is by no means understood. In the context of relativity theory, there is no a priori definition of the simultaneity of two distant events, and it becomes subject to an arbitrary or conventional choice, called a convention of simultaneity and synchronization. The sense of simultaneity (and succession) in LIR is in our opinion metaphysically richer.

6.5 Time in Philosophy

The disjunction between the physics and the phenomenology of time has been apparent since antiquity. It is impossible to even summarize here what has been written on the subject of time as an *aporia.* In any event, the few prior intuitions of a more complex structure of time can be subsumed under the concept of 'both at once', and the medieval *coincidentia oppositorum.* We, therefore, turn to the analysis of time in phenomenology which represents the modern understanding of philosophical temporality.

6.5.1 Time in Phenomenology

One original contribution to the phenomenology and philosophy of *time* is that of Varela [56]: "The familiar account of time inherited from our modern Western cultural background is inadequate". However, he also states: "The experience of temporality addresses head-on the fundamental fact that we exist within a transparent web of time." Husserl's view of subjective time was that of a paradoxical appearance of 'double intentionality', a mixture of passivity and activity, of invariance and change. Temporality is constituted through complementarity between spatial and dynamical ingredients, affect, and trajectories in a phase space landscape. A dynamical 'mutual bootstrap' principle applies such that the trajectories provide the "conditions for an embodied coupling, since through their coupling they shape their dynamical landscape. Metaphorically, the walker and the path are intrinsically linked." Varela refers to his neuro-phenomenological hypothesis that states:

> "Phenomenological accounts of the structure of experience and their counterparts in cognitive science relate to each other through reciprocal constraints." Given two accounts – a process of external emergence with well-defined neurobiological attributes and a phenomenological description that stays close to our lived experience – Varela asks for a 'circulation' between them, that is, their mutual or reciprocal constraints, "*including both the potential bridges and contradictions between them*". (emphasis ours)

We notice that while the concept of time is given substantial development, the standard concept of space (including phase space) seems perfectly adequate to Varela and does not receive additional comment.

A number of questions can be asked here, in relation to the development that follows:

- Does a web of time exist, or is this a metaphor?
- How can 'passivity and activity' be mixed?
- Is 'embodied coupling' real or also metaphorical?
- From where might 'contradictions' come and what is their significance?

From the LIR standpoint, such views fail by postulating an independent subjective temporal flow within which entities and their relations are located. According to Grush, Husserl had the better intuition that it is the relations themselves that constitute the flow of subjective time. The LIR system, that takes words like coupling, contradiction and constraint seriously and asks that their content be defined, is no more (and no less) in our view than well-intentioned critique.

A concept of time being deployed by objects or systems might thus be generally employed as a rule rather than an exception. A significant example is that of biological time, such as circadian rhythms, as a function of biological processes. The primary area of application of what was described above as the *three* contradictorially-related forms of space–time is that of subjective time as a function of human cognition. However, further development of this application must await a detailed discussion of the origin of consciousness and knowledge, which is outside the scope of this book. We will discuss the more philosophical aspects of phenomenology in Chap. 10, as well as the concept of 'Dynamic Being' in Chap. 15.

6.5.2 Derrida and the Complexification of Time

The theme of *Philosophy in Reality* could be summarized, at this stage in our discussion, by stating that something like phenomenology is a necessary part of it, to insure that 1st-person views are given adequate ontological value. However, it is not sufficient part, and ways must be found to harmonize it with science. One way of accomplishing this task is to take the insights of a philosopher and complexify them to establish such a junction. As one example, our approach, here as elsewhere, provides a dynamic link between the terms of an *aporia*, such as that present in 'time'. It can thus both correct and amplify the line of argument on time like the one developed by Derrida.

In a major book, *Of Grammatology* (1997), Derrida [12] explored the complex relationship between the structure and meaning of written language; the existence and role of the writer; and the process of writing, *ipso facto* temporal. The relationship to Lupasco is immediately obvious in his statement that a book, as something repeated and duplicated is more than 'just' a book but a 'text', "constituted by the play of identity and difference. Derrida's idea of the method of philosophy is *processual*: "The method of philosophy is the structure of knowing, an activity of consciousness that moves of itself; this activity, the method of philosophical discourse, structures the philosophical text.

We find in Derrida three philosophical 'reasons' for the unacceptability of the system of signs proposed by Pierce in his deconstruction of transcendence in the dialectic, immanent sense in which we address it in Chaps. 10 and 15. We will follow the approach of Husserl's phenomenology that the "principle of principles" is a radical restoration of the metaphysics of presence—hence of absence, hence a logical dialectics in reality. "The difference between Husserl's and Peirce's phenomenologies is fundamental since it concerns the concept of the sign and of the manifestation of presence, the relationships between the re-presentation and the original presentation of the thing itself (truth)." We take truth here in Lupasco's sense of the truth in reality, not simply propositional truth.

Derrida established the connection between time and the 'absolute past' and writing through his concept of the 'trace': writing results in the trace, that part of the past that does not allow itself to be summed up in the simplicity of a present. Past has in any case always signified past-present. "The concepts of *present, past and future,* everything in the concepts of time and history which implies evidence of them—the metaphysical concept of time in general—cannot describe the structure of the trace. Derrida goes to the heart of the problem by asking: is the temporality described by a transcendental phenomenology as "dialectical" as possible, a ground which the structures … of temporality would simply modify? Or is the phenomenological model itself constituted, as a warp of language, logic, evidence … upon a woof that is not its own?" Our answer is that the 'woof' needs to be augmented by elements of the physics described in this Chapter, without at all detracting from the value of Derrida's insights. Brenner discussed his concept of differ**a**nce and its relation to LIR in (2008). The neologism *différance*, with an 'a' in the third syllable, differs from the word for difference in French which is spelled *différence. Différance* is a kind of dynamism in the LIR sense, which "invites us to undo the need for balanced equations, to see if each term in an opposition is not after all an accomplice of the other. At the point where the concept of *différance* intervenes, all of the conceptual oppositions of metaphysics, to the extent that they have for ultimate reference the presence of a present …(signifier/signified; diachrony/synchrony; space/time; passivity/activity, etc.) become non-pertinent. A new definition for dialectics is necessary [13]. Equally valid in our view is his use of the word of *spacing*, again a process word, to capture "the articulation of space and time, the becoming-space of time and the becoming-time of space". This is a formulation close to that of Lupasco described above. Derrida's intention was to use these concepts to explore the phenomenology of writing, the configuration space-time in which it occurs. In our opinion, Derrida has made accessible a conceptual structure that applies to complex processes and their evolution in general. We suggest that one should see the Derridean visions—the trace, spacing and difference—as Lupascian included middles between the prior either/or terms of the *aporia* of time. We now look further at the problems of space and time together.

6.6 Space–Time in General Relativity

The phenomenon of universal gravity—the gravitational field—determined the new approach to the formulation of a frame of reference in the general theory of relativity. One way of characterizing early 20th Century cosmology is to say that it has been a cosmology of *identity* and non-contradiction, with major effort devoted by Einstein and others to a search for invariants. In special relativity, an absolute Newtonian time and space was replaced with a new invariant (the universal interval s). This approach gave a relative reality to time from which everything that depended on the observer was eliminated. The principles of Special Relativity (SR), pushed to its limits, essentially eliminated energy and dynamics from existence. General Relativity (GR) was introduced to handle the extension of relativity to non-uniform movement and microphysical phenomena, quantum and wave mechanics. For GR, Einstein proposed as the most general invariant the total curvature of the universe. This required a non-Euclidean geometry, from which all heterogeneity of its space–time points at a microscopic level was eliminated by the notion of bodies in continual acceleration (due to the curvature). In other words, through the requirement of general covariance (changes of position with respect to a frame of reference), these space–time points lost what was critical, namely, the really and dialectically necessary component of *partial* individuation at the level of phenomenology.

Two cosmologies were proposed to explain the relation between matter and space–time: the universe of Einstein, in which the geometric structure of the universe was determined by its total material mass. Finite, curved space–time was reduced to matter and absolute time, the absolute time "of the universe" was considered a sort of subjective noumenon with which no interaction could be possible. In the Eddington-De Sitter model, there was no causal relation between the curvature and the total mass; matter was responsible only for local irregularities. In the De Sitter model, a discontinuous intensity takes the position that extensity occupies as an external objectivity in the Einsteinian universe. Both Einstein and Eddington seem to have avoided any *antagonistic* dualism, one throwing into the subject what the other threw into the object, and vice versa; for both one is appearance, the other reality. Heisenberg [19] claimed that the apparent duality arises in the limitations of our language that is incapable of describing the processes occurring within atoms while mathematics does not have this limitation.

Alexander Friedmann [15] in 1924 and Georges Lemaître [30] in 1927 were first to formulate clearly that the universe is non-stationary and is now observed in the phase of expansion, in which, accordingly, the matter of the Einsteinian model of the universe that is becoming rarified, approaching the universe of De Sitter of pure geometric form. The Lemaître model reconciled the opposition between the first two models: there is a dual tendency to expansion and contraction, suggesting that a basic antagonism exists in the universe between the constitutive properties of this expansion and contraction. In the light of current cosmological work, the Lupasco picture of the universe of Lemaître is of interest: the dynamic opposition excluded in the other two models is maintained, and can be a source in principle of instability and becoming. We will now try to describe how this is done.

6.6.1 The Dual Role of the Metric Field

As discussed above, Special Relativity (SR) describes an absolute chronogeometrical structure of a 4-dimensional Minkowski universe with a Euclidean geometry in which is embedded a frame-relative, observer-dependent space–time. 'Distant' simultaneity is defined by convention. General Relativity (GR) is necessary to handle the universal nature of gravitational interaction and aspects of physics at high energies. However, this theory removes all physical objectivity from space–time, and is in direct conflict with the 'apparently real' objectivity of the phenomenological world. There is, accordingly, a need for some kind of frame- and observer-independent description which would ground both the reality of both experience and scientific knowledge. The debate revolves over how to assign physical meaning to the *metric field* which is the central concept in GR. The metric field is a term for the mathematical (tensor) description of the geometrical and gravitational structure of the universe as a 4-(3 + 1) dimensional space–time manifold—the background for, but also possibly a participant in the manifestations of physical events.

The solution proposed by Lusanna and Pauri [34] starts by seeing the metric field as split into two parts: an ontic part corresponding to the autonomous degrees of freedom of the gravitational field in the absence of matter, and an epistemic part to the information encoded in the metric. The latter must be specified in order to get empirical access to the ontic part, which refers to the appearance of gravitational phenomena. A physical individuation of point-events is derived in terms of the ontic part of the gravitational field. The identity of point-events is conferred upon them by a complex relational structure in which they are holistically enmeshed. This relational structure includes all the elements of a so-called complete gauge fixing, supported by a definite solution of Einstein's equations (a definite Einstein "universe"). The characterization of such space-time points includes the fundamental intrinsic properties instantiated at those points including mass, charge, spin and perhaps others. However, it also gives a physical meaning to the coordinate indexing of such which makes point-events as ontologically equivalent to the existence of the gravitational field as an extended entity.

Summarizing, this view holds that space–time point-events do exist as individuals however, their properties can be viewed both as extrinsic and relational, being conferred on them in a holistic way by the whole structure of the metric field and the extrinsic curvature on a simultaneity hyper-surface (see below), and, at the same time as intrinsic, being coincident with the autonomous degrees of freedom of the gravitational field. "In this way both the metric field and the point-events maintain their own manner of existence, so that the structural texture of space–time in this model does not force us to abandon an entity realist stance about both the metric field and its points". This theory supports LIR since for the above description of time and space no background of *dimensionless* points (like those of differential calculus) needs to be postulated in addition to, or apart from, either the causally effective quantum field or the gravitational field and their self-duality.

Here, the thesis according to which metrical relations can exist *totally* without their constituents (point-events as *relata*, some of which may be relations; this point remains open) does not hold. LIR is consistent with this physics that provides a basis for a physical individuation of *some* point-events (entities), while insuring the required indistinguishability of quantum particles. LIR insists on a contradictional relation between identity, which implies indistinguishability, and diversity, which implies individuation.

Lusanna and Pauri [35] proposed solutions to Einstein's equations in which there is a dynamical emergence of 'instantaneous' 3-spaces, the 3-dimensional instantaneous spaces in which ordinary phenomena are observed and described. In their striking metaphor, 'space-time' and the vacuum (matter/energy free) gravitational field are "two faces of the same reality". The 3-spaces are 'embedded' in an Einsteinian 4-D manifold. The appropriate theory is a new kind of structuralism, containing with elements of both the substantialist and relationist points of view, implying a 4-dimensional holism resulting from a foliation or un-folding within it of the instantaneous spaces with three spatial coordinates (3-spaces). The reality of the vacuum space–time of GR, the dynamical instantaneous 3-spaces and their dynamical individuation of point-events described by the epistemic part of the metric field is ontologically equivalent to the reality of the autonomous degrees of freedom of the gravitational field as described by the ontic part of the metric field. What this study brings out is that GR contains the potential for a differentiated description of both the 4-dimensional space–time manifold implied by the existence of universal gravitation and a 'foliation' development into 'sheets' of observable 3-spaces.

In this picture, the individuation of the point-events involves their change in time. This is an attribute whose information is not wholly contained in the 3-dimensional space at a time *t*, but this is not inconsistent with the dual role of the metric field in GR. That a complex material process entity cannot be described as *wholly* present or absent is logically acceptable in LIR, and we see here a basis for the eventual formalization of the LIR conception of simultaneity and succession.

The program of Lusanna and Pauri thus establishes the mathematical basis for a reinterpretation of a 'join' region between phenomenal and non-phenomenal reality, establishing a relation that is both physical and epistemological [54]. LIR is compatible with this interpretation, which can serve as an extension of its foundation in physics as foreseen in Sect. 6.6.2. The Lusanna and Pauri description suggests that the underlying *principle* of the metric field is one of self-duality, expressed by the properties of the gravitational field alone, which the authors describe as more ontologically diverse than any other. The critical insight is that GR is a theory that from the physical point of view is radically different. Its reference is to the space–time that evolves 'within' the gravitational field rather than to some internal mathematical groups, and it leads to a dual role of having both a unique dynamics (reality) and appearance to an observer.

6.6.2 Self-duality

Self-duality in cosmological theory refers to a type of solution to Einstein's equations, for which LIR suggests a physical interpretation in terms of the same kind of opposition between inherent properties as in the quantum field. One can differentiate the descriptions by referring to the respective meanings of dynamics and dynamically: in the LIR 'metaphysical' model, dynamic in dynamic opposition refers to the real physical interactions between any entity and its opposite or negation resulting in an alternating potentiality and actuality of both. Where those are absent, as in the case of three billiard balls, there is no question about their all being on the table 'simultaneously'. In the GR model, as in some forms of structural realism, dynamic refers to a principle of change inherent in the *equations*, Einstein's or others, which describe a non-trivial evolution in 'time'. These are referred to as 'dynamic symmetries', dynamical tensor fields and so on. All of the chrono-geometrical structure of an Einsteinian universe is dynamically determined in this sense. The point-events located on the simultaneity surfaces are achronal (timeless per se), but are characterized by dynamically determined conventions about distant simultaneity. Our conclusion is, however, that simultaneity in GR is still defined as a *convention*, with respect to such large distances. There is no explicit reference to the simultaneity or lack of it in proximal events that may be in just as much need of definition.

The major difference in the LIR and GR accounts of simultaneity is in thus in their treatment of *proximal* events involving entities of greater tangible complexity than those at the quantum or cosmological levels. Another way of saying this is that wherever the process phenomena require formalisms for their description that are dual in the LIR sense, but not self-dual as in the case of quantum and gravitational fields, the LIR picture applies [7]. It describes the contradictorial evolution of biological and cognitive processes as involving a richer structure of simultaneity and succession than provided by GR. Having arrived at a *logical* theory of time and space, or space–time, it might be asked what possible consequences it might have, practical or theoretical. We suggest two preliminary answers: first, this logic could make more accessible, by pointing to the operation of the PDO in them, the contrasting or opposing domains of cosmological physics. It is, if one likes, a restatement of the fundamental duality of the universe into perceiving and non-perceiving domains. The mathematical and physical structures of these domains are separately describable but linked as suggested by the LIR Axioms, in particular, of Functional Association, Conditional Contradiction and Asymptoticity. Second, it could serve as another way of reminding people that the space and time they experience are neither eternal and absolute realities, nor the framework of a priori functions of sensibility, but the consequence of the existence of matter/energy itself. The further role of LIR may be, given the application of the PDO 'across the board', to facilitate the application of significant aspects of the new theories at the quantum and cosmological levels of reality to the levels of intermediate biological and agentive human existence.

LIR considers the role of becoming and change in the comprehension of complexity. The analysis of issues in systems theory (cf. Chap. 16) such as transience and quasi-ness, compatible with the LIR approach can be found in the books by Minati et al. [40] and Minati and Pessa [41]. Our approach further develops the framework introduced in the famous book *From Being to Becoming* of Ilya Prigogine [46].

6.7 Relationality and the Frame of Reference in Physics

The term physics comes from the Greek '$\varphi \acute{\upsilon} \sigma \eta$' which means 'Nature' and refers to natural reality as contrasted to $\mu \acute{\alpha} \theta \eta \mu \alpha$, meaning "that which is learnt", "what one gets to know". In other words, even in original definitions of these words, we are faced with two different aspects of reality, one natural reality and the other learnt. Aristotle's objective in his *Physics* was to discover the real principles and causes of movement or motion ($\kappa \acute{\iota} \nu \eta \sigma \iota \varsigma$) which conceptually resulted in shaping a new field of knowledge, although the basic principle for providing a scientific basis for this concept which would take it away from pure philosophy was found much later. This principle was introduced by Galileo as *inertia*, and on its basis Newton formulated the entire construction of modern physics. The formulation of the inertial frame of reference by Galileo and his principle that the laws of motion are the same in all inertial frames became the first law of Newton's mechanics. This in turn became the cornerstone of modern science.

We consider the mechanics of Newton as a concept based on the assumption of absolute space and time as the preferred reference and thus independent aspects of objective reality. Its background lies in the explicit formulation of a reference frame which places limits on the role of the observer in the description of the physical world. It establishes the basic relativity of motion to the observer which allows one to define a field of externality that is not influenced by the observer exploring (measuring) this externality. At the next step of development of physics, it became evident that the principle of relativity of the external reality to the observer had to be expanded. This resulted in the formulation of the special theory of relativity in 1905 by Einstein and approached by Poincaré and Lorenz. The need was to shape a new objective pattern of externality in the formulation of the General Theory of Relativity (GR) which, in fact, is the general theory of gravitation proposed by Einstein in 1915. The abandonment of absolute space–time meant a change of the preferred frame of reference in modern physics, see above, but it did not mean abandoning *any* frame of reference. This was impossible and would result in the disappearance of physics as a set of general scientific concepts.

In fact, the main challenge in establishing and developing physics as a science was how to formulate externality as an area lying outside of the observer and not influenced by him/her. Physics as science appeared when the frame of reference of external reality in relation to observer was formulated [49]. Aristotle obviously did not go this far, although he approached the establishment of the border between externality and internality by postulating the two types of time in physics: one which is measured and the other by which we measure. When a reference frame is established as it was done by Galileo and Newton, it results in the basis for the absolute space and time of Newton's mechanics. Later the relation of observer to external reality was reformulated in the theory of relativity via communication between observers by the signals possessing the same speed in all directions, and in a different way in quantum mechanics via relationality of actualization simply during observation of external reality. When the relation to the external reality is reduced via consideration of the accelerated movement as relational and by postulating the constancy of speed of light, the result is the space–time of GR. As we show later, the relation of observer to the externality in quantum mechanics includes also the relationality of a process of actualization. This is a new type of relationality which is not compatible with either Galilean relationality or Einsteinian relationality. The basic incompatibility of the fundamental foundations of the general theory of relativity and of quantum mechanics represents the main challenge of modern physics as a complete theory [49]. It has been outlined in many works (*cf.* Lee Smolin [55] "*The Trouble with Physics*"). Before we move to this fundamental contradiction of modern physics, we will have a brief look at the development of physical concepts in Ancient Greek and Modern European philosophies as they relate to philosophy today.

6.8 Substance and Movement in Reality: A Brief Philosophical Overview

After Thales (c. 624–546 BC), who formulated the concept of substance and thus is recognized as the first philosopher, Anaximander (c. 610–546 BC) became the founder of what we now call scientific thought [50]. He eliminated the association of substance with any of its visible manifestations, and defined a primary substance as *apeiron*, thus also introducing the idea of potentiality in philosophical thought. According to Anaximander, "things are transformed one into another according to necessity and render justice to one another according to the order of time" [50]. We can interpret this only documented statement of Anaximander as saying that time orders things by separating them in sequential order so that simultaneous contradiction is avoided. While Pythagoras (c. 570–495 BC) is regarded as the founder of mathematics which we described in the previous section, and Parmenides (c. 540–470 BC) as the founder of logic, Anaximander can be considered as the founder or at least the precursor of natural science [50]. He claimed that nature is ruled by laws like human societies and developed the idea of relationality of the 'upper' and the 'lower' claiming that the Earth floats freely without falling and does not need to rest on something. This idea was considered by Popper [45] as "one of the boldest, most revolutionary, and most portentious ideas in the whole history of human thinking".

While the statement of Anaximander on relationality of the upper and the lower positions in space became the basis of scientific thinking, the relational nature of the world was explicitly put into the philosophical basis of thinking by Anaxagoras (c. 510–428 BC). He postulated the existence of a multiplicity of "seeds" called later by Aristotle *homoiomeroi*— particles having same nature as the whole [16]. *Nous* (mind), in the philosophy of Anaxagoras, orders all *homoiomeroi* and can be related to the philosophical idea of the externality to the world itself forming and organizing its pre-established harmony [27]. Later, the relational nature of the world was developed in detail by Aristotle (384–320 BC) who, in his tractate *De Anima* (On the Soul), attributed the notion of self to an internal determination within living systems (Aristotle, On the Soul, web link), and introduced two types of time in physics, as noted above, one which is measured and one by which we measure. This suggests that that our visible world is a combination of externality and internality generated by a reflexive loop that involves these two types of time (Aristotle, *Physica*). This loop inevitably assumes the postulation of a fundamental quantity that corresponds to a minimum action that cannot be further divisible, which provides a possibility of the physical movement and establishes what was later seen as the quantum nature of the physical space and time. The idea of an indivisible quantum goes back to Democritus (460–370 BC) who, by accepting the atomic structure of the world, escaped the paradoxes of movement formulated by Zeno of Elea (c. 490–430 BC).

The modern concept of thinking arose from the attempt to understand the dual nature of the world by René Descartes (1596–1650). To resolve the duality of the *res cogitans* and the *res extensa*, Baruch Spinoza (1632–1677) considered these two perceivable properties, among an infinite number of others that we do not perceive, as the qualities of the ultimate *causa sui* substance (that which is its own cause). Gottfried Wilhelm Leibniz (1646–1716) challenged this solution and, in fact, revitalized Plato's concept that the "existing one" is actualized as the "many" in the real world (the dialogue "Parmenides" of Plato). The actual *res cogitans* represents the pluralism of monads (Leibniz, 1768), where their coexistence is expressed as a "pre-established harmony", while the *res extensa* appears as the relational space–time of interacting monads and is their intersubjective pattern. The exposition of *res cogitans* into the world of *res extensa* takes place via the common potential field which corresponds to the existing one of Plato's philosophy (Plato, web link). Each monad possesses a kind of subjective being, the subatomic levels with the pilot-wave duality of quantum mechanics to complex living beings having free will and consciousness [42]. The parameters of space–time satisfy the condition of coexistence of monads and correspond to the observability of the world. To repeat, such a representation of the world replaces its objectivity by an intersubjective pattern arising from the relativity of a single picture represented by a monad's 'point of view' having the characteristics common to all individual beings. The temporal evolution of the world is a process aimed at overcoming the limits of its individual representations; it has no external frames and opens into to the infinity of our view of it (for details see Igamberdiev [20–25, 27].

Modern physics, that which appeared two centuries after Leibniz, is based on the relational nature of space–time in its original sense. Space–time in the special theory of relativity was transformed a decade later into the quasi-substantial space–time of the general theory of relativity. Gödel was probably the first to suspect that absolute time may be valid for some formulations of general relativity [52]. We should keep in mind, however, that without at least a limited introduction of relationality as in Galileo's principle of motion, it was impossible to formulate the conceptual apparatus of physics despite the apparent acceptance of absolute space–time in Newtonian mechanics where it appears as the preferred frame of reference.

Logic in Reality talks to this foundational issue. It avoids a dichotomy between a relationality is retained in the picture of the world formed by the observer and an externality free of relationality, both being part of an objective material reality. Relationality will remain in modern and future physical theories as long as they remain physical. Some unification theories of modern physics often indirectly abandon the relational nature of space–time. This is particularly evident in the models of the evolution of the universe evolution which assume uniform time 'flowing' from the Big Bang to the future phases of its expansion. In any case, alternatives to the Big Bang now exist that may be scientifically and philosophically interesting to explore (see Sect. 5.5).

6.9 The Intelligible ("Learnt") Reality of Mathematics and the Actual ("Objectively External") Reality of Physics

The development of mathematical descriptions of the physical world occurs via the imposition of limits of computation that appear as a set of fundamental constants. These limits shape our picture of the spatiotemporal physical world and determine its dimensionality and curvature. The idea that the physical world is shaped by the limits of computation appears in Parmenides and Plato and then in Leibniz. It is introduced in modern science [31] and represented, in particular, as the existence of limits of computation in the universe [1]. The fundamental parameter, which is intrinsic to the action introducing computation, is time which separates contradictory statements [21] and defines the velocity of observation propagation [17]. Physical complementarity corresponds to the non-simultaneous existence of contradictory statements and properties, which, in turn, generates recursion and flow resulting in complexification [7, 28]. In the dialogue "*Parmenides*", where Plato made a profound attempt, while remaining in the framework of pure logic to resolve the question of how ideal forms attain reality and operate in reality, introduced multiplicity via the process of assignment of being to existence. This assignment, according to Plato, takes place via temporality (see above), and the existing being appears in its multiplicity. This the most complex dialogue of Plato describing *his* logic in reality! It is quite different from his other dialogues and transforms the basic ideas of his philosophy into an internal dialectics of reality operating intrinsically, within reality.

From the physical point of view, the multiplicity of the world requires explanation in the frame of an unfolding of the basic components of physical reality into the observed space-time of the universe. Existing models of the universe, to be sufficiently powerful, need to explain the existence of multiple equivalent particles having the same properties such as electrons. The idea of retrocausality which was outlined by Matsuno [37] and appears in Feynman [14], helps to understand the origin of multiplicity via what can be described in terms of motion in forward and backward directions in time. From only a few retrocausal loops corresponding to basic elementary particles appearing as antiparticles in the reverse direction, the whole Universe unfolded in correspondence to the one electron model of the universe [14]. In the flow of time, contradictory statements (in the standard sense of semantic contradiction) appear as the retention (presentation of a temporally extended present)—protention (anticipation of the next moment) relations as defined by Edmund Husserl in his philosophical works. Memorization of retention leads to certain basic values of the actualized structures such as the golden section [26]. Only few geometric atoms are needed if they are implemented in this picture of a physical world moving back and forth in time. In this picture, the physical world emerges when time is introduced into the mathematical world [22], in other words, when numbering comes into being as a result of measurement of reality taken as the external reality. Measurement as the basic underlying process in physics corresponds to the relation in which time and space become connected via the values of the fundamental physical constants.

The principle of retrocausality that assumes at least limited reversibility of time became a powerful principle for the understanding of quantum measurements in the transactional interpretation (see below). It is based on the assumption that the distinction between cause and effect does not appear at the most fundamental level of reality so that time-symmetric systems can be viewed as causal or retrocausal. An assumption that causes could be partly or completely negated by their own effects may be important for understanding of the formation of complex structures associated with retrocausality such as reflexive loops or even biochemical cycles as was suggested by Matsuno [37]. The principle of downward causation in complex systems may be considered as a consequence of physical retrocausality and can be related to the Epicurean *clinamen* (see previous Chap. 5). Downward causation is defined as a causal relationship from higher levels of a system to lower-level parts of that system, *e.g.* mental events as high-level physical events acting to cause lower-level physical events [9].

While both the Special and General Theory of Relativity deal with an external time which is measured, the quantum mechanical concept of quantum reduction or decoherence is associated with an internal time by which we measure and which itself can be measured, provided that quantum measurements are performed in a regular way, with low dissipation of energy [25]. In this case the internal autonomous 'clock' we possess that distinguishes the past (memory), the present (appearance), and the future (anticipation based on the reproducible model) can be seen as a kind of quantum system. Both theories of both relativity and quantum mechanics are relational theories, but they use different concepts of time. Their synthesis is also only relational, based on a dialectic interpretation of the two types of time.

To unify physics in a general theory integrating GR and QM, the question of the relation between external and internal time has to be rethought and further developed. In the frames of external time envisioned in SR and GR, a positive, generative aspect as a consequence of contradictions is not anticipated, however, the generative aspect is present as a latency in internal time grounded in quantum phenomena. Note that this latency is not the potentialization of Logic in Reality, which applies only to the thermodynamic world. It corresponds to the *res potentia* of Kastner (2010). Bridging the gap between external and internal time has been a hard problem since Aristotle, and only the resolution of this problem can open the way for the unification of physics and for its constructive integration with the other fields of knowledge including biology in order to provide the necessary sound basis for philosophy.

6.9.1 Time and LIR as a Theory of Change

Brenner [7] defined the logic of/in reality as, among other things, a theory of change. In it, change follows a fundamental dynamic pattern of alternating movement from a state of actualization to one of potentialization. Emergence during the interactions is the source of new entities or phenomena. The change from predominately one to predominately the other of two opposing elements depends on statistical and probabilistic parameters at some stage of the process at all levels of reality. However, the problem remains as to the appropriateness of the statistical generalizations of standard statistical mechanics for description of this alternating movement between potentialization and actualization. As pointed out by Sklar [53], it is not clear what could ground the introduction of probabilistic and statistical notions even in simple macrophysical systems. Non-equilibrium systems in Logic in Reality follow the rules of entropic asymmetry insofar as they are physical, following the 'arrow of time'. However, at, for example the biological level of reality, anti-thermodynamic processes (of heterogenization or diversification) take place, only part of which is subject to statistical factors. The PDO applies most clearly, outside the quantum world, at the level of mental and social phenomena, when the opposing elements are often close to equal in energy, leading to an emergent included middle (T-state). To the extent that statistical probability issues remain open in all physics, it is clearly beyond the scope of this study to decide how the energy necessary to effect a change at these higher levels 'gets where it should go'.

6.10 The Fundamental Problem of the Origin of the Universe (Big Bang) in Reality

In the frame of reference of a science of relations, where time is viewed as a functional operator separating contradictory or, more precisely, countervalent processes, we need to reconsider the evolution of the universe from the Big Bang, which in a relational perspective appears in a different way than usually taken for granted. The origin of a relational universe as a Big Bang in modern interpretations is taken reductionistically with an assumption of the uniform time flowing from the origin of the Universe to present. In the relational interpretation, Big Bang remains real, but not in a local way (for details see Igamberdiev [23–24, 27]. A "delocalized" Big Bang can be visualized as a fixed point of all intercrossing reflexive loops of observation from which everything emerges in the physical Universe. In the generation of the Universe, the unique set of fundamental constants is non-locally defined via introduction of time and quantization of the elementary action.

Evolution of the Universe with such a delocalized origin would be dependent on the observed velocity of propagation of interactions being different in different systems of observation. Then, assuming an independent time, and this is a major unproven assumption, it could evolve at different rates depending on observation or even be absent as in some relational models of the physical world [51]. Such an assumption is neither necessary nor required in an LIR context: the action of separation of contradictory interacting entities is quantized. Minimal quanta are the units of action which are represented by the loops of quantum gravity while macroscopic quanta correspond to the emergent complex structures of life such as hypercycles, cells, finally conscious beings and finally self, at the highest level of cognitive recursion.

The process of avoiding infinity in natural philosophical pictures of the world grounded in the property, noted by Cantor and quoted by Lupasco, that the physical world is neither finite nor infinite, but *transfinite*. This provides the theoretical basis for the emergence of a quantized physical world, consistent with a quantum of action. Democritus had an intuition of something like this as a necessary operation to escape from the Zeno paradoxes. In quantum mechanics, to avoid infinities, the operation of renormalization is applied. Renormalization corresponds to the process of putting finite limits to make the system computable. The physical world emerges epistemologically as divided into the computable and non-computable parts [10]. The computable part is shaped by the fundamental constants while the non-computable part represents the set of actions that introduce the possibility of computation in the real world.

6.11 The Question of the Speed of Light

As is well known, the speed of light is a universal constant setting an upper limit on the speed of observation. It determines the synchronization of signals between observers communicating via the vacuum and appears as the basis in the concept of the Special Theory of Relativity (SR). In reality, movement at the speed of light is possible in principle only in the absence of a particle's rest and relativistic masses. Particles such as photons represent pure energy, and the concepts of time and space do not apply. The velocity of particles having mass (particles of matter) can never reach the speed of light. The observed propagation velocity of energetic particles can be also slowed down significantly in coherent media. A shielded state with very low temperature is characterized by long relaxation times corresponding to macroscopic scales of interactions of the complex structures of matter. This is explained *via* Heisenberg's uncertainty relation between time and energy, where very low dissipation corresponds to the extended times of relaxation. In coherent physical states, the speed of light appears to have been slowed to very low values.

Biological macromolecules exhibit long relation times of their thermal excited states in their natural environment, primarily explained by interactions with their media. These consist of water and higher level organized elements (filaments, organelles, cells and so on), but also of approaching other macro—as well as small molecules. One can make the hypothesis that the macromolecule thus has the 'time' to recognize substrate and transform them into products with high precision and/or generate a precise signal transduction event in the case of receptor molecules. We will discuss this in more detail in the Chap. 18 on biology and the organization of living systems. It has been suggested that the 3rd Law of Thermodynamics is more important for understanding life than the 2nd Law which is considered the basis of Prigogine's dissipative structures [21]. The 3rd Law establishes the reference state with the lowest entropy in relation to which the order (described as information) can be referred. This state, according to the 3rd Law, is achieved at the temperature of absolute zero. Since living systems operate at temperatures near 300°K, they in fact operate far from this reference state. However, it has been suggested that they maintain a long-lived cold decoherence-free internal state (called the internal quantum state), which is achieved by applying error-correction commands to the internal state and by screening it from thermal fluctuations. Mental processes could be associated with the long-lived internal states maintained within the nervous system, a coherence which is equivalent to the effective temperature being lower than ambient [26]. However, further independent evidence of this phenomenon is not yet available.

In a relational world, but one, let us not forget, of both reality and appearance, external space–time *appears* as the medium ('environment') suitable for the coexistence of individual units having their own substantiality. Classically, these units are "seeds" or monads which actualize a small part of existing potentialities, and the world is such that these actualizations prohibit or inhibit actualization of other events. In any non-thermodynamic world the simplest monads correspond to the quantum loops in the concept of loop quantum gravity (see [51]). Space, according to modern views, represents a fine network of finite loops called spin network. Its evolution over time is called a spin foam. Quantum cells of space are connected with each other via their internal field and can be referred to geometric atoms. This concept also appears in the work of Günther [18], see Chap. 3, as "kenogrammatic structures" (cf. our discussion of ontolons in Chap. 9).

This internal field has a certain apparent "internal time" [27]. The transition from a weak field to a strong field appears as if there was a "past", which affects the "future" (causality). In the macroworld, these cells can be viewed as merging, forming the common space–time. The basic process that underlies complexification and expansion is the process of actualization that appears in physics as quantum measurement. It one way of establishing spatiotemporality and determines its growth via continuous measurement of the system plus environment [25]. This special non-Newtonian causality corresponds to the expansion of the Universe viewed as a consequence of quantum measurement. We propose giving credence to alternate ways of establishing spatiotemporality independently of measurement (see Sect. 6.13) such as that of Lupasco. He saw objects, concentrations or condensations of energy, as developing and being accompanied by their own space–time. He used the word 'unrolling' (*dérouler*) to describe this process. Although this picture is somewhat counterintuitive at first sight, it is useful in cutting through the debate about the nature of apparent or subjective time in human beings. This view of time will be discussed in Chap. 7.

6.12 Relationality in Quantum Mechanics and Apparent Reality

Going beyond the discussion in Sects. 6.7 and 6.8, we note that Leibniz, being a proponent of a relational space–time, kept only the *cogito* of Descartes and not extension as the basic property of substance. Such a representation results in certain interpretational difficulties in physics, therefore, it has been largely ignored in science. However, the understanding of the fundamental relational nature of space–time requires an interpretation of the Leibniz paradigm that is compatible with the foundations of physics. In this interpretation, human observers, acting as measuring and communicating agents, constitute a network of mutual interactions, in which the refinement of the wave function generates intersubjective patterns having universal characteristics and corresponding to human perception of the reality of external world [48, 47]. This understanding leads to the idea of relational quantum mechanics in which all its particular interpretations remain correct [7, 49]. The many-worlds interpretation of Everett is valid, but only as a cognitive model. The Copenhagen interpretation of Bohr in its more recent interpretations relates to matter, which includes ultimate representations of reality and its dynamics which are lacking in the Everett approach. The relational interpretation of quantum mechanics may not be exactly isomorphic to the monadological approach of Leibniz but it corresponds to it in its conceptual basis.

The main challenge in modern physics is thus to link the relationality in quantum mechanics based on the relational actualization of events *to* space–time and the relationality of the special theory of relativity in the interaction of observers *in* space–time. This can be formulated as the paradox of two relationalities. Quantum mechanical interpretations are attempts to resolve the problem of potential/actual, their interaction and their evolution. They are based on a relationality that is beyond physics while Einsteinian relativity is in the basis of the physical theory itself. In the next paragraphs we will describe different interpretations of quantum mechanics and discuss their relation to basic physical space–time formed by the gravitational field beyond entropy. We will show that both relationalities can be complementarily combined in the process of transaction.

In fact, the interpretations of quantum mechanics differ in their assessment of the relation between the actual and the potential. The original Copenhagen interpretation of Bohr and Heisenberg postulates a spontaneous unpredictable transition from the potential to the actual. It has been one of the most attractive interpretations, but Brenner, following Lupasco, has shown where it needs to be modified. The most radical many-world Everett interpretation, does not, in fact, describe actual reality. It represents the view in which reality is only potential and all actualizations are considered as imaginary or even non-existing. This conflicts with the interactive dualism of Logic in Reality.

There is the only one interpretation to our knowledge that considers quantum measurement as a purely physical phenomenon and attributes it to gravity, so the actualization takes place when the curvature of space reaches the value of Planck's mass. This is the interpretation of Penrose which is not widely accepted. The pilot-wave interpretation of Louis de Broglie [8] and David Bohm [4] assumes that the wave function (potential) and its actualization perpetually coexist in a dual reality. There are several advantages of this interpretation for explaning physical reality, obviously including its both potential and actual constituents. The pilot-wave interpretation can be a useful tool to explain the interdependence of the potential and the actual phenomena, from elementary particles to advanced living organisms. Its further development consists in the understanding of the interactions between the pilot-waves as transactions in which a new reality emerges and the pattern of the distribution of potentialities changes (the "implicate order" of fluctuating pilot-waves in Bohm's terms).

Other interpretations such as consistent histories and environmental decoherence, as well as logic in reality do not view the transition between the potential and actual in the abrupt spontaneous way as in the Copenhagen interpretation but tend to avoid the radical view of the many-world Everett interpretation. The notion of "decoherence" instead of "reduction of the wave function" assumes a soft relation of the potential and the actual with their interchange and refinement through self-consistence and environmental fitness but does not put the whole reality into the potential field. The consistent histories interpretation tends to explain decoherence via self-consistence of relative actualization while environmental decoherence assumes a kind of natural selection of actualization towards the fitness to the "environment" in which a previous actualization preceded the current one, allowing such processes to be both continuous and discontinuous without a fixed frame of reference.

6.13 Potentiality and Quantum Measurement: The Transactional Interpretation

From its inception, quantum mechanics faced the problem of actualization that occurs in the course of the measurement process. This process is, in fact, beyond the theory itself but needs to be interpreted in order to understand its power and meaning. The notion of wave function of Erwin Schrödinger was interpreted in a way that the physical nature is divided into "waves" and "particles". The interpretation of "wave" as a potentiality accompanied quantum mechanics from the beginning, but it did not receive a sufficient development mainly because of the difficulty for physicists to accept probability as one of basic foundations of the physical theory.

When, in the classical experiment, a photon passes through two splits, it passes as a potential field and then becomes actualized on the screen. This all shows that the "particle" or the "wave-particle" approaches are not fully relevant. More relevant is an interpretation of a potential field and its actualization. In other words, any measurement is the establishment of a correspondence between the potential and the actual, mathematically corresponding to a mapping from complex to real numbers.

The basic principle of quantum mechanics is, in fact, is probabilistic. It can be expressed in the form of the Born Rule which was introduced by Max Born in 1926. The Born rule claims that the probability of obtaining any possible measurement outcome is equal to the square of the corresponding amplitude. The wave functions, in fact, are considered as potential distributions and the complex numbers that describe these probabilities are associated with the potential reality which is actualized by squaring to real numbers.

The quantum theories of measurement are not strictly physical theories. However, this does not mean that they are not scientific theories. In fact, there is no physical verification for any version of the quantum theory of measurement. In the many-world interpretation, parallel worlds are not interacting or this interaction cannot be verified. In the Copenhagen interpretation, the wave function collapse is also positioned beyond physical causality. Interpretations of quantum mechanics can be compared on the basis of their simplicity in the sense of Occam's razor and Mach's principle and their usefulness for the explanation of phenomena of "measuring" reality such as life and consciousness. We face the question of what in fact is the wave function. Does it represent a real physical entity which undergoes a collapse or it is a derived, auxiliary mathematical instrument for calculating probabilities? In fact, being non-local and having the nature of a "potential reality", it is placed in the area of Kantian *things-in-themselves* before their actualization. The nature of this area cannot be directly referred to phenomena but can help to estimate probabilities in the phenomenal world.

6.13.1 Propensiton Quantum Theory

Maxwell [38] made an attempt to formulate real probabilistic quantum theory. He has suggested the propensiton quantum theory (PQT) as opposed to the ordinary quantum theory (OQT). He interprets the ψ-function as specifying the actual physical state of propensiton quantum entities. This understanding of quantum reality could be traced to the Einsteinian approach to clarify the quantum effects. The difficulties in interpreting quantum mechanics lay in the fact that the physics community has failed to take probabilism seriously. Maxwell suggests the idea that ψ specifies the real physical state of the propensiton system at any given instant by specifying the value of probabilistic properties or propensities possessed by the propensitons at the given instant. PQT provides a natural solution to the quantum wave/particle dilemma by introducing "beables" in the sense of John Bell who suggested this term and does not have reference to observables and macroscopic phenomena or processes.

Probabilism in understanding the quantum phenomena is now re-emerging in the transactional interpretation of quantum mechanics introduced by Cramer [11] and further developed by Kastner [29]. This theory considers quantum measurement as a transaction of the retarded and advanced probabilistic waves attributed to the measured system and the measuring device and claims that the actualization of quantum events takes place outside of the space–time domain. While this interpretation is not dependent upon consciousness, it allows for the existence of life and consciousness as the phenomena characterized by the special organization of quantum measurements. The instantaneous processes of measurement, according to the transactional interpretation, proceed between the hidden possibilities, not actualities. They remain hidden on the "empirical" level but may be experienced in an inner form at the highest levels of quantum phenomena that appears as consciousness. In the transactional interpretation time emerges only as a relative concept, i.e., relative to a given frame of reference which is tied to the objects with non-vanishing rest mass, like atoms or electrons. The transaction in Ruth Kastner's interpretation combines the relationality of the process of actualization with the Einsteinian relationality of space–time. It represents an elementary process that possesses the two complementary relationalities, one being the quantum mechanical and the other relativistic. It appears also at higher levels of organization causing transformations of complex systems from the chemical structures through living bodies to social structures. We expect that the further development of natural sciences will generate concrete and detailed approaches that will incorporate the concept of transaction at different levels.

In the transactional interpretation, the quantum measured system emits an ordinary (retarded) wave forward in time and also emits an advanced wave backward in time, as well as the receiver positioned later in time also emits an advanced wave backward if viewed in time and a retarded wave forward if viewed in time. The nature of advanced and retarded waves can be understood not in terms of the physical time, as the transaction itself introduces time, but as the operation of two components of causality, one acting upwards and the other downwards. Actualization appears as the interference of these two causalities. A quantum event occurs when a "handshake" exchange of advanced and retarded waves triggers the formation of a transaction in which energy is transferred. In this interpretation, the collapse of the wave function occurs along the whole transaction and cannot be attributed to any specific point in time. The waves are seen as physically real and exist as possibilities outside of physical space–time being a part of physical reality (Kastner). Referring back to the *I Ching*, we can say that the transaction is based on the unity of opposites of the advanced and retarded actions, which includes their contradiction and complementarity and generates the actual state as a result of this transaction.

The transactional interpretation of quantum mechanics has the important additional advantage that it incorporates relativistic physics into the apparatus of the measurement theory. While the ordinary Schrödinger equation does not substantiate the advanced transaction during measurement as it was postulated in this measurement theory, its relativistic version does. Kastner [29] has developed a new Relativistic or Possibilist Transactional Interpretation of quantum mechanics in which space–time itself emerges by a way of transactions. She has suggested that the relativistic version of the transactional interpretation provides the quantum dynamics for the concepts of quantum gravity (causal sets). The retardation and advancement of the waves constituting the transaction are based on the relativistic effects that were successfully incorporated in the foundation of the transactional interpretation.

The term "transaction" can be used in place of "quantum measurement". A transaction refers to a process in a state prior to its actualization, while the term transduction designates a transition as we observe it in the actual world, following (in the causal sense) after the transaction. The non-equilibrium thermodynamics approach describes transduction but not transaction. Transduction in biology refers, e.g., to signal transduction when all components are defined before the transduction starts. To understand the basis of not only physical but also biological and social events, we need to understand how the transaction takes place that precedes and initiates the process. We can therefore also consider biological and social transactions also, and we will discuss this further in Chaps. 17 and 18.

6.13.2 Transactional Substantiation of Fundamental Constants

6.13.2.1 Consistency Between the Two Waves (Retarded and Advanced) in Relation to Actualization

It is generally accepted in all interpretations that quantum measurement leads to the appearance of classical features of physical systems. The measurement and its result are connected in such a way that satisfies the condition of consistency and the expression of this consistency is the existence of invariants being fundamental world constants which appear as a correspondence of actualization to the consistency of the actual world. The potential field (vacuum) is a superposition of opposite states. A choice of the definite set of constants is determined by the consistency and optimality of all of them. The three basic physical constants are the most fundamental because they can be used to define all Planck units and, thus, all physical quantities. In the transaction process of measurement those components of the offer wave that satisfy conservation laws and entanglement criteria are permitted to be projected into the final transaction, which exists in three-dimensional space (Cramer).

In the early 1930s, Bronstein (1906–1938), a victim of Soviet totalitarian regime who had excellent ideas on the foundations of physics, wrote: "After the relativistic quantum theory is created, the task will be to develop the next part of our scheme, that is to unify quantum theory (with its constant h), special relativity (with the constant c), and the theory of gravitation (with its G) into a single theory." Considering fundamental constants of the physical world, we return to the fundamental question of Albert Einstein whether God had any choice in the creation of the world. We can consider the quantum measurement as a self-referential transactional process through which the actual part of the world is restricted ("filtered") through the introduction of certain fundamental values (constants). These constants should provide an observability of the Universe and a possibility of free choice at higher levels of self-reference in accordance with the anthropic principle [57].

To illustrate this concept, we introduce the following statements:

(1) The finite gravitational constant follows from the reduction of superposed potential states during the measurement transaction.
(2) The finite velocity of propagation of signals follows from the sequential appearance of dynamic oppositions in the course of measurement transactions.
(3) The finite quantum of action follows from the consistency of the measurement transaction corresponding to causality (conservation laws).

Three most basic fundamental constants and three fundamental parameters: mass, length and time arise from the reflective structure of quantum measurement-based transaction. Therefore, the fundamental constants are the invariants of the process related to the quantum mechanical reduction, appearing as a result of the self-referential process of transaction. The computable part of the world, which is consistently describable by mathematics, is defined within the whole set of potentiality by the fundamental length, time and velocity. These constants appear not as "given" in the apparent harmony of the world in the Pythagorean sense but rather as the parameters of "fixed measures" according to which the Heraclitean fire (*Pyr*) is kindling and extinguishing (Heraclitus: Fragment 30). "An unapparent harmony is stronger than an apparent one" (Heraclitus: Fragment 54), as discussed above.

6.13.2.2 Limitations Imposed on Measuring Devices and Substantiation of the Fundamental Constants

The fundamental constants divide the world into two parts: one being recursive and computable and the other is non-recursive where algorithmic operations are impossible. We repeat that it is to the latter that the dialectics and principles of

Logic in Reality apply, and accordingly to our macroscopic reality. The dynamics of the one part is apparently derivable by the computation from a limited set of physical principles known to be valid in very simple systems whereas the second part cannot be reduced to these fundamental principles. The interface between the computable and non-computable parts is formed through the interference between the two aspects of one transaction, one spreading from the measuring device to the object of measurement, and the other originating from the object of measurement toward the measuring device.

If one (computable) part is taken separately, time (defined as "entropy time" by Boltzmann) is associated with the physical movement of particles ("the time which is measured" according to Aristotle). The other (non-computable) part determines the non-local unity of all points of the Universe and provides "the time by which we measure" (Aristotle). Time in Heisenberg's uncertainty relation "energy-time" is the time of the measuring system by which it measures the energy of a microsystem. The time flow of a microsystem is explicitly contained in its velocity and energy. The correspondence of the time of measuring system and the time of microsystems is realized on the objective-subjective transactional interface, being possible only at the certain values of fundamental constants. Time as a reduction of potentialities is a construction, which therefore includes both these time constituents.

According to the hypothesis of Penrose [43], the quantum vacuum is a superposition of potentially existing contradictory states. The two states existing potentially as a quantum superposition might be judged with regard to their respective gravitational fields, i.e., their respective space-times might be too different from one another for them to be able to coexist in a quantum linear superposition. Accordingly, reduction would have to take place at that stage. The same explanation can remain valid also in the transactional interpretation. The superposed widely different states are unstable and undergo the transactional reduction. The question appears if a superposition of potential states can be related to objective curvatures of the General Theory of Relativity (GR).

The set of alternative potentially existing possibilities is actualized forming dynamic oppositions. In other words, actualization results in the appearance of contradictions reflecting certain opposite actualized possibilities. The process of transaction in measurement remains self-contradictory unless we do not consider it as propagating with finite velocity. Otherwise, it will contain contradictory statements at the same moment of time [17]. The advanced and retarded waves of the potential field in the quantum transaction generate the actualized patterns in the process that satisfies the condition of consistency, *i.e.* of the conservation of energy law. The idea of the energy conservation law as a physical manifestation of consistency is quite old. The minimum time interval for measuring conservation laws internally remains finite and non-vanishing ensuring the conceptual epistemic consistency of physical processes in the Universe. The apparent inconsistencies in the actual processes in reality require the invention of more universal reference frames, which corresponds to the progress in physics and in its description of the Universe external to the observer.

6.14 Transactional Substantiation of Life and Consciousness

If we consider a living system as a measuring device that redistributes quantum mechanical probabilities, we can derive its main characteristics from the particular types of quantum-like transactions that can be described as inherent to it. The specificity of biomacromolecules for strictly determined interactions is grounded in essentially low energy dissipation during their operation, providing for the registration of energetic signals difficult to distinguish from the surrounding noise. According to Heisenberg's ratio between the uncertainty of energy corresponding to its possible dissipation in the quantum mechanical interaction and the time to perform a quantum measurement ($\Delta E \Delta \tau \approx \hbar$), the interactions between a quantum system and a macroscopic measuring device may follow a path that provides, practically to any level of precision, the non-demolition registration of the strictly determined low energy signals [6]. These interactions are characterized by a high precision and certainty of the result of measurement, and a sensitivity of the detector is determined by its relaxation properties in the course of measurement. These measurements were defined as quantum non-demolition (QND) measurements and their main property is low energy dissipation appearing at the expense of a sufficiently long time of measurement to achieve the needed level of low dissipation [5]. According to Matsuno [36], the internal QND measurement is based on participation of entangled quantum states, however, this form of quantum entanglement does not survive during external measurement. The redistribution of probabilities during the internal process can leave them entangled but, in the end, mapping to real numbers will take place and the result will be revealed externally, in classical time-space as Bohr and Heisenberg described in late 1920s. The transactional interpretation can substantiate the flexibility of the potential-actual interface depending on the system which performs the measurements and its relation to other systems.

In general terms, the description of the operation of macromolecules in living systems is described by a concept developed by Lev Blumenfeld in 1970-1980s and profoundly related to the theory of quantum non-demolition measurements developed by Vladimir Braginsky in the same years [6]. Blumenfeld [3] considered the conformational relaxation of macromolecular systems acting as macroscopic oscillators to be an elementary action of the bioenergetic process, in which a fast quantum effect (*e.g.* the capture of electron) is followed by a slow conformational transition during which the energy is not dissipated and remains stored for a lifetime long enough for work to be performed. This corresponds to the ideas of McClare [39] about resonance in bioenergetics and the conformational movements of macromolecules. The conformational motion of biomacromolecules is many orders of time slower than the initial quantum effect, and the rate of a bioenergetic process is, therefore, determined by the rate of the conformational relaxation. The latter takes place only after the action of a force converting the system into the new conformational state, *i.e.* after the generation of a non-equilibrium state resulting from fast initial interaction. From this point of view, the specificity of enzymes is their capacity for the recognition of specific configurations of electron clouds (distributions of electron wave functions) in certain compounds. This specificity can therefore in principle be described by using a quantum mechanical formalism.

The transactional interpretation of quantum mechanics is a powerful tool for the probabilistic explanation of quantum measurements which assumes that actualization may take place in different ways depending on the nature of the transaction. It does not require consciousness for performing measurements but supports a theory of the appearance of life and finally consciousness in systems realizing certain types of quantum measurement transactions, characterized by complex units capable of forming reflexive loops at different levels of organization. On the basis of the above considerations, we consider that we have opened the philosophical discussion of the origin of life to some new scientific, physical considerations. These will be explored further in Chap. 18.

References

1. Bekenstein, J.: Universal bound on the entropy to energy ratio for bounded systems. Phys. Rev.D **23**, 287–298 (1981)
2. Bishop, R., Brenner, J.E.: Potentiality, Actuality and Non-separability in Quantum and Classical Physics: Heisenberg's *Res Potentiae* in the Macroscopic World. arXiv: 1801.01471 (2018).
3. Blumenfeld, L.A.: Physics of Bioenergetic Processes. Springer, Berlin (1983)
4. Bohm, D.: Wholeness and the Implicate Order. Routledge, London (1980)
5. Braginsky, V.B., Khalili, F.Y.: Quantum non-demolition measurements: the route from toys to tools. Rev. Mod. Phys. **68**, 1–11 (1996)
6. Braginsky, V.B., Vorontsov, Y.I., Thorne, K.S.: Quantum non-demolition measurements. Science **209**, 547–557 (1980)
7. Brenner, J.E.: Logic in Reality. Springer, Dordrecht (2008)
8. De Broglie, L.V.P.R.: Certitudes et incertitudes de la science (Certitudes and Incertitudes of Science). Albin Michel, Paris (1966)
9. Campbell, D.T.: Downward causation in hierarchically organized biological systems. In: Ayala, F.J., Dobzhansky, T. (eds.) Studies in the Philosophy of Biology: Reduction and Related Problems, pp. 179–186. Macmillan, London/Basingstoke (1974)
10. Conrad, M., Liberman, E.A.: Molecular computing as a link between biological and physical theory. J. Theor. Biol. **98**, 239–252 (1982)
11. Cramer, J.G.: Transactional interpretation of quantum mechanics. Rev. Mod. Phys. **58**, 795–798 (2009)
12. Derrida, J.: Of Grammatology. Johns Hopkins University Press, Baltimore (Originally published 1967) (1997)
13. Derrida, J.: Glas. Galilée, Paris (1974)
14. Feynman, R.: The theory of positrons. Phys. Rev. **76**, 749–759 (1949)
15. Friedmann, A.: On the Possibility of a world with constant negative curvature of space. General Relativity and Gravitation **31**, 2001–2008 (Originally published 1924) (1999)
16. Graham, D.W.: Science before Socrates: Parmenides, Anaxagoras, and the New Astronomy. Oxford University Press, New York (2013)
17. Gunji, Y.P.: Autonomic life as a proof of incompleteness and Lawvere's theorem of fixed point. Appl. Math. Comput. **61**, 231–267 (1994)
18. Günther, G.: Time, timeless logic and self-referential systems. Ann. N Y Acad. Sci. **138**, 397–406 (1967)

19. Heisenberg, W.: Die Physikalischen Prinzipien der Quantentheorie. S. Hirzel, Leipzig (1930)
20. Igamberdiev, A.U.: Quantum mechanical properties of biosystems: a framework for complexity, structural stability and transformations. BioSystems **31**, 65–73 (1993)
21. Igamberdiev, A.U.: Quantum computation, non-demolition measurements, and reflective control in living systems. BioSystems **77**, 47–56 (2004)
22. Igamberdiev, A.U.: Physical limits of computation and emergence of life. BioSystems **90**, 340–349 (2007)
23. Igamberdiev, A.U.: Physics and Logic of Life. Nova Science Publishers, Hauppauge, NY (2012)
24. Igamberdiev, A.U.: Biomechanical and coherent phenomena in morphogenetic relaxation processes. BioSystems **109**, 336–345 (2012)
25. Igamberdiev, A.U.: Time rescaling and pattern formation in biological evolution. BioSystems **123**, 19–26 (2014)
26. Igamberdiev, A.U.: Time, Life, and Civilization. Nova Science Publishers, Hauppauge (2015)
27. Igamberdiev, A.U.: Time and life in the relational universe: prolegomena to an integral paradigm of natural philosophy. Philosophies **3**, 30 (2018)
28. Kafatos, M.C., Kato, G.C.: Sheaf theoretic formulation for consciousness and qualia and relationship to the idealism of non-dual philosophies. Prog. Biophys. Mol. Biol. **131**, 242–250 (2017)
29. Kastner, R.E.: The possibilist transactional interpretation and relativity. Found. Phys. **42**, 1094–1113 (2012)
30. Lemaître, G.: Un univers homogène de masse constante et de rayon croissant, rendant compte de la vitesse radiale des nébuleuses extra-galactiques. Annales de la Société Scientifique de Bruxelles, série A, **47**, 49–59 (1927)
31. Liberman, E.A.: Molecular quantum computers. Biofizika **34**, 913–925 (1989)
32. Lupasco, S.: Le principe d'antagonisme et la logique de l'énergie. Editions du Rocher, Paris (Originally published 1951) (1987).
33. Lupasco, S.: L'énergie et la matière psychique, Julliard, Paris (1974)
34. Lusanna, L., Pauri, M.: General Covariance and the Objectivity of Space-Time Point-Events. [Preprint] (2005). URL: http://philsci-archive.pitt.edu/id/eprint/2224
35. Lusanna, L., Pauri, M.: The physical role of gravitational and gauge degrees of freedom in general relativity — I: Dynamical synchronization and generalized inertial effects. General Relativity and Gravitation **38**, 187-227 (2006)
36. Matsuno, K.: Forming and maintaining a heat engine for quantum biology. BioSystems **85**, 23–29 (2006)
37. Matsuno, K.: From quantum measurement to biology via retrocausality. Prog. Biophys. Mol. Biol. **131**, 131–140 (2017)
38. Maxwell, N.: Is the quantum world composed of propensitons? In Suárez, M. (Ed.) Probabilities, Causes and Propensities in Physics. Springer, Dordrecht (2011)
39. McClare, C.W.F.: Chemical machines, Maxwell's demon and living organisms. J. Theor. Biol. **30**, 1–34 (1971)
40. Minati, G., Abram, M., Pessa, E. (eds.): Methods, Models, Simulations and Approaches—Towards a General Theory of Change. World Scientific, Singapore (2012)
41. Minati, G., Pessa, E.: From Collective Beings to Quasi-Systems. Springer, New York (2018)
42. Nakagomi, T.: Mathematical formulation of Leibnizian world: a theory of individual-whole or interior-exterior reflective systems. BioSystems **69**, 15–26 (2003)
43. Penrose, R.: Emperor's New Mind. Concerning Computer, Minds, and the Laws of Physics. Oxford University Press, London (1989)
44. Petit, G., Wolf, P.: Relativistic theory for time comparisons: a review. Metrologia **42**, S138-S144 (2005)
45. Popper, K.: Conjectures and Refutations: The Growth of Scientific Knowledge. Routledge, New York (1998)
46. Prigogine, I.: From Being to Becoming: Time and Complexity in the Physical Sciences. W. H. Freeman & Co., New York (1981)

47. Rosen, R.: Drawing the boundary between subject and object—comments on the mind-brain problem. Theoret. Med. **14**, 89–100 (1993)
48. Rosen, R.: Life Itself. A Comprehensive Inquiry into the Nature, Origin, and Fabrication of Life. Columbia University Press, New York (1991)
49. Rovelli, C.: Relational Quantum Mechanics. Int. J. Theor. Phys. **35**, 1637 (1995)
50. Rovelli, C.: The First Scientist Anaximander and his Legacy. Westholme Publishing, Yardley (2011)
51. Rovelli, C.: Reality Is Not What It Seems: The Journey to Quantum Gravity. Penguin Random House, New York (2016)
52. Savitt, S.F.: There's no time like the present (in Minkowski space-time). Philos. Sci. **67**, S563–S574 (2000)
53. Sklar, L.: Philosophy of Physics. Oxford University Press, Oxford (1992)
54. Smith, Q.: The ontological interpretation of the wave function in the Universe. The Monist **80**, 160-185 (1997)
55. Smolin, L.: The Trouble With Physics: The Rise of String Theory, the Fall of a Science, and What Comes Next. Houghton Mifflin Harcourt, Boston (2006)
56. Varela, F.J.: The Specious Present. In: Naturalizing Phenomenology. Issues in Contemporary Phenomenology and Cognitive Science, eds. Jean Petitot et al. Stanford University Press, Stanford, CA (1999)
57. Wheeler, J.A.: Law without law. In: Wheeler, J.A., Zurek, W.H., (Eds.) Quantum Theory and Measurement, pp. 182–213. Princeton University Press, Princeton NJ (1983)

Chapter 7
Mathematics in Reality

7.1 Introduction: Mathematics and Reality

Mathematical concepts and structures, since Pythagoras and Plato, were often considered as separate ideal essences not linked to changing actual reality. Following this thinking, Parmenides claimed the latter was imaginary. This statement, however, revealed substantial paradoxes in the formulation of such an ideal reality, as first was noted by Parmenides' disciple Zeno. Following this revelation, Plato in his dialogue "Parmenides" introduced dialectics to overcome the substantial insufficiency of the original formulation of the concept of ideal reality. Indeed, the dialectical interplay of cognition and space–time, and energetic as well as geometrical principles, are instantiated in the foundations of mathematics. These considerations substantiate the concept of the real world being shaped in a particular way, resembling the (weak) anthropic principle in physics. Inclusion of these principles in the foundations of mathematics increases its power for description of change in reality (Chap. 2). We will discuss here the following main topics: (1) the transition from the Platonic foundations of set theory to the Aristotelian concept of univalent foundations of mathematics based on the theory of types, as a conceptual basis for a Mathematics in Reality; (2) a non-standard set theory, which we relate directly to the Principle of Dynamic Opposition as originally formulated by Lupasco, as well as to the Logic in Reality (LIR) derived from it; (3) further parallels are made to the mathematics of Voevodsky and current issues in the dialectics of rough sets. One of the most important consequences of applying Logic in Reality is the value ascribed to inconsistency, incompleteness, 'roughness', contradiction, and in general the qualitative versus the quantitative. We therefore conclude with an overview of recent developments in mathematics and science which demonstrate the philosophical implications of rigorous qualitative approaches. The basic foundations of mathematics need to be reconsidered to overcome the apparent problems of implementation of LIR in its framework.

J. E. Brenner and A. U. Igamberdiev, *Philosophy in Reality*, Studies in Applied Philosophy, Epistemology and Rational Ethics 60, https://doi.org/10.1007/978-3-030-62757-7_7

7.2 The Generation of Noumena from Potentiality for Defining Phenomenal Reality

In his *Critique of Pure Reason* of 1781, Immanuel Kant [19] claimed that the thing observed in space and time is actualized from a *Ding-an-sich* (see Sect. 5.5.1 above) which cannot be described by consciousness and represents a broader true reality before its actualization in our perception. According to Kant, the *Ding-an-sich* exists beyond space–time, and it is not a thing in the trivial sense because the category of multiplicity is applicable only to phenomena. In the world of *Ding-an-sich*, there is no natural causality, but instead a free causality (reasoning) between phenomena and noumena that is realized through the world of *Ding-an-sich*. The uncertainty of the primary Kantian definition resulted in the identification of the *Ding-an-sich* with the *noumenon*; however Kantian philosophy does not consider these as equal. The *omnium* of the *Ding-an-sich* is rather a transitional state between noumena and phenomena, a kind of philosophical 'purgatory'. By analyzing Kant's doctrine, Teodor Oizerman showed that the noumenon and *das Ding-an-sich* are different concepts in Kant's philosophy [33]. The first refers rather to the Platonic *eidos* (image) while the second represents potentiality before its actualization. However, we need to understand conceptually these two aspects of reality and this may be achieved through Lupasco's principles underlying Logic in Reality [25, 26]. When we grasp potentiality, it becomes represented in consciousness noumenally as a reflection of the original potentiality.

One of the founders of quantum mechanics, Max Born felt a very close relation of the Kantian doctrine and tried to resolve the problem of correspondence between quantum measurement and the Kantian concept of perception [2]. Much later, Nielsen and Chuang [32] also argued that the *Ding-an-sich* corresponds to the Feynman path integrand or, more strictly, to certain fundamental quantities determined from second order effects of the latter integrand. However, we can move away from quantum mechanics and consider any potential reality in natural sciences from the point of view of its associated actualization in on-going naturally evolving processes. Returning to the concept of the *clinamen* in the philosophy of Epicurus, we can distinguish between the *clinamen* itself and the potentiality of decrease of the unpredictable movement through the formation of complex structures in which the degrees of freedom of atoms are reduced. The simplest example of such restriction of movement imposed by the structure in physics is the assembly of elementary particles into atoms and atoms into molecules.

As some readers may have already concluded, there is a striking similarity of form between the *Ding-an-sich* and the Included Middle or Included Third Term in the logic of Lupasco. However, the respective constitutions of the entities are quite different, conceptual in one case and energetic in the other, they both direct thought to 'that something in addition' that enables reality to be discussed. While *Ding-an-sich* represents a Potentiality of Reality, the n*oumenon* appears as a Reality of Potentiality. Ontological (physical) potentiality becomes real when it is grasped through *noumena*. Mathematics, in fact, grasps this potential reality and uses phenomena as the tools to reconstruct the real world via the application of computation. The *Noumenon* as a reality of potentiality and the *Ding-an-sich* as the potentiality of reality are linked dialectically through the dynamic process of the formation of complex dynamic systems in the real world.

7.3 The Noumenal Reality of Mathematics and Its Historical Relation to Natural Language

The formal language of mathematics is suitable for the description of the portion of the world that is computable but it is rather not suitable for common communication. On the contrary, the language of art (e.g. poetry) is non-formalized and suitable for the description of *meaning* in the real world. Both are derived from the basic structure of human language, one by reducing the field of meanings to symbols and the other by expanding it through metaphors and other means. Common communication is more complex and goes beyond mathematics. The link between the mathematical and the physical world refers to the link between the noumenal and the phenomenal, and the link itself cannot be computably ("with certainty") described in the frames of mathematics and formal logic. As once Einstein [7] mentioned, in so far as theories of mathematics speak about reality, they are not certain, and in so far as they are certain, they do not speak about reality. He also noted [8] that "physics is essentially an intuitive and concrete science. Mathematics is only a means for expressing the laws that govern phenomena".

The operational notion of the world external to the observer and of its cognitive acquisition via external activity opens a possibility of counting. Early mathematics is related to the Neolithic revolution and is initially based on invention of the symbols that designate the number of objects used in agriculture [29]. These symbols were clay tokens, and early languages contain counting words in addition to numerals. The counting words appear in the language such as "number of round objects" or "number of triangle objects", etc. These are the symbols of particular classes with a limited generalization as compared to the idea of number itself. The Sumerian and Egyptian mathematics still did not have such generalization of terms as acquired in Ancient Greece. The development of a formal language of mathematics paralleled the development of writing systems for common language.

The origin of mathematics coincides with the recognition of the reflexive capacity of a human consciousness that operates with dynamic oppositions, developing dualism and making much more variety that exceeds the fundamental binary choices described in the model of reflexive consciousness introduced by Lefebvre [21]. This capacity, established on the basis of referral of the subject (self) to its symbolic image, acquires the potential to rationally describe the external world through the structure of human language. It also reveals computable principles in the relation to externality and becomes a premise of the origin of counting and later mathematics [17]. The origin of computability arises from the concept of regulation of potentiality which becomes more powerful with the development of the reflexive structure of the human self. Operational regulation is achieved at this level is a *noumenal* reality.

7.4 The Relation of the Mathematical and the Physical Worlds

Mathematics is thus a reflection of potentiality in the real world that can be transformed into a computable principle of its actualizations and potentializations. Mathematics is a powerful tool for of computing actualizations themselves, while the mathematical entities are separable from the actual reality, behaving like epistemons not ontolons. Mathematics also contains many formulations that are not actualized in the real world. The most striking is the dimensionality of space which is unlimited in mathematics but is realized in physics as a perceived three-dimensional space. In modern physical theories, 3-D space is supplemented by additional compacted dimensions whose number appears to be also limited. The question of reduction of the number of dimensions in actual physical space is not trivial and is approached in different philosophical concepts in different ways. The most important is the strong anthropic principle which states that the possibility of conscious perception is possible only (1) in three-dimensional space and (2) the existence of living systems composed of atoms and molecules in such space. This point was discussed above in Sect. 1.2, and we go beyond it here.

As far as the obvious curvature of space is concerned, in its relation to objects, there is no mathematical basis for a particular value of it. The existence of metrics, according to the basic ideas of Riemann [36], can be explained by the external physical forces of connection that operate in/on space. This is an example of physical principles imposing limitations on mathematical forms. Igamberdiev defined this as essentially a science of the *limits of computation* which appears to us as physical laws [15]. These forces of connection, as suggested by Riemann, are related to the discreteness of space, and form its actual observed structure. In Sect. 6.11 above we described the inversion of this point of view proposed by Lupasco. 'Crystallizations' of energy in the form of objects determine the characteristics of their spatial and temporal environment and its evolution. Mathematics enters the picture in a strictly a posteriori manner. Thus, it cannot provide the proof of validity of the fundamental constants in physics, which can be substantiated, say, as Diogenes proved the existence of movement by walking. As Einstein noted, "God does not care about our mathematical difficulties, he integrates empirically" [18]. In other words, the Nature establishes the limits of real processes that shape the physical world in a manner of 'walking-Diogenes'.

Kant analyzed the problem of dimensionality of space in "*Thoughts on the true estimation of living forces*" [20]. Physical forces can act according to the inverse square law in three-dimensional space and thus the potentiality can be actualized in a consistent manner [1]. In such a space, it is possible to define energy as a countable representation of the potential for realizing a process measured as work. This explanation of dimensionality does not contradict the anthropic explanation, but it provides well-defined physical reasons for a dimensionality that underlies all possible anthropic arguments. Only in real space can processes take place operating in time that separates contradictory statements and generates dynamic oppositions.

Unfortunately, the explanation of three-dimensionality given by Kant depends on assumptions of consistency and semantic truth that beg the question of the origin of the dynamic oppositions it describes. Our world exhibits mathematics in reality, not as a noumenal description of the phenomenal world via actualization of the potentiality of the *Ding-an-sich*, but as a consequence of our own reality as subjects able to cognize the movements of the world and then become capable of transforming them into an epistemic *Ding-für-uns*.

7.4.1 The P Versus NP Problem in Computer Science

The P *versus* NP problem is a basic unsolved problem in the foundations of computer science that has major significance for mathematics, logic and philosophy in general. This problem involves the study of the relation between deterministic problems (P) and non-deterministic problems (NP) in computer science in the sense of computability as well as complexity. For comparison, according to the basic LIR Principle of Dynamic Opposition in reality, complex processes do not go the absolute limits of 0 or 1, and thus have a status which is *both* partly actual and partly potential. Our Principle thus implies a fundamental non-determinism of existence which will be familiar to some readers as the Heisenberg uncertainty principle.

Non-determinism as studied by Liu [22] at the Jules Verne University of Picardie from the perspective of computer science and logic is related to Gödel's incompleteness theorems: in any consistent axiomatic system, as long as it contains a proposition of elementary arithmetic, there must be an undecidable proposition which cannot be determined to be true or false with this system. There would therefore be no basis for a statement of 'truth' regarding the (1) existence of an exact solution to the problem or (2) whether or not it could be found. These concepts are, as these authors state, beyond the reality describable by classical logic.

The additional aspect to which Liu calls our attention is that the current definitions of NP based on a hypothetical NDTM (Non-Deterministic Turing Machine) seem to tolerate the idea that "deterministic algorithms may exist for non-deterministic problems, even though none has been yet found". This implies incorrectly that NP problems are equivalent to P problems (P = NP), because as long as one admits that such "deterministic algorithms" might exist, this implies already that NP problems may have solutions! Thus the current definitions of NP beg the question; the definition is circular and one is stuck in the aporia that implicitly affirmed NP = P in advance.

In summary, we can state that what Liu has now clearly shown is an aporia for standard logic and mathematics is not one for Logic in Reality. This is simply further support of our position that the absence of a deterministic algorithm for solving complex, natural problems, and their 'component' of non-determinism does not place them outside the realm of science and hence of a philosophy that accepts science.

7.5 The Pythagorean Idea of Number in the Foundations of Mathematics

Although Pythagoras is a semi-legendary figure, he still should be considered as the founder of mathematics. In the foundations of his philosophical concepts, mathematics is treated as *real timeless* reality itself as opposed to the *unreal changing* reality of our world. This view influenced the development of the foundations of formal logic by Parmenides and Zeno that in turn determined the entire foundation of Plato's philosophy. Taking number as the primary entity in the foundation of the world, Pythagoras established a frame of reference for mathematics by performing an Husserlian *epoché*, i.e. by bracketing perceptually observed reality from what he considered as the real timeless reality, the mathematical intelligible reality. This point appears in the structure of modern mathematics and formal logic as their incompleteness. This was realized by Parmenides and Zeno and revived in the twentieth century in Gödel's theorems. The limits of computation are a consequence of the same reasoning: through them we establish the impossibility to reach infinities in computation. This concept also obtains outside mathematics, for example that of finite atoms to avoid the paradox of an infinite division of matter. However, the abstracting capacity of consciousness operates with assumed infinities which remain in the epistemic noumenal sphere. Their relation to actual reality is that of abstract operational signs, of which one sees echoes in Pierce.

7.5.1 Number as Substance

The basic idea of Pythagorean philosophy is the concept that all things are made of numbers, where 'made of' means constituted by epistemologically. The number one (the monad) represents the origin of all things; the number two (the dyad) represents matter; the number three was considered an "ideal number" because it has a beginning, middle, and end and defines a plane triangle. The idea of number as the basic foundation of existence dominated mathematics until Georg Kantor introduced the concept of set as a well-defined collection of distinct objects, considered as an object in its own right, as the most fundamental foundational entity. The persistence of this idea has resulted in statements made even today about things being 'made' of numbers.

Regarding the concept of number as the primary "substance", it was a major invention, beyond Greek philosophy, when it was realized that it is possible to have a symbol signifying nothingness, that is, zero. Partridge gives the origin of zero as the Arabic word '*shifr*', empty which also led to the French word for number *chiffre*. This is considered as the invention of Eastern philosophy [23] although zero was also defined by a special symbol in Mayan mathematics. The relation of zero to the numbers designating the existence is a problem that dominated further development of mathematics. It further determined the invention of such mathematical objects as infinitesimals, negative numbers, imaginary numbers, etc. The axiomatic basis of the system of real numbers was developed from the viewpoint of set theory over a hundred years ago. It is rooted in the intuitive Peano axioms for countable numbers and in the Zermelo-Frankel axioms of set theory. We will discuss this in relation to (1) the formulation of the foundations of mathematics and (2) its operational power in reality. There is no evidence at all that any other system can replace these foundational functions.

The original number system was expanded through the invention of calculus. This was done independently by Leibniz and Newton and accompanied the development of classical mechanics. This required the introduction of a new type of mathematical object, the infinitesimal. Leibniz comprehended infinitesimals as having certain values not equal to zero, but only in the 1960s, was the concept revived in non-standard analysis (NSA), although much earlier Kurt Gödel suggested that this view would dominate future mathematics [37]. considers infinitesimals as a special type of number contrary to the previously accepted view on them as variable parameters (see Sect. 4.3.2 and our discussion of the Continuum Hypothesis). NSA allows one to actualize ideal elements making the transition from the finite to the infinite formulated as an actual process. The basic point of the axiomatics of NSA is the idea that each mathematical object contains elements of two types. Elements of the first type are accessible to us in the sense that we can either indicate such elements directly or prove their existence and uniqueness using available objects that are already available. Objects of this type are defined as standard, while others are defined as non-standard. NSA postulates that in any infinite set of objects, at least one non-standard element exists which appears as a "principle of idealization" and forms an internal structure of the mathematical object. The numbers that are infinitely close to the given finite number represent its *monad*. The monads of different standard numbers include all finite numbers. As Brenner [5] showed however, Robinson's infinitesimals and the objects of Bell's Smooth Infinitesimal Analysis cannot be objects of the real world.

Anticipating calculus, a limit of transition can already be found in Galileo's works. For Galileo, a segment is a part cut from the infinite straight line. Calculus is based on the idea that the infinite number of infinitely small values gives finite value. This is profoundly different from the mode of thinking of the Greeks. The term actual infinity was introduced by Spinoza, but current mathematics has moved back again from the concept of actual infinity. Computers work only with finite numbers. This is their difference from conscious beings. However this may not be fully true for quantum computers.

7.5.2 Phenomenological Feasibility of Pythagoreanism

The Pythagorean view of mathematics is possible because of the property of consciousness that Husserl called *epoché* and which is achieved, in his words, via "bracketing". According to Husserl, knowledge of essences would only be possible by "bracketing" all assumptions about the existence of an external world, which means also "bracketing" the whole process of actualization. The paradox of computation cannot be resolved in another way than via "bracketing" our intrinsic ability to compute. The epistemic noumenal world of mathematics consisting of ideal entities such as numbers can exist independently in consciousness as a result of the bracketing operation. But this statement is true of course for all fictional objects. On the other hand, the Pythagorean view of number as the only real essence cannot be complete due to the nature of the process or quasi-process (see Minati [30], Chap. 16) of bracketing. Considering this procedure as really taking place, albeit requiring special analysis is essential for any kind of Pythagoreanism. This feature of the Pythagorean philosophy made Heraclitus dislike Pythagoras as someone learned but lacking understanding (Fragment 40, see also Zhmud [44]). Positioning mathematics in reality means going beyond a Pythagorean paradigm that tends to restrict the domain of its application to epistemological entities (defined in Section 9 as 'epistemons'). The Pythagorean paradigm fits the internal meaning and structure of mathematics that had emerged and developed in this framework. The first substantial move towards overcoming this paradigm was the formulation of geometry by Euclid, which placed mathematics into the observed spatiotemporal reality and determined its subsequent applications, while Euclid's fifth postulate (the parallel postulate) even narrowed the reality to its simplified view challenged later in the non-Euclidean formulations of geometry. In the fundamental scientific revolution in Europe in the 17th century, Descartes built the fundamental bridge between algebra and geometry, which later led to the discovery of infinitesimal calculus and analysis. The Cartesian coordinate system as a cornerstone of analytical geometry was a breakthrough in placing mathematics in reality and led to the formulation of classical physics.

7.6 The Operational Interpretation of Mathematics: From Marx to Twentieth Century Concepts

Karl Marx formulated his views on mathematics in the work "The Mathematical Manuscripts" which is his least known work [35]. It discusses the logical foundations of an infinitesimal differential calculus as then understood. Marx follows here the historical-genetic method as in his critique of Political Economy. He assumes the possibility of derivation of the derivative directly from the process of variation of the function, claiming its algebraic origin and contrary to consideration of differentials as individual entities with substantive content. According to Marx, differentials are as inseparable as numerator and denominator in the differential ratio. The latter is a unitary operational symbol indicating an ordered set of logical operations. These views of Marx anticipate the modern concept of the algorithm, making him a precursor of modern computational mathematics and logic.

Although Marx's contribution to the philosophical foundations of mathematics is not widely known, it is worthwhile mentioning his operational views on the subject. His concept places mathematics in reality, in contrast to the original Pythagorean paradigm. The operational nature of mathematics was developed in the twentieth Century in the logical approaches of intuitionism and constructivism. It resulted in an understanding that the power of mathematics is in the *process* of real computation which also determines the computational limits of reality itself. The operational power of mathematics is realized if and only if it refers to reality. The basic operational aspect can be analyzed using the Zermelo-Frankel Axiom of Choice, considered the most basic foundational axiom underlying the whole construction of mathematics.

7.6.1 The Operational Essence of the Axiom of Choice

The Axiom of Choice refers to a function *f* for any set A that includes in its domain every non-empty subset of A and selects (chooses) exactly one element from each subset. This axiom was first introduced by Zermelo to prove that sets can be well-ordered, and it has subsequently been used, not without controversy, to solve problems in set and number theory [41]. The relation to current logic and philosophy is a consequence of the implications of the axiom for completeness, continuity and demonstrability of mathematics in view of Gödel's incompleteness theorems [24].

Priest [34] showed that his Logic of Paradox (LP) can be used, among other things, to produce a paraconsistent set theory that provides a picture of the "set-theoretic universe". The universe contains all sets, consistent ones and inconsistent ones, i.e., those giving rise to set-theoretic paradoxes. Nothing here establishes, however, that such paradoxes could or could not be resolved at some higher level; in this interpretation, in our view, one simply oscillates from one term to the other. Priest shows that proofs in the metatheory of relevant paraconsistent logics, even if they use essentially classical reasoning, can be interpreted as establishing their results in a way that is consistent in these logics. In particular the orthodox set theory of Zermelo-Frankel can be subsumed by paraconsistent set theory. Yessenin-Volpin sketched a program for proving the consistency of Zermelo–Frankel set theory using ultrafinitistic techniques [43] which refers to the operational power of the axiom itself.

In 1951, Lupasco mentioned a project (see [26]), unfortunately never realized, for a contradictorial set theory, based on the PDO and the logic of the included middle. He did show, however, that the Axiom of Choice, which he considered a difficult problem for classical logic, was a natural illustration of such a set theory. The key insight illustrates the fundamental difference between LIR and all other classical or non-classical logics. This understanding is that each element of a set is a contradictory duality, composed of an element and its anti-element, such that the former expresses primarily an actualized identity and potentialized diversity (or non-identity) and the second the contradictory picture of diversity and identity. Since sets, as well as elements, must be accompanied by their contradictories, the LIR system described above results in the existence of three related sets.

One of these sets is the set M of Zermelo, in which the identity of the elements is actualized and the diversity of the elements is potentialized; one, of maximum contradiction (T-state) which forms sub-sets whose elements are characterized by both identity and diversity—the set P of Zermelo; and the third which is *exactly the set N*—that of choice, in which diversity is actualized and identity potentialized. This is why it can, and must, contain one element and one element only of each of the sets in P (otherwise, if more than one, it would be equivalent to a partial actualization of identity). It is the principle of non-identity, absent in classical logic, but here given its correct place in the scheme of things, which makes possible this result. Lupasco also noted the analogy with the Pauli Exclusion Principle (which partly inspired this approach), which standard logic seems to pay little attention to. The objection that the Axiom of Choice deals with infinite sets and the physical Pauli Principle with, necessarily, finite ones is not valid since the principal quantum number of a particle can be a whole number with values from one to infinity, and that it is the contradictory logical structure of the Axiom of Choice and the Pauli Principle that is the same. This view of the Axiom of Choice should indicate that the LIR could be applied to other mathematical as well as non-mathematical problems. A further philosophical interpretation of the Axiom of Choice has been provided by Badiou in his 'anti-philosophy' (see Chap. 14).

7.7 Mathematics in Reality: Overcoming the Pythagorean Paradigm in the Twentieth Century

It should by now be obvious that positioning mathematics in reality today means overcoming the Pythagorean paradigm. The problem is that this paradigm fits the internal meaning and structure of mathematics well, and it was initially formulated and developed as the area of knowledge in the frames of this paradigm. Two programs of the foundations of mathematics in the twentieth century (Gilbert's formalism and Frege's logicism) were rather Pythagorean and the third (intuitionism) was anti-Pythagorean but it did not develop sufficiently powerful framework for replacing the Pythagorean paradigm.

7.7.1 Three Programs of Foundations of Mathematics in the Early Twentieth Century

The foundational crisis of mathematics of the early twentieth Century (in the German *Grundlagenkrise der Mathematik*) was the most important event in the search for the proper foundations of mathematics. The abstraction of mathematics from reality was now seen to be clearly reflected in its internal paradoxes. In the formalist approach, David Hilbert grounded mathematics on a logical system supported by metamathematical means. This was opposed by the intuitionist school of L.E.J. Brouwer who rejected formalism as a meaningless game with symbols. Hilbert, in turn, insisted this game is not based on the arbitrary rules but develops in the internal necessity of how our thinking proceeds.

Logicism has some points in common with formalism but it is based on the more explicitly formulated thesis that mathematics is an extension of formal semantic and mathematical logic. Gottlob Frege, influenced by Richard Dedekind, initiated this theory, which was further philosophically supported by Bertrand Russell and Alfred N. Whitehead in *Principia Mathematica* [40]. Later Gödel established that logical systems of arithmetic can never contain a valid proof of their own consistency, which led him to the formulation of a set-theoretic Platonism in the foundations of mathematics.

The twentieth century alternative to the pure Pythagorean approach was intuitionism. Instead of referring mathematics to reality, it referred it to the human mind and creativity. Intuitionism, contrary to Pythagoras, considers numbers as merely mental entities which would not exist if there were never any human minds to think about them. A development of intuitionism toward referring mathematics to reality instead of pure subjective mind came with the formulation of constructivism that requires proofs to be "constructive" in nature. According to this view, the existence of an object must be demonstrated in reality. It is not sufficient to infer its existence from a demonstration of the impossibility of its non-existence.

Intuitionism referring to the human mind thus represents only one part of the constructivist program. In a broader sense, constructivism refers to the reality in which human mind is only one of the aspects. Constructive mathematics—and this is a major conclusion of our study—has the capacity of integration with the Logic in Reality of Lupasco to get a foundational substantiation of its ontological principles. Constructive mathematics initially refers to the attempts of David Hilbert, despite its formalism, to substantiate constructively a set-theoretical approach developed by Markov, Shanin and Bishop. The apparent limitations of constructive analysis, which many mathematicians associate with removing the principle of excluded middle, was compared by Hilbert to the prohibition of astronomers to use telescope. The rejection by many mathematicians of any mathematics not based on constructive methods led to the development of new approaches to its foundations. These incorporate the elements of the constructivist approach but search for more broad substantiation of the operational power of mathematics. We repeat here for reference the position of LIR which is that in rejecting the excluded middle, the intuitionist mathematicians and logicians did not go far enough in the direction of reality, since the principle of absolute non-contradiction was maintained (Chap. 3).

7.7.2 Mathematical Continuum and Mathematical Forcing

The continuum hypothesis deals with the possible sizes of infinite sets. It states that there is no set whose cardinality is strictly between that of the integers and the real numbers. Zermelo–Frankel set theory with the axiom of choice included requires that either the continuum hypothesis or its negation can be added as an axiom. The continuum hypothesis refers to the Cantor–Dedekind view of the universe founded on geometry [24], according to continuity ontologically precedes the discrete and the latter appears merely as an accident. This does not conform to the foundational principle in LIR that both continuity and discontinuity are fundamental and are dialectically related. The mathematical approach to understand the fundamental relation between continuity and discontinuity and elucidate the minimum conditions for granularity of space is presented in the dialectics of rough sets developed by Mani [27]. This approach formulates a program that helps in formalizing the concepts and notions that remain untamed at meta-levels in the usual approaches. The foundations of the theory also rely upon the axiomatic approach to granularity for all types (granular operator spaces) obtained from the developed dialectical counting procedures. Cohen [6] demonstrated that a model of set theory could be constructed via the procedure that he called forcing in which the Continuum Hypothesis fails. Forcing introduces discontinuity and provides the formal basis for the freedom of the individual and his subjective response to an event—becoming a subject. The subject 'becomes' philosophically to function with a freedom, inherent in the Axiom of Choice, that is equivalent to "pure chance". Forcing places mathematics *in reality* where discontinuity is inherent to the actual existence and represents a premise of the movement that avoids the paradoxes following from continuity. More detailed analysis of the continuum hypothesis and the axiom of choice is presented in Sect. 14.3.1.

7.7.3 Topoi and the Foundations of Mathematics in Reality

The development of mathematics after the formulation of set theory aimed at overcoming its limitations, and the invention of category theory provided an alternative foundation for mathematics. The notion of category incorporates the unity of objects and relational operators (functors, arrows) in their connection thus introducing a process of interaction between mathematical objects and their axiomatic foundations. The objects and arrows may be abstract entities of any kind, and the mathematical entities together with their relationships provide the explanation for the interactive continuum appearing through the process of actualization. Further development of category theory resulted in elaboration of the idea of topos and topoic logic. The special type of category defined as a topos possesses its own intrinsic logic. It is defined as a space with variable topology in which the inclusion of contradictory statements is possible by means of its that intrinsic logic. In other words, the fundamental logical structure of a topos possesses its own logical calculus with potentially and actually existing elements, in which a certain set of points is stable in relation to topological reconstructions [11]. This is a significant breakthrough in putting the foundations of mathematics in relation to reality.

The application of category theory and the concept of topoi became an important step in formulation of the basic principles of theoretical biology in the relational biology of Rosen [38] and in the concept of internal measurement introduced by Matsuno [28] and developed by Yukio Gunji. This is discussed in more detail in Chap. 17 on living systems. In particular, Gunji [13] stated that the Gödelian incompleteness of formal mathematical systems is incorporated in the structure of a topos: fixation of contradictory statements within a topos is possible by means of its intrinsic logic. Lawvere's fixed point theorem implies that self-definability cannot be proved logically, and this in turn implies that incompleteness is an established principle from yet another perspective other than Gödel's [13].

Living systems can be modeled in part using category theory as if one were an epistemic 'endo-observer', with its own self-referential coding, *operating* on the model in an iterative recursive process. The output of the operation then appears as the development and evolution of the system, with any solution relative to that observer [15]. This methodology clearly shows that, in contrast to the LIR ontology, the topological approach is in the epistemological domain and does not add knowledge about the physical operation of the system. The development of the concept of topoi can be viewed as grounded in Spinoza's general approach to conceptualizing the world [45]. Spinoza's intention was to prove the principles of ethics mathematically by using geometry. The most fundamental geometric relations are topological, and their use in biological and social sciences continues the conceptual paradigm of Spinoza, but it also demonstrates the disjunction with an approach based on energy (classically, force).

The categorial approach to living systems and their behavior established a relation between semiosis in cognition/communication and standard logic. It aimed at covering the complex network of biological and social interactions, as well as a model for the recursive operations necessary for the acquisition of knowledge. The mathematically grounded conceptualization of science in topos theory thus attempted to establish a methodological bridge between philosophy, logic and science and even the development of an ethical system that could clarify the grounds of social behavior. We have concluded that this approach is and could be only partly successful since it has bracketed too much of the actual dialectical aspects of living and cognitive systems.

7.7.4 *Qualitative Equations*

The concepts of topos as well as category theory nevertheless bring out the qualitative relations in the foundations of mathematics and the role of topological relations in relation to energetic ones as the most fundamental. A development in a similar direction is the study of the domain of qualitative differential equations. Kenneth Forbus has suggested a qualitative process theory with real-world applications, with clear references to LIR-type concepts [10]. He emphasizes that the changes in objects over time that we characterize as processes can be described via qualitative reasoning. In the framework of this approach, a new qualitative representation for quantity in terms of inequalities, called the quantity space, was introduced. The basic concepts of qualitative process theory include several different kinds of reasoning that can be performed with them, but they have received little academic attention.

7.8 Using Univalent Foundations of Mathematics to Place Mathematics in Reality

Following the development of the ideas of mathematical foundations in the twentieth century and incorporating the principles that underline the computational process, a new approach to foundations of mathematics was introduced by Voevodsky (1966–2017). In his concept of univalent foundations, all mathematical structures are built out of objects called *types*. Each entity (type) can be considered as a mathematical point, the points create topological spaces in which no definitions of distances available but paths, the process from one point to a next point is path induction (homotopy). Voevodsky's type theory is a deductive system with suitable rules of inference and its goal is automatization of mathematical proofs. In his paradigm, foundations of mathematics are computationally testable (for details, see [12, 42]).

Formal mathematical language, together with an axiom of univalence, fulfills the mathematicians' dream: a language for mathematics invariant under equivalence and thus freed from irrelevant details, capable of merging the results of mathematicians taking different but equivalent approaches. Voevodsky called this invariance property of the language univalence. Following Voevodsky, the unity of cognition and spatiotemporality needs to be incorporated not only in philosophical thought but also into the foundations of mathematics, into which Voevodsky also included geometry. The intrinsic logic of metamathematics generates its spatiotemporal structure. When geometry is introduced to the foundations of mathematics, the world can be shaped theoretically in a way that fits its inhabitability, which resembles the anthropic principle in physics. The limits of geometry are associated with the limits of the computability of the particular world, and in the theory of homotopic types developed by Voevodsky, the basic foundations of mathematics can be verified computationally [12].

Voevodsky's types in univalent foundations do not correspond exactly to anything in the foundations of set-theory, but they may be thought of as spaces, with equal types corresponding to homotopy equivalent spaces, and with equal elements of a type corresponding to points of a space connected by a path. Univalent concepts were inspired both by the Platonic ideas of Hermann Grassmann and Georg Cantor, and by "categorical" mathematics in the style of Alexander Grothendieck. Univalent foundations depart from the use of classical predicate logic as the underlying formal deduction system, replacing it with a version of Martin-Löf type theory. Homotopy type theory thus represents a powerful operational interpretation of pure mathematics, but of pure mathematics only. Its dependence on an abstract, non-contradictorial logic disqualifies it as a mathematics in reality. To paraphrase the term in philosophy, it is mathematics *tout court.*

7.9 The Problem of the Origin of Computation

Up to this point, the main paradox in our understanding of mathematics and its place in reality is to describe the origin of computation by computational means. It may be considered as an expression of the basic paradox of our relation to the external world which generates the dynamic opposition of computable and non-computable aspects of knowledge.

The universality of mathematical laws can be explained by the fact that the same limits of iteration are formed in different generally non-equivalent systems having internal constraints [16]. Any internal choice exhibits a structure resulting from mixing the notion of indicating an element with the act of indicating a set consisting of elements [14]. Closed loops of causation in them appear as objective entities which can be mapped into a single coherent space–time frame [39]. Coordinate systems are established in the sets of interacting ontolons (Chap. 5), and the fundamental parameters may correspond to a consistency of the spatiotemporal landscape formed through communication [16]. A physically embodied recursive loop has parameters that confer on it an objective existence. These parameters include values that can be redundantly repeated in all loops that co-exist, corresponding to a topological inevitability of actualization processes. We observe ourselves as being embedded in the whole world simulated by our internal computation, internal and external to it at the same time. Formal logic and mathematics thus appear as a consequence of a self-identification process, but the description of operational activity in the external world requires the additional logical-physical principles of LIR, as discussed in the next Section.

The book of Brabazon et al. [4] presents a comprehensive and widespread analysis of natural computation that becomes the basis for the algorithms developed by humans. The whole framework of this book is based on an approach similar to the LIR-based understanding of natural computation as derived from reality and serving as a basis for the interaction with reality. The incompatibilities appearing in the course of derivation of computation from natural processes can be referred to the requirement of factorization of numbers for effective computing and the theoretical unavailability of a distribution law for factorial properties, which is equivalent to its non-computability and undecidability. This has far-reaching consequences for the arguments on inviolable general consistency of mathematics. In this regard, in mathematics this could be considered as the end of the formalist approach usually associated with the Bourbaki program [3] and relying on defining to reach a completely self-contained treatment of the entire mathematics. Furthermore, this generates the problems related to incompatibilities between quantum and relativistic theories in the foundations of physics.

7.10 Logic in Reality and the Foundations of Mathematics

7.10.1 The Complex Plane: Initial Considerations

One of the reasons why Logic in Reality has not been accepted as a framework for the foundations of mathematics is the difficulty in visualizing, feeling and comprehending the principles of the dynamics of reality as well as the dynamics of the principles on which reality is established. This difficulty is particularly significant for the interpretation of mathematics because of the explanatory problems associated with the appearance of computational principles. Here we will discuss the interpretation of the complex plane based on Logic in Reality. The relation of complex numbers to the duality of potential and actual realities was assumed in earlier philosophical systems, in particular, by the Russian religious philosopher Florensky (1882–1937)[1] in his book *Imaginaries in Geometry* [9]. For recent developments of understanding of the relation of complex numbers to reality see the book of Nahin [31].

For people trained in classical logic, it is difficult to see implication itself as a process; implication is a simple connective like and, or, and equality. Material implication, or just implication is another name for the material conditional if > then, both applying only to propositional logical proofs and rules of logical inference or their mathematical equivalents. (In Boolean algebra, True and False are assigned the values of 0 and 1.) As shown by Lupasco, however, it is useful to look at implication, as well as the other logical operators as a process as processes themselves capable of degrees of actualization and potentialization, and being the source of emergence.

We begin our alternative discussion by considering the geometric representation of complex numbers as points on a complex plane or in a space of higher dimensions, given that there are other acceptations of the term:

1. A two-dimensional complex vector space.

[1]Florensky was a victim of the communist totalitarian regime. In our pantheon, he joins Benjamin Fondane (Chapter 14), who was a martyr to fascism.

2. A (1 + 1) Minkowski space, also known as the split-complex plane associated with algebraic split-complex numbers that can be separated into two real components that are associated with points in the Cartesian plane.
3. The set of dual numbers over the real numbers which can also be placed in a one-to-one correspondence with the points of the Cartesian plane.

In our opinion, definitions 2 and 3 refer to formal aspects of number theory. Definition 1, however, is part of our preferred conception of the complex plane or solid as describing the properties of real processes which a priori have direction.

We do not, however, wish to push our concept further in the direction of standard topology and mathematics, since one runs immediately into requirements (e.g., for coherence) which are incompatible with the dynamic vision we are proposing. The process view of the dynamics of a complex valued reality that we propose here has an additional feature: it has proven difficult to visualize the dynamic structure of the complex real processes of mind and society. Our picture of this structure is one of the changing, reciprocally defined degree of actuality and potentiality of the opposing elements that constitute it. Although defined by energetic processes, they appear to have a different character from simple processes involving the transfer of energy. For the first time, we propose here that the potential and actual parts or aspects of a process can be represented by the imaginary and real values of a coordinate in the complex plane. Imaginary of course does not mean unreal, but corresponds to a real number times the square root of -1 or i. If **a** is the real axis of the corresponding graph and **b** the imaginary axis, the coordinate is described as (a, b*i*) Many processes, which are 'physical' in the standard way, such as electronics, sound, and geophysics, for example, have an 'imaginary' component in the equations defining them.

The basic postulate of the Lupasco logic is also an implication, but it does not apply to propositions, but to the physical elements of real, dynamic processes. In LIR, the predominant actuality of any such element *e*, logical or physical, implies the predominant potentiality of its opposite, non-e, alternately and reciprocally, without either ever going to the absolute limit of 100% one or the other. We propose that what we might call the dynamic coexistence relation of *e* and non-*e* can be represented by assigning two values to *e*, the real number a to its degree of actuality or presence and the imaginary number ***bi***, that is, **b** times the square root of -1, to its degree of potentiality or absence. We do the same for non-*e*, obtaining ***ai*** and **b**. We can then plot these values on a graph with two axes, one of which is for the real component and one for the imaginary component. The evolution or change in the presence, the real, actual state of the process moving from **a** to **b** on the real axis and the corresponding, linked 'imaginary' state of the process moving from ***bi*** to ***ai***.

As noted, the evolution of a real interactive process involves the (predominantly) actualized component becoming (predominantly) potentialized. Since the value of actuality is the reciprocal of the value of potentiality, the new value resulting from evolution of a process can be represented as (***ai***, **b**). In the simplest case, the line between the two points has a slope of 45°. The values of **a** and **b** can vary between 0 and 1, corresponding to different values of interaction between the elements, and to a required property of non-standard probabilities. However, there are many processes in which the degree of interaction between the real elements is zero to all intents and purposes. Without going into detail here, the limiting case correspond to standard processes without an imaginary component, for example, a standard pendulum.

7.10.2 Initial Complexification of Our Representation

The values of the actuality and potentiality of opposing or contradictional process elements have been defined, in Logic in Reality, as never going to zero. Accordingly, our graph of the evolution of a process in two dimensions will not pass through the origin, the point [0,0], nor intersect either of the two axes. Only the upper-right quadrant will be occupied. On the other hand, many physical processes can and in fact have been described by equations to the solutions to which can take on zero and negative as well as imaginary values. If real numbers are assigned to an x axis and the imaginaries to a y axis, the upper-left quadrant defines the negative real and positive imaginaries, the lower-left quadrant the negative real and negative imaginaries, and the lower-right quadrant positive real and negative imaginary values.

We conclude that a single graphical representation of both kinds of processes is possible by moving to three dimensions, adding a z axis for the imaginaries of LIR kind. Here, we have two intersecting complex planes. Values are defined by a triple [a, b*i*, c*i*], as noted. For all classical physical processes which can take on zero or negative as well as standard imaginary values, c is always zero. However, all quadrants in the x, y plane can be occupied. For complex processes following the LIR logic, b is always zero and the sole quadrant occupied is the upper-right in the x, z plane.

7.10.3 Second Complexification. Identity and Diversity

It was part of the basic insight of Lupasco to have seen the isomorphism between processes at the inorganic macroscopic, biological and cognitive levels of reality. The methodology of the applicable logic involves identifying the logical elements of the processes which are in fact opposing one another and from which a third element may emerge at a higher level. One example is the residual positive or negative charge on an amino acid that determines its possibilities for non-covalent bonding and the secondary structure of proteins which include it. The question is now whether the graphical representation of the degrees of actuality and potentiality of those elements, and their evolution of those degrees toward an opposite one, adds to our analysis of complex processes. Once the concept is understood, our answer would be no unless the further parameter at the core of the Lupasco logic, Identity and Diversity, is taken into account.

It is therefore necessary to add an additional two real variables to our picture which we represent by the additional real values d and e, plotted on a new axis w.

This converts the x, z complex plane to a three-dimensional graph. In words, we state that any point in the 3D graph [a, c*i*, d, e] is characterized, in addition to its degree of actuality and potentiality, by a similar reciprocal degree of identity and diversity which evolves concurrently but independently of the first property. The trajectory of a real process, involving the movements between a and ci, and between d and e, will be a two-dimensional surface, which may be plane or curved, in a three-dimensional space. The function of the graph is to facilitate visualization of a process in which the evolution of the actuality and potentiality of identity and diversity cannot be separated from that of the process as a whole. Eventually, an animation would be required to capture the dual evolution in time.

We started with an attempt to describe information in the complex 2-D plane. Does our '3-D' picture, that is, the representation of a movement between identity and diversity apply to information? It certainly would seem to apply to theories of information, whose proponents seek to achieve some degree of unification or are content with an apparent unending series of different theories. We conclude that at each stage of complexification of our graphical picture, the real nature of change in general becomes more apparent. We would call our representations 'living graphs' but the trope has been overworked. We now will move backwards (or forwards?) toward the original graph. It has become, with the four axes x, y, z and w, a four-dimensional hyperspace, even if only 2-D spaces or 3-D volumes of it are 'active' at any time.

7.11 Concluding Remarks

In conclusion, mathematics, as formulated by Pythagoras and 'bracketed' since then from perceived reality, has fundamental theoretical gaps in its construction. These flaws are substantial; they cannot be overcome within the construction and require referral to a reality that is external to the epistemic reality of mathematics itself. The principles of constructive mathematics were one attempt to improve the foundations of mathematics. In our view, however, further structural substantiations are required to bridge the gap *with reality*. These were formulated explicitly in the program of the univalent foundations of mathematics of Vladimir Voevodsky. The Logic in Reality of Lupasco and Brenner can be seen as a true framework for the foundations of a mathematics 'in reality': mathematics can now be viewed operationally as included in the dynamic processes of reality. It is possible to incorporate the principles first explored in constructivist mathematics and in its univalent foundations in a new fundamental concept based on Logic in Reality. This provides a new understanding of the operational nature of mathematics and substantiates its explanatory power. In this framework, the basic foundations of mathematics can be reconsidered with a reference to their operation in the external world. This resolves some of the problems associated with the application of Logic in Reality, bringing it closer to a synthesis which includes mathematics.

References

1. De Bianchi, S., Wells, J.D.: Explanation and the dimensionality of space. Synthese **192**(1), 287–303 (2015)
2. Born, M.: Natural Philosophy of Cause and Chance. Oxford University Press, Oxford (1949)
3. Bourbaki, N.: The architecture of mathematics. Am. Math. Monthly **57**(4), 221–232 (1950)
4. Brabazon, A., O'Neill, M., McGarraghy, S.: Natural Computing Algorithms. Springer, New York (2015)
5. Brenner, J.E.: Logic in Reality. Springer, Dordrecht (2008)
6. Cohen, P.J.: The independence of the continuum hypothesis. Proc. Natl. Acad. Sci. U.S.A. **50**, 1143–1148 (1963)
7. Einstein, A.: Geometrie und Erfahrung. Springer, Berlin (1921)
8. Einstein, A.: Lettres à Maurice Solovine. Gauthier-Villars, Paris (1956)
9. Florensky, P.A.: Imaginaries in Geometry. Mimesis International, Milan, Italy (Originally published 1922) (2021)
10. Forbus, K.D.: Qualitative process theory. Artif. Intell. **24**, 85–168 (1984)
11. Goldblatt, R.: Topoi: The Categorial Analysis of Logic. North-Holland, Amsterdam (1979)
12. Grayson, D.R.: An introduction to univalent foundations for mathematicians. Bull. Am. Math. Soc. (New Ser.) **55**(4), 427–450 (2018)
13. Gunji, Y.P.: Global logic resulting from disequilibration process. BioSystems **35**, 33–62 (1995)
14. Igamberdiev, A.U.: Quantum computation, non-demolition measurements, and reflexive control in living systems. BioSystems **77**, 47–56 (2004)
15. Igamberdiev, A.U.: Physical limits of computation and emergence of life. BioSystems **90**, 340–349 (2007)
16. Igamberdiev, A.U.: Objective patterns in the evolving network of non-equivalent observers. BioSystems **92**, 120–131 (2008)
17. Igamberdiev, A.U.: Evolutionary transition from biological to social systems via generation of reflexive models of externality. Prog. Biophys. Mol. Biol. **131**, 336–347 (2017)
18. Infeld, L.: Quest: An Autobiography. AMS Chelsea Publishing, New York (Originally published 1949) (2006)
19. Kant, I.: Critique of Pure Reason, translated by P. Guyer and A.W. Wood. Cambridge University Press, Cambridge (Originally published 1781) (1998)
20. Kant, I.: Thoughts on the True Estimation of Living Forces, ed. by E. Watkins. Cambridge University Press, Cambridge (Originally published 1747) (2012).
21. Lefebvre, V.A. 1982.: Algebra of Conscience. A Comparative Analysis of Western and Soviet Ethical Systems. Springer, Dordrecht (2001)
22. Liu, Y. 2020.: NP (Non-deterministic Problem) and Gödel's Incompleteness Theorems. Published on-line January 20, 2020, https://nptheory.blogspot.com/2020/01/npnondeterminis tic-problemand-godels.html
23. Logan, R.K.: The importance of nothing. Zero was the gift from the East that made the scientific revolution in the West possible. Prog. Biophys. Mol. Biol. **131**, 500–503 (2017)
24. Longo, G.: The mathematical continuum. In: Petitot, J., et al. (eds.) Naturalizing Phenomenology. Issues in Contemporary Phenomenology and Cognitive Science. Stanford University Press, Stanford (1999)
25. Lupasco, S.: Logique et Contradiction. Presses Universitaires de France, Paris (1947)
26. Lupasco, S. 1973.: Le principe d'antagonisme et la logique de l'énergie. Editions du Rocher, Paris (1987)
27. Mani, A.: Dialectics of counting and measures of rough set theory. In: IEEE Proceedings of NCESCT 2011, Pune, 1–3 Feb., p. 17 (2011)
28. Matsuno, K.: Quantum and biological computation. BioSystems **35**, 209–212 (1995)
29. Mellaart, J.: Neolithic of the Near East. MacMillan, London (1976)
30. Minati, G.: A conceptual proposal on the undecidability of the distribution law of prime numbers and theoretical consequences. Ratio Math. **37**, 69–84 (2019)

31. Nahin, P.J.: An Imaginary Tale: The Story of "i" (the square root of minus one), p. 12. Princeton University Press, Princeton (2010)
32. Nielsen, M.A., Chuang, I.L.: Quantum Computation and Quantum Information: 10th Anniversary Edition. Cambridge University Press, Cambridge (2011)
33. Oizerman, T.I.: Kant's doctrine of the "things in themselves" and noumena. Res. **41**, 333–350 (1981)
34. Priest, G.: Paraconsistent belief revision. Theoria **67**(3), 214–228 (2001)
35. Ricci, A.: The mathematics of Marx. In the bicentenary of the birth of Karl Marx (1818–1883). Lettera Matematica **6**, 221–225 (2018)
36. Riemann, B. 1868.: Collected Papers. Kendrick Press, Heber City (2004)
37. Robinson, A.: Non-standard analysis. Indagationes Math. **23**, 432–440 (1961)
38. Rosen, R.: Life Itself: A Comprehensive Inquiry into the Nature, Origin, and Fabrication of Life. Columbia University Press, New York (1991)
39. Rosen, R.: Drawing the boundary between subject and object—comments on the mind-brain problem. Theoret. Med. **14**, 89–100 (1993)
40. Russell, B., Whitehead, A.N.: Principia Mathematica, vol. 1–3. Cambridge University Press, Cambridge (1910–1913)
41. Suppes, P.: Axiomatic Set Theory. Dover Publications, Mineola NY (1972)
42. Voevodsky, V.: A1-homotopy theory. In: Proceedings of the International Congress of Mathematicians 1, 579–604 (1998)
43. Yessenin-Volpin, A.S.: About infinity, finiteness and finitization (in connection with the foundations of mathematics). Lect. Notes Math. **873**, 274–313 (1981)
44. Zhmud, L.: Heraclitus on Pythagoras. In: Fantino, E. (ed.) Heraklit im Kontext, pp. 173–187. De Gruyter, Berlin (2016)
45. Zimmermann, R.E.: New Ethics Proved in Geometrical Order: Spinozist Reflections on Evolutionary Systems. Emergent Publications, Phoenix (2010)

Chapter 8
Chemistry: Modeling Activation and Transition

8.1 The Included Third—T-state

Let us first recall the most important axiomatic consequence of the LIR categorial ontology of energy. It is the existence of a point of maximum contradiction or interaction (T-state) in the movement of processes from actual to potential or vice versa, when each is semi-actualized and semi-potentialized to the same extent.

A suggestion of the T-states emerging from contradictions or counter-actions in reality is indicated in Table 8.1, which places chemistry in the context of other entities.

LIR sees as dynamic processes both 'properties' (or the event or states consisting in a system's having a property), and systems and objects as such, seen as emergent T-states arising from dialetheias, real-world contradictions or counteractions. Note that the discussion of contradictions by the para consistent logician Graham Priest (Priest 2002) includes only limited, essentially binary examples of real processes (Sorites, being inside or outside at room when entering or leaving one). In LIR, emergence as a process is not separable or different from its instantiations. It is no more correct to say that emergence 'is' something than that cause or consciousness 'is' something. The only criterion for emergence is the existence of real entities or processes, those which consist of energy-in-change. Emergence does not take place in or between the non-spatio-temporal, epistemological entities described by binary logic.

In LIR, the categorial features of phenomena involve configurational forces, in which significant energy is encoded in potential form. The actual and potential states of quantum particles are present in configuration space, and these potentialities as well as actualities are the carriers of the upward causation necessary for emergence of entities at higher levels. Let us see in more detail how these processes play out at the level of reality of chemicals as real substances.

J. E. Brenner and A. U. Igamberdiev, *Philosophy in Reality*, Studies in Applied Philosophy, Epistemology and Rational Ethics 60, https://doi.org/10.1007/978-3-030-62757-7_8

Table 8.1 Levels of Reality and Emergence

Level of reality	Emergent process/entity
Quantum	Gluon, space–time
Chemical	Transition state or complex
Biological	Life, evolution
Cognitive	Consciousness, reflection
Artistic Creation	Work of art
Ethical	Solidarity, compassion
Sociological	Sustainable development
Cosmological	Quantum vacuum
Philosophical	Ending the infinite regress

8.2 Chemistry or Chemical System?

The locution 'chemical system' is not used routinely, as are physical system and biological system. In some hierarchy of fundamentality, chemistry as a discipline is considered as falling 'between' physics and biology, grounded in the former and grounding the latter. From the point of view of complexity, chemical entities in our view occupy a minimum position. They are less complex than their particle-field constituents, that occupy the threshold to the quantum world, but also less complex than the simplest biological entity built up from macromolecules. The laws governing the behavior of chemical entities are more accessible than those of physics and biology.

We may ask, then, for a definition of the boundaries between the disciplines: between chemistry and biology, it is usually placed at the as yet unknown 'point' or better region of emergence of living systems. We will return to this subject in Chap. 15. Between physics and chemistry, the boundary may be identified with the cosmological emergence of the hydrogen atom: individual quantum particles, including quarks, electrons (leptons) and photons, belong to the domain of physics. Chemistry for us begins, together with thermodynamics, with the hydrogen atom constituted by one proton and one electron. This is not only the simplest entity that cannot be described as a single field; it is also the simplest chemical system. An atom of hydrogen clearly fits the definition of an entity from which a part, say the electron, cannot be removed without destroying the system, leaving behind a bare proton. A hydrogen atom, as a substance is also the simplest physical entity displaying a *potential*. The oxidation potential measures the ease or difficulty of removal of the, in this case, single electron from the atom (ionization). The simplest molecule is constituted by two atoms of hydrogen, H_2.

This chapter will review, from the standpoint of the logical and dialectical principles established above, some of the essential current concepts we believe should be included in the Philosophy of Chemistry. We will show that certain concepts within chemistry have a purport for the understanding of Philosophy in Reality in general.

8.3 The Philosophy of Chemistry. Hylé

The purpose of a chapter about the Philosophy of Chemistry in this book is to call attention to the scientific continuity of chemistry with both physics and biology and the philosophical issues which arise from this intermediate position. The discussion of living systems and human beings capable of high-level cognitive functions generally accepts, without comment, their underlying physical structure composed of atoms and both simple and complex molecules, constituted by atoms in various arrangements. Some of the major questions asked in the Philosophy of Chemistry have been about the reality of chemical elements, the nature of chemical bonds and the nature and function of molecules.

Among the philosophical questions studied we note those of the reduction of chemical substances to molecular species and molecular species to quantum mechanics, that is, physics. The authoritative article on the Philosophy of Chemistry in the *Stanford Encyclopaedia of Philosophy* [13] has a Section on the forms of reasoning involved in the definition of chemical reaction mechanisms—the chemical reaction is the basic process entity. Reference is made to the potential energy or free energy inhering in a chemical system as a guide to the evolution of a reaction from reactants to products. This philosophical literature includes a (for us a rather unreadable) discussion about microessentialism, whether water is or is not "H_2O". This means that a molecule of the substance water can be designated symbolically and statically as composed of two hydrogen atoms and one oxygen atom. Of course, as soon as one starts to talk about real processes involving *change*, for example the interaction of water with sodium metal, it very quickly becomes impossible to refer to water and sodium, H_2O and Na, and necessary to refer to sodium hydroxide, NaOH. It is not even symbolically correct to refer to 'water' by the symbol for a single molecule, since it is fact a dynamic three-dimensional network of H_2O units connected by hydrogen bonds of one molecule to the oxygen of another.

Startlingly, no reference is made to the *form* in which energy is present in a chemical system, that of potential energy. This is a concept with which every student of chemistry is familiar: one talks about the different oxidation and reduction *potential* of the atoms that determines the course of any subsequent reaction, a potential that varies with the environment, in particular, existing chemical bonds or weaker electrostatic interactions, for example with solvent molecules, that can raise or lower these potentials. Neglect of these questions is based, in our view, on the absence of a process view of chemical systems, whose the constitutive entities are considered as a set of identities.

For readers who may not be aware of it, we call their attention to *HYLE: The International Journal for Philosophy of Chemistry* [2]. Its articles deal with "epistemological, methodological, foundational and ontological problems of chemistry and its subfields; the peculiarities of chemistry and relations to technology, other scientific and non-scientific fields; aesthetical, ethical and environmental matters in chemistry, as well as philosophically relevant facets of the history, sociology, linguistics and education of chemistry, as well as ethical issues. This describes an impressive, valuable program and distinguished philosophers as well as chemists serve on its Editorial Board. Our problem with this *Journal*, however, starts with its name! *HYLE* is the classical Greek word for matter or *substance, hylé*. Both authors of this book have direct experience with chemistry, Brenner with Organic Chemistry and Igamberdiev with Biological Chemistry, and the idea of a chemical substance is a perfectly familiar one. For the philosophical objective of this book, however, we cannot allow ourselves to be limited by any implied substance ontology. The subject of Models in Chemistry is a widely studied sub-field, but there is a strong tendency, here as in other fields, to occult the dynamics of chemistry as not philosophically significant. In particular, what are not considered are the key principles of chemistry that in our view are models for other processes! The most important and to us obvious one is that of the transition state or transition complex, discussed in the next Section.

8.4 The Transition State

In his *Logic in Reality* [1] Brenner described the Principle of Dynamic Opposition as a principle of antagonism in energy. We recall that a quantity of energy A can only be prevented from moving in one direction or an other by an opposing quantity of energy, non-A or anti-A, such that the actualization of non-A implies the potentialization of A, and vice versa. At some point, for all phenomena, there will be a point of equilibrium between the two tendencies, 'on the way from one to the other', which can be considered as the third element existing simultaneously with the other two (T-state, from tiers inclus or included 'third' element (cf. Chap. 3).

The transition state, better the activated transition complex in a chemical reaction is a paradigmatic example of a T-state. It is the point at which the 'activation energy', the energetic barrier to a reaction taking place between two molecules, the reactants, has been overcome by an input of energy—heat, light, etc. —from the environment, and the thermodynamically favoured products of the reaction can form. Consistent with the principles of LIR, thermodynamically less favoured products can also be formed, but to a much smaller extent, unless further energy is made available in some form, for example by the use of a catalyst with a unique molecular or surface structure, capable of placing two reactant molecules closer or in a more favorable configuration.

This description needs further elaboration. The transition state is a 'point' on an energy curve, but other entities are involved in the reaction. The first *change* that takes place is the conversion of the reactants into an activated complex which is in a state of equilibrium (quasi-equilibrium) with the reactants. A transmission coefficient measures the probability that an activated complex forms a particular set of products rather than reverting to reactants or forming alternate products. With its additional degree of freedom, the activated complex is a much better model for the situation at higher levels of reality, especially, interactions at the interpersonal and social levels.

It is a curious fact, perhaps a significant reflection of current philosophical thought, that the transition state is mentioned only once, in passing, in a quite recent version of the Philosophy of Chemistry entry in the otherwise authoritative *Stanford Encyclopaedia of Philosophy.* Hooker, on the other hand, cf. Chap. 18 on systems, points out that most chemical reactions are in practice irreversible thermodynamic processes for which an analytical dynamic characterization can accordingly not be made. The whole of chemistry is from this perspective a (the) first 'special science' of complex systems beyond physics.

8.5 Toward a New Chemistry

To support the application of his logic to living systems, as early as 1962, Lupasco pointed, to the necessity for a 'new chemistry' as well as a new physics [11]. The postulate of the real, logical and dynamic opposition at the heart of energy, and consequently of its embodiment in matter and information at all more complex levels of reality, requires that it applies to atoms and inorganic and organic molecules and ultimately the living organisms that are constituted by them.

The principles of LIR applied to the philosophy of chemistry could be stated in the same terms as many of Wimsatt's heuristics for fundamental problems in philosophy and science [14]:

- Look for robust tendencies (e.g., toward identity or diversity), and for conditions under which those tendencies are likely to be realized, rather than for absolute positions.
- Study context-sensitive inferences rather than ones that are context-free, along the lines of Aerts' analysis of non-classical contextuality, in which both system and perturbation have an internal relational structure [4].
- Rather than looking for universal theories or principles which are foundational to all the elements of a given domain, look for the conjoint application of such robust principles which may be heterogeneous in application, but complement each other to give a better fit to the details of the situation.

For our Philosophy in Reality (PIR) we seek no more (and no less) than the status of such a robust principle for the Principle of Dynamic Opposition, exemplified in chemistry and elsewhere.

Reductionism, understood as a metaphysical doctrine that denies or discredits the explanatory and/or causal power of emergent higher level entities or phenomena still has its advocates. We will not enter into this further area of polemics here. We say simply, with the chemist Hoffmann (see Sect. 8.5 below) [7] that vertical understanding, corresponding to classical reductionism and horizontal understanding are involved in any scientific or philosophical undertaking, and human beings combine the two modes of explanation. The process is a typically dialectic one and the rules of alternate actualization and potentialization of LIR apply.

8.5.1 Entropy and Negentropy

The second most fundamental aspect of energy is expressed by the dialectic between its entropic and negentropic properties. Energy moves from diverse, heterogeneous high-level forms toward a single, homogeneous low-level form (heat), governed by the 2nd Law of Thermodynamics. Let us designate such a tendency toward a single undifferentiated state of affairs as exemplifying identity or, simply, identifying or 'homogenizing'—an 'identification'. At the same time, energy, as apparently indistinguishable electrons, shows a fundamental duality governed by the Pauli Principle of Exclusion. Electrons are located in shells around the nucleus of an atom, but two electrons in the same shell cannot have the same quantum numbers for their properties, such as spin. Build-up of a multiplicity of shells is possible, for atoms heavier than helium, in which the electrons will all have, as a consequence of their distance from the nucleus and the degree of completion of the shells, a different capacity (potential) for reacting with other atoms to form different molecules enabling the existence of, ultimately, life and human beings. (A similar phenomenon, due to Pauli exclusion at the nuclear level, results in the existence of nuclear spin isomers [8]. The diversity is real and leads to differences in some physical properties as in, for example, molecular *ortho-* and *para-*hydrogen. These species have been separated and characterized, but the differences in chemical properties are negligible.

Energy and accordingly all existence thus also show a tendency toward or instantiate an opposing process of heterogeneity, or non-identity or diversity, a 'heterogenizing' process, a diversification. Combining this idea with the one in the previous paragraph, we suggest that homogeneity, exteriority and objectivity characterize the process of extensity, and heterogeneity, interiority and subjectivity that of intensity, (cf. Chap. 3).

In any phenomenon, one should always look at the respective tendencies toward homogeneity and heterogeneity in its identifying and diversifying aspects, in order to understand the structure, orientation and the laws applicable to the entities instantiating them. The details of all of these will depend on the level of reality at which these processes are taking place. Further, the 'coefficients' of homogenization and heterogenization define a relation of contradiction or opposition since they imply the coexistence, in the energetic constituents of the phenomenon, of identity and non-identity. A cell (an identity) that has lost its capacity (potential of non-identity) for maintaining its differentiation from other cells is either dead or cancerous, and our thesis is that these can and should be considered *logical* identity and non-identity.

The terms identity or identifying are to be understood as subsuming the notions of rationality, reality in a limited, classical sense, invariance and classical logic and tautology. The opposite process of diversity or diversification refers to notions of irrationality, variance, inconsistency, *non-sequiturs* and negation in general. The essential point is not only that these two groups of processes are connected physically and dialectically, but also that they have equivalent ontological status. The exclusions, negative implications, etc., the 'non-rationalizable' elements of reality that instantiate diversity constitute an integral part of the total logical structure of reality, independently of anything arbitrary or subjective. When the two processes are of equal strength, in a symmetric dynamic equilibrium, one is in the presence of a 'contradictorial' process, expressed most clearly in human mental and social phenomena. The formal structure of these processes has the form of chains of chains of implications, the ortho-deductions, indicated in Chap. 3.

Everything more complex than a single proton or electron that involves this principle is a system, a process of systems of systems, etc. Everything physical is also a chemical system, capable of interacting with other systems in a manner that one can call antagonist. In chemistry, the calcium ion, Ca^{++} is a biologically 'active' system as are the toxic thallium ion, Tl^{+} and the carbon atom with its capacity (potential) for four covalent bonds to other atoms. It is these that make possible complex molecules, an amino acid, a polypeptide, a hormone, a gene, an egg and a human being. The proposed consequence of LIR for biology and philosophy is that its principles are universal in the sense of applying to chemical elements, inorganic and organic materials, macro- molecules, their dynamics, the memory they embody in their three-dimensional structures obtained via folding and to all other constraints that enable self-replication of living systems.

Some early proponents of emergence believed that primitive features of matter could exert a primitive form of causality, involving fundamental 'configurational forces'. LIR gives meaning to this thesis: the 'features' of phenomena, starting with energy, can be described as involving 'configurational forces', in which significant energy is encoded in potential form. It is in configuration space that the bound actual and potential states of electrons are present, and it is both these categorial features that are the carriers of the upward causation necessary for emergence. To take the example of the calcium ion, again, the combination of its size and net positive charge results in different potentialities for interactions with, say, water molecules than that of a lithium ion, Li^{+}, and their biological activity, partly as a consequence of this, is quite different. For example, at the psycho-physical level, the 'simple' lithium ion is highly psychologically active.

We note that the reduction of chemistry to physics is no longer an issue. Every physical entity is a system, unsaturated in its potential for further interactions, the more complex chemical systems that emerge from those interactions will retain part of that unsaturation as higher-level causal properties that we designate as the residual potentialities of the system. These consist, again, in the ability to lose an electron 'completely', to form an ionic bond, say sodium to sodium chloride; to share electrons in a covalent bond, as in the unsaturated ethylene molecule; or to form electrostatic bonds such as those between water molecules and sodium chloride ions. The greater thermodynamic stability of the hydrated versus the unhydrated ions is the thermodynamic basis for the solubility of salt in water. The potential of unsaturated organic molecules like ethylene enables them to become saturated by the addition of two further atoms.

8.6 The Chemistry of Life

While the details of the initial production of biological macromolecules at the origin of life remain unknown, the concept of opposition or antagonism provides a further entry point for analysis of models of these processes. As Lupasco pointed out in [11], the question of whether or not biology and life can be reduced to physics and chemistry is badly posed. The problems of trying to explain the existence of any change, but particularly of the emergence of biological systems in terms of physics and chemistry, were and are still due to the retention of classical notions of cause, time and matter-energy solely in terms of actualities. This is particularly important in regard to living systems, as we will now show.

The Pauli Exclusion Principle for electrons establishes a basis for heterogeneity, in real as well as epistemological opposition to the 2nd Law of Thermodynamics, which is a basis for homogeneity. This principle of heterogeneity, or tendency toward heterogeneity, is what is considered, in LIR as the basis for the existence of increasingly complex macro-physical systems, leading ultimately to those designated as living systems at a biological level of reality. The processes leading, in some as yet undefined way, to entities and their constituents at the biological level—the genome, gametes, other cells, organs, and living individuals—all involve the emergence of new forms, which I have tentatively identified as T-states, included middle elements at an- other level of reality or complexity. Their origin in turn is in properties of molecular and chemical substrates (under-levels) and processes that are less complex, and we have postulated that all processes, at all levels, are characterized by more or less easily identifiable aspects of dynamic opposition, instantiated in those properties. An overly simplistic model of natural selection as a consequence of the 'pressure' of the environment on the evolution of a species is an example of such opposition.

We have discussed elsewhere the on-going attempts to decide whether geometry (position, form) or matter (energy) is more fundamental in the universe. The question is debated today in the terms of information and quantum physics and cosmology. Diaz Nafria and Zimmermann, proposed that information, to which a non-energetic description is often given, Diaz Nafria and Zimmermann [3] suggested that both matter-energy and information are two different, associated aspects of the same underlying and still unknown primordial structure of the world. The best picture is that they emerge together from this substrate, and that the question of the priority of position and energy does not require choosing between them.

Following our own philosophy that a theory, to be valid, should refer to its opposite, we call the reader's attention to a recent geometric proposal by Yurkin [15, 16] of the basis of the Pauli Principle and its expression in existence of the different elements. Yurkin and his colleagues have provided a novel geometrical model of the structure of atomic electron shells, able to assess the main quantum numbers in terms of simple wave trajectories, angles and distances. The geometrical steps necessary to build atomic shells and sub-shells in the form of such trajectories in terms of angles and energy levels are described. Possible applications are the description of cellular metabolic paths and how brain oscillations, frequencies and amplitudes are correlated. Yurkin's approach allows the comparison of various angles and paths on the same grid to evaluate the presence of unexpected correlations among features lying on different systems' energetic levels. This could make possible access to information about previous and future states of the system.

From a philosophical standpoint, this approach would seem to have merit, but it is by no means a 'pure' geometrical interpretation. We retain our postulate of the real, logical and dynamic opposition at the heart of energy, and consequently of its embodiment in matter and information at all more complex levels of reality. It requires that it apply to fundamental particles, protons and electrons, atoms and inorganic and organic molecules and ultimately the living organisms that are constituted by them.

Everything that involves the above principles is a system, a process of systems of systems, etc.; therefore, everything is a system, capable of interacting with other systems in a manner that one can call antagonist. Lupasco's systems theory was originally published in [11] and includes the following statement: "A system is a set of elements that constitute energetic events linked by a relation." We discuss the philosophy of specifically living systems in Chap. 19.

8.7 Distinguishability: The Same and Not the Same

The philosophical topic of identity and diversity was mentioned above to position their dialectic relation at the core of Logic in Reality. The poetic book of the Nobel-prizewinning chemist, philosopher and poet Roald Hoffmann, *The Same and Not The Same* [6] embodies the Lupasco principle of incompleteness of distinguishability corresponding to the identity-diversity duality present throughout nature and natural philosophy. In the logic of/in reality, it is not the truth of propositions about reality but the reality of complex things being partly the same and partly different at one time or the same or different at different times that is grounded in our current physics. The existence of such a principle in nature becomes a logical methodology for analyzing and elucidating many problems of this kind, and we will therefore present briefly some other illustrations of this principle.

There is a substantial discussion in the chemical and philosophical literature as to whether two hydrogen atoms, say, in a molecule of ethane are the same or different. In our view, this apparently scientific and philosophical point does not require such discussion. If the criteria for distinction are the space–time coordinates of an atom, then they are distinct. If the question is about distinguishability by some functional property, they are not. They become distinguishable as soon as some element of asymmetry is introduced within the molecule. In a next stage of complexity, they can be distinguished if the addition of a new atom *changes* the polarization distribution structure of the molecule, resulting in new interactions. Given a ring in which one of the six carbon atoms has been replaced by a chlorine atom (mono-chlorobenzene), there is only one hydrogen atom next to the chlorine. But this hydrogen atom is different in *three* ways from all the other hydrogen atoms in the molecule: sterically, oscillating between axial and equatorial orientations, electrically, its bond strength to carbon is changed and in accessibility to reaction, steric hindrance.

8.7.1 Asymmetry and Interaction

Asymmetry and interaction are necessary for distinguishability, but they may not be sufficient. In the cyclohexane ring composed of six carbon atoms and twelve hydrogen atoms, the three in the equatorial plane of the ring are different from the three axial ones above/below the plane. They cannot be distinguished statically because vibration of the molecule interchanges them. However, a ring with two hydroxyl groups on the same side is different from one in which they are on opposite sides of the plane of the ring, reflected in the ways in which they inter-react.

The philosophical nexus of this brief excursion into organic chemistry is the isomorphism of difference at both this and higher levels of reality, up to and including the complex 'molecules of life'. At the lower atomic level, energy content or level are essentially the only criteria of difference. Two electrons in the first *s* shell around the nucleus are not only the same from an energy standpoint, they are indistinguishable from a positional standpoint; the meaning of position, which involves relations and *change* is equivocal. Energy differences define the differences between electrons at s and p or d shells, that is, distance from the nucleus in an atom. Change between them requires input or removal of energy. However, electrons are not distinguishable in the timeless quantum world of their individual existence.

Individual difference of human beings is, or should be, coupled with their sameness from an ethical standpoint, that is, how they are treated by society: the acceptance of difference (the other) and of his/her having equal rights and responsibilities, In the combined picture of reality in this book, the dialectics of ethics is isomorphic to the physics of matter above the atomic level. The philosophy of chemistry is the basis for the natural processes of natural philosophy in the sense that both operate according to the same logic.

8.7.2 Some Examples from Cognition

8.7.2.1 "I Am You"

One of the false dichotomies in philosophy in our view is that established between bodily continuity *vs.* psychological continuity. In LIR, the two are related dialectically, consistent with Kolak's view [10] that our intuitions that someone can be "the same and not the same" are incompatible "under the received logic". What Kolak calls Alter Subject Identification—the intuition of the subject in the object—is captured by the LIR non-separable category of the Subject-Object.

A language can never remain the same, as Dante wrote, but at any moment, it is incapable of being fully possessed and already partly forgotten. Again in Hoffmann's phrase, like a living being at another time, it is the same *and* not the same.

8.7.2.2 Identity Again

Many of the problems that are associated with the notion of identity in general and personal identity in particular can be ascribed to the need of their proponents to justify their views in terms of bivalent, truth-functional logic. The complex derivations of Kripke amount to no more than a requirement, allegedly for completeness, to assure that catastrophic counterfactual cases can be covered. This strategy is fatal, since it assumes, finally, that there are no real constraints on our universe that guarantee stability except for borderline, pathological cases which cannot be the basis for an existential argument.

Two approaches to escaping from such dichotomies are permitted in the framework of Logic in Reality: (1) go to process language; (2) go to a logic which allows things to be the same and different. In LIR, there is no semantic iteration of the relationship of self to self, no lurking regress, as in one's relationship to oneself, different information is involved in the two terms.

A key chapter of Hoffmann's book is entitled "Identity—the Central Problem". In organic chemistry, the field that includes but is not limited to the understanding of the molecules of life, we see the philosophical issues raised by such questions as the difference between isomers, different structures composed by the same atoms, and of the same atoms at different locations, in different electrical environments, in the same molecule. Such differences, for the same overall composition, are the basis for differences in any subsequent reaction of the molecule as a whole. Inherent in the discussion of these issues is the duality of chemical structure and its representation. What emerges in our view is the more or less conscious ability of the chemist—an ontological included middle—to move between the two epistemological dualities.

The final section of Hoffmann's book has the title "The Dualities that Enliven", in which he refers to the question of identity in chemistry, of *being the same and not the same* as a tension. Is this tension epistemological or ontological? It, too, is both. In 'chemistry in reality', synthesis and analysis are partners, interacting dynamically in a way that has little to do with the analytical-synthetic distinction in standard philosophy. Chemistry is unique—both making things and taking them apart. "Too little attention has been paid by reductionist philosophers to show how this very chemical procedure—synthesis—works." Here again, in our view, it is in chemistry that a foundational principle inheres that applies to science and natural philosophy that can, also, be described by the process logic of Logic in Reality.

Further comments made by Hoffmann almost serve as a charter for our present study. He suggests a dual definition of argument: a simple process of reasoning or statement of fact or disagreement, the confrontation of opposites and he believes that both are essential to science. In this book also, we have combined dispassionate logical reasoning and impassioned conviction we are right, knowing that we must prove our case to the reader's satisfaction. Hoffmann refers to what Wolfgang Pauli called dichotomizing "an attribute of the Devil, too easy and ultimately tiresome". "There is a difference between just listing of oppositional qualities and coming to terms with the tense synthesis of opposites that makes life interesting". "If beauty is to be found, it is surely at the edge where symmetry and asymmetry contend", we add, in dynamic opposition.

For Hoffmann, the usual dynamics of duality resembles the working of Hegel's dialectics, but Hoffmann's approach to chemistry moves beyond austere dualism. He refers to both the activity of the chemist in creating a new molecule and in the relevant theory, the absence of a single thesis and antithesis but a multiple, multidimensional perspective. Opposition is at the core of the living, changing human activity such as chemistry. Hoffmann quotes the poet and philosopher Emily Grosholz to the effect that "metaphysical schemes that accord reality to development *must display the structure of reality in terms of possible changes:* change requires difference, and difference takes the form of binary oppositions in our language and thought". "The venerable binary oppositions of metaphysics are a part of human wisdom; for all their indeterminacy, they stand for something fundamental and inescapable."

In agreeing with the spirit of this citation, we only suggest that the "structure of reality in terms of changes" is better described in terms of *non*-binary oppositions and their dynamics, which Hoffmann sees also in the relation between science and poetry. He was attracted to these not because of their similarities but "rather their dissimilarities, even contradictions: to see the same thing from two *apparently* opposite perspectives and feel the mounting tension between the two".

8.8 Chemistry in Human Beings. Obesity

From an energy perspective, Natural Philosophy—philosophizing—cannot be separated from the chemistry of the synthesis of the major sources of energy human beings use to support their existence. These are the molecules of fatty acids and proteins built up from amino acids and carbohydrates with which we are all familiar. But which human beings are involved? Half are starving and half are too fat, due to a socially unethical distribution of access to energy in its different forms.

Let us consider the similarity between dynamic opposition resulting in an included middle—the LIR 'cybernetics' in general—and biological regulatory mechanisms for maintaining homeostasis. To extend the discussion to a higher level of complete human organisms, an example of current socio-political importance is that of obesity [9].

Human caloric intake is regulated by the interaction of the hormone leptin, produced by body fat and the hypothalamus. Increased levels of leptin act to reduce food intake. Many other factors, conscious and unconscious, genetic and cultural, influence eating behavior and the development, or not, of obesity.[1] However, the statement of the relations between leptin, fat and food intake can be made in LIR terms that are useful for comparison with homeostasis in simple mechanical systems, e.g., thermostats. Actualization of increased leptin production potentializes food intake and vice versa. One can identify the resulting homeostasis when the opposing forces are equal in strength with a T-state, one in which body weight is maintained constant between very narrow limits (0.17%). Is it correct to say of this T-state that it is at a higher level of reality, subject to different or additional laws than those that apply to the chemical substances, hormones and lipids, from which it emerges?

A thermostat also maintains homeostasis, but it has no semantic complexity, internal dynamics or, literally, metabolism. Energy was expended by human beings, as for the construction and operation of all measuring instruments, only for their construction and isolation from the environment. To use the terminology of catastrophe theory, the number of co-dimensions of its space of control is small. The thermostat is not maintaining equilibrium *against* anything; it *is* (more rather than less) *at* equilibrium. This argument has the same logical form as that in regard to real and artificial intelligence. In the biological case, however, a decrease in body fat leads not only to a decreased level of leptin, inducing a powerful unconscious drive to eat after a significant amount of weight has been lost, but also to lower energy expenditure. This requires a disproportionately low caloric intake for the stable maintenance of weight loss. It is the *activation* of this potent behavioral and metabolic response to weight loss that makes successful dieting so difficult.

One author calls for increased understanding of the biological basis of obesity to "win the battle against biology" as well as a different kind of understanding, that of the psychological, social and economic problems of the obese. The desirable consequences of a satisfactory energy balance, among other things, are clear: an improved quality of life or 'wellness syndrome'. Such a complex emergent state can, in my opinion, be viewed as being the resultant of resolved contradictions at both the chemical and behavioral levels. From our point of view, then, it would obviously incorrect to tell people that they can lose weight without an input of energy, mental as well as physical, via simplistic methods. In fact *more* effort is required, in our view, especially in the behavioral and social area. The critical source of such energy is, as indicated in a companion article [12], the time of the health practitioner. "Intervention (in obesity treatment) requires persistence and repeated contact and takes a significant amount of time. This necessary time is missing from the current treatment paradigm."

At another level of society, there is a major contradiction between efforts made to reduce obesity and lessen its effects on individuals, and the interests of economic agents in maintaining and increasing consumption, including that of cars, television sets, etc. that encourage sedentary behavior, as well as of high-caloric foods of low nutritional value.

---

[1]Two other peptides control sensations of hunger and fullness, one peaking before meals and triggering appetite, the other during meals and signaling satiety. One might question, given their reciprocal relationship, whether simple 'blockage' of the former could be an adequate strategy for breaking weight-gain patterns in obese subjects.

Another article in this group [5] attempts an answer to the question of "Where do we go from here?" It is of interest because of the terms involving dynamic opposition in which it is formulated. "We must inspire people to make behavior changes within the current environment that are sufficient to resist the push of environmental factors toward weight gain. This will require conscious effort on the part of most people to make behavioral choices that counteract the environmental pressure." It is not easy to see what kind of resolution might be possible between such opposing phenomena. The conception that there is a *logic* in such opposition might be considered as a source of hope of a potential resolution.

8.9 Why Physicalism is not Reductionist

In this Chapter, we have not approached the philosophy of chemistry from an historical standpoint. There are two main reasons for this: many analyses of this kind have been made, and we have preferred to look at "what chemistry can do for philosophy, rather than what philosophy has done for chemistry". Accordingly, we will not discuss here the philosophy of how a chemist's explanatory and predictive goals can best be achieved, for example, how a proposed reaction mechanism could be confirmed. It is sometimes said that there are 'logics' of discovery in chemistry, but they are basically recipes or procedures, some now highly efficient due to automation, as in the search for vaccines.

Logic in Reality becomes important to the philosophy of chemistry is relation to the issues of physicalism and reduction—the reduction of chemistry to physics. Physicalism is the doctrine that everything in the universe is physical, here, that chemistry is nothing but physics. Chemical entities and their properties have no reality independent of the fundamental particles composing them. We believe that the problem becomes moot if one leaves the (reductionist) language of reduction. What *is* involved is not the reduction of chemistry to physics but the expansion of the ontology of physics to include the ontology of chemistry.

The situation is made clear by comparing two molecules with same quantum mechanical description of their physical states but very different properties. Quantum mechanics can explain why, for example, ethanol and dimethyl ether are stable, how they interact with other molecules and radiation, and so on. But the primary existence of the two structures remains unexplained, and must be seen the result of what is called a 'strong' emergent relationship to their respective substrates. The other issue, mentioned above, was the reduction of molecular species to quantum mechanics. This seems to us primarily an epistemological problem of classification, the question of the reduction of chemical *theories* about chemical entities to quantum mechanics. Whatever else may be true, this has not yet happened, whether it is because of non-computability or other difficulties, such as the accommodation of molecular shape.

The existence throughout nature of emergent, non-computable properties, to be discussed in Chap. 13 on Natural Philosophy is confirmed in chemistry. By judicious reference to the principles of LIR that point to the 'unsaturation' (a term used by Frege without reference to chemistry) of real entities that constitutes their potential for the emergence of new ones, realist philosophy receives further support. We will revisit the concept of physicalism again in Chap. 15 on Philosophy in Reality as Process.

References

1. Brenner, J.E.: Logic in Reality. Springer, Dordrecht (2008)
2. Börsen, T., Schummer, J (eds.).: Editorial introduction: ethical case studies of chemistry, Part III. HYLE—Int. J. Philos. Chem. **24**(1), 1–3 (2018)
3. Diaz Nafria, J.M., Zimmermann, R.E.: Emergence and evolution of meaning. Triple-C Commun. Capitalism Crit. **11**, 13–35 (2013)
4. Gabora, L., Aerts, D.: Evolution as context-driven actualization of potential. Interdisc. Sci. Rev. **30**, 69–88 (2005)
5. Hill, J.O., Wyatt, H.R., Reed, G.W., Peters, J.C.: Obesity and the environment – Where do we go from here? Science 299, 853–855 (2003)
6. Hoffmann, R.: The Same and Not the Same. Columbia University Press, New York (1995)
7. Hoffmann, R.: What might philosophy of science look like if chemists built it? Synthese **155**, 321–336 (2007)
8. Hougen, J.T., Oka, T.: Chemistry. Nuclear spin conversion in molecules. Science **310**, 1913–1914 (2005)
9. Kelner, K., Helmuth, L.: Obesity—what is to be done? Science **299**, 845 (2003)
10. Kolak, D.: Room for a view: on the metaphysical subject of personal identity. Synthese **162**, 341–372 (2008)
11. Lupasco, S.: L'énergie et la matière vivante. Éditions du Rocher, Monaco (Originally published 1962) (1986)
12. Pi-Sunyer, X.: A clinical view of the obesity problem. Science **299**, 859-860 (2003)
13. Weisberg, M., Needham, P., Hendry, R.: Philosophy of Chemistry. https://plato.stanford.edu/archives/spr2019/entries/chemistry/Spring 2019 Edition, Stanford Encyclopaedia of Philosophy (2019)
14. Wimsatt, W.K.: Re-Engineering Philosophy for Limited Beings. Harvard University Press, Cambridge (2007)
15. Yurkin, A.V.: On a descriptive geometrical interpretation of the principle of Pauli, elements of the table of Mendeleyev and the Newtonian laminar current of liquid (*verbatim*). Progr. Phys. **12**(3), 149–169 (2016)
16. Yurkin, A., Tozzi, A., Peters, J.F., Marijuán, P.C.: Cellular gauge symmetry and the Li organization principle: a mathematical addendum. Quantifying energetic dynamics in physical and biological systems through a simple geometric tool. Prog. Biophys. Mol. Biol. **131**, 153–161 (2017)

Part II
Toward a New Natural Philosophy

Chapter 9
Semiotics and Semiosis: The Units and Dialectics of Meaning

9.1 Introduction: Language, Logic and Meaning

In the first Part of this book we have discussed the major disciplines of science and philosophy, especially, in their relation to change. Underlying this discussion were the basic philosophical facts that language is used to describe logic, and this language must have meaning and be understood. A great deal of philosophy is devoted to the content of this last phrase. Since it is obviously impossible even to review it, we will simply restate our objective: to show that some of the major philosophical problems in understanding the world and its meaning for *us* can be addressed by the principles and dialectics of a non-propositional logic of processes (LIR), first outlined in the work of Stéphane Lupasco in the twentieth century [19].

In this period, meaning, and the meaning of meaning were partially codified by the semiotic approaches of Saussure and Peirce, and the latter's conceptions of logic. This work is relatively well-known and will be only summarized below. Logic in Reality (LIR), to repeat, is a *non*-linguistic logical mode of studying both epistemological and ontological phenomena, which for us include meaning, the meaning of and in reality. Reality includes but is not limited to all propositional expressions, including meaningless ones, and the process of their generation. We will begin this chapter, therefore, by looking at the much less known work of Lupasco in this area.

9.1.1 Meaning and the Absurd

As should be clear from the discussions in previous Chapters, our method attempts to consider together, and give equivalent value to, both a concept and its opposite or contradiction, and their conjunction. We will therefore begin by looking at both meaning and non-meaning, the absurd. In fact, Lupasco devoted only a few pages of text, in his *Logique et contradiction* [17], in which the subject of meaning was discussed explicitly. The subject with the title of this Section reference was addressed

J. E. Brenner and A. U. Igamberdiev, *Philosophy in Reality*, Studies in Applied Philosophy, Epistemology and Rational Ethics 60, https://doi.org/10.1007/978-3-030-62757-7_9

in the Second Part of Lupasco's book entitled "*Affective Logic*", and his conclusion is clearly stated: affectivity is absurd, and can have no meaning. Affectivity for Lupasco is a singularity, pure present, relative to nothing else, self-sufficient, without orientation or signification. "These are the characteristics of the affective state. Lived in itself, it is totally devoid of meaning, it is absurd. More, it is the absurd, the only absurd, because everything that is logical, because it is dynamic, possesses a meaning. Affectivity is the only ontological given, hence also the only absurd one".

As others have also said, the classical logic of identity also has no meaning, since its terms are ideal and impossible, but Lupasco went further to postulate three "non-meanings" (*non-sens*), which is not the same as "nonsense", and not equal to the absurd). These correspond to the three possible meanings, two non-contradictory of affirmation or identity and negation or non-identity and one of contradiction. Thus, everything that is logical, that is experimental or existential has a meaning, is meaning. Meaning is dynamic, and there is a transfinite multiplicity of meanings. Meaning cannot be finite or infinite, because then it would be absolute and meaning would cease. One can thus, Lupasco believed, define logic or existence or "non-being" as an incessant activity of signification, as a movement of meaning generation that never stops. The reader may wish to compare these statements, which are not intended metaphorically, with Peircean ontology, which also refers to signification as a process (see Chap. 3 on Philosophers of Process). As we will see further however, the meaning (*sic*) of process is quite different in the two authors.

Faithful to his thesis that affectivity accompanies logic, Lupasco insisted that meaning is coated with a layer of affectivity, which will develop, increase or be eliminated according to the pattern of contradiction in the underlying meaning. "Affectivity has only the meaning of its concomitants, of the enigmatic conditions of its presence and evacuation, which conditions are the meanings of logic relative to logic." The 'shocking' result of this treatment of the ontological cut between affectivity and the absurd on one hand, and logic on the other, is that being, the absolute, and substance are absurd. In addition, the only way Lupasco finds to characterize the "abyssal", non-real connections between logical phenomena or operations and affective givens is to consider them as the "realm of the pure absurd."

The absurd should not be confused with the non-sense of current multi-valued logics suggested as a third logical term between true and false. This non-sense is, rather the irrational, negative meaning that is the inverse of the positive meaning, identifying, linking and rational. The third term is the one defined by Lupasco as the T-state, arising from the contradiction between these two. This third meaning, which is the sense or the signification of the contradiction, can be considered a kind of *a*-sense, which could be confused with non-meaning, but not with the absurd.

In a final comment on affectivity, Lupasco proposes that the affective given might be an enigmatic *catalyst* of logical operations, consequently of change, experience and existentiality, without which nothing would happen and nothing exist, even though it would remain forever "the absurd witness and foreign agent, whose nature would carry no trace of the actions, effects and relations which it catalyzed." It must be the conclusion of this Section that although Lupasco preferred a highly limited vision of ontology, his own "cut" allows the significance of its value as being "logical" to be seen clearly. It is thus our view that meaning, like consciousness, is a *necessary* consequence of the existence of the physical world and its characteristics.

Carnap's comments in his *Meaning and Necessity* (Carnap 1947) show that his idea of meaning is solely the meaning of linguistic expressions, analyzed into intension and extension. Carnap could not be clearer in placing meaning in the binary, semantic domain—meaning is the meaning of language. Any reality underlying the meaning of the expression is separated from the question of new linguistic forms or framework for analyzing philosophical questions: "Above all it (the framework) must not be interpreted as referring to an assumption, belief or assertion of the reality of the entities. An alleged statement of the reality of the system of entities is a pseudo-statement without cognitive content."

As Peruzzi puts it in the previously cited reference [20], it is the furniture of the base macro-world (light, chemical bonds, gravity, temperature) that molds conditions on the earth's surface and living beings, among other things for the emergence of language. Language cannot be separated from its basis in perception and the "Scylla of relativism and the Charybdis of idealism" can be avoided by maintaining the flux of meaning from perception and action to cognition. The consequence, that conforms to the LIR category of Non-Separability, is that there cannot be any cut between perception and cognition without depriving sentences of meaning. The choice of logic is not arbitrary. Both the structure of objects and the dynamics of perceptual interaction with objects contribute to establishing features of both logic and language.

9.1.2 Meaning and Free Will

The concept that actions have meaning in a logical sense is also consistent with our view of free will. If our actions have meaning, there is even more reason to have moral responsibility for them. Non-meaning is not some kind of absolute 'absurd' (nonsense). This non-sense is, rather an irrational, negative meaning that is the inverse of positive meanings that are primarily identifying, linking and rational. Meaning and non-meaning are the cause of one another's existence, and a T-state arises from the contradiction between the two. This third meaning is the sense or the signification of the contradiction. In this theory, meaning, like determinacy, is thus a function of degree of antagonism involved in a pair of elements. As dynamic opposition goes toward zero, in the terms of binary logics, meaning also goes toward zero, and but is a maximum at points of maximum contradiction.

But is this right? We are also told that life is a "tale told by an idiot", full of energy and opposition, "signifying nothing"! We can only conclude that life and the contents of the world do and do not instantiate meaning, sometimes one, and sometimes the other, and we can only hope that partisans of each view may find some basis for *entente* in this theory.

9.2 Toward an LIR Theory of Meaning

Meaning in metaphysics and logic generally refers to the meaning of propositions or sentences, and is closely related, not to say dependent on, the notion of truth. There exists a strong commonsense intuition, however, that events, processes and living beings, especially humans, have meaning. People have always sought the 'meaning' of life and existence. Meaning thus has several definitions, which run from "something that is signified, especially by language" to "something that is felt to be the inner significance of something". We see in this polysemy the same conceptual range that we proposed in logic, that is, from a method for managing linguistic phenomena to a description and explanation of the structure of the world.

There thus appear to be two essential views of meaning—a philosophic/linguistic view of meaning and a processual, dynamic view, which we see as based on the principles of Logic in Reality. In order to grasp more firmly the difference between these two views, let us go back to the philosophy of Deleuze [8]. Deleuze's definitions and description of meaning are consistent with the following position: the existence of meaning in language, at several steps removed from reality, does not establish, although it often seems to imply, that there is no meaning in reality itself. Thus, for Deleuze, "Meaning is the fourth dimension of the *proposition* (emphasis ours)." "Meaning is never just one of the two terms of a duality that opposes things and propositions, nouns and verbs, designations and expressions, since it is also the frontier … or the articulation of the difference between the two, and because it possesses its own impenetrability in which it is reflected, it must develop within itself in a new series of interior paradoxes." Clearly, Deleuze is talking about the meaning of expressions, and not of designated things. In this philosophical world, "one arrives in a region where language has no further relation with the designated, but only with what is expressed, that is, with meaning." This separation is exactly what we contest.

In the LIR view, reflecting one of the usual meanings of meaning, it is what is *felt* that is significant, in other words, meaning-as-experience, and my experience of the world. We do not, however, wish to advocate a conception of meaning based on indexicals (my, our, here, now). The problem with indexicals is that, in a typically binary fashion, they obfuscate the dynamic, contradictorial (or reciprocal, if you prefer) aspects of the relation between the world and us. This enables us to say, non-metaphorically, that if the world has meaning for us, we have meaning for the world.

The approach to a theory of meaning, which emerges from the considerations in this book, corresponds to the "observational and experimental techniques of the scientist" that Dummett thinks should be used in metaphysical research. As we have seen, in LIR the domain of application of logic is extended to real phenomena and events, beyond its limitations, assumed by most philosophers and logicians, to linguistic elements, sentences, propositions and their predicates and their mathematical equivalents. We claim that this should be done for meaning also. As we have seen, realism, in this view, does not require, logically or philosophically, an adherence to classical logic and classical notions of truth and leads, accordingly, to a non-semantic realist position.

Again from the LIR point of view, classical binary logic has no intrinsic meaning, since its terms are abstract or ideal—topic-neutral. However, everything that is logical in the LIR sense that is experimental or existential, *has* a meaning and *is* meaning because it not only emerges from the underlying dualities of energy, but is directed toward the relationship with the other, the other subject—'meaning for' him/her.

In the Lupasco formalism, the elements of LIR have three possible meanings, two (primarily) non-contradictory of affirmation or identity and negation or non-identity and one of contradiction. Meaning is involved in dynamic processes, and there is a transfinite multiplicity of meanings. Meaning cannot be finite or infinite, because then it would be absolute and meaningless. One can define logic, existence or becoming as an incessant activity of partial signification, as a movement of meanings that never stops; this is a kind of Peircean view of process with the additional aspect of contradiction. This theory, sketchy as it is, is nonetheless compatible with *part* of the topological description of the form of meaning in catastrophe theory. All catastrophes are schemes of qualitative opposition, and meaning emerges at the threshold between entities that are the equivalent of the LIR subject and object [21].

9.3 Non-meaning

The concept that actions have meaning in a logical sense is also consistent with the view of free will in the previous Section. If our actions have meaning, there is even more reason to have moral responsibility for them. To repeat, non-meaning is not some kind of absolute 'absurd' (nonsense). This non-sense is, rather an irrational, negative meaning that is the inverse of the above positive meanings that are primarily identifying, linking and rational. Meaning and non-meaning are the cause of one another's existence, and a T-state arises from the contradiction between the two. In our theory, meaning, like determinacy, is thus a function of degree of antagonism involved in a pair of elements. As dynamic opposition goes toward zero, in the terms of binary logics, meaning also goes toward zero, and but is a maximum at points of maximum contradiction.

But is this right? We are also told that life is a "tale told by an idiot", full of energy and opposition, "signifying nothing"! We can only conclude that life has and does not have meaning, sometimes one, and sometimes the other, and we can only hope that partisans of each view may find some basis for *entente* in our theory.

Dummett [10] has indicated two approaches to resolving metaphysical controversies that show the relation between the problems at issue, meaning and logic. The first, more common line is top down; it attempts to resolve a metaphysical problem of, for example, deciding between realism *vs.* anti-realism in regard to an issue, derive from a plausible solution to it a correct model of meaning and an appropriate notion of truth. One then deduces from this an appropriate logic governing the sentences or propositions of the proposed theory. Dummett prefers a bottom up approach that starts by ignoring the metaphysical problems, constructing a theory of meaning based on natural language, and deriving from it a logic and corresponding notion of truth. The controversy can be settled by enabling a concept of truth, applied to statements about physical reality, provide a characterization of the meanings of those statements. This leads to a logical basis for the investigation of metaphysical disputes that avoids assuming the correctness of any one logical system but rather describes how the choice of logic depends on the choice of meaning-theory. One then needs only to "resolve the issue about the correct meaning-theory, and then the best picture (of reality, mind, etc.) will force itself on us, or else we will see that we should reject all the pictures." In this system, logic is considered as completely subsidiary, concerned solely with the validity of forms of argument, represented by inference schemas, and requiring a notion of truth under interpretation. It takes truth for granted and does not inquire into the point or interest of the notion of truth. In direct opposition to this, in our system, it is notions of truth that are subsidiary and, in Lupasco's words, "truth is the truth of reality". In the choice of the correct workable meaning-theory, Dummett gives preference *ex officio* to one that validates classical logic. LIR provides an alternative to this picture.

9.4 The Logical Philosophy of Peirce

There are significant similarities but also differences to be noted between the systems of Peirce and Lupasco. As mentioned in Chap. 3, they are important to point out to illustrate the advances made by Lupasco and support their credibility. In view of Peirce being considered by some people as a precursor of Lupasco, as well as his contributions to the field of semiotics, we provide here a critical overview of the former's work.

Despite his many references to logic, Peirce spoke primarily of logic as semiotics. In speaking of the triadic structure of the sign, in relation to the concept of Thirdness, he distinguishes three relations: (1) affective; (2) energetic and (3) logical, but this latter is defined as "a mental effect bringing about a change in habit." This is not a dynamic logic in the sense of Lupasco, but something like an epistemology. Elsewhere, Peirce defines logical values of 1, 0 and something like ½. This is an extraordinary anticipation of the logic of Lukasciewicz, but this is still a classical, truth-functional if not Aristotelian logic. Peirce does propose a "teridentity" $x = y = z$ in place of the classic identity $x = y$, but he does not describe any form of interaction between the three terms.

In a general way, as a Kantian, it would appear that Peirce was uncomfortable with contradiction, and rejected even Hegel's more dialectic categories and their resulting dynamics. He considered a principle of continuity as "a supreme guide in framing philosophical hypotheses", relegating heterogeneity and discontinuity to second-class status. "Sportings" and pure chance are the sources of evolution in his cosmogony [29].

Peirce said: "Every physical force reacts between a pair of particles, either of which may serve as an index of the other. … On the other hand, we shall find that every intellectual operation involves a triad of symbols". We can see an intuition of the Lupasco concepts of dynamic opposition, involving an included middle at a complex mental level of reality. But Peirce thought the ideas he had elaborated accounted for more, "for the main features of the universe, as we know it—the characters of time, space, matter, force, gravitation, electricity, etc. It predicts many more things which new observations can alone bring to the test."

Part I of this book was devoted to a science of the world that requires attention to the real properties of phenomena which cannot be explained so easily. There is perhaps a kindred spirit in the inquiries of Peirce and Lupasco, and it is perhaps this which is of more importance than the details of their systems. Thus, Peirce's conception of Thirdness and its logic do not correspond directly to the included middle of Logic in Reality. The concepts of dynamic opposition and alternating actualization and potentialization, which would provide the basis for the application of Thirdness to real phenomena are absent in Peirce. However, some of Peirce's other intuitions of dynamic antagonism are either easily interpretable in Lupascian terms and or provide support for Lupasco's approach.

Peirce wrote that "a belief and a habitual connection of ideas are one and the same." a formulation that to us emphasizes the aspect of a belief as a dynamic process—"which will be followed by an action of the nerves when needed." Beliefs are habits of action rather than what one might call terms in an epistemological equation. Peirce would like to avoid absolute abstract concepts such that "even the most highly theoretical belief will be imbued with meaning or rational purport."

The close relation between action and belief is shown by his statement that he knows of no other way of defining a habit (of action) than by describing the kind of behavior in which the habit becomes actualized. Finally, the central role of opposition in reality was clear to Peirce:

> ... The fact fights its way into existence, for it exists by virtue of the oppositions which it involves.... For just as we can only know facts by their acting upon us, and resisting our brute will... so we can only conceive a fact as gaining reality by actions against other realities. And further to say that something has a mode of being which lies not in itself but in its being over against a second thing, is to say that that mode of being is the existence which belongs to fact. [1.432 (c. 1896]
>
> "A thing without opposition ipso facto does not exist." [1.457 (c.1896)].

Other passages show that Peirce grounded his idea of consciousness in the "struggle", the action and reaction between "our soul and the stimulus", but Peirce did not go as far as seeing an overlap, as well as a polarity, between subject and object, ego and non-ego. His realist approach to volition or "free will" is close to the dynamics of consciousness developed by Lupasco. Choice—an exercise of freedom, occurs prior to the volitional act—and in it are where "states of unstable equilibrium" are to be found. In our terminology, choice would be an included middle emerging from the opposing alternatives. Simply, where Peirce says that "chance" is involved in such states, Lupasco would refer to statistical probability at the point of maximum contradiction, that of the T-state, as discussed.

9.5 Toward a Philosophy of Semiotics as Process

9.5.1 Signs, Representation and Semiosis

At first sight, the semiotic approach to information might appear to capture its multiple facets, ordering them into the functional categories proposed by Peirce outlined above, and Brier has provided a complete current interpretation of Peirce in [5]. The standard Peircean definition is used of semiosis as a process of meaning making, of construing a material entity or phenomenon as a *sign*, rather than simply interacting with it energetically: "semiotic interpretation differs from simple physical interaction". One could consider information and meaning as energy here, but the *distinguo* is not trivial; meaning is at a higher level of interpretation in its inclusion of "meaning for". This is the essential distinction between information considered in the sense of Shannon as simple negentropy and what Logan has called instructional or biotic information [16]. Standard logic is applicable to the first since it represents only the non-contradictory aspects of diversification. The second requires LIR since it involves emergence and meaning.

To make this difference more specific, it should be clear that Logic in Reality is a logic of *change*, perhaps the only one. Real, energetic change involves the caused appearance of new entities, and to the extent that higher-level information and meaning emerges from lower-level information or data, such non-Shannon information is a new entity, instantiating "newness". (Such newness may be of interest per se, but this does not concern us here.) As discussed further below, semiotic signs, whatever else they may be, are not required to characterize the informational aspects of dynamically evolving systems.

Thus as might be guessed from our previous comments on dualism and emergence, we do not accept the Peircean theory, which we consider insufficiently dynamic, despite the common interpretation that the relation is dynamic because it leads to "chains of triads". We believe there is no energy that can be assigned to the triadic relation that would give it a basis in reality (physics). We see the same problem with Peirce's categories as with the Hegelian triad of thesis, antithesis and synthesis: there is no deductive basis for the movement from one term to the other or a description of any physical interaction between them. If the argument is made that nothing of the sort is required, our response is that is exactly the problem—the terms are not physically grounded and hence have limited explanatory value other than as a heuristic device for keeping track of the entities involved in biological and informational processes; its use should not make one neglect the real properties of the system.

The Peircean semiotic concept of information has been summarized by Quieroz, Emmeche and El-Hani [23] as a "triadic dependent" process where a form is communicated from an Object to an Interpretant through the mediation of a Sign. Our critique of this approach is that as stated by Peirce himself, it is derived from a *formal* science of signs that provides an *analytical* framework. Thus the QEE approach to information as process is constrained by the abstract characteristics of the Peircean categories, that is, their abstraction from dynamic aspects of real physical phenomena.

In contrast to QEE, we derive the triadic characteristics from the LIR view of the contradictorial evolution of all real processes, providing the physical basis for the QEE differentiation of potential and effective (actual) semiosis and consequent definition of potential and effective *information* as well. In LIR, information is a complex of processual interactions with both binary (dyadic) and ternary (triadic) properties, all of which can be predominantly actualized (effective) or potentialized (not effective) at any time. An example of elements of these processes would be the conflicting messages as to a real state-of-affairs of some catastrophe. This characterization would seem preferable to the concept of a Sign as a Medium for communication of Form.

The essentially static linguistic definition of Form in terms of "conditional propositions" states that certain things would happen under certain circumstances. Strikingly, as quoted by QEE, Peirce said that "*Form can also be defined as potentiality* (*real potential:* (EP 2.388)" (emphasis ours). In LIR, structure and form are also operators, capable of causing change and physical processes, including the physical processes of their conceptualizations, capable of undergoing change. Form is characterized not as "potential" only, but as a process whose elements are both actual and potential at the same time. It is the changing shape of change that we believe the genius of Chinese thought was the first to capture.

LIR confirms the QEE critique of the argument by Jablonka that "for a source to be an information input rather than merely a source of energy or material, its *form*, or variations in its form, rather than any other attribute should affect the interpreter's response in a consistent, regular way". Here, a distinction has been created according to which form is idealized as something non-energetic, but still with causal properties. To say that form is also energy is not to make a physicalist reduction, but to avoid conceptualizing it out of existence.

9.5.2 Semiotics and Representation

Semiotics is a theory of representation—it is things standing for other things which are clearly to be distinguished in some views (see for example Quieroz and El-Hani [24]) from "reactive", that is physical systems, composed of "Dynamical Objects". In a sense this is exactly the problem we address: the semiotic relation is one of "standing", that is inactive and causally and physically inert. The LIR thesis can be illustrated, if not proved, by reference to a citation by these authors from Peirce, p. 15): "...we have to distinguish the Immediate Object, which is the Object as the Sign itself represents it, and whose Being is thus dependent upon the Representation of it in the Sign, from the Dynamical Object, which is the Reality *which by some means contrives to determine the Sign to its Representation*" (CP 4.536) (emphasis ours). For us, this statement is an implied disavowal of any purely semiotic concept of the representational relation.

The problem really arises if representation is supposed to be about a processual change. If it is the sign relation—a member of parliament "officially" represents his constituents, it is trivial. What is not trivial here is the causally efficient group constituted by the elected official and his electorate. In this case, complex interactive processes between them are involved, and their dynamics depend on many other factors, the socio-political context and so on [9]. What is common to all types of signs is that they are all ontologically subordinate to the existence of what they are a sign of, that is represent. In a simple definition of a representation, for example, as a theatrical performance, the term is simply a shorthand synonym for a play involving real actors, fictitious characters, a plot, etc.

Let us make this critique more direct by looking more closely at the kinds of signs defined by Peirce: A sign or Representamen is something which stands to somebody for something in some respect or capacity, and Peirce classified signs, *in his earliest work*, into three groups: icon, index and symbol, based on the different relations between a sign and what it represents. An icon represents something based on the resemblance between a sign and its object; an index represents something by being directly related to its object; and a symbol represents something by convention. In later work, Peirce complexified these distinctions, even collapsing them into a concept of a sign that displays all three characteristics [29].

We will not discuss further the extremely complex taxonomy of Peircean signs and the multiple changes in interpretation and application they underwent. At one point in the Peirce's development of his theory, as mentioned by Short, the only true, genuine or complete sign seems to have been the symbol. Other signs may be incomplete or composite, but the critical relation of representation remains unchanged for all of them. In complex human situations (a lawyer representing his client), the abstract formal relation of representation will always be superfluous. One might say that if a relation moves and changes, it is not, or not only, a representational relation. What we mean by "not only" is that the concept of representation is strong and compelling. It is not going to "disappear" as a consequence of this book. We wish only to suggest the utility of its being understood in dynamic terms, in other words, our representation of a thing must be seen as a "thing" (process) in itself. Deacon (2012) correctly criticizes notions of representation and meaning when they are collapsed into something like a physical pattern as being reductionist, but a physical pattern or structure in LIR terms does not suffer from this drawback, as indicated in Sect. 2.3.

9.6 Semiotics and Semiosis

The views of dialectics and logic that we developed in Part I can be used as tools to reconceptualize the areas of semiotics and semiosis as such. The key concept in semiotics has been the introduction of an ontological and epistemological role for the sign. Signs are elements or units of language that are assumed to convey meaning to living beings capable of receiving and interpreting them, that is, of evaluating their relevance to the on-going processes of survival and growth, physical and intellectual. A name in the thought of the last 150 years to be associated with semiotics as a discipline, in addition to Peirce, is that of Ferdinand de Saussure. The problem with these authors and their countless followers is that their theories are non- or anti-dialectic. Simply, contradiction is given a purely negative interpretation.

As soon as living beings appear in the universe, meaning appears in and as the processes essential to their survival and reproduction. With the arrival of self-awareness of existence, say, at the level of human beings and cats, semiosis describes the active, ontological process of the *discovery* of meaning in and as existence-as-process. It is creative, participatory and relational; semiosis carries an emotional stance or feeling toward existence and the self as part of that existence, which includes the capacity for its perception. It is a phenomenon within the scope of natural philosophy, with its own dynamic 'logic' of processes.

9.6.1 The Semiotics of Peirce

Based on our understanding of the complex Peircean system outlined above, we turn now specifically to his semiotics. For us, his semiotics of Peirce is an epistemological doctrine describing the classification or categorization of the meanings or meaning-processes inherent in all existence. As noted in Chap. 3, Peirce has been considered a precursor of Lupasco, for example by Nicolescu. Browning and Myers include him in their 1988 Compendium as a philosopher of process.

We conclude that Peirce's theory is, finally, reductive and inert, since inserting the concept of sign or representation between a phenomenon and meaning adds no new meaning [3, 4]. According to Peirce, semiotics is a logic and is logical, but logic here is to be taken only in the sense of standard, bi-valent or multi-valent linguistic logics. Peirce's famous 'three-tailed' graphs, designed to illustrate his logic [28] do nothing but provide a visual equivalent of bivalence.

For us, the discovery of new relations between categories is indeed a semiotic process, but only because it takes place *outside* the epistemological categories. To say that it is 'Thirdness' is a tautology, and the entire structure of Peircean semiotics lies in the domain of non-natural philosophy. Our concern with semiotics thus stems from the feeling not that it is wrong in some absolute sense, but that it addresses secondary aspects of the complex phenomenon of meaning and its exchange. The hermeneutic process adopted by Peirce illustrates this: he takes the extant domain and categorizes it without making an ontological commitment as to its basis, something which we consider an unavoidable necessity if the properties of existence are to be correctly represented.

In the realistic approach adopted in this book, we suggest that the only possible hermeneutic process requires starting from a correct assignment of both ontological priority and dynamic structure. We have access to an historical emergence of knowledge, defined as a mental process for insuring the flourishing of living beings, as individuals and species. Since survival, if possible without pain, is the objective, knowledge of how to accomplish this is what is meaningful for the individual; it is constitutive of meaning.

Logic in Reality, which provides a rigorous framework for discussing presence and absence and their interactive relation, provides a non-linguistic definition of a sign as a perceived phenomenon in relation to meaning in the above sense. For a deer, the information received as the smell of a lion, or sight of its droppings, *are* the lion as a potentiality, in potential form. The deer, having interpreted these signs, undertakes actions to insure this potentiality does not become actualized nearby.

Peirce based his theory on a division of phenomena into categories of Firstness (possibility), Secondness (existence) and Thirdness (reality). The 'First' is a 'Sign' or 'Representamen' which is in a genuine triadic relation to a 'Second', called its 'Object' so as to be capable of determining a 'Third', its 'Interpretant' to assume the same triadic relation to its Object in which it stands itself to the same Object'. The term 'Sign' was used by Peirce to designate the irreducible relation between the three terms, irreducible in the sense that it is not decomposable into any simpler relation, such as some form of part-whole relation.

In our view, the Peircean categories invert the ontological priority in the universe and fail to add to knowledge or how to acquire it. The problem was succinctly addressed by Petre [22] in his "Mixing signs and bones: John Deely's case for global semiosis". On p. 412 he writes: "The transformation of physical reality into objective sign reality is the same as the transition between Peirce's categories of Secondness (a binary relation of opposition, impact, cause and effect) and Thirdness (a triadic structure in which one item relates to another *for* yet another, the last one being the 'interpretant' of the dynamic between the first two).

> The trick succeeds only because one has put the rabbit in the hat beforehand only to pull it out later. Secondness is already implicitly Thirdness, and the so-called physical relation is already implicitly objective (italics ours).

When one tries to talk to Peirceans about possible overlaps between the categories when applied to real change, one is told "Oh yes, Peirce was aware of them and they exist". However, there seems to be no more grounding for any physical relationships in these than in the original categories.

9.6.2 On Signs and Meaning in Peirce. The Pragmatic Maxim

Quieroz and Merrell [25] conclude their paper on Peirce with the following statement: "In sum, according to Peirce's pragmatic model, semiosis is a triadic, dynamic, context-dependent (situated), interpreter-dependent (dialogic), materially extended (embodied) dynamic process. It is a social-cognitive process, not merely a static symbolic system. It emphasizes *process* rather than *product, development* rather than *finality* (italics theirs). Peirce's emphasis rests not on content, essence or substance, but, more properly, on dynamic inter-relations." Just before, "Meaning as form-becoming is an emergent inter-active, inter-relational pattern of behavior. The authors describe meaning in much the same terms as Wu does the informosome [3, 4]: "The meaning of a sign depends upon comparable past contexts of what is taken to be the same sign, in the present context and in imagined, conjectured, or hypothetical future contexts; sign meaning is a time bound process." "Meaning is in the entirety of contextualized body-mind interaction…".

In principle, we would like to endorse this doctrine of meaning from the perspective of LIR as a philosophy of real process. For example, referring to Savan, Quieroz and Merrell suggest "an Interpretant I is both the third term of a given triadic relation and the first term (Sign) of a subsequent triadic relation. Forgetting the use of sign here, the concept is similar to a Lupascian third term entering into dynamic relation with other terms to generate a new third at a higher level of reality or complexity. Further, Interpretant, Sign and Object compose a triad whose definition can only be circular; each one of the three terms is defined by the other two". There is not distinct essential or substantive property; at any moment as S can become an O or an I, and the same can be said of O and I. Brenner suggests that in fact Logic in Reality, metaphysics and the ontology developed from them can be seen, non-metaphorically, as a triad, each member of which is the included middle of the other. If one concentrates on the interaction of the logical and metaphysical aspects of, say, the process of implication, as in the Lupasco calculus, an ontology emerges. This formulation is adapted from the 'trialectic view of reality' of Pompiliu Craciunescu, a follower of Lupasco, in which each member of the triad of epistemology, poetry and metaphysics can be the included middle of the other two [6].

However, Quieroz and Merrell also state [26] that "meaning is neither in any correspondence between a sign and its reference, nor is it the case that meaning or sense determines reference"; this is our view would invert the ontological priority of reality. The problem is that here, as in all the work of Peirce and his followers, the notion of sign superfluous, detracting attention from the underlying contradictorial or countervalent elements of the processes in progress. Our authors practically admit this much: "Thought itself is impossible without the material incorporation of some aspect of the world"; our position is, however, that meaning *is* possible without the participation of the Sign, Object and Interpretant because it *is* the 'sign', inherent in the energetic interactive processes underlying it. As stated explicitly, "for Peirce it is impossible to think without signs", but not for us. It is finally the semiosis of Peirce that is a "static symbolic system".

For our authors, Peirce's theory allegedly describes a processual becoming from Firstness as possibility, Secondness as actuality and Thirdness as potentiality in 'mediation' with the prior two. Thirdness in this scheme, whatever else might be said about it, is ontologically posterior. Without any justification in science, this remains an epistemological classification, highly oversimplifying the recursive energetic interaction patterns of reality. By way of conclusion, we propose that the signs identified by Peirce as carriers of meaning are essentially the same as the phanerons he claimed were the units of existence. In our scheme of epistemons and ontolons, signs are epistemons, structures that are non-real and inert, to repeat, not in terms of their *production* by minds, but in terms of their limited, descriptive or classificatory function in knowledge.

9.6.3 The Search for the Meaning of the Sign

If there were any lingering doubt about the ontological status of signs, or rather, their lack of any, it should be dissipated by Peirce's 'maxim' applied to a process of attempting (our addition) to draw meaning from signs in regard to their interactive interrelations. Note that the use of 'interactive' here, which is not explained, begs the question of ontology *vs.* epistemology. The maxim is supposed to be the means for constructing the meaning of a sign as a consequence of practical validation of the sign put in the form of a proposition whose nature is that of conjecture as to the possible meaning of a sign. Quieroz and Merrell [26] cite the first, most commonly quoted version of the maxim: "Consider what effects, that might conceivably have practical bearings, we conceive the object of our conception to have. Then our conception of these effects is the whole of our conception of the object (CP 5.402 *passim*). We consider our authors more honest than Peirce: the maxim has to do with the semiotic subject's construction of *her* world. It is an indeterminately variable (random?!) method for interacting with signs in such as way that the 'semiotic world' with which they interrelate appears to be the case. We repeat our preferred formulation: signs are reduced versions of meaning, suitable only for their classification.

In summary, only physical processes are required for meaning; for us, they can be and **are** meaning. Signs are inert, a posteriori epistemological constructions without explanatory value. Biological 'codes' are not signs in our conception of reality. They are static scientific abstractions. The concept of operators is useful here (Burgin and Brenner 2017); it is the chemical structure of nucleotides in the DNA that 'operates' the genetic process.) Let us now therefore leave the austere world of Peirce and his acolytes and discuss concepts of semiotics and semiosis from other logical and scientific standpoints.

9.7 The Semiotics and Semiosis of Paradoxes

From the earliest known examples of abstract thinking, humans faced the two types of paradoxes, one referred to as semantic and the other as kinematic. The first apparent formulation of the semantic liar (*pseudomenos*) paradox was that of Epimenides of Cnossos who was a semi-mythical thinker of the 7th or sixth century BC. Its later formulations belong to Eubulides (fourth century BC), and in other cultural traditions to Bharthari (India, fifth century AD) and to Nasir al-Din al-Tusi (the Islamic tradition, Persia, thirteenth century), who identified the liar paradox as self-referential. This paradox substantiates the impossibility of non-contradictory application of binary logic to the process of signification (assigning of meaning).

The first known formulation of the kinematic type of paradox in the West was by Zeno of Elea, who was the disciple of Parmenides (fifth century BC). They were formulated independently by the representatives of the School of Names (Míngjiā) in China (5th—third centuries BC) and by Nagarjuna in India (third century AD). The origin of kinematic paradoxes is in the impossibility of a non-contradictory formulation of an infinite space–time and to the incompleteness of any formal representation of it. In such representations, contradiction appears between the description of space consisting of an infinite number of points and the possibility of moving past them in finite intervals of time.

Over many centuries, the semantic and kinematic paradoxes were analyzed independently of one other. The main challenge was to try and unite them in a single dynamic process. This was, in particular, reflected in the designation of Cretans as liars in the Epimenides-Eubulides paradox in which the physical (kinematic) parameter (movement) was introduced. If we remain within the physical domain and observe a moving object (e.g. Zeno's flying arrow) embedded in physical space, we can introduce a semantic constituent by assuming that the arrow, at a concrete finite moment of time, is present at a certain point of space, and at the same time is absent there, beyond the semantic field of the person who launches it (the bowman). The dynamic process thus includes the contradiction that is embedded into the moving system and can be fixed logically. Space–time is structured by physically defined parameters that appear to fulfill the function of avoiding the apparent inconsistency of its representation as infinitely divisible to allow movement in it. These parameters represent the fundamental constants of the physical world, and the physical complementarity which manifests itself, in particular, in the impossibility to define strictly the position and momentum of a particle simultaneously. This arises as the consequence of the process of representation into the represented (measured, observed), which continuously produces an infinite recursion. Quantum measurement appears as a semiotic process in which signification takes place continuously and in which the kinematic and semantic constituents are complementary and emerge in tandem from a still unknown *pleroma*.

9.8 The Units of Knowledge and Existence

There have been many attempts in the literature to respond to a need that, we also have felt to, somehow define knowledge and existence itself in terms of simpler, more stable and graspable entities. One of these is the 'phaneron of Peirce'.

9.8.1 The Phanerons of Peirce

In view of the importance of Peirce's thought, we need to refer to his concept of the phaneron [29] as one candidate for entity of existence. In his highly useful discussion of causal processes, semiosis and consciousness [11] claims the advantage for contemporary biosemiotics, the application of concepts of signs to living systems, is that "it does not force on us a dualist metaphysics that separates phenomena into two distinct worlds or realms which are afterwards difficult to reconnect again".

Emmeche considers that Peirce's system was an ideal combination of semiotic monism, conjoined somehow with an ontological category theory. We noted above how Peirce based his theory on categories of Firstness (possibility), Secondness (existence) and Thirdness (reality), without the requirement for radically different ontological domains.

He states that his "phaneroscopy (phenomenology) has nothing at all to do with the question of how far the phanerons it studies correspond to any realities. It "religiously" abstains from all speculation as to any relations between its categories and physiological facts, cerebral or other. It does not undertake, but sedulously avoids, hypothetical explanations of any sort." Elsewhere, Peirce says specifically that the "one intelligible theory of the universe is that of objective idealism." These positions should not be taken too seriously. Peirce was anxious to avoid being tagged as a naïve realist or nominalist. Nevertheless, this kind of endless philosophical debate is what Lupasco is able to take us beyond.

As might be guessed from our comments on dualism above, we do not fully accept this reading. Despite his deep and anticipatory intuitions, Peirce made no metaphysical commitment regarding his concepts. He wrote specifically that his 'phaneroscopy' (phenomenology) had nothing at all to do with the question of how far the phanerons it studied correspond to any realities. It 'religiously' abstains from all speculation as to any relations between its categories and physiological facts, cerebral or other. It does not undertake, but sedulously avoids, hypothetical explanations of any sort. Peirce also said that the one intelligible theory of the universe is that of objective idealism. In a general way, as a Kantian, it would appear that Peirce was uncomfortable with contradiction, and rejected even Hegel's more dialectic categories and their resulting dynamics. He considered a principle of continuity as "a supreme guide in framing philosophical hypotheses", relegating heterogeneity and discontinuity to second-class status. 'Sportings' and pure chance are the sources of evolution and change in Peirce's cosmogony. These positions should not be taken too seriously. Peirce was anxious to avoid being tagged as a naïve realist or nominalist. The point of LIR is to provide a working alternative to 'naïve' realism and classical dualism, and we have suggested the conditional dualism of Lupasco as such an alternative, capable of taking us beyond previous endless philosophical debate. In the next Sections in this chapter, we will present some more acceptable candidates for the units of knowledge and existence.

9.8.2 Epistemons and/or Ontolons?

In referring to prior work by Barham, Igamberdiev proposed the concept of an 'epistemon' as a description of the sub-systems present in biological functions. Barham [2] regarded living cells as 'epistemic engines', in which a low energy or regulation (epistemic) stroke and a high energy or work (pragmatic) stroke constitute the work cycle. Both phases of the cycle are connected in such a way that the low-energy (informational) constraints act as signs with respect to high energy (pragmatic) constraints, leading to semiotic, epistemic correlations that have predictive value and which can insure a semiotic, epistemic correlation between the measurement of low-energy environmental signals and the response. The recognition, through its active site of an enzyme to an external chemical stimulus, based on its structure, is such an epistemon. This reaction cannot, at the present level of knowledge, be predicted from the structure a priori, but the two components in interaction can be said to be joined by a relation possessing semiotic character. The stability and reproducibility of this relation is made possible via operation of a second semiotic subsystem—the encoding (digital) system. Biological systems therefore include two semiotic subsystems, one based on the structure of imprint and on the recognition of three-dimensional shapes (images), and the other based on the digital linear structure of code.

The interpretation currently preferred by both Authors of this book is based on living systems being first and foremost real, physical, dynamic systems, with a corresponding ontology and hence describable by the 'onto-logic' of Logic in Reality. As in the previous Section, the semiotic descriptions of the meaning-laden biological processes are not false, but are abstractions from the actual flows of energy and molecules in progress. To try to capture these notions, we replace the concept of an 'epistemon' by that of an 'ontolon', an identifiable but dynamic 'individual' [15] that refers to or better implies a portion, recognized as such, of that ontological process. As such, an ontolon is a quasi-individual, individual and non-individual, and is an analogue of a quantum system, but only an analogue, quantum-*like*. The ascription of *essential* quantum properties to brain processes leads only, in our view, to anti-scientific views in which decoherence is conveniently ignored.

Ontolons are thus real, but unlike the epistemological counters of Peirce, we do not require any superaddition of classification or categorization. One can use the concept of ontolons as a device to help focus on the ontological characteristics of processes in relation to other critical aspects such as their ethical valence.

In contrast to the concept of 'monads' (cf. Leibniz) the notion of ontolons as entities or forms of existence—'being in reality'—has received far less discussion. We propose ontolon as our basic term for the essences of beings-in-reality, for the *forms of existence* which assume a multiplicity of dynamic processes in the extant domain, including knowing. The epistemon, then, represents the internal semiotic structure of ontolons, their epistemic representation. Logic in Reality provides the framework for the description of both ontic and epistemic processes and their relations, united via the transcending cognitive logical operations involving potential and actual states in interaction. Taken to its logical (*sic*) conclusion, the notion of ontolon is better grounded in reality than that of monad. It can replace the term of monad and, eventually, that of the *Dasein* of Heidegger in a Philosophy in Reality, in which the number of non-natural concepts is reduced to a minimum.

9.8.3 Ontolons in Relation to Other Units

As candidates for the entities of dynamic processes, let us briefly consider three other attempts to identify or characterize what those units might be. Capurro (2019) sees the clear historical derivation from the Greek *ta onta*, a neuter plural participle form best translated by 'existings' or beings. The singular, *to on*, a being, understood as a verbal, participle hence 'dynamic' form would be the equivalent of the ontolon. This interpretation is only acceptable to us if being here includes *becoming*, the processual activity of reality. In our opinion, any attempt to totally separate being from becoming is incorrect. Support (not proof; we are not in a domain of 'proof') comes from the perhaps little-known fact that in several Indo-European languages, the original word for *be* and *become* as the same.

9.8.3.1 Memes, Holons and Partons

In discussing the evolution of culture and knowledge, Dawkins [7] defined the 'meme' as "a unit of cultural transmission or imitation". Subsequent discussions failed to agree on what a meme 'was', perhaps due to confusion of its epistemological and ontological aspects. Thus memes are supposed to have a genetic character in that they replicate and multiply like genes, and/or are transmitted as a viral phenomenon. In [30], Velikovsky, proposed a holon-parton structure for memes as a unit of culture. The properties of his holon and parton structures are, in our view, static and purely epistemological. The reflect nothing but a part-whole concept of entities with a completely classical mereology, that is, total separation between wholes and parts.

As is clearly pointed out by [1], the concept of memes as the units of evolutionary epistemology should be placed 'on hold' pending more adequate understanding of its basic ontology—the ontology of cultural transmission. We agree and intend to make this a subject of further investigation, but we can say already that the LIR ontology can apply to cultural transmission as an informational process. We will show that it can follow the usual principle of dynamic opposition applied to what XX calls "cognitive and perceptual hardware", beliefs and concepts, the latter having been specifically addressed by [18].

One 'entity' that we must reference in this book is the 'holon' described by Arthur [14]. His simplest definition is that of something that "is at the same time a whole and a part". This is a weak version of the LIR mereology which describes a dynamic, interactive relation between whole and part, but we can now make this reference in describing the ontolon. Koestler further differentiated holons into individual and social, the former being what we would call an identity—discrete, having agency and 'self'-directed behavior. The social holon is a diversity referring to groups—flocks and swarms—which as usually presented look like diversities but because of the absence of LIR/dialectics, remain binary to all intents and purposes. For whatever reason, the holon has not attracted much discussion, but the concept has been used by Luhmann and Ken Wilber, both of whom have been criticized by Brenner. The major difference between the ontolon and the holon is that the ontolon is grounded in LIR hence in a dialectical framework of process. Koestler uses only the familiar 'identity' words.

9.8.3.2 Other Conceptions of Units

In Chap. 6, we mentioned the 'propensiton' as a term for a unit of existence that carried the meaning of potentiality. In addition to ontolon and epistemon, discussed in more detail elsewhere, other terms of interest can be found ("zenon", "symbolon"). The term quanton is used in French, for example by Nicolescu, for a generalized quantum particle. It should be pointed out that the quanton is not a good model for the partly actualized—partly potentialized states of macroscopic phenomena. When a quantum particle is in 'contact' with a measuring device, the wave function is collapsed, and the quanton appears as 100% one form, with only the potentiality to be the other. In the absence of the measuring device, the quanton is a state of 100% epistemic and ontological uncertainty, which is not the case for objects or processes in the thermodynamic world.

We also ask: are there representations of certain appearances of an ontolon originating in its relation to other ontolons, to whole reality, etc.? If we consider the ontolon as an *eidos* in the Pythagorean-Platonic tradition, it appears probably as "zenon". Lefebvre suggested that each living system has its own stable point that governs whole system and remains "unmoved". He called it "eidos-navigator" and associated with the self. The synonyms could be "internal quantum state" or better "zenon". In principle, an ontolon can be reduced to a zenon but by a kind of reductionism which is opposite to the most common reductionism dominating current science. But in this case we position ourselves outside the movement. When the dynamics is introduced, we enter into epistemic or epistemic- symbolic relations.

9.8.3.3 Kykeon

Another possibility is to call the basic stable non-equilibrium cyclic structure of living systems "kykeon". As is mentioned in Fragment 125 of Heraclitus, "Also kykeon disintegrates if it is not stirred". Kykeon was the ancient drink that needed to be continuously stirred to keep its 'magic' properties. In this case the ontolon could be described as having the properties of the zenon and the kykeon, forming the symbolic relation within the unit, with the epistemic relation referring to externality.

The perspicacious reader will have noticed that, although we claimed at the outset to focus on the ontological entities of natural philosophy, we have introduced several concepts in the discussion that are primarily epistemological: (1) the *clinamen* in Sect. 5.4.3; (2) the ontolon above, which retains substantial epistemic character in its relation to the epistemon. We justify the inclusion of these perspectives, as we do that of Peirce, for their hermeneutic value. They can provide additional insight once it is established and clearly understood that there is no direct physical evidence for the existence of the respective entities *qua* entity. In fact, we have returned in restated terms to the dialectics of appearance and reality. The novel feature that we are able to introduce into this dialectics is the availability, offered by LIR, of a dynamic relation between the two—no appearance that is not also partly reality and vice versa, sometimes changing in time. Meaning inheres, implicitly and explicitly, throughout the domain.

9.9 Gödel and Incompleteness

The description of the units of knowledge and existence described above requires taking into account the underlying structures of language and mathematics, as we have seen in Chap. 7. Their codification in the theorems of Kurt Gödel represents a modern interpretation of the semantic paradox but at the same time they have a kinematic constituent which appears in the process of obtaining proofs. The claim that any sufficiently rich formal system is incomplete means that it contains statements which cannot be proved inside the system, but they can be enumerated (encoded) in the representation of the system. A system of axioms can never be based on itself since the statements about the system itself must be used in order to prove its consistency.

The great invention of [12] in his first incompleteness theorem is that its proof is achieved via assigning statements about the whole system (corresponding to meta-mathematical statements) to the elements of the system. Signification is therefore attributed to the dynamic process of (epistemologically) developing the proof of incompleteness itself. Although this dynamic process is not analysed logically (or "dialectically"), it represents the necessary condition for obtaining the proof for the theorem. As shown by Igamberdiev earlier, *a* process of obtaining proof, which necessarily generates the encoding in the system, is continuously realized in living systems and their evolution [13]. However, the latter is not the same as proof of statements or their mathematical equivalents. At a higher level of reflexion, it is present as a dynamic process of evolution of social systems and civilizations (Igamberdiev 2014). A concept that attempts to explain biological encoding as a consequence of metabolic reflexive loops in the system, "closed to efficient causation" in the Aristotelian sense was developed by Robert Rosen [27], discussed further in Chap. 19. Its further development can be based on analysing the dynamics that results in generation and further complexification of these loops.

To understand this dynamic process, we need to analyze the course of obtaining the proof in Gödel's sense which will characterize the process itself and generate encoding in the system such that it becomes a semiotic entity (seme). During this process, growth of complexity takes place and the system acquires *information* about itself and about the external (to it) reality. In other words, Gödel contributed more to the understanding of semiosis than Peirce or Saussure because he explained the necessity of encoding as a required step in the signification process. This process represents a dialectical discourse of the system with other organized systems embedded in the reality external to them. The complementarity between a formal Gödel system and

Gödel numbering is based on the separation between the system itself and its embedding that represents its encoding. It follows or tracks the growth of complexity via the dynamics that is inherent to Gödel's self-reflexive loop and includes a certain freedom as the encoding (Gödel's numbering) can be reached by many alternative, i.e. complementary, ways. Thus the growth of complexity is based on the contradiction between a system and its representation (including its internal "encoded" representation), generating new levels of "gradation" of organizational structure, and includes the relational abolition (negation) of previous steps. Some arrangements that were actualized in previous steps emerge later but at a higher level of complexity, as suggested by the Axiom of Emergence in LIR. These moments of the dialectical discourse loosely correspond to "three laws of dialectics" formulated by Engels. This dynamic process was defined by Heraclitus as the self-growing Logos (*logos heauton auxon*).

When Gödel enumerated metamathematical statements about a dynamical system by encoding them within the system, he incorporated at the same time the paradox in their nature. The next task is to describe in logical terms the procedure of the proof of incompleteness itself. This introduces another paradox: to describe the origin of logic by logical means and the origin of computation by computable means. This means that we enter into a discourse between our own logic and the internal logic of the system with which we interact and which we aim to describe. The dialectical nature of this paradox has not been fully analyzed, except in Plato's dialogue "Parmenides" and perhaps a few other philosophical works. It is very easy to slip from the dialectical discourse to the concept of final truth. This was not avoided by Plato himself (finally in his "Laws") and by Hegel and Marx who abandoned their own well-developed discourse for such a "final synthesis". This synthesis, in fact, rejects the very nature of dialectics and results in total failure for the conceptual field and in social practice. We will return to some of the consequences of Gödel's work in the discussion of information and meaning in Chap. 13. For now, we open a sequence of three Chapters dealing with the major topics in philosophy of metaphysics, metaphilosophy and phenomenology.

References

1. Atran, S.: The trouble with memes: inference versus imitation in cultural creation. Hum. Nat. **12**(4), 351–381 (2001)
2. Barham, J.: A dynamical model of the meaning of information. Biosystems **38**, 235–241 (1996)
3. Brenner, J.: Wu Kun and the Metaphilosophy of Information. Int. J. Inf. Theory Appl. **18**, 103–128 (2011)
4. Brenner, J.: On representation in information theory. Information **2**, 560–578 (2011)
5. Brier, S.: Cybersemiotics. Why Information is not Enough. University of Toronto Press, Toronto (2008)
6. Craciunescu, P.: L'état T' et la transcosmologie poétique. In: Nicolescu, B. and Badescu, H. (Eds.), Stéphane Lupasco ; L'homme et l'œuvre. Monaco: Éditions du Rocher (1999).
7. Dawkins, R.: The Selfish Gene. Oxford: Oxford University Press (1976)
8. Deleuze, G.: Logique du Sens. Paris: Les Éditions de Minuit (1969)

9. Dovi, S.: Political representation. In: Zalta, E.N. (ed.) The Stanford Encyclopedia of Philosophy; Summer 2011 Edition; (Ed.) 2011. https://plato.stanford.edu/archives/sum2011/entries/political-representation/. Accessed on June 1, 2011
10. Dummett, M.: The Logical Basis of Metaphysics. Harvard University Press, Cambridge (1993)
11. Emmeche, C.: Causal processes, semiosis, and consciousness. In: Seibt, J. (ed.) Process Theories: Cross-Disciplinary Studies in Dynamic Categories. Kluwer Academic Publishers, Dordrecht (2003)
12. Gödel, K.: The Consistency of the Axiom of Choice and of the Generalized Continuum Hypothesis with the Axioms of Set Theory. Princeton University Press, Princeton (1940)
13. Igamberdiev, A.U.: Evolutionary transition from biological to social systems via generation of reflexive models of externality. Prog. Biophys. Mol. Biol. **131,** 336–347 (2017)
14. Koestler, A.: Darkness at Noon. The Macmillan Company, New York (1941)
15. Krause, D.: Separable non-individuals. Representaciones **1**, 21–36 (2005)
16. Logan, R.: What is Information? University of Toronto Press, Toronto (2010)
17. Lupasco, S.: Logique et Contradiction. Presses Universitaires de France, Paris (1947)
18. Lupasco, S.: Du devenir logique et de l'affectivité; Vol. 2:Essai d'une nouvelle théorie de la connaissance. J. Vrin., Paris (Originally published in Paris: J. Vrin, 1935) (1973)
19. Lupasco, S.: Le principe d'antagonisme et la logique de l'énergie. Editions du Rocher, Paris (Originally published in Paris: Éditions Hermann, 1951) (1987)
20. Peruzzi, A.: Prolegomena to a theory of kinds. In: Macnamara, J., Reyes, G.E. (eds.) The Logical Foundations of Cognition. Oxford University Press, New York (1994)
21. Petitot-Cocorda. : Physique du Sens. CNRS ÉDITIONS, Paris (1992)
22. Petrov, P.: Mixing signs and bones. John Deely's case for global semiosis. Sign Syst. Stud. **41**, 404–423 (2013)
23. Queiroz, J., Emmeche, C., El-Hani, C.N.: Information and semiosis in living systems: a semiotic approach. SEED J. **5–1**, 60–90 (2005)
24. Quieroz, J., El-Hani, C. N.: Towards a multi-level approach to the emergence of semiosis in semiotic systems. Acta. Biotheor. **54**, 179–206 (2006)
25. Quieroz, J., Merrell, F.: Abduction: between subjectivity and objectivity. Semiotica **153**(1), 1–8 (2005)
26. Quieroz, J., Merrell, F.: Semiosis and pragmatism: toward a dynamic concept of meaning. Sign Syst. Stud. **34**(1), 1–63 (2006)
27. Rosen, R.: Life Itself: A Comprehensive Inquiry into the Nature, Origin, and Fabrication of Life. Columbia University Press, New York (1991)
28. Shin, S.J.: The Iconic Logic of Peirce's Graphs. The MIT Press, Cambridge (2002)
29. Short, T.: Peirce's Theory of Signs. Cambridge University Press, New York (2007)
30. Velikovsky, J.T.: The Holon/Parton Structure of the Meme, the Unit of Culture (2013). https://storyality.wordpress.com/2013/12/12/storyality-100-the-holonic-structure-of-the-meme-the-unit-of-culture/

Chapter 10
Three Major Domains in Philosophy: Metaphysics, Metaphilosophy and Phenomenology

10.1 Introduction. Is Metaphysics Possible?

That we have tied philosophy to change, already in the title of this book, is not innocent. It is a statement that at least part of philosophy cannot be related to or grounded in static individual objects, facts, or statements, but only in change. This is a metaphilosophical position with drastic consequences for metaphysics. It implies the 'end' of a metaphysics that in fact slavishly follows the precepts of binary propositional logic.

10.1.1 From Dichotomies to Processes

Metaphysics as the foundational branch of philosophy deals with the first principles and abstract concepts that include being, knowing, substance, cause, identity, time, and space The major conceptual principles such as causality, continuity, statism and dynamism, determinism, realism and laws of nature constitute the whole structure of metaphysics. The development of these principles led to the appearance of dichotomies of cause and non-cause, determinism and indeterminism, and continuity and discontinuity. These dichotomies were incorporated in the basic framework of formal logic operating with opposing statements. To overcome the limitations and essential incompleteness of this construction, based on such opposing dichotomies and to introduce a basic process of their continuous generation and resolution, the concept of dialectics was introduced in thought. It placed these dichotomies into a single evolving reality that could be either related to a concept of universal mind or to inanimate matter, which led to the dichotomy of idealism and materialism. The concept of Logic in Reality introduced by Lupasco and developed by Brenner became the dialectically based concept that provided a way of resolving dichotomies. In this book, we generalize from Logic in Reality (LIR) to a Philosophy in Reality

J. E. Brenner and A. U. Igamberdiev, *Philosophy in Reality*, Studies in Applied Philosophy, Epistemology and Rational Ethics 60, https://doi.org/10.1007/978-3-030-62757-7_10

(PIR). PIR forms the basis of Metaphilosophy as a separate domain. It dialectically reflects the entire domain of basic Metaphysics and investigates the origin of its fundamental principles and nature. It also substantiates the foundations of Phenomenology as the third major domain that enables the development of a more rigorous form of philosophical discourse. This discourse embodies phenomenology in a way that makes a convergence of science and philosophy possible from an informational perspective. Adding the domain of phenomenology to Philosophy in Reality 'recovers' phenomenology from its weaknesses and allows its humanistic aspects to be brought out and given the proper ontological status.

10.1.2 Every Thing Must Go

Ladyman and Ross [31], later abbreviated as L&R, raised many of the same questions we do in their book *Every Thing Must Go*, and the different answers the two approaches provide are useful to compare. These authors are critical of any metaphysics or philosophy that relies on intuitions or concepts that do not take into account the most recent advances in fundamental physics. In their "Naturalization of Metaphysics", (*Metaphysics Naturalized* is the subtitle of their book) they propose a picture of the world that raises the scientific standard for any theory that purports to describe and/or explain aspects of reality. In noting that much of the philosophy of science is also metaphysics, the authors look at their book as "an exercise in metaphysics done as naturalistic philosophy of science because we think that no other sort of metaphysics counts as inquiry into the objective nature of the world". Provided it is accepted that the existence of subjectivity is included in that nature, we agree entirely with this statement.

There is in some cases an 'unholy alliance' between pseudo-scientific metaphysics and standard logic, where the latter is called upon to prove the former. L&R correctly reject the idea that some non-actual physics can explain metaphysics, where 'non-actual' refers appeals to obsolete features of classical physics or ignores the reality of the non-separability of quantum states and the absence of a background space–time. We insist again on the necessity of non-separability at all levels of reality. Mainstream contemporary analytical metaphysics has become almost entirely a priori, but we place ourselves on the side of metaphysics *informed* by real physics. For us, the *Principle of Dynamic Opposition* (PDO), observable in simple as well as complex phenomena, is a *Principle of Naturalistic Closure* (PNC), of the kind required by L&R.

The mechanisms proposed by L&R for verification of the proposals of the metaphysics we on which we both agree are, however, inadequate, due to their reference to standard, propositional logic and taking the results of semantic analysis as the criterion of validity. As L&R state, one cannot obtain the necessary specificity of a metaphysical analysis "by reference to representational (syntactic or semantic) properties of hypotheses themselves, but by reference to well-understood norms of scientific practice that are identified empirically". One should articulate something with the status of a normative heuristic principle, not that of a logical analysis in the standard sense of truth-functionality. The dialectical method of LIR, taken in conjunction with its physical grounding, is our proposed basis of our form of 'verificationism'. The several hypotheses of physics that are unified within Logic in Reality and the PDO are no more and no less than the antagonisms of electric charge, magnetic polarity and spin, the basis for the existence of electron shells, and then of chemical compounds and molelcules (Chap. 8). LIR further provides a naturalistic explanation of emergence, of entities as well as local or global behavior, without 'mysticism'. We thus have no difficulty in accepting the other major, necessary *Principle of Physics Constraint* (PPC), cited by L&R. We maintain that rejection of the PPC would amount not merely to doing bad science, but to not doing science at all, and, we would add, not doing philosophy either. Most interestingly, in this respect, is the conclusion of L&R that some scientific propositions are sufficiently general for them to be metaphysical: "Our notion of metaphysics is thus recursive, and requires no attempt to identify a boundary between metaphysical and scientific propositions." Our book is an attempt to show, further, that there is no such boundary between such propositions and the *logical* ones of Logic in Reality.

10.1.3 The Basis of a Relational Philosophy of Science

We will not go very far here into the arguments between materialists and empiricists. In LIR, we have a basis for a principle of scientific stance, which L&R call scientistic, because our fundamental principles—the PNC and the PPC, plus PDO are normative. This is an expanded, second-order, stance-relative 'scientistic' metaphysics that reciprocally supports our philosophy of science and our natural philosophy (Chap. 13). We have a new basis for a reasoned scientific realism, with self-subsistence of individuals at all levels of reality, starting from those defined in Relational Quantum Mechanics. Facts about identity and diversity are never intrinsic to entities but are determined for all complex cases by the relational structures into which they enter.

We thus offer an alternative to the attitude of contemporary philosophers who accept as modern a logic and set theory that retain a classical framework of objects, logical variables and constants and invariants for the definition and description of relations, even though it is dynamics of the latter that carry all the ontological weight. As we will see in Chap. 13, Lupasco took this concept one key step further by seeing structures themselves as dynamic and changing, in which changing relations are present ab initio and from then on. LIR and the PDO thus offer a new response to van Fraassen's statement that the difference between mathematical and physical structure cannot be explained in purely structural terms. LIR supports a 'world-structure' that lies beyond mathematical description of all or part of it. The tentative metaphysical hypothesis of L&R is that there is no fundamental level of reality: it is patterns and relations—contradictorial relations—'all the way down'. They suggest that there is nothing more to the existence of structure than what it takes for it to be a real pattern. The contribution of LIR is to explicate further the nature of 'what it takes'.

10.1.4 The Nature and Role of Information

The importance of the nature and role of information in science and philosophy is discussed in detail in Chap. 11. Here, we simply agree with L&R that one need not worry about the apparent insubstantial nature of information because there is no absolute dichotomy between the abstract and the concrete or between the substantial and the structural in the first place. One does not need and should not say that the world is 'made' of information, which is bad, old dualistic ontology, but that information is a fundamental concept for understanding the objective modality of the world—its laws, cause and effect and kinds. Citing Zeilinger, L&R repeat that "it is impossible to distinguish operationally in any way reality and information—the notion of the two being distinct should be abandoned." Our logic and dialectics enable this statement to be rationally applied, that is, to cases and situations whose complexity requires our approach. Attempts to counter it by reference to two types of limiting, idealized cases should be rejected, here as elsewhere. Examples are simple binary events at one end of the spectrum, and constructed catastrophic counterfactual 'examples', with zero probability of actualization, at the other. It is part of reality that two human beings are physically, and immunologically separate, but their minds are or can be in a joint state of communicational 'overlap' (cf. our discussion of communication in Chap. 12.

In the work of Asher Peres, however, the formal similarities between thermodynamics and information theory are *not* coincidental, without going to the excessive concept of Wheeler that all things physical are information-theoretic *in origin* (italics ours). L&R use a neologism that nevertheless expresses clearly our position that our metaphysics is a metaphysical dynamicism. We thus recover, in a more modern context, the frequent use by Lupasco of *dynamism* to refer to the same process. Both are ontic notions involving real process properties in contrast to the epistemic notion of dynamicity of Johanna Seibt (see Chap. 15).

The key statement by L&R is that structural information is to be attributed directly to the real patterns found in nature, and not just to representations of them. To be is to be a real pattern, and real patterns are ones that are projectable and carry information about other patterns at a lower logical depth. Their 'Rainforest Realism (RR) is grounded firmly in a perspective of patterns that support informational projection physically as opposed to one only mathematically or logically possible, where logic, again, means standard logic. The major addition that can be made using our theory is that the use of a formalism for modeling a given range of phenomena can be empirically *and* logically motivated. A formal basis exists for valuing qualitative as well as quantitative knowledge.

Further and in conclusion, the verificationism of L&R extends to both philosophy and science, as in the convergence of science and philosophy under the influence of information proposed by Wu Kun (Chap. 11). The verifiability criterion endorsed by L&R is based on a relation, into which observation and the standard empiricist relation of being an observer do not enter per se, of 'being informationally connected to': x is informationally connected to P iff there exists an information channel from P to x. Exactly because our notion of inference and implication themselves involves dynamic ontic as well as epistemic processes, informational connectedness does not imply the absence of mediation by inference.

10.2 Lupasco in Relation to Physics and Metaphysics

Lupasco based his logico-philosophical system on some of the foundational principles of both classical and quantum physics. These included

- the macroscopic potential and actual properties of energy;
- the dualisms of forms of energy: two opposing magnetic poles; two kinds of electrostatic charge;
- the Heisenberg uncertainty principle;
- the corresponding minimum Planck length.

10.2.1 Lupasco, Modern Physics and Cosmology

In Chap. 6, we have provided some of the detailed implication of the Lupasco system to the interface between the so-called classical macroscopic world and the world of quantum entities. Lupasco was also aware of the cosmologies being developed by Einstein, De Sitter and Lemaître. He saw in the key principles of the theories an analogy to each of the three kinds of elements of the logic of dynamic opposition. These concepts apply to the actual physics of the universe in the sense of a Lemaître's expanding universe. The model of expanding universe suggests that a basic antagonism exists between the constitutive properties of expansion and contraction of space.

At the other end of the spectrum of size, Lupasco did not demonstrate an application of his logic to quantum systems as such. His logic of energy describes the evolution only of thermodynamic systems, not of microphysical ones. In the microphysical realm as it is now visualized, changes in the form and level of energetic particles occur, and systems can be constructed in which something like information can be transferred, stored and even used for computation.

Later, Nicolescu proposed an equivalent to the principle of superposition of states in quantum mechanics as describing a physical model for the emergence of an included third, epistemological and ontological, from the interaction of opposing process elements at a lower level of reality. On this basis, one can construct a general theory that can describe the quantum-*like* behavior of some complex systems [23]. For the purposes of this book, the existence of a fundamental physical structure for interactions at all levels of reality above that of one hydrogen atom provides the basis for emergence of complex systems, up to and including human individuals and societies which demonstrate the operation of the Principle of Dynamic Opposition.

The English emergentist Broad developed a synchronic, non-causal, co-variational account of the relationship of emergent features to their pre-conditions. Broad was interested not only in resolving the debate between mechanists and vitalists, but also in answering the question of whether biology and chemistry were reducible to physics. The unpredictability associated with Broad's emergentism does not present a problem for LIR/PIR, given our contradictorial view of determinism and indeterminism. Unpredictability is not constitutive of emergence, but rather a consequence of the metaphysical irreducibility of emergent properties.

Broad's ontological description of emergence is generally compatible with the LIR view: in both, emergent laws are not totally irreducible to laws characterizing properties at lower levels of complexity (or reality); otherwise there would be no basis for the discontinuity between levels. Both concur that since emergent features have not only same-level effects, but also effects in (or on) lower levels; they accordingly accommodate the concept of downward causation.

10.2.2 Lupasco and Metaphysics

We will attempt to position the logic of Stéphane Lupasco, which we have called a *philosophical* logic [3] in relation to the major recent developments in philosophy as such in the last century. We remind the reader that Lupasco did not look for parallels to or references for his "new theory of knowledge" in the work of contemporary philosophers, but rather in that of scientists active in the fields of quantum mechanics, developmental biology and cognition. From the standpoint of this book, the demonstration by Lupasco of the generality of the fundamental Principle of Dynamic Opposition, applicable to both the epistemological and ontological aspects of knowledge, was a profound statement about what philosophy is or should be—Philosophy in Reality.

Our task is made more difficult by Lupasco's own overextension of his fundamental principle by his assignment of a purely transcendental character to affect. In way that is no longer acceptable, given the work of Damasio and others on brain-body dynamics, Lupasco claimed that affect operated independently of energy and the logic of energy. Affect, in some magical way, accompanies phenomena and 'ordinary' mental and emotional processes. We therefore need to repeat to those readers who may have seen the full title of Lupasco's early major study (in French), that the authors of this book do not subscribe to this theory. Lupasco also drew consequences from the PDO for cosmology, based on some doubtful work by the Russian Blokhintsev for which we find no evidence whatsoever. Further details of this work will not be provided here.

There is a sense in which these ideas of Lupasco coincide with the transcendental aspects of the philosophy of Husserl and Heidegger. A study of such relationships might be of the kind that Husserl himself would reject, but a more subtle reading of Husserl, in Chap. 20, suggests an unsuspected correspondence. We now leave this subject for the time being and return to metaphysics as such.

10.2.3 Dubito Ergo Sum

We believe that the principles enunciated by Lupasco have an additional until now completely unformulated value: they provide a scientific basis for the *cogito*. Thinking does not take place in some sort of mental configuration space: it follows the same rules as do the other processes that are the persistent "furniture" of the world. The final argument, allegedly irrefutable, of the solipsist is that everything is an illusion, including our writing of these words. We have two answers: one is the 'classical' one that the difference between illusion and reality becomes meaningless and the question is badly posed. The principle offered by the dualism of Lupasco, that the world is real *and* illusion (to sentient beings) at the same time, expresses the same duality positively.

One aspect that is real is our ability to move, mentally, between considering one aspect, reality or illusion, as dominant or more completely actualized at one particular moment than another. This places us squarely in the domain of belief, which ipso facto is not decidable scientifically. I believe in a deity that is real, and all else is illusion and vice versa, not only this morning but permanently. If I accept for a moment that any part of my reality is illusion at any time, accepting in part the other point of view, mine is refuted. Thus belief establishes the only certainty we have, namely that there are incompatible beliefs, including of course what is 'science' and what it is not. Nicolescu [36] has synthesized the duality in a phase that for us captures the 'best of both worlds': "The sacred is what is irreducibly real." Metaphysics is neither more but also not less that this. It is also possible that some believers (or non-believers) are not sincere about their beliefs and are lying. This possibility has been explicitly expressed by Franz Kafka, whose insights are so deep that we consider them part of science. In referring to a short story—parable—that is embedded in his novel *The Trial*, entitled "Before the Law", one of the protagonists concludes the parable with a linguistic, self-referential paradox. K.: "I bet that is also a parable." The Priest: "You have won, but only in parable. In reality you have lost." K.: "A melancholy conclusion, it turns lying into a universal principle."

This passage further emphasizes the value that Lupasco placed in *doubt* as compared with his *cogito* a belief or knowledge that could not be grounded. He therefore stated: "*Dubito ergo sum*": I doubt therefore I am. One is then a firm realist; skepticism is limited to the value of one's own beliefs, not in the reality of the world. The principles of Logic in Reality thus provide a response to the radical nihilism of Nietzsche: what is true is my doubting, and its existence makes truth possible [30]: "My act of doubting implies that I believe *something* to be true, but (even if) I am unable to decide what that something is. If we get rid of truth, doubt becomes impossible.)". As Kolakowski puts it, simply but accurately, our direct experience is not of being but of incessantly losing our existence in the irretrievable 'has been', 'having past'. The philosophical-psychological-value of the dialectic view of time, indicated in Chap. 6, should now be apparent. In the 'activation complex' in our mind, as in the activation complex in a chemical process (see Chap. 8), the present persists, even if it specious from a completely external point of view.

To say that a self-referring human existence is never self-identical is as much a non-sense as to say that it is *always* self-identical, without the need for an unchanging *One* as a basis for our change. The search for an Absolute in the context of the Absolute's non-being already being a contradiction is a vicious circle, paradoxical as are all such attempts to 'put things into words'. In a major difference from the standard logic which has been the reference for philosophy, in LIR logical terms, we can say that the circle is *also* a virtuous one. In fact, the Principle of Dynamic Opposition cuts through the entire, endless debate about Truth and the existence of what is contingently real depending on the existence of something that is self-grounded, real non-contingently. Let us utter the test phrase: everything is real in part contingently and in part non-contingently. This statement may be at best 'inconsistent', in a Gödelian sense, and at worst unprovable, but it is a place in which we can live in our finiteness.

If one starts from LIR, many major issues can be summarized quite rapidly: if the Absolute is, as alleged, pure actuality, not susceptible to change, it is timeless and cannot exist for us. Kolakowski's 'metaphysical horror' sets in with the recognition that we cannot find a One or Absolute to justify our existence. We are convinced that like beliefs as a class, one's view of this state-of-affairs is personal. One can feel quite comfortable lying in a hammock strung up on the inside of the virtuous circle above rather than constantly sliding down its convexity. The problem of the co-existence of a 'Many' and 'One' is 'solved' by our principle that they are connected or overlap *physically* and logically, sharing one another's properties.

The One is as defined classically by Damascius [40], is perfectly simple, having no relationship to anything else, even to itself and is unknowable; only the fact that it cannot be known is known. It cannot be named, it cannot be 'higher' or 'lower', it is one of two 'nothings', one that lies above the One and Being, and one below' matter. Kolakowski gives the standard + scholastic arguments about the fallacies in this reasoning, also in its summary by Hegel: "Pure Being and pure non-Being are the same". This Absolute provides the ultimate support for the existence of anything, but can never be conceptually reduced to anything. It is *Nothing*, it explains nothing and is, according to Kolakowski, the *horror metaphysicus*. Kolakowski undertakes the task of finding in the *ego cogito* the meaning of the idea of existence, but he fails to discard completely the notion of "something else besides" the universe. Lupasco shows how the concept of the *ego*, the 'me' that is usually defined as *different* from the experience is not *totally* different or incommensurable with it. As noted above, Lupasco then throws his result to the wind and defines a 'realm' of affectivity as noted in Chap. 15 that is one of pure Being, without energy or logic, but which 'accompanies' our existence, not to say that it is parasitic on it.

10.3 Experimental Metaphysics and Metaphilosophy

For the purposes of this book, we use Michael Redhead's definition of metaphysics as the critical examination of the principles, concepts and fundamental presuppositions that lie behind modern physics. From time to time, in relation to specific topics, we will refer to the *Metaphysics* of Aristotle where a useful synergy appears. In our discussion, we cannot reproduce all the problems that have arisen in the first hundred years of quantum mechanics (QM) let alone provide definitive solutions. We can only suggest a few points at which an interpretation that would make reference to the LIR Principle of Dynamic Opposition (PDO) would provide an alternative approach.

10.3.1 Dynamic Opposition and Metaphilosophy

The metaphysical world-view that is implied by the Principle of Dynamic Opposition (PDO) is compatible with the metaphysical revision that has been "engendered by quantum mechanics" [41]. One does not have to have a prior 'orthodox' concept of reality in order to define the best possible active role for what is observed, namely, that dualities are present at all levels of reality, starting with that of the quantum field. The dualities in question have a kind of part-whole relation to the world, but one need not assume that at the end of this analysis, one will have captured all the essential aspects of the world. As noted above, one will not have, as a consequence, a 'Theory of Everything' (which was not an objective in the first place), but one will have a framework that can evolve in parallel with further development in the physical understanding of our universe. Our view is consistent with the work of Vedral [52].

Logic in Reality, as outlined in Brenner [3], includes a 'New Energy Ontology' which includes categories but redefines their characteristics. The absolute, binary concepts, exclusivity and exhaustivity are eliminated in favor of a dynamic relation between 'objects and forms' of thought. The role of such categories in ontology is essential in defining LIR as a conceptual structure that has additional explanatory power. In a categorial realist conception, as suggested by Thomasson [51], "providing a system of categories can be seen as a, or even *the* central task of metaphysics". We believe a robustly realist position is made more plausible by the principles of LIR, since they improve our ability to discern intrinsic divisions and redefine changes or movements in physical reality as the "entities" in the category of existence. For our purposes, it is not necessary to decide for an ontological or metaphysical reading of the term 'category': both can be used as they complement one another.

Summarizing outrageously, we claim that the principles of QM eliminate the possibility of analyzing wholes in terms of their parts or reducing wholes to them. At face value, if true, this statement implies a profound change in our metaphysical picture of the world. Both the orthodox and so-called hidden variables approach to the reduction fail, but as Redhead cogently points out, for very different reasons. "In the orthodox interpretation, there are not enough local properties of the constituents to sustain the reduction, while in the hidden-variable interpretation there are in a sense too many 'local' properties, but owing to the required ontological contextuality that we have argued for, these properties are not really local properties at all."

We agree with Redhead's first consequence—accepted in today's theory—that one cannot found the entirety of physical science on the properties of individual elementary particles. However, a second consequence which emerges from our theory breaks with all potential extensions of standard quantum mechanics of macroscopic phenomena. We offer an enhanced description of the properties necessary to sustain the reduction by including those implied by the PDO of coexisting actuality and potentiality. One clear advantage of our approach is that there are *no* aspects of the classical world that are somehow false in the basic theory. Further, we avoid the dangers of *scientism*, applying the methods of science to the subjective content of human thought by concentrating on the *processes* involved rather than on content *qua* content and the arbitrariness that would imply. These considerations provide an occasion to say what our approach, in the 'language' of ontolons and epistemons, is *not*. It is not an 'experimental metaphysics' in the concept of Redhead [41], as the latter still embodies the tautological assumptions of classical, propositional logic and its notion of truth in the methods of valid argument. Redhead's (well-intentioned) movement "from physics to metaphysics" moves further away, not towards, reality.

The role of categories in ontology, independently of the formal mathematical uses to which they can be put, is essential in defining LIR as a conceptual structure that has additional explanatory power. In a categorial realist conception, as suggested by Thomasson [51], "providing a system of categories can be seen as a, or even *the* central task of metaphysics". We believe a robustly realist position is made more plausible by the principles of LIR, since they improve our ability to discern intrinsic divisions and redefine changes or movements in physical reality as the "entities" in the category of existence. For our purposes it is not necessary to decide for an ontological or metaphysical reading of the term 'category': both can be used as they complement one another.

10.3.2 Anti-realism and Metaphilosophy

As we design our realist approach to the logic and dynamics of social systems, we should not forget that such approaches have been and probably always will be subject to anti-realist attacks. These say something like "the processes that we see operative in the society are purely mental constructions that have no proper existence". The concept of the reality of a social operator, outlined in the paper of Burgin and Brenner [9] would be contested, despite the reasonable argument for it (to us). Let us reproduce this argument briefly.

The contemporary American philosopher John Searle is perhaps best known for his so-called 'Chinese Room' thought experiment which attempts to offer a way of distinguishing a response to a question by a person as opposed to a machine. His [45] book, *The Construction of Social Reality* is useful in countering the above anti-realist argument. First of all, he is interested in a class of institutional facts which are those deriving from simple but real social interactions, those which he considers has a "huge invisible ontology" that constitutes social reality. The origin of this ontology is the capacity of the human brain to operate with characteristics that are both 'mental' and 'physical'. Searle supports a complex view of human rationality that goes beyond sharply delineated, and in our terminology binary, intentional content. The resulting picture is that of a reality independent of human representations—"external realism" (ER). This is the position that we have taken earlier in the discussion of "representation in information theory" in Chap. 11. Searle goes further toward a concept of realism that is compatible with the LIR dialectics: "Realism does not say how things are, but only that there is a way that they are". 'Things' includes processes as well as objects in a universe without an absolute independent space–time to worry about. These considerations are directly relevant to a social reality and its dynamisms. Giving the human brain the capacity to construct, as some anti-realists such as Maturana do in order to support their concept of autopoiesis, is a self-referential monster. Searle: "It is just a *non sequitur*, a genetic fallacy to infer from the collective neuro-physiological causal explanation of our knowledge of the external world to the non-existence of the external world." In standard philosophy, it is possible to counter arguments based on perception, but they require what Brenner has referred to as "catastrophic counterfactual" arguments whose probability is zero. Again, it does not follow from the fact that all cognition is within a cognitive system that *no* cognition is ever of a reality that exists independently of cognition.

Searle suggests that realism is attacked by people who would like truth, an epistemic principle, to coincide with reality, an ontological principle. Our thesis does not discredit naïve notions of truth and reality but redefines their relationship, following Lupasco, who stated that "truth is the truth of reality". External realism, again, says that there is a way that things are that is independent of representations of how things are. Reality is thus not constructed either by individuals or groups of individuals—society. Public reality also exists independently of representations or statements about that reality. According to Searle, *this* statement is not a truth condition but a condition of the *form* of intelligibility that such statements have as properties. Searle's thesis applied to money rather to facts about money illustrates our dualistic approach: facts about money can be both epistemically objective—money is a social construct—and ontologically subjective (internal reality). Searle considers that a socially constructed reality presupposes a non-socially constructed reality is a "transcendental" argument, but we find it a logical consequence of the coexistence of individual minds and society, immanent to them. Culture is thus a 'large' ontolon that describes collective intentionality, functions of phenomena that go beyond the content of the underlying physical reality. This work of Searle is most congenial to our study as well as important because he equates the rejection of realism and the denial of ontological objectivity with attacks in general on "epistemic objectivity, rationality, truth and intelligence in contemporary intellectual life". We unfortunately have to emend this to "all contemporary social and political life". Arguments against external realism go hand in hand with irrational extremism. Let us now look at some ways of combating it.

Some of the most pertinent examples of the Principle of Dynamic Opposition and the related property of non-linearity are to be found in the social sphere. In the Greece of Thucydides there was a regular alternation between democracy and tyranny in the city-states. In the West of today there is a similar regressive polarization trend toward autocracy in parallel with continued advances in technology and science. The French sociologist Michel Henry summarizes the current situation in the one word title of his [20] book—*Barbarity*. Henry's thesis is that since one cannot expect tyrants to have a proper understanding and appreciation of life, scientists must. If they abdicate this responsibility and fail to take into account subjectivity as a valid part of the real, they contradict and self-negate their own existence. As science loses its own foundations, culture which was excluded by has nowhere to go but underground. The nexus of the new ideological barbarity lies in the behavior of political leaders to reproduce, at a social level, the contradictions in science for selfish purposes. Scientific understanding of the dynamic oppositions appearing in social dynamics should help in bringing new relations to human societies. This understanding is present in the reflexive psychology of Vladimir Lefebvre [34] who defined opposite types of human reflexive behavior continuously reproducing themselves in social evolution. A principle of social evolution that is instantiated by generation of opposite references to external reality is discussed by one of the authors [22].

10.4 The Problem of Free Will in the Context of Philosophy in Reality

10.4.1 Free Will, Appearance and Reality

Philosophy in Reality can contribute to the free will problem in relation to the question of its compatibility (or incompatibility) to the determinism of natural phenomena. At the level of atoms Epicurus introduced the property of indeterminacy that he defined as *clinamen*. In modern physics, the free will theorem of Conway and Kochen [13] states that if we have a free will in the sense that our choices are not a function of the past, then, subject to certain assumptions, so must some elementary particles. Its physical representation is quantum uncertainty (or "inherent ambiguity" in the terms of Niels Bohr and David Bohm) which spans from the level of elementary particles to biological evolution and to the phenomenon of free will and consciousness (reviewed in [21].

The debate on whether or not human beings possess free will, defined as the capacity for choice not determined by prior events, continues unabated in the literature. One does not have to be a psychologist to understand the importance of this question to individuals in their search, conscious or consciousness, for an understanding of the world and their place in it. In both Eastern and Western antiquity, similar formulations of the existence of another, 'unseen' world co-existing with that of day-to-day conscious experience is to be found. The existence of the other as a reflection of a real dialectic interaction between appearance and reality in the context of our Philosophy in Reality provides a new framework for understanding the problem of free will.

Since the advent of quantum mechanics, the debate has taken on another dimension, since the inaccessibility of the quantum domain to direct experience places it in a similar philosophical situation to the world of religious belief. Starting from the original ideas of Werner Heisenberg and Niels Bohr on quantum uncertainty, the problem of fundamental indeterminism of the quantum reality for the explanation of the problem of free will was brought to modern interpretations of quantum mechanics. Ruth Kastner has analyzed the role of quantum logic and philosophy in illuminating the question of free will in several recent publications [28–28]. Kastner's transactional picture of quantum mechanics points to an additional conceptual structure for a real noumenal world beyond that of experience and space–time. She proposes that it is constituted in part by incipient transactions of which one, 'chosen' probabilistically will collapse to one actualized transaction. However, as the author clearly states, in a world of classical fields, in which energy can be distributed in arbitrarily small or large portions, there is no single obligatory outcome of the process.

Kastner's concept of the situation for a human subject is that his/her capacity for choice is governed by a 'mix' of quantum propensities but with an additional degree of (or for) freedom due to the complexity of the biological system involved. Deviations from the Born rule for the probabilities of the evolution of the system are possible because the relation between subject's physical state and the relevant choice available are continually changing; no violation for the quantum statistics for the quantum part of the process is required. The system that introduces several levels of the reflexive relation to externality according to Lefebvre [33] provides a continuous choice between the different types of reflexive behaviour and thus active interference into the probability of evolution of different paths at the quantum level. This choice is realized in the process of interaction with the external reality mediated by the communication with other subjects. The reflexive subject is not a free agency in the sense of Kane [25] who considers it in the way of causal indeterminism. On the contrary, a real choice appears via the reflexive control in complex systems that can perceive their potentiality as a particular phenomenon. The phenomenology of free will can investigate the feeling of potentiality which is based on the perception of the potential choice that is, in fact, grounded in the physical indeterminacy of quantum behavior.

10.4.2 Free Will as Self-forming Willings

Kane's postulate of the existence of a minimum set of self-forming willings (SFW, called also self-forming actions) in a free agent is only a postulate. It is not clear why any given SFW itself is immune to a vicious infinite regress. Further, the concept of a free agent or free agency itself begs the question of the existence of free will (the free agency principle: free agents have the power over the determination of their own wills). It is based on the possibly superfluous and misleading notion that 'we' are something separate from and in addition to our 'will', attitudes, beliefs, etc., as in the expression "attitudes the agent has". For, (1) either all that constitutes us as agents is determined, in which case it is impossible to see how that part of us which is our will has not also been determined, or (2) nothing that constitutes ourselves as agents has been determined, in which case determinism and the *appearance* of indeterminism are the only issues and not free will. The assumption of separate categories of event-causal factors and reasons to act in all free agent models [37] is accordingly unjustified. The entities designated as reasons-to-act are mental entities that have themselves causal powers; they *are* the causes of any subsequent choices.

Although Kane is an incompatibilist (claiming a dichotomy between determinism and free will where anyone must choose one or the other), in regard to the question of determinism, he provides a description [26] that is very close to the view that we develop in the frames of Philosophy in Reality. This view can be placed in the context of compatibilism claiming that free will and determinism are mutually compatible and that it is possible to believe in both without being logically inconsistent:

> One has to imagine a different kind of reality altogether – one in which physical systems are described in ways (say, in terms of multidimensional vector spaces or quantum wave functions) that translate into observable properties only probabilistically – not a familiar world with chance added, but a world in which indeterminacy is woven more deeply into the fabric of reality.

It is not only that, in LIR, (1) the absolute separation of determinacy and indeterminacy is avoided, but also (2) the separation between functional and causal categories of the psychological entities postulated as being involved in free will. Our description of the origin and structure of conscious phenomena is in terms of the contradictorial changes resulting from increasingly complex, recursive physical interactions. These processes have conscious and unconscious components, of which the conscious ones are the only mental entities involved in thought and knowledge, without separate mental representations or symbols. In the LIR view, the existence of such entities is as sufficient for an explanation of any concept of free will, or free intentional action, as being no different in kind from any other partially self-referential thought. 'Free will' in our view, is a term for a particularly vivid process of mentation that is fraught with contradictory elements, rational and irrational and also a strong tendency toward non-contradictory identity or diversity.

People believe 'absolutely' in their existence as free agents, or, just as absolutely, refuse to accept that such beliefs have *any* basis in reality. It is in the availability of alternative explanatory pathways that introduce a *relational* component that appears more abstract than that involved in a potential behavioral action. The LIR position is that the appearance of an abstract structure is also an appearance, but it would be better to start from the position that it is an energetic process. The LIR view is that it is potentialities, called here 'more abstract', that are the carriers of causal sufficiency, as well as the conscious appearances of freedom. Any abstracting capacity operates on the potentiality, and constructs the new noumenal essence on the basis of reduction of this potentiality. In other words, it is not the abstractness of the potentialities that are critical; in fact, they have all the real characteristics that are required in current views of free will: the possibility of non-actualization (failure) and biological functionality. In the LIR picture, there is no need to consider these characteristics purely functional to the exclusion of any causal relevance, since cause is also relational in the same sense as the entities: the entities are cause-(effects), and, of course, mappings between the organism and external world are categorized as processes in the ontology of LIR.

10.4.3 Free Will as Intuition

Our preferred way of looking at the phenomenon of free will, as a particularly strong and lasting intuition, is one that inverts the usual relation between intuition and knowledge-as-such. The resulting sense of modal insufficiency at the conscious, potential level is accompanied, dialectically, by the largely unconscious, actualized basis for choice and behavior.

Taking all the above into account, the phenomenon of free will can be said to instantiate several processes explicable in the system outlined in this book:

- Apparent free will exists, but only as an appearance in the conscious mind of an individual in opposition to and because of the unconscious knowledge of the lack of total 'freedom', that is, isolation from other individuals. The issue of compatibility is therefore a false problem; individual responsibility for one's actions *does* exist, but its source does not lie in free will, or the absence of it.
- We are in the presence of a reciprocal mutual instantiation of appearance and reality that corresponds to the contradictorial, LIR interpretation: appearance and reality, as real processes, can never both be fully actualized at the same time, in line with the basic principle of dynamic opposition defined in Chap. 3.
- If we as knowers are not totally external to what is known by us, not completely different from it, then we conclude that there are other knowers that are part of our known and vice versa.
- The source of our dignity *is* then in ourselves as knowers, but if we avoid the error of solipsism, the origin of the sense of moral responsibility can only come from the relation to other knowers, in other words, all human beings, and by extension, other beings. *A contrario*, one cannot find responsibility in oneself as an isolated agent. Since we are both 'not-other' *and* 'other' in the two-tier structure of LIR, a self-interest argument for morality holds.

The LIR position is thus that, paradoxically, two or more human individuals are entities in the categories both of Subject-Object and Non-Separability. An individual is no more isolated logically, psychologically and morally than he or she is economically. It is because our will is *not* free that we must try to insure the viability of the environment and so on. Issues of altruism vs. self-interest arise in model conflict situations such as the Prisoner's Dilemma. By far the majority of analyses start from the assumption that maximization of self-interest determines behavior. This is binary (identity) logic in yet another form.

The world described by LIR is one in which determinacy and indeterminacy are both woven deeply into the fabric of realty; they *are* reality. Kane also quotes Popper to the effect that "freedom is not just chance, but rather the result of a subtle interplay between *something almost random or haphazard and something like a restrictive or selective control* (emphasis his)." This "subtle interplay" is, of course, what we propose the dynamic logic of reality explicates. Kane proposes a 'fusion principle' in connection with the proposed feedback between neurons involved in consciousness and choice (the parts) and the neural net (the whole). The process is viewed as follows ([26], p. 151):

> One does not therefore have merely indeterminism or chance at the microlevel *followed* by a determinate effort, or a determinate effort *followed* by indeterminism or chance. Rather the indeterminism and the (determinate) effort are 'fused'; the indeterminacy is a property *of* the effort and the effort *is* indeterminate.

10.4.4 Free Will and Moral Responsibility

Dennett [15] has expressed the best compatibilist view: we have ultimate moral responsibility because the natural causes (in our brains) involved in determining our behavior do not *control* us. However, this is somewhat of a tautology that the antagonistic formulation, which does not require separation of 'causes' and 'us', and also permits indeterminacy at one level of description, can avoid.

In LIR, one can retain the libertarian belief that persons are most fully human and have greatest worth when they are ultimately responsible for their own destinies, but enable it to be reconciled with modern science (and logic) by adding 'and others' to 'own'. The origin of responsibility thus lies in the still unexplained nature of individual ontological personality, existing at the same time as that of others. The idea that that we are 'independent' sources of activity is as untenable here as it is in the economic world. Free will is both the consequence and expression of the "struggle and conflict" within, but also between people. It has a place in a scientific picture of the world that gives adequate ontological status to appearance as well as reality, and to their dynamic relations. LIR provides a way of understanding the fundamental relations of contradiction in existence as the origin of the phenomenon of free will.

10.5 Phenomenology: Understanding the World Through Human Experience

10.5.1 Appearance and Reality

Phenomenology as the philosophical study of the *structures* of experience and consciousness is concerned with the systematic reflection of consciousness and the phenomena that *appear* in acts of consciousness. This understanding of phenomenology goes back to Hegel; he considered phenomenology as an approach to philosophy that begins with an exploration of conscious phenomena as a means to grasp the absolute, logical and metaphysical essence that is behind phenomena. Phenomenology as a separate domain of philosophy was introduced by Edmund Husserl. It deals not with objects as they are in themselves, but as they appear in the intentional acts. This becomes possible by "bracketing" all assumptions about the existence of an external world and the subjective aspects of how the concrete object is given. This bracketing procedure Husserl called *epoché*. Like *being* and *becoming*, and as discussed in Sect. 10.4 in relation to free will, we include *appearance* and *reality* in the group of philosophical dualities which can receive improved interpretations in our logical and dialectical framework. One also cannot discuss 'reality' as we have tried to do in this book without reference to the philosophical positions referred to as realism and anti-realism (and we do this later in this book Chap. 15). The application of Logic in Reality to phenomenology as a domain of knowledge suggests a separation, for discussion, into *two* domains, one within natural philosophy (Chap. 13) and integrating science, and one in which what we believe is a misleading separation from science results in restricting it to that of philosophy *tout court*. Our strategy in the remainder of this chapter will be to introduce a dualism of method: we will study (1) the recovery of essential concepts that can be made *by* phenomenology and (2) the recovery that can be made, in part as a consequence of the preceding *of* phenomenology as a valid partner in the development of new knowledge.

10.5.2 The Recovery by Phenomenology

The objective of our study is the recovery of both dialectics and semiotics from reductionist interpretations and to their reunification in a synthetic paradigm centered on meaning. To a certain extent, the entire enterprise of phenomenology as it emerged in the twentieth century was also an effort at recovery, in this case, of the value of first person experience as an integral part of knowledge. Unfortunately, but understandably in view of the overemphasis on hard science and mathematics to which it was a reaction, the practitioners of phenomenology adopted an anti-scientific stance that reduced the value of their contribution. Among other things, 'science' was referred to as a monolithic methodology and body of knowledge which clearly is not. The authors of this monograph have direct personal experience of the use of the scientific method (and thus some understanding of what science is not).

The approach that we adopt is to identify the valuable insights developed by phenomenology and recast them in a context, which hopefully will become more familiar, of dynamic interaction between a first and third person phenomenology. In this approach, the scientific value of the 'view from within' gains in validity by accepting the interactions that exist between it and the 'view from without' [3].

10.5.3 *The Recovery of Phenomenology*

The value of an approach based on Logic in Reality lies not so much in its capacity to generate new *information* as such. That is the proper domain of science, especially, today of information science. What LIR does provide, however, are contexts and structures for evaluating and naturalizing qualitative knowledge that is contradictory, fuzzy, incomplete and inconsistent. We therefore consider it a necessary step for progress in its application to real problems of the individual and society. As we have seen, LIR is a process-ontological view of reality, applying to theories, to trends and tendencies, rather than to objects or the steps in a state-transition picture of change. LIR does not replace classical binary or multi-valued logics, but reduces to them for simple systems and situations. The critical categorical feature of the LIR process ontology is the non-separability of opposing dynamic elements of a process and extends to ideas, people and societies. This recovers phenomenology by "naturalizing" its main principles and ideas, as noted in the next Section.

10.6 The Naturalization of Phenomenology

The Naturalization of Phenomenology is case of a general process of ascribing philosophical properties to phenomena which in our sense of natural philosophy must be in conformity with the principles of the logical evolution of natural processes described by LIR. This form of phenomenology is thus a core part of our overall conception of Natural Philosophy integrated in Chap. 13. The concept of a Natural Philosophy, which has evolved into science, raises the further complication of what *it* is and how it differs from philosophy *tout court*. Whitehead's focus, anticipating-phenomenology, is on the "texture of human experience". Coherence is also defined as the body (*sic*) of our theoretical knowledge, but at the same time it is "a basic inventory of concepts", resembling an inert Peircean classification of signs. Much closer to natural philosophy, which Brenner [2] discussed, is Whitehead's concept of 'concrescence', extended in his *Process and Reality* [53] toward real systems: "The coherence, which the system (organism) seeks to preserve, is the discovery that the process, or concrescence, of any one actual entity involves the other actual entities among its components. In this way the obvious solidarity of the world receives its explanation."

As one example, let us look at the Heideggerian principle of the 'clearing', that is, a specifically human domain in which people experience themselves and the world as a part of existence that is irreducible to physics and chemistry, as the expression goes. There is no difficulty in seeing what Heidegger's intent was in creating this epistemological domain: it was to confer on it characteristics that would emphasize its importance as a piece of the intellectual 'furniture' of human cognition. In our approach, the logic and dialectics of cognition that are grounded in the logical-physical properties of reality simply confer a further ontological status on this domain that places it not in conflict with 'science' but in interaction with it.

LIR differs in its treatment of the subject-object duality from all of the vast discussion of this topic, including that of standard phenomenology, and can be characterized as a non-naive dualistic realism. The LIR view, critical for the discussion of free will and the origin of moral responsibility, is that the world is ontologically deterministic and epistemologically indeterministic, in the contradictorial relation suggested above.

The reworking of some additional subjects within philosophy seems necessary to eliminate what amount to category errors, ascribing to natural phenomena logical properties which are essentially fictions. These subjects were thus the main targets for the naturalization proposed by Brenner (2015). We will now discuss how this grid might apply to them, in other words, 'move' them to Natural Philosophy where possible. The precursor to naturalization as a process Feldman [17] was 'scientization', defined as the incursion of empirical science into areas of knowing previously the purview of theology and philosophy. An example of this is the attempted naturalization of intentionality Jacob [24], which has been only partially successful. If one looks explicitly for precedents to naturalization in philosophy, one finds that the term is generally used to describe a kind of grafting of philosophy onto science studies. This dead end suggests that the entire domain requires reconceptualization.

Naturalization of phenomenology was the focus of the special issue of the journal Progress in Biophysics and Molecular Biology in 2015. The need of phenomenology for natural science for understanding non-classical phenomena was discussed in the paper of Rosen [43].

10.6.1 Husserl and the Naturalization of Phenomenology

In contemporary philosophy, phenomenology occupies a strange position: on the one hand, it seems to place major ontological value on appearance while at the same time denying access to it by science. Phenomenology could thus have the not very felicitous designation of anti-scientific realism. Perhaps it is best seen from an historical standpoint as a reaction, perfectly understandable in the twentieth century, to the hegemony of reductionist scientific views of a universe in which a transcendental deity no longer had a place.

The'naturalization of phenomenology' might however be considered an oxymoron to the extent that phenomenology was designed by Husserl to exclude physical reality. The most natural view of physical reality is simply the world independent of our thought processes (which are nevertheless part of it and without which it would not have meaning). Husserl's approach of focusing on the unique lived character of experience available to individuals has led to the familiar definition of phenomenology. Familiar, also, is Husserl's later bracketing of the question of the existence of the natural world and its relation to experience *vs.* a realist ontology.

As discussed below, the informational approach leads to alternate, and from an LIR standpoint preferable, descriptions of the co-existence of meaning and non-meaning in the world.

Roy [44] confirms that Husserl's phenomenological theory of intentionality is based entirely on the assumption that "truly adequate characterization of intentional phenomena can only be achieved by renouncing all forms of naturalism, both ontological and epistemological." We agree with Roy that this seems to be too high a philosophical price to pay, but it is important to see in what sense the role of nature is to be understood. Although in Husserl every intentional state is conceived as one entity mediated by another, his relation of interpretation resembles that of Peirce with an ontological 'cut' between the interpretation and its intentional correlate. There is semantic 'identity within difference', but no real interaction between them. This is the kind of standard inert conception that we refuse to accept.

10.6.2 The Naturalization of Heideggerian Phenomenology by Capurro

Capurro and his associates [11], see also [6], have sought to formulate the essential phenomenological insights of Heidegger in a way that makes a concept of progress in philosophy as accessible as progress in science. In such a doctrine, a relation to ethics is always natural and present. It 'is' ethics, a doctrine whose implementation as a moral program in the social arena might be facilitated if its compatibility with science were more visible.

Capurro [11] suggests that phenomenology in Heidegger's *Being and Time* [19] can only be understood if one pre-understands Husserl, particularly what his '*epoché*' means, i.e. putting the question of reality in brackets, in order to see the 'essence' (*eidos*) of things as they appear to consciousness (*noesis*) as their correlate (*noema*). Husserl's discovery was that there is no consciousness without a correlate and this correlate he called 'intentionality'. Husserl was arguing 'with and against' Descartes but instead of asking if things are out there (*ex*-ist) at all, he 'forgot' this question methodologically in order to analyze the 'whatness' of things. Husserl was a mathematician and not a natural scientist, so his logic is 'Logic outside Reality', in our terminology. Heidegger's 'existential turn' was to understand intentionality as the basic phenomenon and to start with being-in-the-world itself. This is the reason why Heidegger does not use the word/concept of consciousness, but remained all his life a phenomenologist '*sui generis*' with regard to Husserl.

Heidegger's step beyond Husserl's '*epoché*' is a logical one in terms of 'Logic in Reality': 'reality' is the world which precedes its explicit interpretation within the framework of natural science. Husserl was well aware of this 'revolutionary' step with regard to his *epoché*. He revised his foundation and introduced the concept of *Lebenswelt* as a foundational horizon also for the specific perspective of natural sciences. While Husserl developed a phenomenological model of pure consciousness, similarly to Plato, the link to being as presence, formulated in the ancient philosophy by Aristotle, became the central point of the philosophy of Heidegger. We can say, however, that Heidegger's role is not comparable to Aristotle's and his influence on natural sciences including physics is much less pronounced than it could be. He admitted that his *"Sein und Zeit"* is a completely imperfect attempt to enter into a temporality (*Zeitlichkeit*) of being (*Dasein*) after Parmenides [18]. This definitely imposed certain limitations on the development of the metaphysical basis of modern natural science.

In Heidegger, the question of what 'reality' or 'being' means must be analyzed from the outset in order to avoid dualisms such as the Platonic ones to which Husserlian phenomenology remained 'ideologically' bound. In other words, starting with a doctrine dependent solely on physical science or, in Heideggerian terms, with being-in-the-world, is not the answer (or 'solution') to the question of what reality or being mean. It is rather a new 'original' way of recognizing the contingency or the 'fact' that we 'are' in the world and that the world 'is' our 'correlate' as Husserl would say. Further, however, the interpretation of what things are and of what being itself means is no less contingent or 'historical' (*geschichtlich*). If we want to be critically or phenomenologically aware of this we have to start with questioning our own place 'in' the world and not with a pre-established theory. In the first chapters of *Being and Time,* Heidegger describes the 'crisis' of the (natural) sciences regarding their foundational concepts which he calls, following Husserl, 'regional ontologies'.

The standard anti-realist view is that science cannot put us into direct contact with reality due to our physical limitations and consequently cannot be validated. In contrast, Heidegger's view of science, in Capurro's interpretation, is that science is a valid method of inquiry, provided it is not allowed to contaminate our thinking with its Promethean programs. These include the search for machines with greater than human intelligence, placing humans in suspended animation for centuries, etc. These concepts still contain references to a subject-object duality (correlationism) which is *prima facie* unable to establish a satisfactory relation to *things-in-themselves.*

In discussing Heidegger's concept of science, in particular modern science, Capurro assigns its ontological ground as human being-in-the-world rather than a transcendental constitution in subjectivity [10]. We claim simply that Lupasco, carefully considered at this level, makes the question moot. When 'subjectivity' is part of a dialectical ontological ground of subjectivity and objectivity, as it is in the Lupasco and LIR systems, there is no relation of exclusion required, for any physical reality.

10.6.2.1 The Noema

The noema (Fr. *noème*) is usually defined as an intentional object of thought, but there is no consensus about its ontological and/or epistemological character. Solomon, discussing Husserl's concept of the noema, repeats the dogma the *Sinne* (senses) stand independent of the meaning (*Bedeutungen*) of linguistic expression.: It has become all but axiomatic among analytical philosophers that that there is no meaning apart from language and the noema links language and sense experience. "The noema embodies both the changing phases of experience and the organizing sense of our experience. But these two 'components' are not separable, for all experience requires meaning, not as an after-the-fact luxury in reflective (JEB/AUI: Kantian) judgments but it order for it to be the experience of anything. Logic in Reality and our concepts of the units of reality in thought as well as more familiar processes clarify this standard view by referring to the real processes underlying thought. Meaning for us as subject is present in phenomena independently of language, although is required for operation on it.

10.6.3 The Naturalization of Phenomenology and Logic in Reality

A discussion such as ours of the terms 'naturalized' or as 'admitted by the natural sciences', from the viewpoint of the phenomenology of Husserl and Heidegger means restraining the horizon of understanding of being and beings to the kind of outline (*Entwurf*) of a reality, standard in modern thought, still based on subject-object duality (excluded third). That phenomena 'answer' correctly the questions asked by modern science from this perspective was seen by Husserl as one possible perspective within the larger *Lebenswelt* horizon with important consequences, also 'political' ones. It became not only a leading horizon of understanding but also an 'imperative' one against all other possible perspectives with regard to what is finally 'true' or not and of what 'is' truly means. While Husserl developed the model of consciousness by making a "phenomenological reduction" of consciousness from the temporal materialized world, Heidegger aimed to return to the question of how consciousness operates in the finite world shaped by time as a horizon of being. It shapes the being as presence (*Dasein*) and manifests itself as the horizon of being (*Horizont des Seins*) [19]. The Being-in-the-World (*Dasein*) emerges from the abstract Being (*Sein*) through time and appears as a multiplicity of the forms of existence for which we have proposed the term ontolons.

The LIR expressions "degrees of actualization and potentialization of processes" then include history (not only natural evolution but human history as well) into a new conception of what 'reality' means when trying to understand 'things' from themselves (Husserl's maxim: "Go to the things themselves, not from our methodological pre-conceptions of them, i.e. give them the opportunity to question us.) The term 'phenomena' means, according to Heidegger letting things 'announce' themselves, before one starts to tell them that, in order to have a rational conversation, they have to obey our rules; this is impolite. '*Après vous*,' Levinas, another *sui generis* phenomenologist, used to say. The logic of the included third is a polite and humble logic in a foundational sense in which politeness and humility is enlarged to our relation to ourselves, to others and to the world—opposed to a logic of domination and mendacity.

10.7 Positioning Phenomenology

A new trend in philosophy called Speculative Realism by Sparrow [49] and others makes a necessary critique of certain stubbornly idealistic aspects of 'modern' phenomenology. Sparrow talks about the 'end' of phenomenology, stating in essence that phenomenology does not establish the connection to reality necessary to ground science. His Speculative Realism attempts to solve the problem by replacing subject–object correlationism by a form of reasoning that emphasizes the correlation between thinking and being. Speculative Realism avoids positioning phenomenology "underneath the natural sciences" as they are naïvely understood in order to safeguard the philosophical primacy of the former.

On the other hand, the most obvious applications may be in the areas of cognitive science and the further naturalization of phenomenology. The questions of the existence and nature of free will may be amenable to new readings, providing the support for a more realistic basis for this intuition. In the philosophy of knowledge and belief the disappearance of the barriers of standard bivalent logic may also offer advantages. We have given a few examples in this book of advances in theoretical and experimental science that also imply the operation of the PDO, and one can expect that further ones will be made. It is the *absence* of antagonistic interaction between man and machine, seen as elements in the category of *Separability*, which confirms the uniqueness of the human, living condition.

Two aspects of the LIR approach indicate possible new directions for philosophical and metaphysical research:

- LIR avoids the dichotomy of chance and necessity as the driving principles of change.
- LIR is grounded in both physics and experience and provides a principled separation of phenomena into pre-semiotic, without internal representation, and semiotic, involving emergent qualitative complexity, in which the equivalents of both efficient and final causes are functional.

LIR as logic in this construction is able in principle to naturalize the operations that are involved in internal mental phenomena and their expression in behavior and action. LIR can thus be considered as a framework theory of mental phenomena, including intuition, considered as an essential dynamic property in the world, a part that is not separated from the rest of it. It could support an emergent, interactive model of normative function and representation that might resolve, in a novel way, the opposition between the so-called factual, non-normative world and the mental, normative world.

The line of dialectical discussion that we have adopted in this book is grounded in the antagonisms of physics and their logical interpretation of logic at higher levels of reality. At the cognitive and social level of communication, the areas of most interest are those of interpersonal communication, and these may be approached directly also from a perspective that starts from a phenomenology of opposites, the oppositions of *yin* and *yang* and Heraclitus, as noted above. Among the Russian precursors of

Lupasco, one should also list Mikhail Bakhtin and his ideas of a (phenomenological) 'tension' between unity and difference (see Bakhtin [1]).

For Heidegger, ontology and phenomenology are not two distinct philosophical disciplines. These terms characterize philosophy itself with regard to its object and its way of treating that object. Philosophy is universal phenomenological ontology, and takes its departure from the hermeneutic of *Dasein*, which, as an entity—'analytic' or what we call proper part of existence—establishes a guide-line for all philosophical inquiry "at the point where it *arises* and to which it *returns*." LIR provides the framework for a parallel interactive interpretation of ontology and phenomenology.

10.8 Phenomenology and Semiotics

As discussed in Chap. 9 on the 'units' of knowledge', Peirce coined the neologism 'phaneron' to designate something like the *idea* of Locke forming the 'immediate content of awareness' as opposed to phenomenon. "The phaneron is what appears, as it appears, even if, as it appears, it appears to be more than mere appearance". The difficulty is that Peirce made no ontological commitment regarding his concepts. He wrote specifically that his 'phaneroscopy' (phenomenology) had nothing at all to do with the question of how far the 'phanerons' it studied correspond to any realities. Elsewhere Peirce says that the phaneron is the totality of what is "in any sense present to the mind, quite regardless of whether it corresponds to any real thing or not" [46].

10.8.1 Brier's Claim that Semiotics is More Powerful Than Phenomenology

As noted first in Chap. 9, the semiotician Sören Brier has claimed that the semiotics of Charles S. Peirce can deliver the missing philosophical framework for phenomenology through his semiotic conception of a fundamental triadic structure of the universe. Semiotics is a transdisciplinary doctrine that studies how signs in general—including codes, media, and language, plus the sign systems used in parallel with language—work to produce interpretation and meaning in human and nonhuman living systems such as pre-linguistic communication systems. For Peirce, a sign is anything that stands for something or somebody in some respect or context.

Brier [7] has pointed out the weaknesses in much of standard philosophical and sociological thought in general and phenomenology in particular. He thus writes that Husserl, Heidegger, Merleau-Ponty, and Luhmann were unsuccessful in developing a proper philosophical framework for phenomenology [7]. They did not offer any adequately deep picture of things in themselves in relation to appearance. Thus, both Wu Kun (see below) and Brier state that Husserl's transcendental idealism makes no contact with the world or the natural sciences. In particular, Wu [56] has provided a unique metaphilosophical analysis of Husserl from an informational standpoint. We will simply reiterate here his key conclusion, namely, the complexity of human individual and social existence and experience cannot be captured by reference to a "life-world" and "intersubjectivity" that exclude the detailed functional role of information, information processing and the operation of the physiological and psychological structures necessary for that processing.

Peirce's introspectional method leads him to ascribe a foundational role to chance and spontaneity as causal factors. The subsequent classification of reality into "Firstness, Secondness and Thirdness" follows. We argue that Peirce's system remains *only* one of classification of phenomena in terms of representation by signs and symbols. It adds nothing that helps us understand the ontological basis of the way the world evolves. Signs obviously refer to existing therefore natural objects, but they are abstractions from them, and the corresponding relationship is not natural-philosophical as no dynamic, ontological relations obtain between the entities involved. The above considerations are the basis of Brier's thesis outlined in his major 2008 book *Cybersemiotics* whose sub-title is *Why Information is not Enough.* Our reply to him is that semiotics is not enough since it does not incorporate in its 'flesh', to use the Merleau-Ponty and Lakoff concept, the dynamic, energetic changes from actual to potential, present, and absent that the term 'information' refers to.

10.8.2 The 'Flesh': Merleau-Ponty and Lakoff and Johnson

The contribution of Merleau-Ponty and his existentialist followers can be seen as a necessary reaction to the transcendentalism of Husserl [47]. In his *Phenomenology of Perception*, Merleau-Ponty emphasized the role of the body in human experience. In the terms of Logic in Reality, his body-image is a kind of "included middle" between the mental and mechanical-physical domains. The human subject is inseparable from both his body and the world. Kauffman and Gare [29] have provided an authoritative discussion of Merleau-Ponty's trajectory which led him ultimately to embrace natural philosophy as the framework for his thought.

Unfortunately, by focusing on the human body (the 'flesh') as the primary philosophical entity, Merleau-Ponty effectively eliminated any foundational role for the properties of the underlying physical components of the 'flesh'. In our view, these properties that are not only consistent with consciousness and life but underlie their emergence as real and not epiphenomenal. It is perhaps more than anecdotal to note that in a competition for a key position in the *Collège de France* in the 1950s, Merleau-Ponty was chosen over his contemporary rival—Lupasco. The marginalization of Lupasco can be dated from this event. As Brenner [4] has suggested in a paper in French, it is high time for this 'noble' marginalization to end. Finally, as discussed in another reference to Capurro's work [6], phenomenology should never, *pace* Husserl, have been conceived of as being a science in the first place.

In the light of information theory, the weaknesses of modern philosophy, from Kant through Husserl become apparent [54]. It is the existence of information, even more than, but in concordance with, the logic of and in reality (LIR), that breaks the traditional absolute separation of subject and object. Phenomenology in the informational terms of Wu Kun can be included in Natural Philosophy, but in its initial explicit formulation by Husserl it cannot.

10.9 The Reconstruction of Phenomenology

The domains of phenomenology and the sciences of perception and cognition are closely linked. Smith [48] appears to subsume all of these disciplines under that of an ontology of mind, and we agree with this philosophical concept, with the proviso that an adequate physical and logical foundation be defined for that ontology. Our approach is developed further in this Section.

10.9.1 Husserl and the Phenomenological Representation of Consciousness

Unfortunately, in our opinion, because the details of perception and subsequent cognitive processes presented in this approach are not grounded in physics, phenomenology fails as a science of cognition. In Smith's delicate phrase, Merleau-Ponty, as well as Heidegger and Sartre, seem to have sought "a certain sanctuary for phenomenology beyond the natural sciences". [12] points out that both Merleau-Ponty and Blanchot attempted to replace Husserl's transcendental idealism with a radical empirical idealism but wound up reproducing the former's metaphysics due to a lack of a proper picture of the dynamics of a reciprocal "mind–body" interaction. Derrida, on the other hand, discovered, in the sense that Brenner has discussed [3], a meta-phenomenology in the "quasi-transcendental", a difference *(difference)* which is neither transcendental nor empirical but the interval between them (an LIR "included middle") as the condition of possibility of phenomenology.

The semiotician Sören Brier has pointed out, as noted above, the weaknesses in much of standard philosophical and sociological thought in general and phenomenology in particular. He thus writes [7] that Husserl, Heidegger, Merleau-Ponty and most recently Luhmann were unsuccessful in developing a proper philosophical framework for phenomenology, because they did not offer any adequately deep picture of things in themselves in relation to appearance. Both Wu and Brier state that Husserl's transcendental idealism makes no contact with the world or the natural sciences. In particular, Wu Kun has provided a unique analysis of Husserl from an informational standpoint [56]. We will simply reiterate his key conclusion, namely, the complexity of human individual and social existence and experience cannot be captured by reference to a "life-world" and "intersubjectivity" that excludes the detailed functional role of information, information processing and the operation of the physiological and psychological structures necessary for that processing.

The phenomenological representation of consciousness developed by Husserl bears a certain complementarity or even similarity to the semiotic representation of perception developed by Peirce, which has been mentioned by several scholars [50, 35, 8]. It brings forth the structure of consciousness and provides a kind of non-formal model of it. Husserl proposed a phenomenological theory, which is quite different from semiotics but has similarities in development of the semiotically-oriented logic of relations. These relations can be modeled by Peircean algebras [42], which combine sets, relations and various operations linking the two in a unifying framework. Peirce and Husserl both anticipated the linguistic turn of modern philosophy with its challenges and shortcomings, but with very different implications for an adequate Philosophy in Reality.

Brier claims that the semiotics of Peirce can deliver the missing philosophical framework through his semiotic understanding of the fundamental structure of the universe. Unfortunately, as discussed earlier, Peirce's own framework is based on several assumptions about that structure derived ultimately from introspection. This leads him to ascribe a foundational role to chance and spontaneity as causal factors. As noted above, his subsequent classification of reality into "Firstness, Secondness and Thirdness" follows, but it remains just that, a system of classification of phenomena in terms of representation by symbols which adds nothing that helps us understand the ontological basis of the way the world evolves. The claim that Peirce made an enormous contribution to logic is true, but on close inspection his logic, including his system of graphical representation, is an elaboration of standard linguistic, truth-functional logics that is still incapable of describing real phenomena without excluding their essential dynamic properties. Let us now see, therefore, what a picture of phenomenology in relation to a philosophy of information might bring to the table as science.

10.9.2 Wu Kun and the Informational Standpoint

Briefly, in the work of both philosophers and neuroscientists such as Searle and Deacon as well as Wu, the basic worldview of natural science, namely, that consciousness is part of nature is upheld. The advantage of the informational standpoint, or "stance", is that information serves as the unifying concept between the fields of physics, biology, neuroscience and mind. In this sense, the philosophy of information is a more scientific and reasonable explanation of the mechanism of human understanding than in phenomenology.

Wu's criticism of the method of Husserl and others is that due to the classical phenomenological suspension of natural objects and the human body, its description of the mechanism of human understanding is one-sided and incomplete. The philosophy of information provides a framework for a mechanism of awareness of natural objects, a role for material devices and instruments, human physiological and cognitive and an historical fifth, intermediary-based dimension. In a word, the philosophy of information enables the actors in the cognitive process to return from their phenomenological suspension. The new integrated model of cognition includes the classical phenomenological interpretation, so that the philosophy of information goes beyond the phenomenological one.

Note, however, that due to the fact that phenomenology does not have a natural ontological foundation, we do not claim that the metaphilosophy of information is in and of itself either phenomenology or a naturalized phenomenology. Perhaps in the sense of transcendence and inclusion the philosophy of information could unify phenomenology into a "meta-phenomenology", but, in any case, the philosophy of information is not determined by phenomenology. Rather, the theory of Wu [55], and the logic of Brenner [3] are, together, candidates for the ontology (the metaphysics), called for by Depraz [16], "compatible with the most acute phenomenological insight, an ontology which can be used to reject reductionist tendencies by exhibiting their fundamental non-duality."

10.10 Two Alternatives to Standard Phenomenology

Wu [54] expressed his views of the critical role of information as the basic concept for metaphilosophy. In the light of information theory, the weaknesses of modern philosophy, from Kant through Husserl become apparent. It is the existence of information, even more than, but in concordance with, the logic of and in reality (LIR), that breaks the traditional absolute separation of subject and object. Although Husserl found a way of beginning to describe the reality of consciousness, his in part one-dimensional phenomenological reduction maintains, in another form, the disastrous (for human society) polarization of standard bivalent logics. As a hermeneutic process, Husserl's bracketing requires a redefinition of essence itself and in relation to existence.

In place of standard phenomenology, we first propose an informational ontology in which we as humans have (self-evidently) access to "things-in-themselves". The philosophy of information of Wu Kun and logic in reality of Lupasco at the first sight are not phenomenology because phenomenology is the subjective intent of interpreting the structure of the world. We live, however, also as indicated in the dialectics of Lupasco, by adhering to route on which "the natural noumenon's own movement explains the world" [5].

We have come to the conclusion that while functional and operational definitions of information have their role to play in practical applications, they also fail to capture both the intrinsic dynamics of complex processes and the nature of information itself which is instantiated in them. It is thus, fortunately or unfortunately for some, that, e.g., for the understanding of knowledge and knowledge propagation, drastic modifications of points in standard epistemology, also foreseen in LIR, have to be made. Information in the new epistemology that is to be developed provides the indication of indirect existence even as it exemplifies indirect existence, and its consequences for the emergence of new entities and meaning, in the contradictorial relationship that is formalized in LIR.

Using an informational paradigm illuminates work such as that of Lakoff and Johnson [32] on "The Embodied Mind" (cf. 10.8.2). Here, the physical and physiological structures of the mind and body interact in an informational complex which Wu designates as the informosome, discussed further in Chap. 11. To talk about information at any but the lowest computational level requires attention to the entire objective dynamics and subjective idiosyncratic patterns, consistencies and inconsistencies, styles of the human actors involved in its generation and reception, and its historical dimensions, and so on.

This is a difficult task for science, but it is a more correct position from which to start than by eliminating the complex informational properties of existence from consideration. To quote Wu: "Informational activities have their origin not in the pure "life world" of an idealized subject, but in the objective world of their own interactive existence and evolution." One must maintain in the forefront of one's mind the synergism between the physical form and the informational form and the rules of their evolution to fully understand their unified relationship.

Application of the philosophy of information thus brings out the ontological domain of indirect existence as part of total existence, something that is objective, complex, has meaning and value and thereby constitutes the elusive thing-in-itself that does not require further empirical proof in the reductionist classical sense. We can relate Wu's work both to prior standard theories of information and to new promising work on the dynamics of complex real processes by Deacon [14]. As we have suggested elsewhere, the theories described in this book may constitute part of a new transdisciplinary paradigm, in which information has a central role, together with but more general than approaches from the systems and complexity standpoints. Application of Wu's Metaphilosophy of Information, supported by Logic in Reality and Deacon's complexification of dynamics could contribute to resolving critical outstanding issues in the field of information. In our view, it is this perspective, proposed by Wu as early as 1987, that "fills the vacuum" of speculation in previous dialectical philosophy.

10.11 The Human Mind: The Biology of Phenomenology

The LIR Principle of Dynamic Opposition and its categorial features are instantiated at all levels including the mental level. Lupasco presented a theory of mental activity and consciousness that is grounded in the most well-established dualisms of human nervous systems of activation and passivation, polarization and depolarization of nerve cells. The dualism of LIR does not involve 'flirting' with a classical, indefensible dualism of body and mind, one neurological and the other mental. However, why should the sole presence of a dynamic, interactive dualism, a non-reductive physics involving potentialities as well as actualities, insure preservation of the specific qualities of human thought, that is, provide an adequate explanation for its operation?

A successful scientific theory of human cognition should account for its subjective phenomenality, the fact that things have appearances, and that appearances can also be shared.

The ultimate objective is to bridge the explanatory gap between a phenomenological mind (consciousness) and brain and to naturalize phenomenology, bringing subjective conscious experience within the purview of natural science. Any view of consciousness and mind must account for both external events as they are cognized—phenomenological data - and their internal processing. However, phenomenology cannot be taken into cognitive science as such without substantial modification. To be scientific, phenomenology thus requires some form of 'naturalization', but there are some inherent limitations in both cognitive science and other current approaches. Most of these theories involve a kind of realism and objectivism that either eliminates all subjective, 'irrational' dimensions of the phenomena under study as 'unscientific' or assign them to a second-rate logical category. Biologically, a phenomenon can be considered as the subjective consequence of a physical interaction with an individual organism that requires detection or measurement. What is detected is selected as potentially functional for the organism as a self or subject [38]. Thus naturalization of phenomenology expands the original reference of Husserl to consciousness to living systems and potentially all functional complex systems operating as "measuring devices" in their environment.

Smith's strategy for the naturalization of phenomenology was to extend the concept of intentionality associated with human consciousness, viewed as a physical process to the natural world [48]. However, in his 'Unionism', the unity of the mental and physical is to be understood as the product of a categorical constitution and not as a factual reality. Naturalization in this way is alleged to avoid "reduction to causal or computational processes along the lines envisioned by current cognitive science," but it is difficult to see how intentionality defined in this way would not be epiphenomenal. LIR challenges the structure of both cognitive science and Smith's critique of it as embodying classical concepts of cause and separability, *e.g.*, between internal and external. Like standard cognitive science, LIR can propose a "close and explicit relationship between brain mechanisms, their existence within an organism, and a surrounding world with which there is an unceasing coupling" [39], but the basis for such coupling needs to be spelled out.

10.12 Conclusion. From Essence to Existence

The issues discussed in this chapter converge on a new concept of phenomenology. In antiquity, Pyrrho and his followers developed the concept, similar to the *epoché*, of suspending all judgments about non-evident matters, referring mostly to ethics. Husserl in his definition of *epoché* referred to consciousness which reveals its own inherent system of meaning through this process. In our view, the process of bracketing (*epoché*) refers not only to our consciousness but it can be considered a natural informational process which characterizes the abstracting capacity of complex systems. The appearance of the *Ding für uns* from the Kantian *Ding an sich* can also be viewed as a natural *epoché*. Physical laws appear as the computational principles bracketing all other possibilities and, epistemologically, bringing events to existence. There is a place for the *epoché* in the structure of natural philosophy *if* it becomes a part of the process in which dynamic oppositions appear and resolve. Phenomenology arose as a study of the structures of consciousness in relation to the phenomena that appear in acts of consciousness. We can move from such anthropocentric appraisal to the natural dynamic process and reconstruct phenomenology in relation to natural phenomena. Naturalizing phenomenology is an important challenge for the future development of the entire structure of philosophy. Metaphysics dealing with the fundamental first principles of being, metaphilosophy in its understanding of the informational connection between perception and reality, and phenomenology as the study of the structures of experience and consciousness become tied together in the synthetic construction of Philosophy in Reality.

References

1. Bakhtin, M.M.: Collected Writings, 6 vols. Russkie Slovari, Moscow (in Russian) (1996–2012)
2. Brenner, J.E.: Process in reality: a logical offering. Logic Log. Philos. **14**, 165–202 (2005)
3. Brenner, J.E.: Logic in Reality. Springer, Dordrecht (2008)
4. Brenner, J.E.: Stéphane Lupasco: Vers la fin d'une noble marginalité. Sociétés **1**, 71–78 (2013)
5. Brenner, J.E.: The naturalization of natural philosophy. Philosophies **3**, 41 (2018)
6. Brenner, J.E.: The 'Naturalization' of the philosophy of Rafael Capurro. In: Kelly, M., Bielby, J., (eds.) A Festschrift in Honor of Rafael Capurro; Wiesbaden, pp. 45–64. Springer Fachmedien, Germany (2016)
7. Brier, S.: Cybersemiotics. Why Information is Not Enough. University of Toronto Press, Toronto (2008)
8. Broekman, J.M.: Firstness and phenomenology—Peirce and Husserl on attitude change. Prosp. Legal Semiot. **1**, 37–77 (2011)
9. Burgin, M., Brenner, J.E.: Operators in nature, science, technology, and society: mathematical, logical, and philosophical issues. Philosophies **2**, 21 (2017)
10. Capurro, R.: What is information science for? A philosophical reflection. In: Vakkari, P., Cronin, B. (eds.) Conceptions of Library and Information Science, pp. 82–96. Taylor Graham, London (1992)
11. Capurro, R.: Angeletics—a message theory. In: Capurro, R., Holgate, J., (eds.) Messages and Messengers: Angeletics as an Approach to the Phenomenology of Communication, vol. 5, pp. 5–15. ICIE Series, Munich, Germany (2011)
12. Chung, C.Y.: The relation of Merleau-Ponty and Blanchot to Derrida. LiBRI. Linguist. Literary Broad Res. Innov. **1**, 9–23 (2010)
13. Conway, J., Kochen, S.: The free will theorem. Found. Phys. **36**, 1441–1473 (2006)
14. Deacon, T.: Incomplete Nature: How Mind Evolved from Matter. W.W. Norton & Co, New York (2012)
15. Dennett, D.C.: I could not have done otherwise—so what? J. Philos. **81**, 553–565 (1984)
16. Depraz, N.: The phenomenological reduction as praxis. J. Conscious. Stud. **6**, 95–110 (1999)
17. Feldman, R.: Naturalized epistemology. In: Zalta, E.N. (ed.) The Stanford Encyclopedia of Philosophy Summer 2012 Edition. The Metaphysics Research Lab Center for the Study of Language and Information, Stanford University, Stanford (2012). Available online: https://plato.stanford.edu/archives/sum2012/entries/epistemology-naturalized/

18. Heidegger, M.: Ponderings II-VI. Black Notebooks, 1931–1938. Indiana University Press, Bloomington (2014)
19. Heidegger, M.: Sein und Zeit.. English translation: Being and Time. Harper Perennial, New York (Originally published 1927) (2008)
20. Henry, M.: La Barbarie. Grasset, Paris (1987)
21. Igamberdiev, A.U.: Time, Life, and Civilization. Nova Science Publishers, New York (2015)
22. Igamberdiev, A.U.: Evolutionary transition from biological to social systems via generation of reflexive models of externality. Prog. Biophys. Mol. Biol. **131**, 336–347 (2017)
23. Iurato, G., Khrennikov, A.Y.: Entropy, externality and human evolution. Biosystems **191–192**, 104130 (2020)
24. Jacob, P.: In: Zalta, E.N. (ed.) Intentionality, The Stanford Encyclopedia of Philosophy (Winter 2014 Edition). Available online: https://plato.stanford.edu/archives/win2014/entries/intentionality/
25. Kane, R.H.: Free Will and Values. State University of New York Press, Albany (1985)
26. Kane, R.H.: The Significance of Free Will. Oxford University Press, New York (1996)
27. Kastner, R.E.: The Transactional Interpretation of Quantum Mechanics: The Reality of Possibility. Cambridge University Press, Cambridge (2012)
28. Kastner, R.E.: Understanding Our Unseen Reality: Solving Quantum Riddles. Imperial College Press, London (2015)
29. Kauffman, S., Gare, A.: Beyond Descartes and Newton: recovering life and humanity. Prog. Biophys. Mol. Biol. **119**, 219–244 (2015)
30. Kolakowski, L.: Metaphysical Horror. University of Chicago Press, Chicago (1988)
31. Ladyman, J., Ross, D.: Every Thing Must Go: Metaphysics Naturalized. Oxford University Press, Oxford (2007)
32. Lakoff, G., Johnson, M.: Philosophy in the Flesh. In The Embodied Mind and its Challenge to Western Thought. Basic Books, New York (1999)
33. Lefebvre, V.A.: The fundamental structures of human reflection. J. Soc. Biol. Struct. **10**, 129–175 (1987)
34. Lefebvre, V.A., 1982.: Algebra of Conscience. A Comparative Analysis of Western and Soviet Ethical Systems. Springer, Dordrecht, Netherlands (2001)
35. Mullin, A.A.: C.S. Peirce and E.G.A. Husserl on the nature of logic. Notre Dame J. Formal Logic **7**, 301–304 (1966)
36. Nicolescu, B.: Nous, la Particule et le Monde. Éditions Le Mail, Paris, France (1985)
37. O'Connor, T.: The agent as cause. In: Kane, R., (ed.) Free Will. Blackwell Publishing, Victoria (2002)
38. Pattee, H.H.: Cell phenomenology: the first phenomenon. Prog. Biophys. Mol. Biol. **119**, 461–468 (2015)
39. Petitot, J., Varela, F., Pachoud, B., Roy, J.M. (eds.): Naturalizing Phenomenology: Issues in Contemporary Phenomenology and Cognitive Science. Stanford University Press, Stanford (1999)
40. Rappe, S.: Skepticism in the sixth century? Damascius' «Doubts and solutions concerning first principles». J. Hist. Philos. **36**, 337–363 (1998)
41. Redhead, M.: From Physics to Metaphysics (Tarner Lectures). Cambridge University Press, Cambridge (1995)
42. De Rijke, M.: The logic of Peirce Algebras. J. Logic Lang. Inform. **4**(3), 227-250 (1995)
43. Rosen, S.M.: Why natural science needs phenomenological philosophy. Prog. Biophys. Mol. Biol. **119**, 257–269 (2015)
44. Roy, J.: Saving Intentional Phenomena: Intentionality, Representation and Symbol. In: Petitot, J., Varela, F., Pachoud, B., Roy, J.-M., (eds.) Naturalizing Phenomenology, pp. 117–147. Stanford University Press, Stanford (1999)
45. Searle, J.R.: The Construction of Social Reality. The Free Press, New York (1995)
46. Short, T.: Peirce's Theory of Signs. Cambridge University Press, New York (2007)
47. Singer, J.D.: "Menaced Rationality": Husserl and Merleau-Ponty on the crisis and promise of science. Prog. Biophys. Mol. Biol. **119**, 245–256 (2015)

48. Smith, D.: Intentionality Naturalized? In: Petitot, J., Varela, F.J., Pachoud, B., Roy, J.-M., (eds.) Naturalizing Phenomenology. Issues in Contemporary Phenomenology and Cognitive Science, pp. 83–110. Stanford University Press, Stanford (2011)
49. Sparrow, T.: The End of Phenomenology. In: Metaphysics and the New Realism. Edinburgh University Press, Edinburgh (2014)
50. Spiegelberg, H.: Husserl's and Peirce's phenomenologies: coincidence or interaction. Philos. Phenomenol. Res. **17**, 164–185 (1956)
51. Thomasson, A.: Fiction and Metaphysics. Cambridge University Press, Cambridge (1999)
52. Vedral, V.: Decoding Reality: The Universe as Quantum Information. Oxford University Press, Oxford (2010)
53. Whitehead, A.N.: Process and Reality. An Essay in Cosmology. Gifford Lectures Delivered in the University of Edinburgh During the Session 1927–1928. Macmillan, New York. Cambridge University Press, Cambridge (1929)
54. Wu, K.: Fundamental questions of philosophy with the fundamental shift of philosophy. Hebei Acad. J. **4**, 11–21 (2011) ((In Chinese))
55. Wu, K.: The complex characteristics of multidimensional emergence of human knowledge: a Critique of the one-dimensional, simplistic features of Husserl's phenomenological reduction theory. Hebei Acad. J. **4**, 25–30 (2014) ((In Chinese))
56. Wu, K.: The basic theory of philosophy of information. In: 4th International Conference on the Foundations of Information Sciences, Beijing (2010)

Chapter 11
Information; Convergence of Science and Philosophy

11.1 Introduction. A Philosophical Triple

Information is a field for which both formal and non-formal descriptions are possible. Formal approaches are not incorrect, but they can become so when they equate to bivalent or multivalent linguistic logics and standard category theory which in our view have limited explanatory value. Non-formal or partly formal approaches to information have been much less developed, but they offer interpretations of phenomena in terms of dialectics and a non-standard logic of processes, Logic in Reality (LIR), as outlined in previous chapters.

The two lines of thought are present throughout the human and social sciences and philosophy. Again, the second group has received less attention as being allegedly unscientific and non-rigorous, as most of the time it is. In this chapter, we will address the nature of information, especially in its relation to meaning and to the convergence we see between science and philosophy. We will provide a summary of the standard well-known interpretations and then of ours based on the parallel use of dialectics and LIR.

Meaning and information are often discussed separately, and in Chap. 9 we have pointed out some of the difficulties with standard conceptions of meaning in relation to semiotics or semiosis. Even placing them in conjunction has significant consequences—in what sense are meaning and information the same or different? In this section we study what can be gained in the understanding of both meaning and information by looking more closely at the dialectics and logic of the relation between them.

We suggest that in fact meaning, information and communication constitute a *philosophical triple*. It possible to discuss information separately in abstract terms, but as stated earlier our major interest is in information for human beings, which immediately implies a process of informing and hence of communication as communicating between them. These process concepts will inform our approach in this and the next chapter.

J. E. Brenner and A. U. Igamberdiev, *Philosophy in Reality*, Studies in Applied Philosophy, Epistemology and Rational Ethics 60, https://doi.org/10.1007/978-3-030-62757-7_11

11.1.1 The Origin of Meaning

In a way quite similar to the question of the emergence of life from inanimate matter, the question of the origin of meaning from apparently meaningless physical and chemical signals is still the subject of major debate. We summarize the concept developed in LIR in the following three points:

1. Information generation and transfer occur in nature as a natural process that is constituted by and requires energy.
2. Information is present throughout nature, but its conversion to meaning requires organisms capable of recursive processing of it.
3. Meaning is constituted by the all information valuable to a living organism for its survival, reproduction and well-being.
4. Information processes follow principles applicable to energy and energetic processes in general.

Many authors have noted the complexity of information and the difficulty of giving a 'single, clear' definition of it. Attempts to do so are typical of standard substance ontologies, where firm definitions—identities—are automatically given preference. The failure of such attempts suggests that a major categorial error is being made. We therefore make the following lapidary statement which we will try to justify in what follows:

11.1.1.1 Meaningful Information is Reality in Potential Form

It is derived from the Lupasco/LIR conception of consciousness outlined in [31] which basically looks at the real dialectical interactions in and between internal and external, and internalizing and externalizing processes as they move between potentiality and actuality.

In order to understand how our approach to logic and dialectics adds value to the discussion of meaning, information and communication and their interrelation, we need to summarize briefly the major developments in these fields made in the 20th Century and in the last twenty years, both in the West and East, especially in China. The concomitant formulation of theories of information in which it was related to the clearly developing information society resulted in a new rationale and methodology for information studies. From the point of view of this paper, the overall movement was from non-dialectical mathematical theories to dynamic ones, including critical aspects from biological science. This is to be welcomed, yet each of the kinds of theories outlined below suffers in some way from the absence of explicit recognition of the dialectical/logical principles underlying the operation of 'Information in Reality' [7]. This term was introduced by Brenner in his initial study of the logic and metaphysics of information, based on his Logic in Reality.

11.1.2 The Prelude in Communication

In this section, we present a further part of our reconceptualization of Natural Philosophy in terms of meaning. We bring together the essential insights of Chap. 9 on the units of meaning—ontolons—and define the contours of a new concept of meaning. If a natural philosophical statement means something at all, has or conveys meaning, this fact does not and cannot stand in isolation. The meaning in question is not a solely epistemological concept, but puts Natural Philosophy in a dynamic relation with a dialectics and a logic. In this 'meaning in motion', we move from semantics to dynamics where meaning functions as an operator that can effect change. The parameters of this process of structures 'meaning in action' necessarily involve those of information and its communication. It is fascinating to note that the nature of information, often viewed today as 'just' a philosophical question, in fact requires the best available concepts of physical science of the origin and fundamental structure of the universe to be correctly addressed. The modern origin of a theory of information is often ascribed to the work of Shannon on theories of communication, in which information is related to the removal of redundancy in a formal manner. This approach was specifically intended *not* to address or define meaning, but it nevertheless gave rise to many other formal theories of which the most modern and comprehensive is that of the mathematician Burgin [12].

In the 1950s, in our view, the discussion of information was almost fatally polarized by the statement of mathematicians, Norbert Wiener and his follower John Wheeler, to the effect that information was not matter or energy. This statement gives the sense that there is something truly 'different' about information, but scarcely gives an indication of what it might be. Such theories are not wrong but they are incomplete in that the scientific and logical origin of the dialectical, that is, in our terms real physical interactions described is not specified. Faced with the multitude of theories of information of all kinds, Brenner gave his 'personal synthesis' [8] of several competing theories, from which a few key benchmarks for philosophy can nonetheless be taken as discussed further in this section.

Given their origin in communication science and its significant dependence on technology, it is not surprising that theories of information emphasize formal concepts and standard bivalent logics. From the human, *natural* philosophical standpoint—we do not separate these—communication as the transfer of information is a human activity, and its nature and changes must follow the rules and logic of the latter. In the next Chap. 12, we will discuss the theory of communication in detail, in particular that of Loet Leydesdorff. For the time being we just note the dialectic aspect of the understanding of information in Chinese, as pointed out by Leydesdorff [28]. There are two words for "information". Both words contain two characters: one, 'sjin sji', corresponds to the mathematical definition of information as uncertainty, 'sjin' means letter of reliability, and 'sji' means message. The second, 'tsjin bao,' means information but also intelligence; 'tsjin' means situation or status, and 'bao' means report. In other words, it means information which informs us, and which is thus meaningful.

11.1.3 Relational Dialectics

At the end of the 20th Century, Leslie Baxter and Barbara Montgomery proposed a theory of interpersonal communication in terms of patterns of interaction resulting from such dialectical tensions and designated it as 'Relational Dialectics' (RD). The major difference between RD and our presentation is that the former takes its data from individual experience and, totally correctly from our standpoint, seeks a structure in the human relationships which are in constant change. As the authors put it [2] "we all experience different tensions in different ways. The classification of kinds of tensions is a combination of basic existential properties—openness and closedness, equality and inequality, autonomy and connectedness and others involving affect—favoritism and impartiality, instrumentality and affection. The language of this theory is quite simple, but it can be appreciated above all for its description of contradiction is positive terms, that is, terms which assign it an essential ontological status and value. We summarize briefly the key approaches of RD:

- The monologic approach sees contradictions in binary, either-or, mutually exclusive terms, but does include the concept of movement from one extreme to the other.
- The dualistic approach defines contradictions as separate entities, unrelated in nature. An example would be the study of a human relationship limited to one partner.
- The dialectic approach is directed toward the operation of multiple point of view possible within a binary relationship.

The underlying concepts assumed in RD are ones to which we have called attention to earlier: non-linearity (here, of relationships); change; contradiction (or counteraction) as a fundamental fact of existence; and in the human sphere, communication as central in the evolution of relational contradictions. We find this theory congenial as it is basically a process approach; it includes a concept of contradictions as a dynamic interplay between opposing forces that condition and define one another. However, applications of Relational Dialectics will not be discussed further in detail, as its concept of contradiction as "grounded in human nature which forces ethics" is too superficial. On the one hand, RD could benefit from a broader approach to actual human exchanges as its 'data' as an empirical base. On the other the concepts of logical dialectics in LIR and *their* grounding in nature—the Principle of Dynamic Opposition—could provide a sounder theoretical base. Critical sensibility and aesthetic moments in relationships are real processes that in our view absolutely deserve a more rigorous approach. We encourage the reader to take a 'break' so to speak from the 'heavy going' in this book, look at some of the references to RD that we have provided, and make his or her own judgment as to the relations between it and LIR.

11.2 Logic in Reality, Meaning and Information

We have shown above that meaning and information are not identical but also not separable at the cognitive level, the one of interest here. They follow the logic of real processes that is the thread running throughout this entire current paper. It is thus an integral part of Philosophy in Reality, which includes the Philosophy of Information of Wu Kun. In the changes in stance or perspective, dialectical movement **is** both epistemological and ontological. Knowing not is not totally separate from the Known, what we know. It is true that the "map is not the not territory", but stated in this simplistic way, the relation restored or recovered by LIR is obscured.

From our newer ontological perspective, the actors are the same but the relation is logical and contradictorial in the sense of Logic in Reality. The following consequences can be deduced from this consideration. Epimenides as an individual can be regarded as an element of the set which signifies this set, having *discovered* it in the sense of Capurro. The set (the society of Cretans) acquires its own dynamics having acquired the property of being 'liars' in conflict (antagonism) with, but not totally separate from, Epimenides. Different possibilities for the dynamic behavior of a system arise from this. What is important to note is that the 'dynamic behavior' with which we are concerned is that of real persons, individuals and groups, and their relations, dependent on their deep psychology

11.2.1 Geometry/Position or Energy/Force; Change

In seeking the grounding for philosophy and knowledge and their communication, one is confronted by at least two age-old, unresolved arguments. One is whether (1) space, defined by position and geometry or (2) energy, defined by movement and change is more fundamental – has ontological priority—in the universe. Another is whether matter is somehow constituted by our consciousness of it, the doctrine of anti-realism, or whether consciousness evolved from or together with matter. Those who prefer the first elements of these two pairs, including the major figures of Norbert Wiener and John Wheeler mentioned above, also favor the position that information is *sui generis*, neither matter nor energy, present in some way in the form of digital bits, the so-called It (things, processes) from Bit position.

The proposal of Logic in Reality outlined above and in [4] is that what emerges from the still unknown ground of the universe, position and energy, or statics and dynamics, can be rigorously described in dialectical terms as two different, opposing but non-separable aspects of that ground, a position similar to that taken by Diaz Nafria and Zimmermann [19]. Digital bits, now considered part of knowledge, are an emergent epistemological phenomenon—the 'Bit-from-It' position, and the overall situation can be described by an 'onto-epistemology', whose components are the 'ontolons' and 'epistemons' described in Chap. 8.

If, as we argue energy is fundamental, as outlined already in Chap. 1, so in our view are its properties, in any real system, of actuality in a non-separable relation to potentiality. Change is a consequence of our existence within an overall energy gradient moving locally from higher to lower grade forms, finally to heat (low energy photons). The movement of elements itself is sinusoidal, from predominantly actual to predominantly potential and the reverse. No element goes to the absolute, idealized limit of 100% one or the other, except in trivial cases in which there is no interaction between the initial and final states, as in snooker balls falling into pockets. The hypothetical pre-thermodynamic universe is supposed to be something like a quantum vacuum, with elements moving into and out of existence, without reference to time. As attempts to show a direct coupling of such a world with ours, we feel that they are not required for our subsequent analysis. The desire to demonstrate such coupling must be placed in the same domain as that of religious beliefs which by definition cannot be the subjects of scientific proof, except of their *own* existence.

As proposed by Lupasco and discussed by Brenner in [4], whether a dialectics of energy and position, or of consciousness and non-consciousness or of a polarized position of absolute opposites is preferred is a question of individual psychology. Lupasco saw individual human beings as instantiating change and hence preferring the stability conferred by belief in identities, stasis, rather than the dynamics of diversities. This dialectics, the dualistic oppositional properties of energy is reflected in the properties of information. The concept of information as an 'immaterial' substance, involving meaning in some way, has been available since antiquity [15]. For us information is as material as energy. Information and energy are both the same *and* different, as the focus of the mind moves from one aspect to the other.

The major difference is, of course, that information is alleged to be and to convey meaning, and the emergence of meaning from a meaningless substrate has been impossible to explain. In our view, the question of the emergence of meaning is equivalent to that of consciousness as a higher order of mental processing by animals capable of that processing, related to their survival. We do not admit any tautology here; meaning and the emergence of meaning are not identities, and the fact that no explanation of the emergence of meaning exists does not mean that the ascription of dialectical properties to information is invalidated.

11.2.2 Why Information is Enough

The cybersemiotic approach of Søren Brier is outlined in his major book, *Cybersemiotics* (2008). The title reflects is program of defining a new semiotics that in his view emerges from the computational processing of knowledge. The subtitle is *Why Information is not Enough* and refers to the limitations of standard logical and computational theories of information. Brier refers here to the lack of an adequate rigorous framework for philosophy, now including phenomenology, in an adequate relation to the natural sciences.

As noted in Chap. 9, Brier among other semioticians has based their underlying world view on proposals of Charles Sanders Peirce for logic and semiotics, in fact of a theory that implies their equivalence. Peirce made a classification of phenomena into those instantiating categorial properties of Firstness, Secondness and Thirdness, roughly related to the degree of interaction present. The problem for us is that the properties selected have, by Peirce's own admission, no grounding in any physical reality. They thus rejoin the positions generally referred to as anti-realistic. The semiotics of Peirce, and accordingly of Brier, is a description of the world, including living systems, in terms of 'signs' to which human beings can give an interpretation = meaning. Biosemiotics has become a highly active field of inquiry designed to explicate the role of complex signs in science and knowledge in general.

In our view, the approach of Brier and other semioticians displaces the emphasis from the ontology of the world to its epistemology without giving an adequate reason or substitute for it. Peirce did not claim that his system was based on any ontological commitment as to the ground of his signs, but that did not prevent him or his followers from assigning major hermeneutic value to them. Let us restate here the advantages of the LIR physical view of the energetic aspects of the origin and propagation of information, without reference to signs. Information inheres in both the physical and epistemological evolution of real processes and hence serves as a concept unifying the physics, biology and neuroscience of mind.

The somewhat apodictic statement by Lupasco cited above that physical processes, and hence their informational content *are* meaning is supported by the Kaufmann-Logan notion of biotic or instructional information [24]. It is also a dualistic approach in which intrinsically meaningless Shannon information, the lowest level that characterizes a physical system without the self-reference present in living systems, is contrasted with biological information which always entails meaning. It emerges from the material-energetic structure of the latter in a process of which the 'structure' is a dynamic part. As originally emphasized by Lupasco (see Chap. 16) "everything is structure" or better a *structuring* (*structuration*), and Logic in Reality describes the evolution of these structurings.

11.2.3 Information in the Presence- Absence Dualism

The basic philosophical position of Logic in Reality requires a dynamic interaction between opposites at and in respect to metaphysics (Chap. 10). We have seen above the dialectics and logic of the relation between the degree of actuality and potentiality of opposing process elements, as well as the new entities which can emerge from them (included middle or third term states—'T-states'). A key pair of physical/metaphysical opposites, to which we ascribe a physical dialectical meaning, are presence and absence. We have noted that LIR ascribes value, in ways that are more explicitly scientific as well as logical, to concepts generally considered negative—contradiction, inconsistency and vagueness. This approach is echoed by the work of the biologist Terrence Deacon of a metaphysics of incompleteness [16]. Deacon defined what is missing from theories of information in a paper with that title [17] and that this absence is an essential part of its content. For us, also, information refers to something that is not totally present, or is not present yet.

Deacon's key concept is that information is a relational property of systems that emerges from (1) constraints of signal probability, discussed by Shannon; (2) signal generation, by Boltzmann; and (3) those required for an apparently teleological dynamics, essentially those of Darwin in regard to evolution. We refer the reader to Deacon's major work, *Incomplete Nature* (2012). The three properties reflect three levels of entropy reduction that is an informational 'architecture' of recursivity. There is a rough parallel here to the notions of syntax, semantics and pragmatics, here, the pragmatics of Darwinian survival. The hierarchy of levels is that of data, content and significance, and the principles of LIR permit an interaction between them, rather than an absence of interaction as between quantum and non-quantum levels of reality.

At the cognitive level, according to the Lupasco Principle of Dynamic Opposition (PDO), presence and absence correspond to consciousness and unconsciousness and actuality and potentiality in a non-intuitive contradictorial manner. Forgetting as also an active process, and in the complex dynamics of mind, a 'fact' or concept that was present and then relegated to the unconscious is an actuality with a potential for reappearing in consciousness.

11.2.4 Information and the Laws of Thermodynamics

The discussion of complex real processes still suffers from the lack of an appropriate language that takes into account both intrinsic and relational properties of a system at the same time. Deacon defines constraints as relational properties, but LIR amplifies this by the rules for their evolution of their potential as well as actual aspects. Ulanowicz [41] has made an extension of the Deacon approach by connecting concepts of entropy to the Third as well as the Second Law of Thermodynamics. He defines entropy in a relation to a degree of system constraint (actuality) and its conjugate state which, in the residual freedom interpretation of the Third Law, can deliver meaning. A picture of entropy and information whose terms are always relative implies that, like those of quantum physics, they do not commute [36].

In several papers, Igamberdiev further develops the idea that the Third Law of Thermodynamics is more important for understanding information and life than the Second Law, which is considered the basis of Prigogine's dissipative structures. The Third Law establishes the reference state with the lowest entropy in relation to which the order (described as information) can be referred [20]. This state, according to the Third Law, is achieved at the temperature of absolute zero (0 °K). However, living systems operate at temperatures near 300 °K, in fact far from this reference state. It has been suggested that they maintain a long-lived cold decoherence-free internal state (the internal quantum state), within macromolecular structures which is achieved by applying error-correction commands to the internal state and by screening it from thermal fluctuations [22]. Iosif Rapoport was the first who suggested describing the stability of genetic structures by introduction of special thermodynamic principles explaining the persistence of low entropy in living systems [35]. Mental processes could be associated with such long-lived internal states present in the nervous system [23]. This could be sufficient for Lupascian T-states, and a corresponding non-Boolean logic, as applicable physical preconditions.

In our picture, the difference between non-Boolean domains and those in which classical (Boolean) logic and the law of the excluded third operate is determined by the degree of interactivity and sinusoidal movement between them (primarily actual to primarily potential and vice versa, alternately and reciprocally); communication occurs between actual and potential states on a macroscopic level. Stated in this way, the Principle of Dynamic Opposition (PDO) goes beyond thermodynamic principles. It requires implications in our expanded definition in addition to mathematics for description of its operation in complex processes, given their non-Boolean evolution. The operation of the PDO thus functionally replaces the notion of a 'temperature' in metabolic cycles. According to Nicolescu [32], the Principle of Dynamic Opposition is not and should not be described as a thermodynamic principle. The thermodynamics of energy underlies all phenomena in opposition, but it does not characterize all of them. Ontolons, as described above in our picture, are constituted by predominantly actual states, but these are always partially potential states. Together they can lead to the emergence of new entities (T-states). Further complexification of ontolons at the interpersonal level generates perpetually evolving socially organized structures [21]. In the next sections, we will summarize a further approach to the science and philosophy of information. We will then characterize it as part of Natural Philosophy in our new synthesis construction outlined in the Chap. 13.

11.3 Information as an Operator

11.3.1 Burgin's General Theory of Information

In a recent book [12], Mark Burgin has presented an extremely complete approach to a General Theory of Information (GTI) based on a thorough analysis of information processes in nature, technology and society. In particular, Burgin gives an extensive treatment of how information is modeled by mathematical structures. As Burgin and Brenner have shown elsewhere [13], Logic in Reality supports many of the conceptions of the properties of information in the GTI. Brenner offered a "non-representational" logical (in the LIR extended sense of logic) theory of information that complements the semiotic, "representational" theory. The essential binding of the two terms is indicated in Quieroz and El-Hani [34]. With Burgin, we claim that it is necessary to see information, better the information generation process, not as the output of a symbolic operator but of a natural and social operator. In the GTI, information is characterized by a system of principles which are both mathematical and philosophical. We will not list all of them here, but the second of his Ontological Principles, the General Transformation Principle O2, describes the essence of information in a broad sense as the potential (capacity) of things, both material and abstract, to cause changes (transform) other things. When this capacity (potential) is actualized, it becomes a nature or technology operator, acting on systems. The operational essence of information is further emphasized by his Ontological Principle O5, the Interaction Principle, which states that transaction/transition/transmission of information takes place only in interaction. Thus, it is reasonable to distinguish potentialized and actualized components of information, whose evolution follows the pattern of Logic in Reality, as discussed above. Energy *is* information in a broad sense, according to our reading of Burgin's Ontological Principle O2, and is thus the most fundamental natural operator. From this point of view, natural information operators are present in all natural systems. Also, information exists in the form of portions of information. A portion of information, for Burgin, is or can be considered (treated) as a separate entity. For instance, information in a word, in a sentence or in a book is a portion of information, and each such portion is an operator in its own right. This is consistent with our picture of the ontolon as an active unit of knowledge, with a causal capacity for change.

In the language of LIR, it would be more appropriate to say that since process elements are the information themselves, "portion" refers to the energy of the element that is more or less actual or potential, as the case may be. It is important to state in this connection, however, that the categorial feature of Non-Separability operates both within and between ontological levels of reality or complexity. The rules of the LIR logical framework are the same, and they will apply to both basic information-as-process and to the emergent higher-level meanings that can involve different receivers of the communicated information.

A further consequence of the LIR process approach to information is that words or semes are not looked at in isolation but in a context of their actual transmission and reception as non-abstract entities by specific actors. By the LIR Axiom of Non-Identity, no real phenomenon at one time can be identical to that phenomenon at another time. The possibility of differences in the meaning of the allegedly "same" word follows naturally.

In the book indicated above, Burgin's extremely complete approach to a General Theory of Information (GTI) is based on a thorough analysis of information processes in nature, technology and society. In particular, Burgin gives an extensive treatment of how information is modeled by mathematical structures. As Burgin and Brenner have shown elsewhere [13], Logic in Reality supports many of the conceptions of the properties of information in the GTI. Brenner has offered a "non-representational" logical (in the LIR extended sense of logic) theory of information that complements the semiotic, "representational" theory. The essential binding of the two terms is indicated in Quieroz and El-Hani [34]. With Burgin, we claim that it is necessary to see information, better the information generation process, not as the output of a symbolic operator but of a natural and social operator.

In the GTI, information is characterized by a system of principles which are both mathematical and philosophical. The GTI, in Burgin's terms, treats information at a very high methodological and mathematical level. It provides tools for specifying the characteristics of all existing information theories which, together, constitute information science, and of types of information, ethical and semantic, for which semantic meaning is the essential feature.

From another standpoint, Burgin's theory is not so much a General Theory of Information as a Theory of General Information. For us, what are ontologically primary are the phenomena themselves, which have meaning embodied as the information necessary for the survival and reproduction of living beings. These phenomena follow rules that are a combination of principles that are digital and analogue, mathematical and non-mathematical, Boolean and non-Boolean.

Burgin further states that his General Theory of Information "encompasses all existing information theories only potentially". Accordingly, one must have both it and "a variety of special information theories". If we accept this most interesting point as a basis for discussion, a further modification of Burgin's conception suggests itself: the GTI he has defined is in fact *Meta-Theory* of Information, a theory of *Theories* of Information. As we have frequently pointed out, however, there is not and does not have to be an absolute separation between a theory and its associated meta-theory, which coexist dialectically. There is a clear, intriguing parallel here with the approach of Wu Kun and his definition of the Philosophy of Information as a *Meta-Philosophy.*

We conclude that what Burgin has proposed should be called a *Theory of General Information.* It extracts, or abstracts, the general characteristics of information processes independently of their substratum of physical (energetic) properties, that is, their reality. All of the mathematical aspects of his GTI then apply to that abstraction, and a 'meaning' of those aspects exists, but it is tautological. It is a closed, epistemological system, which, like classical logic, is topic-neutral. It cannot found or impact on an ethics. It is a part, but only a part, of the Foundations of Information Science and of an eventual Unified Theory of Information in the sense of Hofkirchner.

11.3.2 Information as a Natural Operator

We claim that it is impossible to provide a Philosophy of Reality that does not include a reference to how information is *actually* transferred in and between subjects. This follows a very useful definition of information by Kolmogorov to that effect [27]. We have referred earlier to the LIR view of probability as non-Kolmogorovian, that is, the limiting probabilities of 0 and 1 are not included in the distribution. When this capacity, which is a potentiality in the LIR sense, is actualized, it becomes a nature or technology *operator*, which acts on different systems. The operational essence of information is further emphasized by the Burgin Ontological Principle O5, the Interaction Principle, noted above which states that transaction/transition/transmission of information takes place only in interaction. Thus, it is reasonable to distinguish *potentialized and actualized* components of information, whose evolution follows the pattern of Logic in Reality.

11.3.3 Energy as Information

Among the many authors cited by Burgin in support of his formal approach to information, some appear close to our position from a completely independent standpoint. Scarrott [38], cited by Burgin, considers energy and information as the two fundamental causal agents, i.e., natural operators, acting in the natural world. Overextending this approach, Bekenstein [3] and others have claimed that the physical world should be seen as being made of information itself. However, we reject this and the even more radical point of view expressed by Wheeler [42] who claimed that every item of the physical world is information-theoretic in origin. In this view, all such information would be indeed be composed of a multitude of information operators, e.g., information in an instruction is an information operator, a *system* or *function* operator. Brenner [5, 6], however, points out that views such as those of Wheeler and Bekenstein can lead to misunderstandings about the correct ontological relation of priority between information and matter-energy. It is the latter that is primitive, and failure to recognize this has often led to unnecessary idealizations of the concept of information. The issue of the 'physicality' of information is the subject of intensive on-going debate (information as a "physical essence"). Thus, we can conclude with Kaye [25, 26]: *"Information is not merely a necessary adjunct to personal, social and organizational functioning, a body of facts and knowledge to be applied to solutions of problems or to support actions. Rather it is a central and defining characteristic of all life forms, manifested in genetic transfer, in stimulus response mechanisms, in the communication of signals and messages and, in the case of humans, in the intelligent acquisition of understanding and wisdom". In other words, natural information operators are pervasive in all walks of life.*

11.3.4 Information in Natural Objects and Processes. Self-regulation

Information present in natural objects is a natural operator, *e.g.*, the genetic information stored in DNA. According to Loewenstein [29], information, in its connotation in physics, is a *measure of order*—a universal measure applicable to any structure or system. It quantifies the instructions that are needed to produce a certain organization. Burgin and Simon [14] also demonstrated that information has been and is the currently prevailing force for evolution both in nature and society. Smith and Szathmary [40] discuss evolutionary progress in terms of radical improvements in the representation of biological information. All these processes are initiated and controlled by information present as a natural operator.

Information as a natural operator is very important for self-regulation of various systems in nature. Self-regulation in a broad sense is the property of a system to regulate its internal environment (state self-regulation) and external behavior or functioning (phase self-regulation) in order to maintain a stable, constant condition.

Any self-regulating system is an operator, specifically, a *self-operator, or natural self-operator*. An important peculiarity of biological systems, such as organisms, ecosystems or the biosphere is that most parameters of these systems must stay under control within a relatively narrow range around a certain optimal with respect to existing environmental conditions. Thus, to achieve stability in its functioning, a biological system performs self-regulation, becoming a self-operator. In this process, the impact of the environmental information operators provides information about outer changes. The impact of the organism information operators provides information about the current state of the self-regulating system, and the self-regulation module of the system applies its information operators to maintain the functioning of the system.

Information has a pivotal role in the self-regulation of a system seen as its feedback. It is possible to understand self-regulation through the interplay of positive and negative feedback cycles in which some variations tend to reinforce themselves, while others tend to reduce themselves. Both types of feedback are important to self-regulation: positive feedback because it increases parameters of the system (up to the point where resources become insufficient), while negative feedback because it stabilizes these parameters. Feedback is central to operation of various biological mechanisms, such as genes and gene regulatory networks. In essence, repressor and activator proteins, acting as operators, create genetic operons, which we can also class as informational ontolons (Chap. 9) which function as feedback loops. These feedback loops may be positive or negative. A similar situation exists in psychology, when the body receives a stimulus from the environment or internally from its parts, causing the release of hormones.

11.4 The Causal—Compositional Concept of Information

11.4.1 The Operation of the Gödel Theorems

An essential component of our synthesis is the Causal-Compositional Concept of Information of Luhn [30]. In contrast to many physicists, Luhn starts from the implications of the Gödel theorems for both the definition of information and the application of that definition. "Information" here is used to describe the kinds of possible interactions between systems, including a process of completion, involving a new interpretation of Gödel's incompleteness theorem from a physical and informational perspective. Physical laws are given through arithmetical/mathematical equations, but Gödel showed that any such system of equations is incomplete. The Completion Theorem states that any system of physical equations can be completed by new equations, which cannot be derived from the existing equations. Further information must be added. This concept lays the foundation for the newness, in part or in whole of any information as an emergent phenomenon. It is the operation of this Completion

Theorem which we believe should be added to the basic laws of physics and to the foundations of ontology.

This interpretation of Gödel's theorems shows their close relation to Logic in Reality. In LIR, they are simply another expression of the fundamental functional duality of the universe. It is not surprising that Luhn uses the completion theorem at the highest cognitive level in a new interpretation of so-called free will. From within this information concept, we can derive that our mentally experienced free will is physically grounded in a system which instantiates a highly recursive physical phase space—a 'natural language'. The system is subject to the 2nd law of thermodynamics such that a physical force instantiates a maximum number of possible states. This is the physically based driving force in the direction of enabling or approaching freedom. People who suppress other people's search for freedom are acting *unnaturally* from a very deep perspective. The universe, including the entire biosphere, evolves in the direction of maximization of the number of possible states.

Our explanation of why some people act 'unnaturally' is the inevitable existence, together with any force, of another force or forces opposing it. In LIR, it is thus 'natural' that some people behave unnaturally and seek a *reduction* of possible states, We do not think that any reader of this book will have been so fortunate as not to be able to confirm this view from personal experience. As noted elsewhere, the term free will, even in this interpretation, is somewhat unfortunate since when the relation between two people is proceeding correctly, in order to achieve the overall goals for both, some restriction, based on the information of the other's states, will operate to reduce some of one's own. In the acceptance of this may lie what is commonly called 'wisdom', but we will not discuss this further here.

11.4.2 The Term 'Causal Compositional'

Luhn's approach can be compared with the General Theory of Information proposed by Mark Burgin (see above). The main difference is that Burgin's descriptive theory still focuses on a strong relationship between what is to be understood as "information" and what is to be understood as "structure". Structures are conceptualized, classically, as quantitative descriptions of some arrangements of parts (elements) of the entire potentiality of possible compositions of those elements. Burgin focuses more on the local changes of a system and its structure ("In a broad sense, information for a system R is the capacity to bring about changes in the system R.").

Luhn's Causal Compositional Concept (CCC) of information [30] states that the overall (set of) rules that define any possible and valid arrangement of elements is "information". In other words, what is defined as information, using the CCC, must incorporate the overall rules applicable to the evolution of elements as fundamental entities. A given structure displays a "projection" of a law of physical regularity, which describes the unity and physical reason of that structure. Thus information is the causal-compositional process:

- which describes as initial values a particular physical system, its actual state, its potentially possible state space and accordingly, its region of stability, as well as the lawfully regular (and thereby) macrophysically-measurable description of the system,
- which describes as further initial values the description of an energetic-material, structural influence on the system,
- describes as the resultant values the physical system, its actualized state, its actualized potential phase space, and accordingly its region of stability, as well as the lawfully regular and thereby macrophysically-measurable actualized description of the system,
- which also describes further possible output values (often designated as output signal); brought into a semantic context, *information is a message that is understood.*

If the environmental conditions change, then the system rearranges its elements and their connections through a new overall pattern. This rearrangement is undertaken by a compositional non-algorithmic superposition. For this reason the structure of the emergent pattern cannot be calculated. This compositional concept of non-algorithmic superposition describes newness as a fundamental quality of information.

One can say that it is the task of our conscious mind to mediate and drive this process of informational creation. Given this, any newly created information acts as the efficient cause of structurally new chains of action—an inherent solution of the mind–body problem. The CCC overcomes the problems and inconsistencies of non-physical information concepts like Shannon's (while never the less integrating Shannon's quantitative basis), and puts primarily subjectively and semantically oriented concepts of information (i.e. Floridi) into a unique physical and logical framework, as proposed by LIR.

One can perceive an ethical implication of this approach. If the CCC in correspondence with LIR strives for such envisaged deeper evidence, then the *knowledge* about those circumstances may deliver portions of energy *as* potentiality to us. In simple chemical reactions, for example, the CCC says that there is a physical basis for the fact that the formation of new molecules will, essentially instantaneously, deliver the *information* about the newly created, enriched phase space. If similar processes describe and enable the course of social interaction, than we are committed to accepting this as logical and ontological evidence of a deeply grounded concept of information exchange. One can envisage the application of this form of compositionality to moral behavior and motivation patterns.

11.5 Wu Kun and the Metaphilosophy of Information

The 35+ years of pioneering work of the Chinese information scientist and philosopher Wu Kun first became available in English in 2010 [48], at a Conference on the

Foundations of Information Science in Beijing. Subsequent papers in English by Wu, including some in collaboration with Brenner [46], have influenced recent approach to information, especially in the context of the International Society for Studies of Information (IS4SI). Conferences of the IS4SI between 2011 and the present have all included major sections devoted to the Philosophy of Information.

Basically, Wu's key contribution was to see that the fact that information is not immediately given as matter or energy, even though always requiring them, implies an emerging crisis in philosophy. Information is required as an additional philosophical category, and Wu redefined the philosophy of information as a *metaphilosophy*. The implications of this change are only beginning to be realized.

A key aspect of this approach is that it *defines* a *stance*, the Informational Stance. Wu and Brenner consider that the LIR "opposites" in information are not captured by the classical concept of a classical, static "unity of opposites", but by their dialectical interaction. This has been classified by Wu based on his general philosophy of natural ontological levels that capture the essence of the properties of information. The resulting doctrine of objective information, subjective information and human information in society constitutes Wu's information theory and establishes it as a unified philosophical foundation for information science.

His concept of the Informational Stance is a philosophical position and attitude that is most appropriate for, and above all not separated nor isolated from, the emerging science and philosophy of information itself. The Informational Stance is a more 'active' formulation of Wu's concept of Informational Thinking, which offers an alternative to standard 'Systems Thinking' in which most standard views of logic are retained. The Informational Stance is an attitude that requires attention to the informational aspects of complex processes as a methodological necessity, starting from the level of an existence theory for information and a methodology for its investigation. Especially, the Informational Stance supports and generalizes the recent work of leaders in the area of information ethics, including Floridi, Pedro Marijuan, Rafael Capurro and Wu himself, grounding the attribution of ethical value to all existence in informational terms. Wu's Philosophy of Information combined with LIR yields a *philosophical* structure of information that is compatible with its dynamic physical and logical structure and has no obvious direct precursor, either in or outside of the field of information.

11.5.1 Convergence of the Science and Philosophy of Information

Wu's definition of the role of information in philosophy is the critical *first step* in the characterization of the complex phenomenon of information and information processes. Further, Wu's classification provides a basis for an understanding of a current development, the convergence of Information Science and the Philosophy of Information as the precursor of en emergent Unified Science of Information [47].

This convergence is obviously not intended to imply an 'end' to philosophy or its conflation with science. Philosophy will continue to explore issues that arise, in particular, in relation to language and knowledge in their aspects as unique cognitive products of the human condition, with a substantial abstract content. But the question of the relation of that condition to the rest of the world logically requires retaining the scientific properties of that world to insure the validity of the comparison.

The new and unique relationships that are developing between the classical disciplines of science and philosophy are a consequence of new understandings of the science and philosophy of information. The overall movement is that of a philosophization of science and a scientification of philosophy leading to their convergence. Wu and Brenner have used the term 'Unified Science of Information', but this is not strictly accurate, as our convergent theory also includes the Philosophy of Information as a proper part, without conflation. Wu and Brenner proposed, despite its awkwardness, the term Unified Science-Philosophy of Information (USPI) as the best possible description of the field of endeavor. We believe we are witnessing the emergence of a new system of science, a metascience in a complex, dynamic reciprocity with philosophy that amounts to a paradigmatic revolution in thought.

In a paper in *Information* [44], Wu describes the current situation as follows. We cite this passage *in extenso* as an example of the kind of new paradigm we referred to as the objective of this book. "As a result of establishing the fundamental role of information in the existential domain, the Philosophy of Information provides a kind of dual-existential and dual-evolutionary theory of matter and information which describes information as a general phenomenon existing in everything in the cosmos. This leads to the acknowledgement of the dual dimension of matter and information in all forms of research. Because the lack of an informational dimension in traditional philosophy and science, it is necessary to transform them completely to take into account the new scientific paradigm provided by the current Science and Philosophy of Information. By means of that transformation, all scientific and philosophical domains become involved in an integrating, developing trend of paradigm transformation, which Wu and Brenner has called the "informational scientification of science" [47].

The current interaction and convergence of the Science and Philosophy of Information represents a fundamental and basic path for the development of scientific and philosophical knowledge. This "philosophization of science" and "scientification of philosophy", anticipated in the progression from ancient philosophy to modern science and philosophy, now represents a completely new way of thinking that is that distinct from that in the contemporary Western philosophy of consciousness. It resists an absolute separation between science and philosophy and establishes interactive, mutually defining feedback loops between science and philosophy which emphasizes their interrelation.

Given the properties of information outlined above, we claim that the relationship between its science and its philosophy is one of non-separability, leading to the convergence described. We assume that the Philosophy of Science is not identical to a Science of Philosophy, which remains to be defined (cf. Chap. 1). However, if as Wu and we suggest science and philosophy are converging, under the impact of the science and philosophy of information, the two cannot be considered as totally independent, separated or separable. The emergence of a revolution in philosophy, as suggested in [46] implies a revolution in the philosophical perception of science. In our paper that is the *Prolegomenon* to this book [10], we have defined the parameters of and paradigm applicable of the Revolution in Philosophy and its implications for the Philosophy of Science. We see the two as a pair of doctrines in opposition in the sense of Lupasco. One may talk about a "New Kind of Philosophy" by analogy with the "New Kind of Science" proposed by Wolfram [43].

11.6 The Philosophy of Information as a Metaphilosophy

Let us now discuss the essential role in knowledge played today by the logic and physics of information, based on the concept of Wu Kun that the philosophy of information is a metaphilosophy. Wu's innovation in philosophy is that information is not immediately given as matter or energy, even though always requiring them, implies an emerging crisis in philosophy. Information is required as an additional philosophical category, and Wu redefined the philosophy of information as a metaphilosophy [48]. The implications of this change are only beginning to be realized.

Through papers in English by Wu and jointly by Wu and Brenner, especially in *Philosophies,* many of Wu's conceptions are now broadly available in English. A key aspect of this approach is that it defines a stance, the Informational Stance. Wu and Brenner consider that the "opposites" in information are not captured by the classical concept of a classical, static "unity of opposites", but by the dialectical interaction of opposites, classified by Wu based on his general philosophy of natural ontological levels that captures the essence of the properties of information. The resulting doctrine of objective information, subjective information and human information in society constitutes Wu's information theory and establishes it as a unified philosophical foundation for information science. At the same time, it constitutes a theoretical framework for all organized justifying its appellation as a metaphilosophy.

The Wu Metaphilosophy of Information positions information as a critical component of all disciplines, beyond the scientific content specific to them. It describes an attitude or stance, which Brenner has termed the Informational Stance, which requires attention to the informational aspects of complex processes as a methodological necessity, an attitude that he called Informational Thinking. A meta-philosophical view of the central role of information in the reform and naturalization of philosophy itself is supported by the recent work of Deacon on the dynamics of informational processes. Brenner has discussed this elsewhere as a further contribution to the understanding of what information *is.* The focus of this Wu—Brenner paper is on a new foundational *role* for information in science and philosophy. The theories described here may thus constitute part of a new informational paradigm, in which information has a central, more fundamental role as compared to other major current approaches from the systems and complexity standpoints.

The discussion of a metaphilosophy of information is perhaps an indication of an approaching maturity of the field of information, based also the contributions of Wu and Floridi. One of the consequences, however, is that the comprehensive nature of such a metaphilosophy establishes a socio-political role of those involved, involving them in the social and ethical aspects of the informational components of existence. The Metaphilosophy of Information, then, requires attention to the informational aspects of complex processes as a methodological necessity, in a process that Wu calls Informational Thinking. Informational Thinking (*IT*), as conceived of by Wu, refers to a way of grasping and describing the essential characteristics and attributes of things by reference to the structure and dynamics of the information involved in their evolution, from their historical origins to future possibilities and probabilities. This strategy involves something like a Husserlian bracketing (Wu of the details of any complex process to consider the ways in which information functions in its dynamics, as well as the dialectical relations between its logical elements as proposed by LIR. However, the difference between Wu Kun's theory and that of Husserl is obvious: the purpose of Wu's original Philosophy of Information is to clarify the nature of the dual existence and dual evolution of material and information in the objective world, starting from the logic of the existence and dynamics of the natural human self. The doctrine of Wu, unlike that of Husserl, does not have to be "naturalized", that is, brought into the domain of natural science. It is already there. Wu then discloses directly the mechanisms of the processes involved in an individual's understanding at the level of the integrated object and subject, with internal and external interactions providing the necessary multi-level objective and subjective mediation.

11.6.1 The Informational Stance

The Informational Stance (IS) that can be adopted is consequently a philosophical position and attitude that is most appropriate for, and above all not separated nor isolated from, the emerging science and philosophy of information itself. The IS is a more 'active' formulation of Wu's concept of Informational Thinking that offers an alternative to standard 'Systems Thinking' in which most standard views of logic are retained (see below and Chap. 16). The Informational Stance is an attitude that requires attention to the informational aspects of complex processes as a methodological necessity, starting from the level of an existence theory for information and a methodology for its investigation.

Especially, the Informational Stance supports and generalizes recent work of leaders in the area of information ethics, including Floridi, Capurro and Wu himself, grounding the attribution of ethical value to all existence in informational terms. Wu's Metaphilosophy of Information combined with LIR yields a philosophical structure of information that is compatible with its dynamic physical and logical structure and has no obvious direct precursor, either in or outside of the field of information.

As noted above, Wu has characterized his Philosophy of Information as a Metaphilosophy since it discusses ways in which all theories are constituted and operate in the evolving informational world. The broader role of a Metaphilosophy of Information and by implication for all other disciplines has been given by Wu: "Now, my research is still basically limited to elucidate the general basic theory of philosophy of information from the angle of a metaphilosophy. At such a level, there is a lot of work we should do to define the philosophical essence of information, the philosophical shape and form of information; the nature of the different levels of information; a philosophical measure of information; the relationship between information and various prior definitions of its scope; information ontology, information epistemology, information methodology; the evolution of the informational world; information in material and social evolution; informational sociology and psychology; informational esthetics: an informational theory of value; there are still have very abundant and a large number of branch issues in the each area field of that listed above [49]." In our view, this conception of "meta-" does not conflict with that above of a recursive conceptual analysis. Philosophy and metaphilosophy are not totally disjoint in the process of philosophical ascent.

An article by Sebastian Sequoiah-Grayson, a collaborator of Floridi, entitled "The Metaphilosophy of Information [39] is one of the few on this subject. It supports Floridi's concept of strongly semantic information, but in fact indicates only two items of content of a "metaphilosophy of information": (1) that it should include "Shannon's Premonition" that there will always be a multiplicity of theories of information and (2) explications of the pre-theoretical notion of information are to be judged by their usefulness. These ideas are more or less acceptable in the context of this paper, but they do not say very much. We have found no other serious references to a Metaphilosophy of Information. Wu, through the substantial exposition and explicit references in [48] should be considered the pioneer in this field.

Our view thus offers a novel dialectic perspective on philosophy, metaphilosophy and their recursive relation. It naturalizes this relation, eliminating any implied circularity, since it does not require the total independence of premises and conclusion of standard logics. Starting from the fundamental properties of matter-energy, my theory permits a new approach to critical issues in both philosophy and science that is both logical and metaphilosophical. Wu's metaphilosophical view of information is not directed toward the codification of a Metaphilosophy of Information as yet another static discipline or body of knowledge. It is rather an attitude toward the position of an adequate Philosophy of Information as encompassing a critical component of all disciplines, beyond the scientific content specific to them. Metaphilosophy is not something "more abstract" than philosophy and it must be able to deal with the essential aspects of all disciplines and their theories.

11.6.2 Informational Thinking and the Metaphilosophy of Information

The Metaphilosophy of Information requires attention to the informational aspects of complex processes as a methodological necessity, in a process that Wu calls Informational Thinking. Informational Thinking (IT), as conceived of by Wu, refers to a way of grasping and describing the essential characteristics and attributes of things by reference to the structure and dynamics of the information involved in their evolution, from their historical origins to future possibilities and probabilities. This strategy involves something like a Husserlian bracketing of the details of any complex process to consider the ways in which information functions in its dynamics, as well as the dialectical relations between its logical elements as proposed by LIR. However, the difference between Wu's theory and that of Husserl is obvious: the purpose of Wu's original Philosophy of Information is to clarify the nature of the dual existence and dual evolution of material and information in the objective world, starting from the logic of the existence and dynamics of the natural human self. The phenomenology of Wu, unlike that of Husserl, does not have to be "naturalized", that is, brought into the domain of natural science. It is already there. The naturalization of Husserlian phenomenology was the subject of a major study [33]. Wu's approach eliminates the arduous task of finding natural equivalents for Husserl's transcendental intuitions. Wu discloses directly the mechanisms of the processes involved in an individual's understanding at the level of the integrated object and subject, with internal and external interactions providing the necessary and sufficient multi-level objective and subjective mediation. But to a certain extent it goes too far. We will explain in our concluding Chap. 20 how a partial convergence with Husserl might be effected, based on his complex concept of the role of science.

11.6.3 Towards an Informational Metaphilosophy of Science

We can conclude from the above that a proper new philosophy of science is an informational metaphilosophy, in that (the science of) information defines both new content of the philosophy of science and new dynamics of the relations between a science and its philosophy. A question of interest to readers will be that of the consequences of this convergence for the Philosophy of Science itself. In the spirit of this paper, we suggest that as the special sciences become more and more informational, their respective philosophies can be included as part of the philosophical aspects of the Unified Science-Philosophy ofInformation (UPSI) [47]. At the same time, logically, the informational-philosophic aspects of the USPI can become part of the general Philosophy of Science. All of these considerations, however, are clearly metaphilosophical, since they touch on the content of the respective philosophies (of information and science). We therefore tentatively conclude that the dynamics of the processes can be captured in the concept of a trend toward an Informational Metaphilosophy of Science as the most appropriate model of knowledge. This Philosophy-Science justifies a new non-reductionist view of the world in which it is ethically impossible to maintain any scientific basis for economic, social or ethical exclusion. The use of this doctrine to promote the development of an information commons and the common good is thus a moral as well as a methodological imperative. We consider this, therefore, a necessary area for further research. Our interim conclusion is that making an invidious distinction between philosophy and metaphilosophy, which is itself a metaphilosophical question with an opening toward an infinite regress, is not an essential one. The intuition of this, which has been noted earlier, is simply supported by our dialectical framework which provides for their interaction logically and cognitively. The distinction may be made for pragmatic reasons, to focus on groups of closely related issues.

We have shown that the new kind of logic we have described, Logic in Reality (LIR), offers a novel dialectic perspective on philosophy, metaphilosophy and their recursive relation. LIR naturalizes this relation, eliminating any implied circularity, since it does not require the total independence of premises and conclusion of standard logics. In the next Chap. 12, we will consider further aspects of information and its relation to a convergence of science and philosophy.

References

1. Adriaans, P.: Information. In: Zalta, E.N. (ed.)The Stanford Encyclopedia of Philosophy, Spring 2019 edn. https://plato.stanford.edu/archives/spr2019/entires/information/
2. Baxter, L.A., Montgomery, B.M.: Relating Dialogues and Dialectic Method. Guilford Press, New York NY (1996)
3. Bekenstein, J.D.: Information in the Holographic Universe. Sci. Am. **289**(2), 58–65 (2003)
4. Brenner, J.E.: Logic in Reality. Springer, Dordrecht (2008)
5. Brenner, J.E.: The Philosophical Logic of Stéphane Lupasco (1900–1988). Logic Logical Philos. **19**, 243–284 (2010a)
6. Brenner, J.E.: Wu Kun and the metaphilosophy of information. Int. J. Inf. Theories Appl. **18**(1), 103–128 (2010b)
7. Brenner, J.E.: Information in reality: logic and metaphysics. Triple-C **9**, 332–341 (2011)
8. Brenner, J.E.: Information: a personal synthesis. Information **5**, 134–170 (2014)
9. Brenner, J.E., Burgin, M.: Information as a natural and social operator. Int. J. Inf. Theories Appl. **18**(1), 33–49 (2011)
10. Brenner, J.E., Igamberdiev, A.U.: Philosophy in Reality: Scientific Discovery and Logical Recovery. Philosophies **4,** 22 (2019)
11. Brier, S.C.: Why Information is not Enough. University of Toronto Press, Toronto (2008)
12. Burgin, M.: Theory of Information. Fundamentality, Diversity and Unification. World Scientific, Singapore (2010)
13. Burgin, M., Brenner, J.E.: Operators in nature, science, technology and society: mathematical logical and philosophical issues. Philosophies **2**, 21 (2017)
14. Burgin, M., Simon. I.: Information, energy and evolution. In: Preprint in Biology 2359 Cogprints, Electronic edn. https://cogprints.ecs.soton.ac.uk (2001)
15. Capurro, R.: Foundations of information science. review and perspectives. In: Proceedings of the International Conference on Concept ions of Library and Information Science. University of Tampere, Tampere, Finland, 26–28 Aug 1991 (1991.)

16. Deacon, T.E., Cashman, T.: Steps to a metaphysics of incompleteness. Theol. Sci. **14**, 401–429 (2016)
17. Deacon, T.: What is missing from theories of Information? In: Davies, P., N.H. Gregersen (eds.) Information and the Nature of Reality: From Physics to Metaphysics. Cambridge UniversityPress, Cambridge, UK (2010)
18. Deacon, T.E.: Incomplete Nature How Mind Evolved from Matter. W.W. Norton & Co., New York, NY (2011)
19. Diaz Nafria, J., Zimmermann, R.E.: Emergence and evolution of meaning. Triple-C **11**, 13–35 (2013)
20. Igamberdiev, A.U.: Physical limits of computation and emergence of life. Biosystems **90**, 340–349 (2007)
21. Igamberdiev, A.U.: Objective patterns in the evolving network of non-equivalent observers. Biosystems **92**, 120–131 (2008)
22. Igamberdiev, A.U.: Biomechanical and coherent phenomena in morphogenetic relaxation processes. Biosystems **109**, 336–345 (2012)
23. Igamberdiev, A.U., Shklovskiy-Kordi, N.E.: The quantum basis of spatiotemporality in perception and consciousness. Prog. Biophys. Mol. Biol. **130**, 15–25 (2017)
24. Kauffman, S., et al.: Propagating organization: an enquiry. Biol. Philos. **23**, 27–45 (2007)
25. Kaye, D.: The nature of information. Library Rev. **44**(8), 37–48 (1995)
26. Kaye, L.: How to avoid holism and draw the analytic/synthetic distinction. https://evans-experientialism.freewebspace.com/kOaye01.htm (1995)
27. Kolmogorov, A.N.: Logical basis for information theory and probability theory. IEEE Trans. Inform. Theory **IT-14**, 662–664 (1968)
28. Leydesdorff, L.: The Challenge of Scientometrics: The Development, Measurement, and Self-organization of Scientific Communications. DSWO Press, Leiden University, Leiden.http://www.universal-publishers.com/book.php?method=ISBN&book=1581126816 (1995)
29. Loewenstein, W.R.: The Touchstone of Life: Molecular Information, Cell Communication and the Foundation of Life. Oxford University Press, Oxford UK and New York NY (1999)
30. Luhn, G.: The causal compositional concept of information. part i: from decompositional physics to compositional information. Information **3**, 151–174 (2012)
31. Lupasco, S.: L'énergie et la matière psychique. Editions le Rocher, Paris (1987) (Originally published in Julliard, Paris, 1974)
32. Nicolescu, B.: Personal E-mail Communication to J. E. Brenner (2019)
33. Petitot, J., Varela, F., Pachoud, B., Roy, J.M. (eds.): Naturalizing Phenomenology; Issues in Contemporary Phenomenology and Cognitive Science. Stanford University Press, Stanford CA (1999)
34. Quieroz, J., El-Hani, C.N.: Towards a multi-level approach to the emergence of semiosis in semiotic systems. Acta. Biotheor. **54**, 179–206 (2006)
35. Rapoport, I.A.: Microgenetics. Nauka, Moscow, Russia (1965).(In Russian)
36. Rosen, R.: Complexity as a system property. Int. J. Syst. Evol. Microbiol. **3**, 207–232 (1977)
37. Sawa, K., Igamberdiev, A.U.: The double homunculus model of self-reflective systems. Biosystems **144**, 1–7 (2016)
38. Scarrott, G. G.: The nature of information. Semantic information. Br. J. Philos. Sci. **4**(3), 147–157 (1989)
39. Sequoiah-Grayson, S.: The metaphilosophy of information. Mind. Mach. **17**, 331–334 (2007).
40. Smith, J.M., Szathmary, E.: The Origins of Life: From the Birth of Life to the Origin of Language. Oxford University Press, Oxford UK (1999)
41. Ulanowicz, R.: Towards quantifying a wider reality: Shannon exonerata. Information **2**, 624–634 (2011)
42. Wheeler, J. A.: Information, physics, quantum: The search for links. In: Zurek, W. (ed.) Complexity, Entropy and the Physics of Information. Addison-Wesley, Redwood City, CA (1990)
43. Wolfram, S. A New Kind of Science. Wolfram Media (2002)

44. Wu, K.: The interaction and convergence of the philosophy and science of information. Philosophies **1**, 208–244 (2016)
45. Wu, K.: The basic theory of the philosophy of information. In: Proceedings of the 4th International Conference on the Foundations of Information Science, Beijing, China, 21–24 August 2010
46. Wu, K., Brenner, J.E.: Philosophy of information: revolution in philosophy. Towards an informational metaphilosophy of science. Philosophies **2**(4), 20 (2017)
47. Wu, K., Brenner, J. E.: A unified science-philosophy of information in the quest for transdisciplinarity. Chapter 8. In: Information Studies and the Quest for Transdisciplinarity. World Scientific, Singapore (2019)
48. Wu, K.: The Complex Characteristics of Multidimensional Emergence: Critique of Cutting-edge Issues of the Contemporary Philosophy of Science and Mind, Guangzhou—Macau, 13–17 Dec 2010
49. Wu, K.: Philosophy of Information—A New Spirit of the Time. Shaanxi Normal University Press, Xi'An, China (1989) (In Chinese)

Chapter 12
Communication

12.1 Introduction

12.1.1 Positioning Communication

Although we have not said so specifically, there is a sense in which theories of all subjects which are not machines are part of Natural Philosophy. This statement, as one of Philosophy in Reality, applies to communications theories or descriptions. Although in Chap. 10 we identified epistemological structures in human communication which had the status of fictional objects not involving energy and hence without change, all other concepts of communication in social interactions are natural processes. From our view of information in Chap. 11, we therefore go to the next level of complexity which is that of meaning and information exemplified in human communication.

If a natural philosophical statement means something at all, has or conveys meaning, this fact does not and cannot stand in isolation. The meaning in question is not a solely epistemological concept, but puts natural philosophy in a dynamic relation with a reality described by a dialectics and a logic. In this 'meaning in motion', we move from semantics to dynamics where meaning functions as an operator that can effect change (see Chap. 8). The parameters of this process of 'meaning in action' necessarily involve those of information and its communication, in what was termed in Chap. 11 a philosophical triple. We will therefore now look at aspects of the science and philosophy of communication as such.

We differentiate, for discussion, between two aspects of the *transmission* of meaning, that is, what is usually called its communication. We reserve the term communication for the active transmission of meaning, essentially, between two human beings or other animals. All such communication involves transfer of information, but some transmission of information is passive, the consequence of human awareness, say, of external objects. Only internal processes are involved initially, here, in the assignment of meaning, although their subsequent elaboration can obviously be communicated externally. At this level, we will also pay close attention

J. E. Brenner and A. U. Igamberdiev, *Philosophy in Reality*, Studies in Applied Philosophy, Epistemology and Rational Ethics 60, https://doi.org/10.1007/978-3-030-62757-7_12

to the concept, also due to Capurro, of "angeletics" or messaging theory [10] as an extension of information theory to communication.

12.1.2 The Question of Dynamics. A Joint Theory

As we approach the promised synthesis of the various parts of this book, one aspect should be stated explicitly: if our Logic in Reality applies to description of the changing phenomena in question, then that description belongs to Philosophy in Reality. Other theories of them, e.g. one of those that will concern us here, is the sociological theory of communication of Loet Leydesdorff [26]. *A contrario*, that theory as described by its author in an analytical theory (of structure, communications and social networks) and hence does not make explicit reference to the logical principles of LIR as they apply to processes. Our point is that domains of intersection can be identified; one can imagine a joint theory in which that of Leydesdorff, for example constitutes the epistemological part, using epistemons as 'units' (Chap. 9) and LIR the ontological part using ontolons. We acknowledge here the positive reaction of Leydesdorff (personal communication) to our categorization of his work as 'epistemological dynamics'. The clear identification and valuation of changes in the concepts, the fact that we are not limiting the range of discourse to static entities, establishes the relation between the two domains of thought. In a forthcoming book, Leydesdorff (below, LL) will focus on the *quantitative* study of science, technology, and innovation in relation to communication [28]. Our *Philosophy in Reality* is also a study of dynamic aspects of 'knowledge, information and meaning', but it is intended primarily as a contribution to the *qualitative* study of science, together with philosophy. We claim that it is possible to define a theory of communication in which both aspects operate jointly.

The cognitive processes involved in human communication are so familiar that their complexity tends to be forgotten. Philosophical issues immediately arise, however, in any discussion of communication since they depend on the existence of at least two individual human or animal consciousnesses, memories to retain verbal or other messages and common codes, which need not be linguistic ones, for one individual to understand the meaning of the other. In our discussion of information, we described it as a process of informing, constituted by both actual and potential energetic elements, such that energy was present both as the carrier of meaning as well as meaning itself.

Communication theories focus on the different aspects of communication, broadly the generation, transmission, reception and interpretation of meaningful information, where meaningful as noted above refers to the value of the information for the well-being, physical and mental, of the individual. However, information, as we have already seen has both simple scalar and complex vector components. The evolution of the former is describable by truth-functional, propositional logics or their mathematical equivalents, but the latter requires a non-propositional logic of real processes, based on the physical-philosophical principle of dynamic opposition, the alternating dominance of actual *versus* potential aspects of the cognitive systems in the communicating relation. The process elements, real and interactive or not, with which one chooses to begin the analysis and synthesis are thus critical. It is therefore not a criticism of existing communication theory to say that it conforms to the rules of bivalent, propositional logic since no valid alternative can be said to have received broad acceptance. LL's basic thesis is that above a certain level of complexity of interactions, recursive and hyper-recursive processes take place essentially involving only the communications systems themselves. In these, the "links" of the network, "process" information differently from human action systems considered as the "nodes". The three scare quotes placed by LL represent his view that not only should such links and nodes not be discussed as if they were independent, the interactions between links and nodes involve the principle of dynamic opposition, in which one or the other feature predominates. The corollary of this is that a purely mathematical theory of communication is adequate only for some statistical aspects of knowledge-based systems.

12.1.3 The Question of Meaning

The similarities and differences between the approach of LL and ourselves can be illustrated by reference to what we define as the two major forms of meaning. They are the further expression in LIR terms of meaning first discussed in Chap. 9, but the basis for their potential synthesis will emerge in what follows.

Meaning I
Meaning I *inheres* in all existent entities. It has *ipso facto* value for conscious entities. Changes in it can be characterized as ontological flows of information (energy). Its logic is the non-propositional logic of energy. Its units are complex dynamic structures—'ontolons'.

Meaning II
Meaning II is *generated* in the codified, interpreted communications between conscious entities. Its dynamics are epistemological, without energy change. Its logic is propositional, bi- or multivalent, paraconsistent and/or intuitionist, modal, etc. Its units are linguistic structures—'epistemons', or in LL's term 'kenes', which refers to units or blocks of knowledge.

Corollary A *For any real system including conscious entities, both forms of meaning are present and influence one another dialectically.*

Corollary B *Both forms of meaning are causally efficient for conscious entities, Meaning I directly and Meaning II via Meaning I.*

The two forms of Meaning do not only *refer* to the Interactions involved in a real process, they *are part of* the Interactions (or are constituted by them). Thus, there are two different kinds of Interaction corresponding to the two forms of Meaning:

Interactions II Epistemic (as described by Leydesdorff in [26]).

Interactions and information of Meaning **II** processes can be measured in bits.

Interactions **II** can be between codes.

Meaning **II** originates from and feedbacks on Interactions **II** between or among communications.

The dynamics of innovation is a case of Interactions **II** between or among communications.

Interactions I Ontic.

Meaning **I** processes and Interactions **I** cannot be measured in bits nor described by an algorithm.

Meaning **I** processes originate from and feedback on Interactions **I** between living agents.

One example of such interactions is the dynamics of innovation: it involves Interactions **I** between living agents (inventors and venture capitalists). The frames of reference are different but not totally disjoint, since the living and communicating agent is present more or less actualized or 'in the background' in each.

In fact, the communicative act involves a complex series of mental and physical events and processes. Some involve actions directly, while others define a 'space' of attitudes and anticipations that are the pre-conditions for/to subsequent communication. In addition to the spatial dimension, all the mental pre-processing as well as operation take place in a time dimension that extends from the domain of historically available options, grounded in memory, to imagined and anticipated future events. The properties of human anticipation have been well studied by Dubois [13] and Poli [34] as well as Leydesdorff. Roberto Poli has developed the concept of anticipation and brought into the domain of the political arena. In our 'real' phenomenological view, imagination precedes anticipation, which involves a greater degree of presence of a concept, albeit still at the level of an actualized potentialization. (We see here a first example of the necessity, pointed to by Nicolescu, of a nine-fold matrix of actuality, potentiality and T-state, in which each of the three can exist in a state which is predominantly one of the others. Some of the pairs, such as potentiality of potentiality have more meaning than others, but not always in the same way. In any case, it is to this area of cognition where they appear to be most applicable.)

We will refer several times to the use, especially by LL, of the term historic or historicity. We will emphasize the *reality* of 'historicity' as the four-dimensional object in which the dynamics of real processes are instantiated. History—and its philosophy—should be understood in ontological terms, as in the intuition of some neo-Kantians (see Chap. 15). We should again emphasize that our designation of some doctrine as "ontological" is to insist on its dynamics, an acknowledgement of the history of a system in its changing form (morphogenesis).

12.1.4 The Floridi Interpretation

A statement by Floridi, whose concept of information as well-formed data we have critiqued above, illustrates this point. Writing in the *Stanford Encyclopaedia of Philosophy*, [15], Floridi equates a mathematical theory of information to communication "because it is a quantitative approach to the analysis of information that has been most influential among several philosophers". Floridi developed his approach in support of a mathematical theory of communication (MTC) in opposition to alternative information incorporating the semantic dimension. Donald M. Mackay [30] proposed a quantitative theory of qualitative information that has interesting connections with *situation logic* (see below).

According to Floridi himself, however, the majority of philosophers agree that MTC provides a rigorous constraint to any further theorizing on all the semantic and pragmatic aspects of information. Disagreement concerns the crucial issue of the *strength* of the constraint. He further states that an MTC defines information in terms of a probability space distribution. "Along similar lines, the *probabilistic approach* to semantic information defines the semantic information in p in terms of logical probability space and the inverse relation between information and the probability of p. This approach was initially suggested by Bar-Hillel [2]. The original idea is simple. The semantic content in p is measured as the complement of the a priori probability of p. It is also compatible with a probabilistic approach, although it does not require a probability measure on sets of states. The informational content of p is not determined a priori, through a calculus of possible states allowed by a representational language, but in terms of factual content that p carries with respect to a given situation. Information tracks possible (or probable) transitions in a system's states space under normal conditions. Both Dretske and situation theorists require some presence of information already immanent in the environment (*environmental information*), as nomic regularities or constraints. This "semantic externalism" can be controversial.

Perhaps surprisingly, while we disagree with a MTC in general, our approach to reality is like Floridi's also probabilistic, perhaps 'more so'; the values of the actual and potential components of real processes are non-Kolmogorovian in not including the limit values of 0 and 1. The phrase "some presence of information already immanent in the environment" as a lawful regularity *and* constraint can be read as fully equivalent to LIR with the term 'immanent' replaced by 'potential'.

Unfortunately, Floridi's description remains at an epistemological level. The contemporary theory of LL is much more useful, despite its also being primarily epistemological. We will now look at this theory in detail and pick the areas of overlap and exclusion with the approach in this book.

12.2 The Philosophy of Human Communication

In Chap. 11, we indicated some recent developments in the science, logic and philosophy of information in relation to meaning. In particular, we have mentioned the dialectical convergence of science and philosophy under the influence of the philosophy of information. The convergence is particularly visible in the field of communication. We have also mentioned briefly the origin of the science and technology of communication in the twentieth century and its enormous increase with the advent of computer technology. In numerous publications, Rafael Capurro has identified progress in communications theory as part of the current digital ontological paradigm. He calls attention, also to their social consequences [8]. From our standpoint, progress in communication is by no means univocal. In the case of natural languages, for example, the basis for human communication, their evolution in the direction of increased efficiency and reduced redundancy is clear. The latter is off-set in many instances by context, as in Mandarin Chinese, where a single tone-symbol may have a dozen meanings, but this may be an exception. It is clear that such developments are accompanied by an impoverishment in the quality and depth of language and communication. Routine errors in English are made and become codified in defining—otherwise responsible—media. The loss of redundancy in individual languages is paralleled by the disappearance of entire 'native' language groups. The destruction of the languages of indigenous populations, as slaves, was part of the political strategy of the Catholic Church in Mesoamerica, the conquest of the North American continent by Europeans and is part of that strategy in Asia and Africa today. The society is impoverished by such losses as much as it is by the loss of animal species, such as the moas, literally eaten to death in Australian the nineteenth century.

Under these circumstances one can only welcome movements in the direction of the saving of surviving languages and the recomplexification of our own. Such a process cannot be formal and artificial; we propose what is essentially a further case of recovery, by the reintroduction of references to classical portions of our cultural heritage, not classicism for its own sake. We are thus proposing a rehumanization and de-digitalization of communication theory which is not intended to eliminate digital perspectives but to provide some balance to them. Capurro correctly points to the prevalence of a 'Digital Ontology' today [7], but this does not mean that a non-digital ontology is not active, admittedly to a lesser degree.

12.2.1 Messaging Theory (Angeletics)

Following McLuhan's seminal work, further emphasis has been placed on communication as the exchanges of messages—'messaging' involving the triple of message, its sender and its receiver. Capurro and Holgate coined the term 'Angeletics' to replace that of information for the description of the exchanges of messages between human beings, going back to the ancient Greek word for messenger $\alpha\nu\gamma\varepsilon\lambda\lambda o\sigma$ [6]. The choice of the term Angeletics for the study of messages and messaging implies a philosophical framework, closest to that of Heidegger, which is ultimately based on Being of which the irreducible uniqueness of the individual human consciousness is a part. It is this central functional role of philosophy in messaging theory that calls for definition of a new specific field for the clarification of residual problems at the interface of the domains of messages, communication and information, in which the notion of Being plays a central role. A major discussion by Capurro and others of the concept of Angeletics is given in [10].

The word message is derived from the Latin *mittere*, to send. For us, this describes a 'packaged' portion of a complex process of information in movement that can be also designated as an ontolon. This is consistent with the concept of Holgate [20] that science should continue "the epistemological revolution of our time". It should extend relational, that is ontological concepts to every field of science and the mind (relativity, pluralism, polarities, information exchange, and ethics).

Communication involving the sending and reception of a message is a creative act, an energetic process that involves the exchange of energy, overcoming resistance to doing nothing, or not sending any message. Messaging theory or Angeletics must reflect the creative, value-laden characteristics of communication, their reference to the physical survival of the receiver, or more indirectly to his/her mental and spiritual well-being.

As shown in Chap. 11, meaning and information are not identical but also not separable at the cognitive level which is the one of interest here. Their evolution follows the logic of real processes that is the thread running throughout this entire current paper. It is thus an integral part of Philosophy in Reality, of which part of our proposed synthesis is the Philosophy of Information of Wu [42]. In the changes in stance or perspective, dialectical movement **is** both epistemological and ontological. Knowing is not totally separate from the Known, what we know. It is true that the "map is not the not territory", but stated in this simplistic way, the relation restored or recovered by LIR is obscured.

The current usage of media as a singular noun is incorrect from the standpoint of Latin grammar in referring to several kinds of medium—press, TV, cinema, etc. Oddly, it is more correct if 'media' is taken to refer to the complex dynamic properties of the messaging process, "the media is the message". Process and process_as_meaning then have the proper ontological relation and value. A good example of our current approach is our reading of the 'liar paradox', which was referred to and analyzed above. In the usual linguistic perspective, it is an epistemological dead end: its terms oscillate between ideal limits of one or the other of two idealized alternatives. In the original perspective of one of us (AUI), the system subdivides into levels, with two actors. In the epistemological mode, as pointed out by Igamberdiev [22] Epimenides and the Cretans are separated by non-existence: Epimenides is a signifier, and the Cretans are the signified. The standard logical contradiction, which appears when we realize this one-level formalization of this system, is to be expected in this mode.

From our newer ontological perspective, the actors are the same but the relation is logical and contradictorial in the sense of Logic in Reality. The following consequences can be deduced from this consideration. Epimenides as an individual can be regarded as an element of the set which signifies this set, having *discovered* it in the sense of Capurro. The set (the society of Cretans) acquires its own dynamics having acquired the property of being 'liars' in conflict (antagonism) with, but not totally separate from, Epimenides. Different possibilities for the dynamic behavior of a system arise from this. What is important to note is that the 'dynamic behavior' with which we are concerned is that of real persons, individuals and groups, and their relations, dependent on their deep psychology

Capurro stated in [7] that the ontic difference between a sender or messager (messenger is too passive) and a receiver as separate entities presupposes the original unity of Being as sender, the world as message and humans as messagers/messengers. We can distinguish this original unity analytically, but we must be aware that any ontic separation (at whatever level of reality and concerning whatever kinds of beings) presupposes our being-in-the-world (to use Heidegger's formula). At the risk of overdoing the form but to make the point, we may use the expression Being in Reality in place of this formula. From its first formulation, LIR has assumed a synthetic non-separation of subject and object, grounded on the original relationship between man and world. In the angeletic perception of Capurro, whatever we perceive AS being this or that, whatever the kind of relation (causal or not, etc.), is achieved on the basis of our being originally open to the message of the world that we process, at a higher level, as Being. One of the problems posed by discussion of Being in non-transcendental terms is whether one can talk also about it in terms of proper parts, as one can for ontic and epistemic phenomena, using the terms of ontolon and epistemon respectively, as was discussed in Chap. 9.

12.3 The Cognitive Basis for Human Communication

We have previously defined information as the knowledge necessary for a human being to survive and grow. We define communication as the processes of emission, transfer, reception and comprehension of information between two or more adult, healthy individuals. Such a definition, for reasons that are not entirely clear, is often attacked by anti-realists on the grounds that it fails in some cases. It obviously has not failed in *all* cases, in which case our species would be extinct.

12.3.1 Personal Identity

The problem, which is a real one, is the *basis* for the mutual understanding of communicated information or messages, since if there is one incontrovertible phenomenological fact it is the non-identity of two human beings, and the independence of their physical structures with the exception of the immune systems of identical twins. This view, however, occults the fact that two or more human beings *do* understand one another, even if the process results in occasional or systematic errors. The mistake lies in the view of individuals as totally isolated mentally. Human beings—and this is perhaps the essence of their humanity—are perfectly capable of internalizing the attitudes, opinions and above all the simple existence of the 'other'.

The questions of the origin, persistence—resistance to change—and functioning of personal identity dominate the philosophy of personal identity and accordingly the social purport of all human interactions. In the LIR epistemology, we as knowers are not totally external to what is known by us and not completely different from it. We must know, then, that if there are other knowers, as there are, they must be part of our known and vice versa. The source of human dignity *is* in ourselves as knowers, but if we avoid the error of solipsism, the origin of the sense of moral responsibility can only come from the relation to other knowers, in other words, all human beings, and by extension, other beings and perhaps even, as suggested by Magnani, [31] certain non-living entities. *A contrario*, one cannot find responsibility in oneself as an isolated agent. Since we are both a "not-other" *and* an "other" at the same time, a self-interest argument for morality holds. Two or more human individuals and their relations constitute interactive systems in the LIR categorial sense of non-separable subjects and objects, sharing in part one another's characteristics. An individual is no more isolated logically, psychologically, or morally than he or she is economically. The fact that potential or potentialized states exist does not, in a deterministic universe, mean that we have the capacity to make a choice among them that is independent of our genetic and experiential background.

To repeat, every individual is indeed unique, but this should not be taken to mean that his or her mind is independent, since each incorporates a portion of the subjective experience of other brains. As Bennett and Hacker point out [4], our ability to know the states of other persons' minds is not folk psychology, but a natural consequence of the evolution of our species. LIR simply adds the logical consequence of this capability as the origin of individual moral responsibility.

12.3.2 Group Identity

Logic in Reality suggests a new definition of what constitutes an individual, a group, and the relationship between them. It is in line with the proposal by Ladyman and Ross [24] of the elimination of the classic abstract a priori philosophic notion of an individual thing and its redefinition in terms of dynamic patterns. The consequent naturalized metaphysics is compatible, as is LIR, with fundamental physics.

In the LIR two-level framework for analysis, groups can no longer be considered, at the social level, as the equivalent of a set composed of individuals, equivalent to members of the set. Classical set theory requires that sets and their members be completely disjoint; set theory is essentially bivalent logic in another form. As indicated above in the discussion of free will, human beings cannot be considered as isolated from one another, nor, we propose from any group that is truly collective and not accidental. Collective groups share, even in standard philosophy, some cognitive properties such as opinions or objectives. LIR sees this sharing as a dynamic process, in which there can be differences between individual and group views such that at one time it is the former and at another the latter that predominates (is actualized) and is determinant for behavior.

The characteristics of collective moral responsibility, and the basis of that responsibility, remain a problem for classical reasoning based on classical logic. Even assuming that individual human responsibility is clearly understood, the existence of collective responsibility poses the following key questions that relate to both groups and their individual members:

- Can an individual be responsible for what a few of the group members do?
- Is it possible for a group, as distinct from its members, to cause harm and/or have intention to cause harm?

As Smiley puts it [39], groups or collectives do not have "full-blown mental lives" and cannot make choices or hold beliefs as the basis for intentions. On the other hand, groups are certainly *perceived* as having such competence, and the competence can have causal properties. Further, criteria can be established to limit the relational conditions under which it is legitimate to talk of collective responsibility: the individuals should be related so that each can act in ways that they could not on their own, and they would have agreed that some of them could represent their actions as the actions of the group.

Group intentions can be seen as arising from relationships between members of the group which are not trans-individual or collective in any sense that stands *totally* (emphasis mine) above individuals; the intentions are group-based and can be treated "as if they were collective", both collective *and* individual. The degree of moral blameworthiness of dissenting group members *vs.* that of the group as a whole seems to depend, in this picture, on how actively such members resist or fight against them, a parameter not easy to measure, to say the least.

The clarification that Logic in Reality brings to the controversy, as noted above, is relax the requirement that individuals and groups are a priori totally separate entities, and allow each to share part of the others properties, including intention. However, the shared properties do not have to be 100% actual or actualized at one time; they can be present as potentials of which individuals are more or less vaguely conscious. Proceeding to actions, however, involves the individual having a rather clear consciousness of what is being done, and the degree of responsibility increases accordingly. On this basis, in a 'potential way' so to speak, we are responsible for the destruction of the rain forest by others, as there might be things we might do to resist it that are more or less (rather less) accessible to us. To the extent that one is a member of a society that includes criminals, one is responsible for knowing about their existence. In a sense, everyone is responsible for understanding his or her single or group opponents. Understanding one's opponents' psychology is not a question of our all becoming amateur psychologists, as a skeptic might argue. It is an accepted strategy in war, business and sports, to which we can add environmental policy-making.

Absolute cultural relativism, the idea that all cultural practices have equal rights to existence and execution, is no longer seriously entertained or defended as a doctrine, given its obvious self-contradiction. However, we believe that discussions of individual and group moral responsibility are still influenced by the *apparent* humanistic focus of ironist thinkers such as Rorty [36] and his followers, and a tendency remains to give some cultural practices more credit than they deserve. We will return to further issues of individual *versus* collective behavior in our discussion of the philosophy of social systems.

12.4 The Ontology and Epistemology of Communication

The major categorial features of communication as a human cognitive process are the following: (1) it takes place in time, involving the past as memory, the present as operation and the future as anticipation; (2) it requires two or more individual human beings as senders and receivers, who process the information communicated or to be communicated; 3) structures can be ascribed to the different phases of communication which refer to both the actual physical process, an ontic Information I in the definition in Sect. 12.2 above, and an epistemic Information II; a final categorial feature is that communication processes instantiate self-organization. As we have shown in Chap. 11 on Information, the use of 'self-' is either incorrect or misleading, to the extent that no ontological process can be separated in reality from its real, potentialized components which exist within the system under scrutiny but on which attention is temporarily not focussed. Systems which are epistemic to all intents and purposes can be viewed *as if* they organized themselves, in spite of the fact that it is the conscious agent that is 'managing' the self-organization.

As discussed by Loet Leydesdorff (LL) in his *Sociological Theory of Communication* [26], the existence of communication processes both characterizes a society and enables it to function. We consider it essential, to include in the discussion of its philosophy some of its formal epistemological aspects in addition to the ontological/phenomenological properties referred to in the above sections. Both standard logic as well as our non-standard logic in reality has a place in this theory. LL's theory provides an explication of communication within social systems, and therefore will act as a bridge to our Chaps. 18 and 19 on Social Systems. It is also an example of what and where to look for the essential features and dynamic structures of a scientific or philosophical domain and the way in which they operate—are operators. This discussion of communication as process is thus parallel and an extension to our discussion of information as process.

LL does not address the principles of our Logic in Reality as such. The differences as well as the relations between ontology and epistemology are real, and we believe we have shown that they are; LL is concerned primarily with the epistemological structure of communication and *its* expression in the social sciences. The units of the corresponding processes have been designated above as epistemons. However, the methodology of LL in going to the detailed pattern of evolution of communication processes is similar in spirit to the dialectomethodology of Lupasco referred to above, that is, to always seek out the elements of a process that are involved in the interaction. In other words, LL's approach implies recursive structures of change that complement and are complimented by the cognitive structures defined by LIR. We therefore will attempt to bring out the relations between the approaches, which are unified of course by their reference to meaning (Meaning I and II above).

Following LL, we have organized the remainder of this section in terms of epistemological categories, but we see an ontological thread between them that allows, if not a complete naturalization in the sense of a natural philosophy (see Chap. 13), a sense that communication involves real energetic and temporal and well as epistemic change.

12.4.1 The Time Dimension: Recursion, Incursion and Hyper-Incursion

Communications are processes occurring in time, and hence instantiate change in many different ways. Theorists of communication are therefore justified in seeking 'responsible invariances' in communications, while avoiding as far as possible their reification into static concepts. The capacity for the human mind to operate in past, present and future modes can be captured formally, as discussed by LL.

A recursive equation refers for the computation of the current state of a system to a previous state (t – 1), but an incursive equation refers also to states in the present among its independent variables, and a hyper-incursive one refers to future states as co-constructors of present states. A model contains a prediction of a next state of the system. Hyper-incursive equations show how models may additionally feed back on the present state of the modeled system. The total system—that is, the system including its modeling subroutines—develops over time and thus generates entropy. The model, however, reverses the time axis and thus can be expected to generate redundancy. In other words, a model provides a form of meaning to the modeled system. Meaning can then be communicated in a reflexive discourse entertaining different models: a discursive model enables us to feed back on the system under study. This hyper-incursive meaning processing—formulated in terms of possible future states—can be instantiated incursively by a reflexive observer who herself operates with an expectation in the present and historically with reference to a previous state. The model is thus instantiated.

12.4.2 Expectations and Contingencies

Few people would disagree that, as Chalmers has proposed, explaining the existence of human consciousness is the hardest problem for science and philosophy. The, for us, even harder problem is one that is scarcely addressed in standard philosophy, which is why I am *this* consciousness and not some other. The idea that it is the sum of all experiences, actualized and potentialized, of the individual beginning, with fertilization, that is somehow necessary and sufficient to constitute an *individuality* seems correct, but simply restates the problem.

In his extremely comprehensive approach, Leydesdorff starts appropriately with the observation that communication must involve a minimum of two individuals, each of whom will be assumed to have their own irreducible identity and consciousness. If we assume that the existence of each individual at a particular time and place is entirely contingent—random, then the system constituted by the two individuals together constitutes a double or second contingency. This double contingency results in mutual expectations which precede interaction. Consequent interactions imply an instantiation of them.

A critical concept in LL's theory, which underlies everything that follows, is that of expectations. The following comments, which touch only a few of the details of this theory, demonstrate that an alternative interpretation of it is possible. Leydesdorff distinguishes clearly between cognitive processes dealing with past and future events, for which the key term is expectations. Expectations epistemically interact, recursively generating interobjective or subjective knowledge with a new dynamics either with or against the arrow of time. These emerging levels or concepts have "a different logic, but not an independent existence".

The expectations that human beings have can be formulated by human beings, but they are themselves not directly observable. As Giddens [17] formulated it: "Structures exist paradigmatically, as an absent set of differences, temporarily 'present' only in their instantiations, in the constituting moments of social systems." The constitutive moments in the historical genesis are relational, the structures are co-relational.

In the theory of anticipatory systems, inversion along the arrow of time is considered as a degree of freedom of a knowledge system [13]. First, an event at time t can be provided with meaning at $t + \Delta t$. In other words, one looks reflexively backward, while moving historically forward within (and potentially changing) the stream of events.

In their "sociology of expectations" cited by LL, Brown and Michael formulated this tension between (1) forward cognitive movement along the arrow of time and (2) backward interpretation and control as a dynamic between "retrospecting prospects and prospecting retrospects." A reflection in the time domain stands orthogonally to substantive reflections at each moment of time.

The interobjective reality of expectations thus generated has the status of a construct that can be entertained, unpacked, and revised. Husserl would call this *cogitatum* transcendental, but this domain for LL is the second contingency. Whereas the first contingency is a given in *res extensa*, the second is constructed as a *res cogitans*. Since they are not hard-wired, the spheres can interpenetrate one another. The warp and the woof of meaning generation and self-organization are not integrated harmoniously, as in textiles, but with differentiation and disturbing one another. Horizontal differentiation is a necessary complement of vertical differentiation and self-organization. The layers are not only vertical and hierarchical, but operate in parallel. This interweaving of horizontal and vertical dynamics can lead to a fractal manifold in different directions.

Readers will note yet another formulation of phenomena as interwoven or intertwined, terms which we insist on replacing, also for phenomena which are primarily epistemic with the interactive formulations of LIR. For Leydesdorff, unlike the *res extensa*, the *res cogitans* is a functional mental entity (mindful). We have reflexive access to the *res cogitans*, but only via the latter to the *res extensa* in terms of reconstructions in models and metaphors. Luhmann [29] called this "a reality which remains unknown." Logic in Reality, as we have indicated and will see again in our discussion of systems, is 'designed' to eliminate such separations. We therefore resist Popper's formulation that one has no direct access to an assumed "nature" as a given through discourse, as Popper formulated in The Logic of Scientific Discovery [35]. The access may not be complete, and its absolute accuracy is not guaranteed, but we are not cut off from nature.

Popper further stated that the empirical basis of objective science has thus nothing 'absolute' about it. "Science does not rest upon solid bedrock. The bold structure of its theories rises, as it were, above a swamp. It is like a building erected on piles. The piles are driven down from above into the swamp, but not down to any natural or 'given' base; and if we stop driving the piles deeper, it is not because we have reached firm ground. We simply stop when we are satisfied that the piles are firm enough to carry the structure, at least for the time being [35]." This is a perfect example of a, rather Western, failure of thought, due again to a misguided search for some metaphorical identity—the 'firmer ground'—that would be considered more reliable, rather than process and change.

12.4.3 Communication in Society

Socio-cultural evolution thus has a dynamic of organization and self-organization that is different from biological evolution. Biological selection is based on genotypes that are hardwired (e.g., as DNA). The "genotypes" of cultural evolution are codes of communication which can further be developed because they are not hard-wired. They are structures of expectations operating at a level above the hardware—that is, the human agency involved. Interactions among the codes can be positive or negative given historical constraints. Information theory enables us to measure whether new options (redundancies) are being created, and in which relations. It is necessary to remind oneself, however, that these are theoretical epistemic constructs, and the interactions are not real apart, from the cognition of those constraints. Theorizing about social systems has thus become more complex than the biological metaphor can carry. Culture operates incursively (see above) on nature and functions have to be specified at the systems level, that is, in terms of expectations.

Unlike the hyper-incursive dynamics of interaction and self-organization which operate vertically and against the arrow of time, organization structures communication at specific moments of time by using incursion. These instantiations also provide room for supra-individual (e.g., institutional) agency. Like (but different from) double contingency as the fundamental operation at the level of the *cogitans,* organizations can synchronously entertain different expectations because they are both interfacing different expectations and looping into the present.

As discussed by LL, both uncertainty—Shannon-type information—and meanings circulate in interhuman relations. Languages enable us both to provide a communication with meaning *and* to distinguish the expected information content of the message. This dual processing can be considered as the evolutionary achievement of developing human language. Carrying the complex dynamics of cultural evolution requires symbolic in addition to linguistic mediation. Thus social communication develops in substantive and reflexive layers at the same time. Reflection is possible from various perspectives, and furthermore reflection is a recursive operation: meanings can be provided with new meanings. A social system thus emerges as a dynamic and flexible coordination mechanism among different levels and dimensions of expectations. The communications can be differentiated both horizontally and vertically: horizontally in terms of different codes and vertically in terms of historical, evolutionary, and anticipatory dynamics. These communicative structures pervade actions to various extents. Society exists inside the individual in the form of language and thoughts; action is based on specific selections among the options available.

The (second) contingency of mutual expectations is *the proper domain of the social sciences and humanities*: in addition to investigating the events, one can always ask "what things mean?" Any attempt to make this domain of expectations the subject of the natural and life sciences in the name of "the unity of science," "monism," or a "grand synthesis" vulgarizes the social sciences or makes them relatively irrelevant as *sozialwissenschaftliche Begleitforschung* (social-scientific assessment). In his *Incomplete Nature*, Deacon [12] called for a focus on "the absent:"

> What is absent matters, and yet our current understanding of the physical universe suggests that it should not. A causal role for absence seems to be absent from the natural sciences."
>
> [...] This something-not-there permeates and organizes what is physically present in these phenomena. Its absent mode of existence, so to speak, is at most only a potentiality, a placeholder.[...] Zero is the paradigm example of such a placeholder. (p. 9)

Brenner has called attention to, rather, a dialectic between absence and presence, in that both co-exist in complex evolving processes. From his biological perspective, Ulanowicz [40] compared "redundancy" with "negative" or "apophatic" theology: in apophatic theology, one can specify God only in terms of what He is not. From a monistic perspective, the second contingency of expectations operating on expectations is "absent" and cannot be studied. There are no data; data are in the first contingency. Both in the philosophy of science and in the philosophy of biology, this limitation to analyzing only observables has been reflected, for example, by Deacon and Ulanowicz, respectively. Ulanowicz argued that the *apophasis* (A)—of redundancy—cannot teach us anything about historical events, unlike the observable information which he labels D (as an abbreviation of *didactic* instruction). In biology, however, the *apophasis* is limited by the constraint that $\Phi = D - A$ should be positive. A biological system with more options than realized ($A > D$) would be vulnerable to such perturbations that a catastrophe would be unavoidable [40].

The priority of a second contingency, to repeat, the probability of interaction, changes the status of the social sciences by taking the place previously reserved for the study of transcendence (e.g., theology and philosophy). What things "mean" is no longer only a metaphysical question, but also an empirical one. This viewpoint is, of course, explicit in the earliest work of Lupasco.

12.4.4 Redundancy

Culture operates incursively on nature and functions have to be specified at the systems level, that is, in terms of expectations. In the information theory of Leydesdorff, redundancy is defined as the complement of the information given the maximum entropy—that is, the total number of options [38]. Redundancy provides a measure for the options that were not realized historically but which could have been realized [5]. By adding redundancy, one adds to the maximum entropy. One can further distinguish between redundancy as the not-yet-realized options (in general systems theory) and the generation of redundancy as new options in the potential synergy among codes in interpersonal communications. Such interpersonal intentionality can be expected to generate entropy in the first contingency *and* redundancy in the second.

In LL's theory, *information is communicated* in interactions among senders and receivers, and at a higher level, *meanings can be shared* to variable extents and thus meaningful information is organized into a vector-space. However, this vector space is constructed and therefore remains subject to reflexive reconstructions. The reconstructions, in terms of different weights of the codes of communication, open horizons of meaning which are self-organizing, but in our view only as epistemological structures, epistemons.

LL suggests that historical and evolutionary processes operate as feedbacks on each other but with opposite signs. The net result of the interactions between information and meaning processing can be measured in bits of information. If this net result is positive, historical organization prevails; if it is negative, self-organization of the communications is indicated. In this sentence, the approaches of LIR and LL can be clearly distinguished: the former is natural, the latter not. The use of 'self-' here is particularly questionable. For us, communications are not grounded, as ontic processes, not only anchored by linguistic codes. The codes are the (unintended) results of repeating patterns of communication. The patterns develop in terms of selections over time operating upon selections at each moment. Some selections are selected for stabilization along trajectories; some stabilizations can be selected in a next iteration for globalization at the regime level. As it emerges, the regime restructures the meanings of the communications from which it emerged. In other words, the communication gains a degree of freedom and can handle more complexity by developing the reflexive capacity to rewrite its history and content at the supra-individual level—that is, discursively. This would be a reasonable description of communication processes if it were not personified, reified by ignoring its process aspects.

Information flow in communications can be considered as a linear flux, whereas expectations can circulate. The cycling generates redundancy on top of the linear flow. Redundancy is generated when two (or more) perspectives on the same information are operating at an interface. Whenever information is *appreciated*, a meaning is generated. The same information can be appreciated differently by other agents or at different levels. Whereas information can be communicated, meanings can be shared.

12.4.4.1 Sharing

Sharing is an eminently dynamic, Lupascian concept that applies in the epistemic part of LL's communication theory. First, sharing can generate an *intersubjective* layer with a dynamic different from that of information processing. The redundancy in the overlaps can be measured as *negative* bits of information; that is, as reduction of uncertainty. Meanings refer intersubjectively to "horizons of meaning" that are instantiated in events. Whereas the events are historical and generate entropy along trajectories following the arrow of time, appreciations are analytical and can add redundancy or negative entropy from the perspective of hindsight—that is, against the arrow of time. One can also consider this redundancy as feedback (e.g., error correction) against the arrow of time. LL thus adds, to the properties of the *res cogitans*, those of non-material expectations and intentions as things which remain uncertain. This is an understandable and possible position, but is not a necessary one; our LIR approach focuses on expecting and intending as cognitive processes, encompassing degrees of uncertainty and *ipso facto* of certainty.

LL believes that meanings cannot be communicated, but they can be shared and organized depending on positions and perspectives, even without requiring a direct communication relation. Semantics are based not on relations, but on patterns of relations or, in other words, *correlations*. However, sharing has to be, in our view, a dialectical process. When one assumes the materiality (as energy) of mental processes that instantiate both actuality and potentiality, as LL would formulate it, the organizational level is historical and thus material. At this level, entropy is generated (and energy is needed) to operate the "communication in reality" in the LIR view.

12.4.5 Weak and Strong Anticipation

Rosen [37] defined anticipatory systems as systems that entertain models of themselves. The model represents a future state that is available in the present and can be used for further development. Dubois [14] added the distinction between weak and strong anticipation. As psychologies, we ourselves can be considered as *weakly* anticipatory systems: we are able to construct and entertain different models of ourselves, but we are always historically constrained by our current state ("the body"). We construct our present state (at t) with reference to both our past ($t - 1$) and our mentally envisaged states at future moments in time ($t + 1$).

As against weakly anticipatory systems, strongly anticipatory ones construct epistemically their next state exclusively from expectations ($t + 1$). However, one cannot expect a system other than systems of expectations to operate in this mode. All other systems take the ontological components (potentialities) of past and present states into account as a basis for further development. The ontological constraint of *having had* to be historical disappears at the supra-individual level of expectations, since the social system of communications is *not* a living or even "existing" system. The anticipatory mode can be considered as a logic operating within the system. The evolutionary dynamics of this strong anticipation in terms of interacting expectations is meta-historical; from this perspective, the history of the system is its *morphogenesis*.

12.4.6 Structuration

In his "structuration theory," Giddens [17, 18] offered another way of discussing these structures in terms of the expectations of individual agents. According to Giddens, structures exist only as memory traces that are instantiated in action [18]. Structures can be considered as providing rules and resources which can be instantiated. However, structure, according to Giddens, exists outside "time and space" as "absent differences" that, in his opinion, cannot be studied empirically. This contrasts with the concept of structures as processes as conceived by Lupasco which will be outlined in Chap. 16 on Systems.

In Giddens' "structuration theory," the "duality of structure" is considered as a "virtual" operation. Thus defined, a *methodology* is suggested for relating institutional analysis to the analysis of strategic conduct: the one narrative can be used as a *context* for informing the other; structure is present in action and actions can be aggregated into structures. However, the two narratives remain juxtaposed by "bracketing" the one perspective when focusing on the other. This is exactly the mental process of actualizing and potentializing described by Lupasco.

In Giddens' opinion, the focus should remain on the individual agents, institutions, and instantiations of structure. Although structure is implicated in the reproduction of social systems, it cannot and should not be studied as such *because* it is absent. However, why would one not be allowed to formulate hypotheses about a second contingency in social structure? Would not the absent ("the zero") be equally interesting? [12]. LIR confers legitimacy to the study of structures: structures can be expected, and observations then serve to test these expectations as hypotheses.

For human beings, a second contingency of expectations is constructed on top of the biological expectation of "living together." For Leydesdorff, the micro-operation of this second contingency is the double hermeneutics in the relation between *Ego* and *Alter*: *Ego* expects *Alter* to entertain expectations about *Ego* just like her own about *Alter*. The second contingency evolves as interacting expectations. A reflexive dynamic of meaning and intentionality is thus added to the contingency of observable actions. As against biological systems in which operational closure can be structural, translations among codes remain possible across boundaries in social systems. Since the communication of information and the sharing of meanings operate in terms of recursive selections, the historical origin of the variation may no longer be relevant after a series of further rewrites. The relevant variation is absorbed and organized into formats such as trajectories which can be overwritten radically when a regime changes. A social system of expectations continuously loops into itself at different levels and from different perspectives.

Both historical trajectories and evolutionary regimes can be expected to change, but at a different pace. The two momenta of historical development and evolutionary change relate as woof and warp, but in a dynamic trade-off, not simply a positional framework. In his famous article about "objectivity" in the social and cultural sciences, Max Weber expressed this change in the dynamics of reflection at the supra-individual level in his concluding paragraph as follows: The Second Law of thermodynamics states that entropy increases with each operation. Shannon [38] defined information as probabilistic entropy—$H = -\Sigma_i pi * \log(p_i)$—and accordingly Shannon-type information is by definition positive.[4] Whereas historical developments thus unfold with the arrow of time, models enable us to anticipate future states from our position in the present, that is, to use future states (x t + n) represented in the present (x t) against the arrow of time. The dynamics of expectations are very different from the historical dynamics [13]. For example, the interactions among differently coded expectations can generate redundancy. Redundancy enriches a system with new options that are available for realization, but have hitherto not been used. In information theory, redundancy R is defined as the fraction of the capacity of a communication channel that is *not* used.

LL argued against the reification of the cognitive process, but not against its operationalization. As Luhmann [29] formulated it: "The most important consequence of this analysis is *that communication cannot be directly observed, only inferred.*" Luhmann added: "The basic problem in the theory of communication lies in the general reluctance of the social scientist to deal with what is not directly observable." The various equations enrich our mental models about the dynamics of expectations, and LL states that different ontologies are indicated by using the different about the actual and potential status of what is not directly observable.

12.4.7 Meaning and Information in Communication

As discussed by Leydesdorff, Habermas and Luhmann shared a focus on "meaning" as the central concept of social communications. In order to make this concept ("meaning") clear, Luhmann first distinguished "meaning" from "information": "We must distinguish between *meaning* and *information.* Sociological theory is not used to working with a pregnant concept of information, let alone making an explicit distinction between information and meaning. In other disciplines too—in linguistics, cybernetics, information theory—there is a lack of clarity, or at least no uniform opinion, regarding these concepts and the way they are related.

The difference can be made clear by using a practical criterion. On being repeated, a message or report loses its information value, but not its meaning. In our view, the strategy used by LL is pseudo-dynamic using a pseudo-dynamic logic such as the logic of forms of Spencer-Brown. Concepts of recursion and feedback are in our view insufficient to model the knowledge-based society because they are tied back to standard binary logic through abstraction of the entities, which is the real underlying dynamics. By putting logic in reality back into systems theory, we undercut the abstract reflexivity of the communication systems picture.

Shannon's counter-intuitive definition of information as uncertainty induced a call for a more intuitive definition of information as "reduction of uncertainty" or neg-entropy. Several authors, summarized in Capurro [9], defined "information" in accordance with the semantic root of the word of "in-formare." The anthropologist Bateson proposed defining information as "a difference which makes a difference" [3]. However, a difference may make a difference for one system of reference but not, or differently, for another. In other words, information would then no longer be defined *analytically*, but in terms of what information means for a receiving system. In second-order systems theory, this receiving system has been denoted as an "observer". LL confirms that binary codes as used by Luhmann are too abstract for sociological analysis, and information theory enables one to specify the uncertainties in a model of communications.

When the sending and receiving systems are considered as the systems of reference (and information is defined as "a difference which makes a difference" for them), the definitions of information, meaning, and discursive knowledge become entangled. The more abstract perspective of information theory on the number of options and the measurement of realized ones in terms of (e.g., bits of) information tends to be lost. For us, this result is exactly what our critique is about! In human communication, it would be *unnatural* for them not to be 'entangled', but the applicable LIR terms for the interactions and their evolution ('disentanglement') are a little more rigorous. The original passage might simply suggest that abstract perspectives are to be avoided!

Hidalgo and Hausmann [19] have defined "information" with reference "to the order embodied in codified sequences, such as those found in music or DNA, while *knowledge and knowhow* refer to the ability of a system to process information." However, codified knowledge can be abstract and—like music—does not have to be "embodied". Floridi [16] proposed "a general definition of information" according to which "the well-formed data are *meaningful*" (italics of the author). Luhmann [29] posits that "all information has meaning." In his opinion, information should therefore be considered as a selection mechanism. Kauffman et al. [23] added to the confusion by defining information as "natural selection."

LL repeats that as Weaver [41] emphasized: "Information must not be confused with meaning." When "something" is communicated, the exchange can *mathematically* be described as information: the distribution of what was communicated has changed Shannon's H is a content-free statistic of the distribution; it remains meaningless without the specification of a system of reference. Specification of a system of reference provides the information with meaning. *The communication is specific in terms of "what is exchanged" when the system operates.* Here, we simply note that whatever in a communication can be described mathematically does not exhaust its ontic, functional content. In LL's communication epistemology, the logic of an emerging system is different form that of its genesis. In our approach, there is no logical or ontological 'cut' between them.

12.5 The 'Intentionality' of the Social System

Leydesdorff starts from the position that the market and the political system contribute two sub-dynamics to the dynamics of our social system. Organized scientific and technological knowledge production and control have added a third sub-dynamic. Interactions between these sub-dynamics lead to a complex model due to differing time scales of evolution, historical (for technology) or instantaneous (for the market). Leydesdorff [25] sees the sub-dynamic of knowledge as analytically different and orthogonal to older economic systems and their communication and control structures. The large and rapid influxes of knowledge create an emergent dynamics of expectations or anticipations that determine the further evolution of the information society. In addition, these interactions in and of themselves contribute to the generation of the knowledge base. LL calls this a "first codification". Knowledge enables codification of our 'Meaning I' of information, and this knowledge can be further codified to Meaning II.

The intentionality of the social system is contained in the distribution of selections of future events and is therefore different from and even orthogonal to those of human beings. However, the two types of intentionality remain interwoven because of the structural coupling between communications and the carriers of communications.

In LL's conception, the processing of information (Shannon-type) and the processing of knowledge exist in a "two-layer" relationship, but dependent on each other because of couplings. The former can be slowed or accelerated by the processing of meaning and knowledge. When the various processes and sub-dynamics are distinguished analytically as independent operations, the respects in which a development is increasingly knowledge-based can be studies. The relations of reflexivity and recursivity [21] that exist define a set of orthogonal relations. Knowledge-based strongly anticipatory systems construct and continuously reconstruct their own bases in potentially self-reinforcing loops. This is a new level of complexity in which it is the knowledge-base itself that is the origin, analytically speaking, of further complexification. To repeat, inter-human communication takes place at two interacting levels. At the first level information is exchanged and provided with meaning, and at a second level, meaning can be communicated reflexively. The processing of meaning is structurally coupled to the processing of information in a co-evolutionary scheme.

Luhmann said that "communication cannot be observed directly, only inferred" has given rise to much discussion about the equivalent of an "observer" in a social network system of communications. Expectations operate as the substance of communication at the level of social systems. We can now specify that and how meaning is generated, communicated and reproduced in communication at an abstract level of the social system. The two systems, the individual and the social, can be considered as structurally coupled, since they provide complexity to each other, however, the axes of the time development of these systems are orthogonal because they are *substantially* different. Over time, the operation is neither exclusively historical nor completely recursive.

The relation system/environment changes when the system of reference is functionally differentiated into sub-systems. Functionally differentiated sub-systems are *both* operationally closed and dependent upon one another, for example, with reference to the reproduction of the next-order system. In a sense, LIR is the sought-for meta-biological model than can capture the complexity of the reflexivity in inter-human communications. To repeat, a knowledge-base develops as the consequence of the two anticipatory mechanisms—functional and reflexive. If the knowledge based is globalized, the social system can be reconstructed in terms of a techno-economic evolution.

Social systems reconstruct reality in terms of meaning that is provided to the system. Meaning processing adds an evolutionary feedback mechanism to the system that inverts the time axis locally. As cited by LL, Luhmann defined social systems as consisting of communications and their attributions as actions. A system of communication reproduces itself by linking communications to one another over time, while the attribution of these events as actions serves the observability of the system in historical time. Communications are operations which cannot be observed directly, but one can make inferences about them by testing hypotheses against the observable interactions among the agents. The communications and agents are strongly coupled and therefore the states of the agents can be used as indicators of the evolving communication processes among them.

Communications can thus be organized non-linearly into systems and systems of communication are self-organizing, where communication is the unit of operation of social systems. The new communications media reinforce direct relations between science, technology and the market as different communication mechanisms. In another specifically social context, the social organization of the exchange of knowledge-based expectations, in addition to the institutional mechanisms of political economy, gradually transforms national economies into knowledge-based economies.

The result is that with less and less central control, hyper-incursive dynamics reconstruct the social system evolutionarily, with the effect that the cultural system becomes increasingly orthogonal—less 'real'—to the natural one. At the level of social systems, it is no longer the facts which matter, but what they mean. Meaning is not given naturalistically, but constructed socially and continuously reconstructed. For us, however, this is an idealization which results in losing focus on the entire process of communication which includes what is being communicated and the *changes* in the senders and receivers and as well as its potential impact on the society.

According to Habermas, intersubjectivity cannot provide meaning to events at the supra-individual level when the sociological model is defined only in terms of social systems theory. The human subject and the social system are then only coupled in an objectified interaction among these systems, where objectified means here, also in our view, blocked at an epistemological level. Both Luhmann and Parsons would thus have reduced intersubjectivity to interpenetration as a systems operation without sufficient appreciation of the dynamic normative and affective components that bind people together at the communal level.

In the network description of society, relations include non-linear interaction terms at various levels and generate dynamics other than those involving the intentionality of the interacting agents. The latent structure of relations operates over time as a virtual (potential) structure and can be considered as a sub-dynamic which interacts with the local sub-dynamics but partly beyond their control. These latent dimensions can structure events or situations. *This is exactly the operational role that LIR assigns to potentialized phenomena.*

12.5.1 Knowledge-Based Systems in Society

For Leydesdorff, knowledge enables us to codify the meaning of information. Knowledge can be considered as a *meaning which makes a difference.* In LL's three-dimensional social system [27], the three "dimensions", as they are instantiated by people are knowledge, geography and economy, and the three corresponding interactions between the pairs are the knowledge infrastructure, the political economy and innovation processes.

A knowledge-based system involve a two-layered network, one of institutional relations in which the carriers constrain each other's behavior and one of functional relations in which the shape each other's expectations. The knowledge base is a second-order interaction term which is the historical result of the first-order interactions in the knowledge infrastructure. The self-organization among the functions involved has a different dynamics from the organizations of relations among the institutions (and people).

Complex social systems thus exhibit both integration (stabilization) and differentiation (globalization) of the various sub-dynamics to enhance or permit further developments. These can operate concurrently, but not necessarily in a synchronized manner. In fact, they are not states, but operations. Stabilization and globalization can be considered as different sub-dynamics operating on each other and on underlying selections. Analytically different functions accordingly couple evolutionarily as different codified expectations, while institutions couple historically. (We suggest that, in this case also, LIR brings a greater degree of rigor than can be conveyed by the term 'couple'.)

12.5.2 Summary of the LIR Perspective. Epistemological Dynamics

We conclude this analysis of communication by summarizing the key conclusions about the dynamic structure of the combined concept of information, meaning and communication. In LIR, none of these can be separated from the real content to which they refer; they are dialectically connected. It is via dynamic systems of relations that structures emerge that can operate as selection environments and thus *determine* the meaning(s) of the information. Better, they *are* the meanings.

It is thus correct, we believe, to refer to Leydesdorff's field of study as "epistemological dynamics"; this is a term with which he has expressed his agreement to us privately [28]. The interobjective reality of expectations thus generated has the status of a construct that can be entertained, unpacked, and revised. The subjective and the inter-objective were not sufficiently distinguished the original form of the first and second contingency. We suggest however, that a more ontological reading of them as incorporating real processes removes the non-scientific flavor of 'subjective' and supports the subjective-objective dialectics.

In a review article mentioned by LL about "social interaction," Parsons [33] related his own [32] formulation of "double contingency" in human interactions to roots in American pragmatism. In addition to the first contingency of observables, a second contingency among expectations can thus be expected. In another formulation, Giddens [17] denoted a "double hermeneutics" between the analyst's and the participant's level of action and accounting. The participant and the analyst can be the same person embedded in different or similar discourses. However, what a communication means is not directly observable. Meanings II as defined above can only be specified as expectations.

For LL, innovation and change are based on specifications of what has hitherto *not* been the case, as in the Deacon theory of information based on absence. We argue that on top of the existing relations, a second contingency of possible relations and expectations can be envisaged. This second contingency of absent possibilities cannot be studied from a natural-science or life-sciences perspective, unless it explicitly includes, imports from ontology, the necessary additional dynamics. It is no different, then, from what LL calls the proper domain of the social sciences.

Different from the first contingency in which one can touch, feel, observe, or directly relate to other human beings, the other's contingent expectations shape a layer of expectations which can be shared to varying extents. In this next-order process one provides meaning to things and events by *partaking* in the social (re)construction of meanings among humans. Meanings can be expected to loop in cycles on top of the entropy flow. Whereas biological systems can gain in complexity by closing themselves operationally—for example, by shaping a membrane—expectations can disturb and penetrate one another "infra-reflexively" and across domains in the second contingency. Neither the systems nor the codes "exist" as hardware (*res extensa*). The reflexive layers (*res cogitans*)—at the individual and the supra-individual levels—can be expected to operate with other selection criteria upon each other and over time. Because of these reflexive couplings in terms of expectations, cultural evolution can be much faster than biological evolution which operates in terms of realizations (over generations).

12.6 The Socio-economic Polarization of Communication

In their natural habitat, but under conditions of stress due to a more or less lengthy need for food, some animals resort to a variety of devices to hide food sources from others. The falsification and/or of blocking information for personal gain is of course a highly developed skill of some human beings. It is therefore philosophically valuable to make a brief digression into the area of *disinformation*, or, better, *disinforming*. For the purposes of this exercise, we define information as a process of informing, a transfer of knowledge from one human being to another that is meaningful in the sense of having value for his/her survival or pleasure. It supervenes on the definition of information as data (Floridi). The theory of information includes its communication or messaging, Angeletics in the term of Capurro, discussed above.

Misinformation is false information that has been generated and transferred by accident, without any intention on the part of the sender. Any negative consequences, even if they are disastrous, does not imply negative intent, but the sender may still be held responsible for them. Negligence, at least in a somewhat decent society, cannot be allowed to go without suitable reaction.

Today, disinformation has become a major topic of concern at the level of the European Union as evidenced in this March, 2019 article.[1] For us, disinformation—disinforming—is an intentional process whose objective is to subvert information for criminal and/or selfish purposes. It is characterized by having no meaning, since there is no dialectical relation between message and intent, and any meaning, for the disinformer, is subordinate to his/her underlying—lying—objective. In other words, disinformation is a lie, characterized by the logical properties of semantic, mathematical and visual paradoxes, namely, the perceivable oscillation between limiting binary logical states of yes or no, truth or falsity, 0 and 1. In the social domain, disinformation is a tool, a method of attempting domination by any means, *ipso* facto

[1] https://www.europarl.europa.eu/RegData/etudes/STUD/2019/624279/EPRS_STU(2019), "Regulating disinformation with artificial intelligence. Effects of disinformation initiatives on freedom of expression and media pluralism".

immoral or unethical. This definition can be compared with that of the EU study: "false, inaccurate or misleading information designed, presented and promoted to intentionally cause public harm or for profit". The difference with misinformation is as in the above in its intentionality.

Typical forms of disinformation consist of messages that are incomplete and misleading as well as directly false. Disinformation in this sense is close to lying by omission, and in fact one could consider disinformation as describing lying in the social sphere. People who withhold information about their physical condition in connection with their employment are 'engaging' in this form of disinformation, and I point here to the utility of using the verb form instead of the noun. Disinformation in all walks of life is so prevalent that it becomes—almost—taken for granted. This is becoming an increasingly greater danger for the society in view of the influence of social media, some of which can now only be described as in part anti-social media. In fact, the only question may be to what extent political and narrow economic objectives can be maintained *without* disinformation.

There is no obvious solution, as we are very close here to the domain of belief, from which science is excluded. There is no overlap or interaction possible in the information/disinformation content of the following two statements: "Climate change is an impending disaster for which there is almost no remaining time to avoid," and "Climate change is a hoax propagated by Communists to weaken the U.S. economy." Philosophy and the social sciences in general benefit in principle from the vast capacities for identification of sources that are now available. On the other hand, these are more than compensated by the information explosion, such that finding all relevant references is still a difficult process. Disinformation can come down to a very specific, at least partly intentional process of ignoring easily available references. Other methods include swamping of new results by overemphasis on classical sources of only historical value.

In general in science, disinformation becomes roughly equivalent to fraud, the dissemination of data not obtained by actual experiments. However, for data with major social implications, such as data on climate change, its misuse is a clear example of disinformation including a major ideological component. In addition, false accusations of fraud or plagiarism are usually supported by a mass of disinformation which can become auto-catalytic. There are two or three levels on which disinformation can be combated: (1) on the personal level, correcting false information in one's personal network; (2) on the institutional level. Many further details on regulatory and technological responses to disinformation are provided in the EU study.

We do not consider this phenomenon, which might be termed a social polarization of communication, as a topic worthy of further philosophical discussion. We note the existence of the phenomenon and relegate it, together with that of selfishness in general, without regret to the domain of mental pathology, one that is outside the scope of this book. This is, we believe, a recognizable philosophical position. As the numbers of people who believe in a flat Earth grow and as more join them to propagate that belief, the only stance we can take, again, is a scientific one, observing the pathology but unable to 'treat' it directly; Flat-Earthers use the same language and conceptual structures that are used by the authors of this book. We face here a rather unique situation. There are no 'lower' animal species that routinely and consistently misinterpret their environment. Somehow, messages about such beliefs, e.g., in a divinity have frequently been exchanged in the past with not only potentially devastating consequences but actualized ones (religious wars).

The basis of our Philosophy in Reality requires that we practice it to allow a dynamics between ontological and epistemological opposites of all kinds, *provided however*, that there is mutual recognition of one by the other. This requires of course communication and the level of cognitive development of a society can be measured by the degree of consensus within opposition, as Hofkirchner's identity-in-diversity (or unity-and-identity-of -opposites), that prevails in it.

12.7 Machine Communication and 'Behavior'

Given our definitions of communication as involving human beings directly, we will not discuss the growing subjects of human–machine and machine-machine communication. There is no substantive question that is part of natural philosophy as we see it. We are dealing with a phenomenon that is a binary artifact of the digital ontology defined by Capurro, and that we have relegated to philosophy *tout court*.

Lorenzo Magnani has asked us to consider the status of human beings with major artificial implants or components of which a major aspect is their computerization [31]. Such entities, for Magnani, are first of all new and second require consideration of their composite status of part-human, part-machine. Again, as with fully human beings, corresponding communications relevant to machine and human components must exist, and the (human) society must develop mechanisms for meeting the requirements of such—individuals? We express here a view attempts to avoid, if nothing else, all possible *fuites en avant* regarding man–machine entities. The only process possible, in this view, is a loss of human independence *with respect to human objectives*. If a runner who has lost a leg now participates in competitions of similarly handicapped people he or she may find this physically and emotionally rewarding—the overcoming of obstacles, etc. But nothing in the human reaction involves the machine as such. One does not love one's eyeglasses.

12.8 Conclusion

As pointed out in a bibliographical article [11], the formal philosophy of communication is fragmented among several knowledge streams. Whether some contribution is 'philosophical' in nature is a problem here, as it has also been in this book, regardless of whether the authors are designated as philosophers, linguists or semioticians. This was also true of an earlier compendium [1]. Partly for this reason we have concentrated on the work of Leydesdorff since it seems to us embody what is most unique to communication from a philosophical standpoint, namely, its epistemic and ontological aspects. In Chap. 16 devoted to systems theory, in particular that of Gianfranco Minati, we will see how communications and systems theory converge in a social context as the basis for a complete logical and scientific structure for Philosophy in Reality.

We will now leave the domain of formal components of philosophy and specific philosophies of disciplines. In the next Chap. 13, we propose a more global view of knowledge as natural philosophy.

References

1. Arneson, P. (ed.): Perspectives in Philosophy of Communication. Purdue UniversityPress, West Lafayette, IN. (2007)
2. Bar-Hillel, Y.: An examination of information theory. Phil. Sci. **20**, 86–105 (1955)
3. Bateson, G.: Steps to an Ecology of Mind. Ballantine, New York NY (1972)
4. Bennett, M.R., Hacker, P.: Philosophical Foundations of Neuroscience. Blackwell Publishing, Malden, MA (2003)
5. Brooks, D.R., Wiley, E.O.: Evolution as Entropy. University of Chicago Press, Chicago IL (1986)
6. Capurro, R., Takenouchi, T., Tkach, K.L.M., Iitaka, T.: On the relevance of angeletics and hermeneutics for information technology. In: Capurro, R., Holgate, J. (eds.) Messages and Messengers. Angeletics as an Approach to the Phenomenon of Communication, ICIE Series, vol. 5, pp. 145-154. Wilhelm Fink, Munich (2011)
7. Capurro, R.: Hermeneutics and the phenomenon of information. In: Mitcham C (ed) Metaphysics, Epistemology and Technology. Research in Philosophy and Technology, vol. 19, pp. 79–85. JAI/Elsevier Inc., Amsterdam, The Netherlands (2000)
8. Capurro, R.: Interpreting the Digital Human. https://www.capurro.de/wisconsin.html. Accessed 3 Mar 2015 (2008)
9. Capurro, R.: Foundations of information science. Review and perspectives. In: International Conference on Conceptions of Library and Information Science, University of Tampere, Tampere, Finland, 26–28 Aug 1991. https://www.capurro.de/tampere91.htm (1991)
10. Capurro, R., Holgate, J.: Messages and Messengers: Angeletics as an Approach to the Phenomenology of Communication; ICIE Series: Wilhelm Fink, Munich (2011)
11. Cooren, F., Bencherki, N.: Having to be. The possessive constitution of organization. Hum. Relat. **64**(12), 1579–1607 (2011)
12. Deacon, T.: Incomplete Nature: How Mind Evolved from Matter. W.W. Norton & Co, New York NY (2012)
13. Dubois, D.M.: Computing anticipatory systems with incursion and hyper incursion. In: Dubois, D.M. (ed.) Proceedings of the First International Conference on Computing Anticipatory Systems (CAYS), vol. 437, pp. 3–29 (1998)
14. Dubois, D.M.: Mathematical Foundation of Discrete and Functional Systems with Strong and Weak Anticipation. In M.V. Butz et al. (Eds), Anticipatory Behavior in Adaptive Learning Systems. Lecture Notes in Computer Science 2684. Springer, Heidelberg (2003)
15. Floridi, L.: Semantic conceptions of information. In: Stanford Encyclopaedia of Philosophy. https://plato.standford.edu/archives/win2019/entries/information-semantic (2019)
16. Floridi, L.: The Philosophy of Information. Oxford University Press, Oxford UK (2010)
17. Giddens, A.: Central Problems in Social Theory. Macmillan, London UK (1979)
18. Giddens, A.: The Constitution of Society. Polity Press, Cambridge UK (1984)
19. Hidalgo, C.A., Hausmann, R.: The building blocks of economic complexity. Proc. Natl Acad. Sci. USA 106, 10570-10575 (2009).

20. Holgate, J.: *Homo loquens* meets *homo informaticus*; exploring the relationship between language and information. In: Paper presented at the 3rd International Conference on the Philosophy of Information, Gothenburg, Sweden, 12–14 June 2017 (2017)
21. Hui, Y.: Recursivity and Contingency. Rowman & Littlefield, London/New York (2019)
22. Igamberdiev, A.U.: Physics and Logic of Life. Nova Science Publishers, Hauppauge, NY (2012)
23. Kauffman, S., Logan, R.K., Goebel, R.E.R., Hobill, D., Shmulevich, I.: Propagating Organization: An Enquiry. Biology and Philosophy 23, 27-45 (2007)
24. Ladyman, J., Ross, D.: Every Thing Must Go. Metaphysics Naturalized. Oxford University Press, Oxford (2007)
25. Leydesdorff, L.: The communication of meaning and the structuration of expectations: Giddens' "Structuration Theory" and Luhmann's "Self-Organization." J. Am. Soc. Inform. Sci. Technol. **61**(10), 2138–2150 (2010)
26. Leydesdorff, L.: A Sociological Theory of Communication. The Self-Organization of the Knowledge Society. uPublish.com, Universal Publishers (2003)
27. Leydesdorff, L.: *The Knowledge-Based Economy*: Modeled, Measured, Simulated. Universal Publishers, Boca Raton, FL (2006)
28. Leydesdorff, L.: Personal E-mail Communication. J. E. Brenner (2020)
29. Luhmann, N.: Social Systems. Stanford University Press, Stanford CA (1995)
30. MacKay, D.M.: Information, Mechanism and Meaning. MIT Press, Cambridge MA and London, UK (1969)
31. Magnani, L.: Morality in a Technological World. Knowledge as Duty. Cambridge University Press, New York NY (2007)
32. Parsons, T.: The Social System. The Free Press, New York NY (1951).
33. Parsons, T.: Social Interaction. In D. L. Sills (Ed.), International Encyclopedia of the Social Sciences 7, 429-441. MacGraw-Hill, New York NY (1968)
34. Poli, R.: Introduction to Anticipation Studies. Springer, Dordrecht (2017)
35. Popper, K.R.: The Logic of Scientific Discovery. Hutchinson, London UK (1959)
36. Rorty, R.: Contingency, Irony Solidarity. Cambridge University Press, Cambridge, U.K (1989)
37. Rosen, R.: Anticipatory Systems. Philosophical, Mathematical and Methodological Foundations. Pergamon Press, New York NY (1985).
38. Shannon, C.E.: A mathematical theory of communication. Bell Syst. Tech. J. **27**, 379–423 (1948)
39. Smiley, M.: Collective responsibility. In: Zalta, E.N. (ed.) The Stanford Encyclopedia of Philosophy, Fall 2008 edn. https://plato.stanford.edu/archives/fall2008/entries/collective-responsibility/ (2008)
40. Ulanowicz, R.: Reckoning the nonexistent: Putting the science right. Ecol. Model. **293**, 20–30 (2014)
41. Weaver, W.: Some Recent Contributions to the Mathematical Theory of Communication. In C. E. Shannon and W. Weaver, The Mathematical Theory of Communication. University of Illinois Press, Urbana IL (1949).
42. Wu, K.: The basic theory of the philosophy of information. In: Proceedings of the 4th International Conference on the Foundations of Information Science, Beijing, China, 21–24 Aug 2010 (2010)

Chapter 13
Natural Philosophy

13.1 Introduction

13.1.1 Relations as Principles of Reality

The starting point of this chapter is that philosophers have something important to say about the reality of our world as well as about appropriate moral rules for living in it. It has been possible to challenge that importance from the standpoint of standard science and the classical binary logic that supports it. Our strategy is thus to apply logical and dialectical tools that can enhance the capability of philosophy to describe complex, changing real phenomena.

In his approach to a new paradigm for Natural Philosophy, Igamberdiev [27] has suggested that it should ultimately be based on relational principles. Relational ideas in physics and biology, logic and mathematics are explored in this paradigm. Space appears as a relational order of co-existences and time as a relational order of sequences. The relational paradigm was recognized in physics as a dependence of thespatiotemporal structure and its actualization on the observer. In the foundations of mathematics, the basic logical principles are united with the basic geometrical principles that are generic to the unfolding of internal logic. Reality in this framework can be represented as a set of self-maintaining reflexive systems capable of a continuous process of complexification. In the framework of the upcoming synthetic natural philosophical paradigm that is discussed here, the observed structure of the world is a result of a perpetual solving process activity rather than given a priori.

In the context of this book, the key dynamic principle of Logic in Reality (LIR), the Principle of Dynamic Opposition (PDO, cf. Chap. 3), can be seen as a relational principle in this sense, with the internal logic of its unfolding based on energy rather than geometry. As it is more closely linked to physics, the 'philosophy of LIR'—Philosophy in Reality—is less dependent on the 'observer' as an idealized human agent and more on the structure of the world itself and human beings as subjects.

J. E. Brenner and A. U. Igamberdiev, *Philosophy in Reality*, Studies in Applied Philosophy, Epistemology and Rational Ethics 60, https://doi.org/10.1007/978-3-030-62757-7_13

13.1.2 The Quasi-Set Theory of Dieter Krause

Throughout this book, we have tried and will continue to try to maintain a thread that links our concepts, when possible, to those of standard philosophy and mathematics. We thus present already for consideration the approach of Krause to relational structures in terms of set theory [31]. For Krause, a relational structure in the usual sense is a collection of sets (or quasi-sets) and the relations among them. Quasi-sets are relational structures where the relations involved do not depend on the particular objects being related. They are collections of elements of which one cannot say that they are identical to or distinct from one another. Formally, this is equivalent to saying that classical identity in its sense as indistinguishability does not apply to the objects in the domain. In still other terms $x = y$ and $x \neq y$ are not well-formed formulas in the logic of this theory. In LIR, entities are, by the fundamental axioms, *both* the same and different, both distinguishable and indistinguishable. This seems perfectly consistent with the interpretation of Krause for quantum cases. But one needs to distinguish in some more formal way between macroscopic process elements involved in an 'active' process and objects for which the dialectical relations are 'frozen'. Subject to a new input of energy, they are to all intents and purposes in the 'classical' part of the LIR theory. This is similar to the quasi-set situation: in macro elements that *are* distinguishable, the set-theoretical description has a classical part.

13.1.3 The Fundamental Nature of Relations

Most contemporary philosophers retain the classical framework of individual objects and static logical variables for the description of the relations between objects. The advantage in Lupasco's theory [33] is its development of logical and philosophical thought on different grounds: changing relations are present ab initio in the universal process of generation of individual objects and structures. In this approach, a 'world-structure' is fundamentally relational and it lies beyond its mathematical description and the whole reality is represented by it is patterns and contradictorial relations.

The approach to reality taken by Ladyman and Ross [32] in their book "*Every Thing Must Go*", which we discussed in detail in Chap. 10, represents an attempt to naturalize metaphysics based on a relational paradigm taken empirically. However, their approach still refers to standard, propositional logic and is based on the criterion of semantic analysis. Relational analysis is fundamentally relational when it takes individual entities not as simple independent objects but as the basic interactive ontological units called in our context "ontolons". The relational communication of ontolons includes the relational representation of the partly actualized and partly potentialized in the system of interactions between these individual units. This can be clarified in the metaphilosophical approach based on ultimate understanding of information as the fundamental category that connects potential and actual realities (see Chap. 11).

Floridi [16] distinguishes digital ontology from informational ontology which interprets reality as the totality of structures dynamically interacting with each other. In this framework, digital *versus* analogue appears as a Boolean dichotomy which reflects not reality itself but the mode of its presentation. Berto and Tagliabue [2] discuss Floridi's argument and suggest that digital and analogue properties are indeed related to the objective logical properties of identity. In fact, the digital description appears within ontolons having a complex structure with an embedded digital description (see Chap. 17) that allows them to be put in relation to external reality. A digital property thus is not an immanent feature of reality taken as a set of individual objects, but as a process that generates the relational continuum, in which relations acquire digital properties as projections at a lower binary level of reality, to which a binary logic applies after further development.

13.2 Non-natural Philosophy

13.2.1 The Grounding Problem

We should further begin this chapter by explicitly stating what we consider is not natural philosophy—standard philosophy *tout court*. It is impossible to list all of the subjects of standard philosophy which, because they are based on binary propositional logic, fail to give an adequate picture of reality. Bringing order within non-natural philosophy remains an important task, and the question of the nature of grounding is one of the most important epistemological problems currently being addressed.

It is possible to show, however, that the point at which philosophy confesses to an inability to proceed is the point at which our non-propositional logic of real processes begins to apply. The question then changes to the difference between nature and reality as a whole, including fictions, non-verifiable beliefs and intangible objects of thought. Since the idea that classical Natural Philosophy evolved into science seems correct, we are left, for the domain of Natural Philosophy, only a speculative interpretation of nature viewed in its entirety.

This interpretation is, *ipso facto*, at a lower ontological level than the science which has largely replaced it. Much of the twentieth century linguistic turn, expressed in both analytical and phenomenological and residual transcendental traditions, is still visible in contemporary philosophy. The most important point for us is that Natural Philosophy tells us something real about the world that is consistent with our best science, physical, biological and cognitive. Speculative philosophy can always re-illuminate 'eternal' questions such as what it means to be a thinking being in a non-thinking environment, but it cannot in itself be other than part of philosophy *tout court*. This non-Natural Philosophy, to repeat, exists for 'natural' reasons: it is a natural necessity for human beings to create it, by a natural process, but it is not part of nature *qua* content.

Natural philosophy or philosophy of nature is still considered by some authors as a precursor of natural science, as referred to the study of nature and the universe as practiced before the development of modern science. Overcoming the simplistic views grounded in positivistic concepts requires creating a modern version of natural philosophy that recasts the transitions between the potential and the actual in natural processes, and substantiates the relational approach to understanding the reality described by the natural sciences.

13.2.2 Reasoning

In considering the appropriate scope of Natural Philosophy, we are driven to the study of reasoning as a critical part of the human activity of thought, better thinking. Let us say that thinking refers to a processing of information common to all animals which may follow some internal rules. In the case of human beings, these have been called 'laws of thought' or something equivalent from Boole [3] to Jerry Fodor and his 1975 Language of Thought Hypothesis [35], often however without making a clear distinction between internal and external. Reasoning is a uniquely human activity that follows external rules given the capacity of humans for higher-level abstract thought. Logic then means, in the heavy phrase of the analytical philosopher Jacquette [25] the science of correct reasoning. Current leaders in the logical philosophy of reasoning include Dov Gabbay, John Woods and Lorenzo Magnani.

As an epistemological, linguistic concept, reasoning has been described as correct or incorrect, that leading to conclusions that are true or false. The logic of correct reasoning is usually understood as classical bivalent propositional and predicate logic or its modern multivalent, modal, deontic or intuitionist versions [26]. Their machinery—symbolism, syntax and semantics—is essentially the same in the related disciplines of standard set and category theory. Logic has been thought of as the study of reasoning and the construction of adequate, formal descriptions of the modes of reasoning, e.g., deduction, induction and abduction. Smarandache's Neutrosophic Logic [38, 39] provides a model of human reasoning that seeks to capture the imprecision of knowledge and linguistic inexactitude, uncertainty due to incomplete knowledge and vagueness, lack of clear contours or boundaries of mental phenomena. However, it remains for us a formal, epistemic logic without the necessary grounding in physics and real change.

13.2.3 Model-Based Reasoning

Model-based Reasoning (MBR) is major relatively recent addition to concepts of reasoning, introduced by Magnani and Nersessian [30]. The perspective on MBR taken by Brenner [5] suggests that if models are used in reasoning to achieve practical results, the process of reasoning with models is neither topic-neutral nor context-independent, while standard logics are virtually required to be both. For this reason, and others, these logics are incapable of describing or capturing any but the simplest, abstract, syntactic and consistent aspects of reasoning, not of MBR in particular and reasoning as a real process in general. In relation to the added concept of 'Reasoning-Based Models', Brenner proposed [7] that LIR was the appropriate theory for capturing the ontological properties of reasoning as a real process, in which, as in information, meaning cannot be separated from its dynamic evolution.

Magnani calls attention to the need, in creative scientific reasoning, to allow two rival theories not only to coexist but to compete, as epistemological and "non-logical" inconsistencies. We will claim that it is possible to describe all of these concepts of interaction in a more rigorous manner by extending the domain of logic to real phenomena, including theories and their "competition".

13.2.4 Abductive Logic and Abduction

A second development is the application by Magnani and Nersessian and their colleagues of the concept of abduction to reasoning, especially, model-based reasoning that goes well beyond standard conceptions of sentential abductive logic. The complexity of the interaction between abduction, considered as equivalent to abductive reasoning, models and reasoning as such is illustrated in Magnani's important paper "Inconsistencies and Creative Abduction in Science" [29]. Abductions of many kinds are model-based if they involve a reasoned problem solving process. The further definition of "model-based reasoning", by Magnani and Nersessian, is "the construction and manipulation of various kinds of non-verbal representations, not necessarily sentential or formal". For the purposes of this paper, let me emphasize the attribution of a process character to models. It is primarily non-physical models that are, of course, most easily seen as processes, whereas most formal models have the mathematical structure that Batterman has called "applied mathematics" [1].

Formal models have inherent limitations in accounting for inconsistencies inherent in reasoning, and accordingly, in sentential abduction itself. There is a problem, however, with the expression "models of abductive reasoning", which are, presumably, cognitive models. Standard logical accounts of abduction fail to capture, in Magnani's words, "much of what is important in abductive reasoning", i.e. abduction, including, and we here exit a threatening circularity, the existence of model-based abductions, in which empirical, as well as theoretical inconsistencies should not be eliminated arbitrarily.

13.2.5 Errors in Reasoning

The above discussion gives a basis for deciding whether reasoning and the logic of reasoning should or should not be included in Natural Philosophy. We have tentatively concluded that Model-Based Reasoning meets our criteria of correctly addressing some of the illogical', in the standard sense, aspects of reasoning. In contrast, since it illustrates this thesis *a contrario*, let us look at recent work of John Woods on errors in reasoning [44]. As clearly stated by Woods, reasoning for Woods is premise-conclusion reasoning which covers both correct reasoning and errors or fallacies in and of reasoning. To approach the problems resulting from the existence of errors, Woods proposed what he calls a 'Third Way Mandate':

> THE THIRD WAY MANDATE: It falls to an epistemically and epistemologically sensitive logic of error to describe the character of third way reasoning and to propose performance standards appropriate to its nature.

This is reasoning which turns out to be 'good' based on criteria other than validity or inductive strength. Of most interest to us is Woods' suggestion of drawing on the resources of a *naturalized* logic to implement this third-way mandate. In turn, a naturalized logic is one which is empirically sensitive to "what people are like, how they are put together and what they get up to when they reason".

This line of argument leads Woods naturally (*sic*) to the advantages of *inconsistency* as permitting as much broader semantic basis for reasoning. However, logics applicable to such states of affairs are now known; they are paraconsistent logics, to which references are provided in Chap. 3.

It is noteworthy that even this approach, which we consider incorrect because limited to propositional logics, by Woods own admission is not widely accepted in the 'informal logic community. Woods attempts to close this gap by proposing new, not totally idealized models of human practice (thought) referring to the 'Rule' of L.J. Cohen: "In the absence of particular indications to the contrary, formalization of an empirically sensitive theory should not introduce idealizations that greatly exceed those required for the theory's descriptive adequacy."

Naturalization (Woods goes on to say) was a radical step in epistemology, and it is no less so in logic. The naturalization of logic has sizable consequences for the intuitionist approach to conceptual analysis ...). Woods proceeds correctly to critique the forms of consequence relations which fail to aid in finding the conditions for judging that right reasoning is right and introduces the need to "investigate the cognitive makeup of the human reasoner". The approach is continued by reference to defeasibility, that is when premises-conclusion reasoning is correctly performed, there is a relation R (we: epistemic relation) on what is reasoned from to what is reasoned to. If the R-relation is subject to change on the addition of true premises consistent with both the old ones and the conclusion, it is a non-monotonic relation and any reasoning which implements it is defeasible: a defeasible consequence rule, which may help in grasping the nature of defeasibility. A logic of defeasible consequence is thus required to specify the properties of non-monotonic consequence which reflect the distinctive character of the reasoning in which it is embedded, that is, reasoning that is vulnerable, aletheically benign, and generically impervious to breakdown. These are disjoint properties of R-relations.

At this point we must leave Woods' development of the epistemic relations between linguistic, propositional logic and human reasoning and go to the conclusions relevant for our position. His self-critique is most instructive: in connection with error-making, epistemological and logical naturalism are totally underdeveloped. He does not have "much of an idea about the natural go of these things" and "most of the work in the logic of reasoning has yet to be done—to be formulated, never mind achieved". We entirely agree with Woods when he describes the current transformation of logic into a branch of theoretical mathematics and then into one of theoretical computer science as the "targets of logic's would be neutralizers". The objective must be new concepts of experimental intent" to avoid "paradigm creep".

As we have tried to show in preceding chapters, nothing can replace the radical step of replacing standard truth-functional logics, which Woods continues to use, by a logic of real processes of reality. This is the first and necessary step in naturalizing *philosophy*, including natural philosophy as it has been generally understood. Let us now return the area of most interest to us, namely, to explore additional areas of *natural* philosophy as *we* have defined it.

13.3 The Naturalization of Natural Philosophy

13.3.1 Defining Natural Philosophy: The Relation of Human Beings to Nature

For the authors of this book, Natural Philosophy is a non-technical domain of knowledge that tells us something real about the world consistent with our best science, physical, biological and cognitive. Speculative philosophy can always re-illuminate 'eternal' questions such as what it means to be a thinking being in a non-thinking environment, but it cannot in itself be other than part of philosophy *tout court*.

In the conception of a major contemporary philosopher Capurro (2016), Natural Philosophy deals with the question about nature as a whole stated by beings (ourselves) who find themselves in nature without having the possibility of a holistic view of being since we are ourselves in nature and not beyond it. The fact that we are able to ask this question means that we have some kind of pre-knowledge about nature as a whole while at the same time this pre-knowledge is problematic, otherwise we would not ask the question and would not be able to become natural philosophers. Much of the twentieth century linguistic turn, expressed in both analytical and phenomenological and residual transcendental traditions, is still visible in contemporary philosophy. The reaction to this unsatisfactory state of affairs has been the reinstatement of realisms and materialisms of various kinds, associated today with the names of Derrida, Badiou, Žižek, and others.

The 'ontological turn' in philosophy is a term of art that designates dissatisfaction with descriptions of reality based on analytical, semantic criteria of truth. Starting with Heidegger's critique of hermeneutics and the basing of philosophy on human life, the ontological turn is a challenge to neo-Kantian epistemologies, and looks to what the structure of the world might be like to enable scientific, that is, non-absolute knowledge. Unfortunately, ontological theories have been hobbled by the retention of static terms whose characteristics are determined by bivalent logic. Priest [34] suggested that an ontological turn in philosophy was taking place, away from language in the direction of a contradictory nature of reality. However, Priest proposed paraconsistent logic as appropriate to this turn. Lupasco anticipated this ontological turn by some 60 years, but his logical system avoids the epistemological limitations of paraconsistency. Anecdotally, we may say that contemporary philosophy is largely natural and realist: we note the date of July 2012 and the Summer School in German Philosophy at the University of Bonn: "The Contemporary Turn in German Philosophy". We have no difficulty in associating this group of thinkers with the term 'natural philosophers', but we do not believe that the issue of what is natural has been exhausted by them.

13.4 Natural Philosophy from the Historical Perspective

13.4.1 Conceptual Precursors of Natural Philosophy in Reality

To support our new approach to Natural Philosophy, reference needs to be made to its precursors to see where the LIR Principle of Dynamic Opposition might apply. Precursors include both non-contemporary and contemporary thinkers, and both philosophers and scientists. We will not attempt to review the entire history of knowledge from 'Heraclitus to Hegel to Husserl to Heidegger', recognizing that significant elements of Logic in Reality are to be found in all of them. To repeat our thesis, to the extent that a component in these theories is present explicitly or implicitly equivalent to Logic in Reality, the relevant doctrine is *ipso facto* in the domain of a Natural Philosophy.

LIR can thus be a method of distinguishing between Natural Philosophy and philosophy *tout court*. It is based on the assumption that energy deploys itself in all existence, in particular in human existence and complex cognitive and social processes, in a movement from actuality to potentiality and vice versa, alternately and reciprocally, without either totally disappearing, except in trivial static cases. This logic, which we call Logic in Reality, supports and is supported by parts of doctrines from many different and disparate sources, including some of the informal perspectives of Eastern, in particular Chinese world-views. This logical approach establishes non-separability as a basic ontological principle and, among other things, supports the role of 'the other' in society, an argument in favor of social-economic justice and the common good.

The domain of philosophy-as-such, 'just' philosophy, as a separate discipline is best directed toward the study of general principles, such as the unity of knowledge and speculative, 'fundamental' questions such as "why is there something rather than nothing?" A characteristic feature of such questions is that they do not change, although an individual's interpretation of them will obviously be a function of his or her context. In this domain, we include apparently dynamic theories such as category theory, semiotics and quantum-type logics applicable to macroscopic phenomena which in fact retain the principles of standard propositional and mathematical logics. Their capability of describing the 'interesting', that is, interactive, changing and moral aspects of phenomena is accordingly limited.

13.4.2 Natural Philosophy in Ancient Greece. Change

The ultimate relational nature of scientific knowledge means that the establishment of the frame of reference to reality as the starting point of any scientific reasoning. After Thales' introduction of the concept of *substance* as the basic principle of our comprehension of Reality, the understanding of reference in this comprehension is due to Anaximander who envisioned the concepts of both scientific knowledge and a frame of reference. Anaximander realized that Earth cannot be this reference because it floats unsupported. Thus, he introduced the concept of *relational order* in which all things have their origin and end according to necessity and in conformity with the 'ordinance' of time [36]. He also rethought the concept of substance as an *apeiron* which refers to a potential reality from which all things originate. His follower Anaximenes located the apeiron of his teacher Anaximander into reality and explained the transitions from potentiality to actuality by the qualitative transformations of the primary substance that he associated with 'air'.

Parmenides, who can also be considered the founder of bivalent logic, argued that all change is a logical impossibility: nothing can go from nonexistence to existence. This logic could be valuable for pure mathematics taken independently from natural reality, but the framework of natural philosophy could and perhaps should be based on different grounds: change evolves not from non-being to being but from one kind or degree of being to another, i.e. from potentiality to actuality. This background was outlined by Heraclitus and developed in some of Plato's works, but it was elaborated in detail by Aristotle. It forms the real basis of a philosophical approach that can be applied to the comprehension of natural phenomena. It allowed Aristotle, in particular, to formulate, although in preliminary terms, the idea of code as the basis of biological organization (see Chap. 17).

13.4.3 Approaches to Natural Philosophy from Fichte to Whitehead

We appreciate the importance of the concepts of Descartes, Spinoza and Leibniz in the development of foundations of modern European natural philosophy. The fundamental role of Immanuel Kant will be analyzed in Chap. 16, but we will focus here on the approaches to natural philosophy of Fichte and Schelling and then turn to Whitehead. German *Naturphilosophie* was inspired by Romanticism and represented an attempt to achieve a philosophical unity of nature and spirit. This was especially noted in the works of Johann Wolfgang von Goethe with his great achievements in arts, philosophy and natural science. This movement was opposed to a more mechanical view of the world arising to Newton and British empiricists, and it is in the work of Fichte we see some of the foundations of our own logical and philosophical approach. Fichte [15] stated that the "thing-in-itself… is a contradiction though as the object of a necessary idea it must be set at the foundation of all of our philosophizing, and has always lain at the root of all philosophy and all acts of the finite mind, save only that no one has been clearly aware of it, or of the contradiction contained therein. This relation of the thing-in-itself to the self forms the basis for the entire mechanism of the human and all other finite minds. Any attempt to change this would entail the elimination of all consciousness, and with it of all existence." It is this positioning by Fichte and Lupasco of the concept of active, ontological contradiction that defines its place in Natural Philosophy.

Arran Gare points out that it was dialectics as developed by Schelling that "provided the forms of thinking required to develop natural philosophy" [16]. Schelling developed Fichte's notions of the appreciation of subjects as activities (we: processes) rather than objects and of cognition as the process by which nature has come to comprehend itself. The notion of synthesis rather than analysis is central, and so, as in Lupasco, is opposition. "Thought is inherently synthetic, Schelling argued, and begins with a genuine opposition between thought and something opposing it, or between other factors within thought." As in Lupasco, the dynamic basis is provided for emergence from a state of contradiction of a "new synthetic moment that can be treated as a product or factor in the next level of development". We only suggest, based on the principles of LIR, "product *and* factor". Emergent entities, as 'products', enter into oppositional relations as 'factors'. Lupasco [34] specifically refers to his method as a 'dialectomethodology' which he saw as requiring the identification of the dialectically opposing forces operating in a process, their reciprocal degree of actualization and potentialization and the direction of the trend toward predominantly one or the other, as noted in Sect. 13.3.

Schelling is an essential figure in defining a relation between natural philosophy and a logic, even if it remains an essentially standard logic. According to Schelling, since the resultant of coming into being is not a finished entity but one still becoming and thus always contingent, existence and movement cannot be a logical category [16]. For the purpose of describing a process of actualization, we need an instrument that can in some way overcome the frames of purely logical description of the process of coming into being, and dialectics represents a substitution for descriptions using classical logic. Gare, analysing Schelling's views, comes to the conclusion that Schelling developed a theory of emergence and a new concept of life relevant to current theoretical biology. This theory is grasped in Logic in Reality in the Principle of Dynamic Opposition (PDO) operative at the most fundamental physical level as well as all higher levels of reality. The PDO and contradiction or counteraction, and a robust notion of potentiality, is required for understanding the relation between substances, events and processes. (We note in this connection the debate in Buddhist logic on whether dynamic (al) opposition was or was not real [40], involving, like LIR, an included middle.

The concept of emergence introduced by Schelling was further developed by Whitehead [43]. Following Whitehead, it is only events that are the actual entities of the physical world which need to exist in order to have a physics. This leads to a new idea of physical substances, that of a distribution of potentialities, "powers" or propensities, that are "reasonable consequences of a theory of processes" as Thompson [41] suggested. He talks of eternal objects with only a 'pure potential' for ingression into reality in which potentiality is realized (actualized). However what Whitehead designates as Categories of Explanation include highly relevant statements of the ontological foundations of Logic in Reality as a logic of process [8]. Thus

1. (That) the actual world is a process and (that) the process is the becoming of actual entities.
2. (That) *how* an actual entity becomes constitutes *what* that actual entity *is*. ... Its being is constituted by its becoming. This is the principle of process.

Another important philosophical concept of organizational evolution of matter is tektology of Bogdanov [3], analyzed by Gare [15]. In this concept Bogdanov put process and organization at the center of a perpetually evolving reality seen as an organizational process of the Universe. Science itself, according to Bogdanov, is comprehensible as a development within and of nature. His 'tektology' claims that processes are the primary reality rather than things or substances and their attributes, which assumes that a non-formal language is more appropriate for characterising the basic characteristics of nature, and mathematics should be seen as having a derivative status.

The aim of modern synthesis in science is to explain the phenomenon of complexification which lies at the foundation of evolution of the Universe [19, 20], in evolution and morphogenesis of living systems [18, 22] and in social progress [21, 23]. For this purpose, the dialectical ideas of Plato and Aristotle, revived in modern times by Schelling, Whitehead, Bogdanov and Lupasco represent an essential alternative to the positivist interpretation of reality that dominated science for many years.

In a paper in *Information* [46], Wu Kun describes the current situation as follows. We cite this passage *in extenso* as it is a textbook example of the kind of new paradigm we referred to at the objective of this paper. "As a result of establishing the fundamental role of information in the existential domain, the Philosophy of Information provides a kind of dual-existential and dual-evolutionary theory of matter and information which describes information as a general phenomenon existing in everything in the cosmos. This leads to the acknowledgement of the dual dimension of matter and information in all forms of research. Because the lack of an informational dimension in traditional philosophy and science, it is necessary to transform them completely to take into account the new scientific paradigm provided by the current Science and Philosophy of Information. By means of that transformation, all scientific and philosophical domains become involved in an integrating, developing trend of paradigm transformation, which Wu and Brenner have called the "informational scientification of science" [47].

Currently, several thinkers have developed ideas of essentialism that revive (recover, to use our term) the Aristotelian tradition and overcome the positivistic attitude toward the natural sciences. Ellis [11] and Oderberg [33] should be mentioned in this respect. Maxwell [31] emphasizes the necessity to create a modern version of natural philosophy and introduces novel vision of the transitions between the potential and the actual in quantum mechanics and in other areas of natural science.

13.5 Natural Logic, Logic of Nature and Logic in Reality (LIR)

We place the discussion of natural logic, the logic of nature, and LIR together in this chapter because we consider them essential components of Natural Philosophy. Indeed, it will be one of our conclusions that the terms are substantially equivalent. We could have used the term Natural Logic for the Lupasco logic of complex natural processes that grounds our development of a logic of real processes. We have not done so because, as Brenner [6] pointed out, the term Natural Logic has been used by Grize [17]. In his *Logique Opératoire,* Piaget defines it as a logic used spontaneously for customary reasoning performed by means of everyday language.

The paper "Natural Philosophy and Natural Logic" of Wu and Wang [48], defines the latter basically as the logic of nature, which is very close to Logic in Reality. A much earlier book [45], unfortunately published only in Chinese, has the title *Logic of Nature.* Wu's Natural Logic has the same concern for the natural ontological aspects of the logic of nature as Logic in Reality, and both are relevant to the construction of scientific rationality. As we will see later, Lupasco's approach emerged concurrently with the early concepts of systems of von Bertalanffy, while that of Wu emerged in his deep analysis, starting in 1980, of the properties and dynamics of information.

Language as well as nature is frequently considered to have its own logic, but one with its own rules, much more vague and difficult to formalize and axiomatize. Natural logic is thus always situated in a social context and is not subject or topic neutral. Formal, classical logic and the formal view of reasoning was specifically criticized by Grize as taking place in a closed domain, elaborated at some point from facts, but without further relation to reality. An absolute concept of truth is implicit in that premises are stated as such and their establishment is not required, and the only rule of deduction is *modus ponens*. Natural logic on the other hand is referred to not only as a "logic of subjects", but also as a "logic of objects", and care is taken to differentiate natural logic from formal logic also designated as a physics or logic of objects. This idea confers the particularity of context to objects and raises their ontological status, so to speak. In the Lupasco logic of dynamic opposition, subject and object are dialectically and contradictorially related.

LIR statements look like what are termed synthetic statements, that is, ones whose truth depends on matters—in particular, contingent facts about the world—to which we have ascribed a certain dialectic structure. Such statements are to be distinguished from analytic statements that are true by virtue of the meaning of their constituent terms alone. LIR thus provides support to a naturalistic, causal-role theory of mental content and a naturalistic means of drawing the analytic/synthetic distinction. This can be part of Natural Philosophy, while a 'pure' analytic theory cannot be. LIR always defines a real relation between the intensional notions or aspects of a phenomenon and the extensional ones. Analytic claims can provide insight into external reality, but only if coupled with a non-semantic theory that provides some basis for explanation of the coincidence between our concepts and the properties of the real phenomena of the world. By starting from the side of the phenomena, LIR permits progress toward a new 'synthetic' philosophy that if not entirely is more within the domain of Natural Philosophy.

13.6 Realism and Anti-realism

13.6.1 Is Natural Philosophy Realist?

In doing experimental or theoretical scientific research, scientists are involved on a daily basis in the inconsistencies and antagonisms in reality, both epistemological and ontological. Examples are the tension between their partial knowledge and ignorance, as well as the frustrating intractability of matter—the 'refusal' of a chemical compound to crystallize from solution. It is not surprising that these complex processes are perceived and conceptualized in an equally complex fashion. Faced by the diversity of the world as uncovered by science, philosophers tend to reject its metaphysical importance in the name of a perhaps laudable but dangerous strategy of simplification, dangerous if it confers a lower ontological value or significance to diversity as such.

Scientific realism is the stance that best captures the general validity which the activity of science has acquired, but its acceptation of entities that are not directly observable has led to its refusal by some philosophers. This is in my mind another instance of the dynamic opposition that *is* observable in all phenomena, physical and mental. Realism must, in my theory, always be accompanied by anti-realism, and both will alternately predominate as more or less valid in specific cases. Structural realism describes a group of relatively recent approaches whose objective is to respond to anti-realist and other challenges to scientific realism. I will show that LIR also supports and explicates aspects of some forms of structural realism that are relevant to science at the microphysical and macrophysical levels (this chapter and Chap. 6).

The varieties of realism on the market are all pertinent to the proposal of LIR as *logic* of and in reality. In a sense, all have been developed as attempts to answer the question "Is science reliable?" The answers given have tended to focus on the microphysical or quantum domain, in view of its 'wealth' of unobservable entities of which only the intrinsic properties (see Chap. 3) are accessible to measurement. Like Ladyman and Ross [27], we wish to support a program of a principled unification of science, in which the special sciences (those other than fundamental physics) exemplify the principles or patterns of physics while also involving emergent ones of their own. As in our discussion of quantum physics, quantum phenomena such as quantum entanglement and quantum coherence cannot be operative at the mental level, but this does not mean that the patterns of interaction at the two levels can not and do not follow a similar logic.

The most important contribution that we see LIR as making is to provide a *non*-mathematical element of structure (see Chap. 16) to the various forms of scientific and structural realism. The core thesis of LIR is that LIR and the PDO apply to both the entities described by scientific theories and the theories themselves (or more generally the epistemic and ontic aspects of theories) in the *first* and *third* of the three relevant levels or domains: the quantum level; the 'inert' macrophysical level; and the biological and mental level. It is for us rather odd to note how often examples used to illustrate philosophical positions about reality and science are taken from the *second* domain. To discuss issues about fundamentalism in laws of nature, examples often used are simple, reversible 'to-all-intents- and purposes' physical changes of phase. We see LIR and the PDO as making accessible for analysis, that is, to science, more dynamic and interesting cases, for example, where appearance and reality are involved as in the psychology of lying or cheating.

13.6.2 Realism and Anti-realism

We find ourselves in an interesting position: realism is positioned inside Natural Philosophy, as one of its major components, but its opposite, anti-realism which by definition lacks a natural component, is located conceptually in Philosophy in general, philosophy *tout court*. We have thus created an example of the kind of dichotomy which it has been our intention to avoid. Where is the 'emergent included middle' that we have raised, earlier in this book, to the level of a universal principle characterizing the outcome of logical, in our sense, dialectical processes, both epistemological and ontological?

Our approach to this aporia is to analyze the (metaphilosophical) intentions behind the realist and anti-realist positions. Our conclusion, again with the recognition that, by definition, it cannot satisfy 'everyone', is that the realist accepts the complexity of the world while the anti-realist looks for forms of simplicity that will permit an eliminativist attitude to that complexity. Some philosophical challenges raised against scientific realism rest on intuitions that beg the question against empiricist antirealism: there can be no evidence that rationally distinguishes between two empirically equivalent scientific theories. Such an approach implies the existence of two such theories, and pending their appearance, inconsistent with the dialectical LIR view of identity. We consider this objection void of content. Anti-realists tend to use arguments based on counterfactuals and highly unlikely states of affairs that carry strong anti-scientific subtexts. They are examples in themselves of dynamic opposition to theories that eschew such approaches.

LIR provides a basis, accordingly, for realist theories to accept a connection between natural kinds and the conceptual machinery of the sciences. Extra-linguistic and mind-independently existing natural kinds, in our view, are metaphysically fitted for explanation and induction. Any version of something like an *objective idealism* is not required. LIR, in contrast to standard naturalism and metaphysical materialism, provides the physically grounded dialectical basis for such a connection. LIR thus opposes and argues against anti-realism in philosophy and science. LIR accepts as reproducible, quasi-scientific evidence that people do defend one position or the other and makes the reasonable assumption that 'psychological factors' of some sort must be at work. But it thus says something further and perhaps more interesting and important about these two opposed positions: they are *inevitable.*

As we have seen in Chaps. 6 and 8, the principles of LIR permit the formulation of a physics and chemistry that includes a locus for the potentialities necessary for emergence. They will be the basis for our discussion of living systems below. A categorial interpretation of the related issues of closure and downward causation is developed, using the LIR notions of time, simultaneity and succession outlined in Chap. 6. LIR is presented as a logical system that can compensate for the inability of standard logics to address general issues in biological science. The application of LIR to the major problems of the origin of life and evolution and natural selection is suggested, and the essential role of the Pauli Exclusion Principle as the physical basis for the emergence of diversity and living systems will be discussed in Chap. 19.

13.7 Toward a Metaphilosophical Rejunction

13.7.1 The Origins of Natural Philosophy

The concept of a Natural Philosophy, or Philosophy of Nature, has been formulated in Western thought since antiquity. At its origin, there were no hard distinctions between the disciplines of philosophy, logic and science. Brenner has pointed out that a metalogical 'rejunction' of them can be found in the contemporary work of Lupasco referred to above (Chap. 3) as the basis for our Logic in Reality. Today, the history of Natural Philosophy, what we can call its cliodynamics [42], demonstrates its partial evolution into science, which raises the complication of what *it* is and to what extent it differs from philosophy *tout court*.

In this complex hermeneutic process, what has been called Natural Philosophy, acquired some of the properties and concepts of the bivalent, propositional logic of European scholasticism. This was in our view a process that *de*-natured Natural Philosophy by its elimination of the inconsistent, incomplete and contradictory phenomena of reality.

The domain of philosophy is also now divided by its relation to science. Analytical philosophy recognizes the central role of science, but applies to it an inadequate conception of the logical operation of complex processes and living systems. Continental philosophers and anti-realists in the area of 'Constrictive Empiricism' such as Bas van Fraassen [38] are fundamentally anti-scientific and thus lack a necessary link to reality. Both kinds of are thus limited for similar reasons.

In Brenner and Igamberdiev [9], the Authors referred to the need for logical *recovery* of "dialectics and semiotics from reductionist interpretations and to their reunification in a new synthetic paradigm centered on meaning and its communication." Our vision was of a unification of science, logic and philosophy in a common meta-thesis to provide the basis for a more realistic picture of nature and civilization. We stated that a logic of real processes—Logic in Reality—was necessary to redefine the ontological relations between meaning, communication and language. This fundamental task forms the background of any description of nature that might accompany the new functional convergence of science and philosophy that we saw in progress, the scientific *discovery* that accompanies the recovery.

13.7.2 Metaphilosophical Rejunction

From a methodological standpoint, continued in this book, two operations had to be effected: the first was to identify the weaknesses in natural philosophy as generally understood due to the retention of inapplicable binary logical concepts. The second was to effect a *second* naturalization of philosophy and natural philosophy that established Natural Philosophy in our dynamic sense as the central, essential domain of Philosophy in Reality. In such a view, what we would have effected is a *meta-philosophical* rejunction of philosophy with its origins in Western and Eastern antiquity, and with many aspects of current Eastern thought. The term Metaphilosophical Rejunction can be considered as descriptive of the new *Book of Changes* that is in the sub-title of this book. It is thus possible to see, as close readers of this book may have already, that our 'new' *Book of Changes* is also a *Book of Non-changes* at the same time; both are 'epistemons' in a dynamic cognitive relation in our minds.

Let us therefore go back to the first step in the process we have identified of the naturalization of 'un-natural' aspects of natural philosophy itself:

(1) the demarcation of Natural Philosophy by a non-standard, non-propositional philosophical logic which we have called Logic in Reality (LIR) [6], (2) the application of LIR to stubborn problems within the philosophies of process and science,

as well as in Natural Philosophy itself as previously delimited; and (3) the role of recently developed philosophies of information, consistent with LIR, in bringing a new ontological dimension to philosophy.

It seems essential to again outline our positions as to what philosophy and Natural Philosophy are and how they differ. Before addressing the status of Natural Philosophy, let us therefore try to answer the question posed by Deleuze and Guattari: "What is philosophy?", "*Qu'est-ce que la philosophie?*" [12]. In their view, philosophy is neither contemplation, reflection nor communication; it is the cognitive activity of creating concepts in a domain of pure immanence, in contrast to science and logic which involve functions and observers, and to art which operates with percepts and affects. Most importantly, philosophy does not operate with propositions: the relations that compose the concept are not those of comprehension or extension but of ordered variation, processual and modular, pure events, real without being actual, ideal without being abstract. We can agree with these authors that standard logic has an 'infantile' conception of philosophy.

Concepts, hence philosophy, should not be confused with the energetic cognitive states-of-affairs in which they are found. "There is no energy; only intensities" in philosophy, whereas energy involves intensity in an extensive context. Taking this line of reasoning one step further, the level of immanence itself is pre-philosophic, becomes philosophic under the influence of the concept and then evolves is a philosophic relation with non-philosophy. Finally, while the two domains—of immanence and event—are inseparable, philosophic concepts do not 'intervene' in scientific functions or functionalities and vice versa. I will introduce a similar demarcation, to use a term familiar to philosophers of science, to distinguish between philosophy and Natural Philosophy.

The philosophy of Deleuze illustrates the results of applying the concepts of immanence and transcendence, without defining and including any dynamic dialectical relation between them. It constitutes a domain, governed by a binary logic of undetermined, idealized entities, Humean in its absence of reference to effective interactions. In the domain of reality to which LIR applies, the existence of all beings depends and is defined by that of others. Infinities and infinitesimals do not exist, but are replaced by transfinite values, and immanent and transcendent aspects of phenomena are mutually and alternately actualized and potentialized. Thus, LIR can discuss philosophical issues in physical, dynamical terms that do not require recourse to any imaginary, abstract structures which to separate aspects of reality. The aspects that are considered 'virtual' or 'possible' in Deleuze are so 'in philosophy', but 'in reality' are instantiated as potentialities. We will review issues of immanence and transcendence in Chap. 17 below on philosophy as process.

We envisage the rejunction as a conceptual movement in which a change is accepted in the purport of logic that *includes* Logic in Reality. This change is by no means accepted today. The major proposals of what logicians such as Jean-Yves Béziau and He Huacan define as 'Universal Logic' does not include in LIR. From the perspective of this study, we can argue that the rejunction is also affected with the most significant principles enunciated in the *I Ching* as well.

Logic in Reality, in our view, can be a method of distinguishing between natural philosophy and philosophy *tout court*. Natural philosophy must be in conformity with the principles of the logical evolution of natural processes described by LIR (see Chap. 10). In this view, natural philosophy describes the natural process in which dynamic oppositions appear and are predominantly resolved, and thus it reformulates the natural aspect of *epoché* and has a phenomenological dimension. Natural philosophy is grounded in metaphysics in its concepts of substance, frame of reference and relationality. By substantiating the concept of information as grounded in nature, it has a metaphilosophical dimension. Thus, all three domains are present in the structure of natural philosophy making it the basis for understanding the foundations of scientific knowledge.

13.7.3 Other Aspects of Natural Philosophy

In Brenner and Igamberdiev [9], we limited the scope of our Philosophy in Reality to what is also the content of *Natural* Philosophy as it is currently being redefined by Schroeder and Dodig-Crnkovic [10] as well as elsewhere by the Authors of this book. Now that we have expanded our view of Natural Philosophy, however, Philosophy in Reality, by our own theory, cannot be separated from a potential *Metaphilosophy in Reality* in which the standard notions of transcendental philosophy are present as real cognitive objects (ontolons) *qua* their production. We claim that the concepts noted at the beginning of this Chapter of *Non*-natural Philosophy are not to be excluded from the domain of philosophy but only if the co-existence and relative roles of natural and non-natural philosophy are acknowledged. One of these roles is the pre-eminence of natural philosophy in the understanding of the science and philosophy of systems as presented in Part III of this book. We conclude this Chapter here, dialectically, with a further perspective on what has been discussed in the literature as *anti*-philosophy, to determine to what extent it functions as a counter-part of or opposition to philosophy/natural philosophy as we propose.

13.8 Conclusion. Homo Sui Transcendentalis

Our natural Philosophy in Reality is a new way to handle the existence of the *homo sui transcendentalis*, the ‘transcendental man’ of Nicolescu [32], who exemplifies transdisciplinary relationships in philosophy and science. In our view, the life of any individual establishes a new dynamic relationship between immanence and transcendence. To the extent that an individual human and perhaps a cat have at least one component that is transcendental—one does not know where one’s own first-personness comes from, one is transcendent. One’s self-awareness is not (except in pathological cases) the same as any other. Brenner has suggested the existence of individual consciousness as an example of the operation of the Axiom of Choice (see next Chap. 14) at the psychological level, but this concept remains to be explored. In any case, even if the basic principle is established, we cannot and must not 'stay' here. We need to go back to science to see where the immanent-transcendent duality will take us. It is that it is, also, a logic and a dialectics that makes science relevant for a New Natural Philosophy, that is relevant in turn to Society.

The next Chap. 14 is about a subject particularly congenial to our method and approach, that of anti-philosophy. It exemplifies our attempt, as far as possible, to give if not 'equal' time, adequate attention, as a metaphilosophical 'necessity', to opposing or contradictory points of view and doctrines.

References

1. Batterman, R.W.: Idealization and modelling. Synthese **169**, 427–446 (2009)
2. Berto, F., Taglabue, J.: The world is either digital or analogue. Synthese **191**, 481–497 (2014)
3. Bogdanov, A.A.: Essays in Tektology: The General Science of Organization. Intersystems Publications, Seaside CA (Originally published 1921) (1984)
4. Boole, G.: The Laws of Thought. On Which Are Founded the Mathematical Theories of Logic and Probabilities. Dover Publications, Inc., Mineola, NY, USA (Originally published 1854) (1958)
5. Boyd, R.: Scientific realism. In: Zalta, E.N. (ed.) The Stanford Encyclopedia of Philosophy, Summer 2002 edn. https://plato.stanford.edu/archives/spr2002/entries/scientific-realism/ (2002)
6. Brenner, J.E.: Logic in Reality. Springer, Dordrecht (2008)
7. Brenner, J.E., Igamberdiev, A.U.: Philosophy in reality: scientific discovery and logical recovery. Philosophies **4**, 22 (2019)
8. Brenner, J.: Process in Reality: A Logical Offering. Logic and Logical Philosophy 14, 165–202 (2005)
9. Brenner, J.E.: A transconsistent logic for model-based reasoning. In: Magnani, L. (ed.) Model Based Reasoning in Science and Engineering. Cognitive Science, Epistemology, Logic. Kings College Publications, London, UK (2006)
10. Brenner, J.E.: Reasoning-based models. In: Paper presented at the Conference on Model-Based Reasoning. Sesto Levante, Italy (2015)
11. Brenner, J.E.: The 'Naturalization' of the philosophy of Rafael Capurro. In: Kelly, M., Bielby, J. (eds.) A Festschrift in Honor of Rafael Capurro. Springer Fachmedien, Wiesbaden, Germany (2016)
12. Deleuze, G., Guattari, F.: Qu'est-ce que la philosophie? Les Éditions de Minuit, Paris (1991)
13. Dodig-Crnkovic, G., Schroeder, M.: Contemporary natural philosophy and philosophies. Philosophies **3**, 42 (2018)
14. Ellis, B.D.: The Philosophy of Nature: A Guide to the New Essentialism. Montreal: McGill—Queen's University Press (2002)
15. Fichte, J.G.: Foundations of the Science of Knowledge. Cambridge University Press, Cambridge UK (Originally published 1794/1795) (1982)
16. Floridi, L.: Against digital ontology. Synthese **168**, 151–178 (2009)
17. Floridi, L.: The Philosophy of Information. Oxford University Press, Oxford, UK (2010)
18. Gabbay, D., Woods, J., Thagard, P. (eds.): Philosophy of Psychology and Cognitive Science. North-Holland Press (Elsevier), Amsterdam (2006)
19. Gare, A.: Aleksandr Bogdanov and systems theory. Democ. Nat. **6**, 341–359 (2000)

20. Gare, A.: From Kant to Schelling to process metaphysics: on the way to ecological civilization. Cosmos History **7**, 6–69 (2011)
21. Grize, J.B.: Logique naturelle et communications. Presses Universitaires de France, Paris, France (1996)
22. Igamberdiev, A.U.: Semiosis and reflectivity in life and consciousness. Semiotica **123**, 231–246 (1999)
23. Igamberdiev, A.U.: Semiokinesis – Semiotic autopoiesis of the Universe. Semiotica **135**, 1–23 (2001)
24. Igamberdiev, A.U.: Physics and Logic of Life. Nova Science Publishers, Hauppauge NY (2012)
25. Igamberdiev, A.U.: Time, Life, and Civilization. Nova Science Publishers, Hauppauge NY (2015)
26. Igamberdiev, A.U.: Evolutionary transition from biological to social systems *via* generation of reflexive models of externality. Prog. Biophys. Mol. Biol. **131**, 336–347 (2017)
27. Igamberdiev, A.U.: Time and life in the relational universe: Prolegomena to an integral paradigm of natural philosophy. Philosophies **3**, 30 (2018)
28. Igamberdiev, A.U., Shklovskiy-Kordi, N.E.: Computational power and generative capacity of genetic systems. BioSystems **142–143**, 1–8 (2016)
29. Jacquette, D.: Ontology. McGill-Queen's University Press, Montreal, Canada (2002)
30. Jacquette, D.: Philosophy of logic. In: Handbook of the Philosophy of Science Series (2007)
31. Krause, D.: Separable Non-Individuals. Representaciones **1,** 21–36 (2005)
32. Ladyman, J., Ross, D.: Every Thing Must Go. Metaphysics Naturalized. Oxford University-Press, Oxford, UK (2007)
33. Lupasco, S.: Le principe d'antagonisme et la logique de l'énergie, p. 1987. Éditions du Rocher, Paris, France (1951)
34. Lupasco, S.: Le principe d'antagonisme et la logique de l'énergie. Editions du Rocher, Paris (Originally published 1951) (1987)
35. Magnani, L., Nersessian, N.: Model-Based Reasoning. Kluwer Academic/Plenum Publishers, Dordrecht, Netherlands (2002)
36. Magnani, L.: Inconsistencies and creative abduction in science. In: AI and Scientific Creativity. Proceedings of the AISB99 Symposium on Scientific Creativity. University of Edinburgh, Edinburgh, UK (1999)
37. Maxwell, N.: In Praise of Natural Philosophy. A Revolution for Thought and Life. McGill—Queen's University Press, Montreal, Canada (2017)
38. Monton, B., Mohler, C.: Constructive Empiricism. Stanford Encyclopedia of Philosophy (Summer 2017 Edition), E. N. Zalta (ed.) (2017) URL = https://plato/stanford.edu/archives/sum2017/entries/constructive-empiricism/
39. Nicolescu, B.: Manifesto of Transdisciplinarity. State University of New York Press, Albany NY (2002)
40. Oderberg, D.S.: Real Essentialism. Routledge, Abingdon UK (2007)
41. Priest, G.: Inconsistency in the empirical sciences. In: Meheus, J. (ed.) Inconsistency in Science. Dordrecht: Kluwer Academic (2002)
42. Rescorla, M.: The language of thought hypothesis. In: Zalta, E.N. (ed.) The Stanford Encyclopedia of Philosophy, Summer 2019 edn. https://plato.stanford.edu/entries/language-thought/ (2019)
43. Rovelli, C.: The First Scientist Anaximander and his Legacy. Westholme Publishing, Yardley PA (2011)
44. Schelling, K.: In: Schelling, K.F.A., Abtheilung, I. (eds.) Friedrich Wilhelm Joseph Schelling's Sämmtliche Werke, vols. 1–10, II Abtheilung vols. 1–4, pp. 1856–1861. Cotta, Stuttgart (1856)
45. Smarandache, F.: Neutro algebra is a generalization of partial algebra. Int. J. Neutr. Sci. **2**, 8–17 (2020)
46. Smarandache, F.: A unifying field in logics: neutrosophic logic. Neutrosophy, neutrosophic set, neutrosophic probability and statistics. https://fs.unm.edu/eBook-Neutrosophics6.pdf (2003)

47. Stcherbatsky, F.Th.: Buddhist Logic. Dover, New York, NY, Originally published in 1930. Academy of Sciences of the U.S.S.R., Leningrad (1962)
48. Thompson, I.: Process Theory and the Concept of Substance. Gener. Sci. https://www.generativescience.org/ph-papers/subst5c.html (1990)
49. Turchin, P.: Arise 'cliodynamics' . Nature **454**, 34–35 (2008)
50. Whitehead, A.N.: Process and Reality. An Essay in Cosmology. Gifford Lectures Delivered in the University of Edinburgh During the Session 1927–1928. Macmillan and Cambridge, New York, NY, USA, Cambridge University Press, UK (1929)
51. Woods, J. 2013. Errors of reasoning. naturalizing the logic of inference. In: Studies in Logic. Logic and Cognitive Systems, vol. 45. College Publications, London UK
52. Wu, K.: The essence, classification and quality of the different grades of information. Information **3**, 403–419 (2012)
53. Wu, K., Wang, Z.: Natural philosophy and natural logic. Philosophies **3**, 27 (2018)
54. Wu, K.: Logic of Nature. Northwest University Publishing House, Xi'an, Shaanxi, China (1990)
55. Wu, K., Brenner, J.E.: Chapter 8: a unified science-philosophy of information in the quest for transdisciplinarity. In: Information Studies and the Quest for Transdisciplinarity. World Scientific, Singapore (2017)

Chapter 14
Anti-philosophy

14.1 Introduction

A specific field of anti-philosophy has been proposed by some philosophers as a necessary adjunct to philosophy. Despite the wide variation in content and level of analysis, and regardless of whether or not the positions expressed have been designated by the authors themselves as anti-philosophy, its existence and its relations to philosophy raise issues which are interesting and can be usefully analyzed from the standpoint of Logic in Reality. The concept of Wu Kun that traditional philosophy was incomplete [30] and needed to be complemented by a science and philosophy of information led as we have just seen to the concept of the philosophy of information as a metaphilosophy (Chap. 11). We do not consider metaphilosophy in and of itself an anti-philosophy, since its dualism with philosophy is transparent and dynamic.

Anti-philosophy has thus not been well-defined. It clearly co-exists with philosophy, and there is interest in determining what sort of relationship may exist between the two domains. Is it an 'antagonistic one' and if so in what way is that antagonism is expressed? This is what should be expected based on our inclusion of the opposition between other philosophical positions as instances of LIR dynamic opposition. Alternatively, is anti-philosophy simply a collection of metaphorical positions, generally antithetical to philosophy in some standard view? For the purpose of this discussion, we have concluded that it is the latter. There is no 'single-clear' definition of anti-philosophy or answer to the question "What is Anti-philosophy?" The absence of such definitions is not necessarily a drawback; simple identities are not appropriate for complex cognitive phenomena, but are rather an indication of the existence of something possibly new and meaningful.

In Chap. 10, we have seen how the basic contradictorial principles of Logic in Reality and levels of reality can lead to metaphilosophy. Metaphilosophy is an example of recursive cognitive processing of 'philosophy' that establishes it as an emergent entity, here a process structure, a new entity. On the other hand, one may directly challenge the validity, applicability or value of philosophical concepts, and

J. E. Brenner and A. U. Igamberdiev, *Philosophy in Reality*, Studies in Applied Philosophy, Epistemology and Rational Ethics 60, https://doi.org/10.1007/978-3-030-62757-7_14

propose others that directly contradict or change them. Such a process provides a first rough definition of the content of a potential *anti*-philosophy.

It should be clear that the principle of non-separability should apply to any anti-philosophy—philosophy dualism, as it does to philosophy-metaphilosophy; it is impossible to discuss one without reference to and use of concepts from the other. This already suggests that dynamic logic of the relation must resemble that of LIR in both cases. The next step we take is to avoid using the abstract noun 'philosophy' and talk in terms of philosophers and anti-philosophers and their respective attitudes. On this basis, we could designate Nietzsche at once as an anti-philosopher, together with other iconoclasts as a group, but let us not get ahead of our argument.

Anti-philosophy is not a *major* segment in current philosophy but, to repeat, we find it congenial because, like our Philosophy in Reality (PIR) it challenges many precepts of the perceived and repeated wisdom and offers reasoned alternatives. This may be clearer after we have seen further applications of our 'system' in later chapters, especially, that about systems themselves. For now, we simply repeat that all innovative philosophy will be *ipso facto* partly an anti-philosophy. This term could well be applied to our discussion of Aristotle in Chap. 5, as it brings in concepts that are normally not used in philosophical analysis. The most anti-philosophical position toward standard philosophy one could take, therefore, is to accept an on-going dialectics between the two domains as one of their essential components.

We have found a few forms of anti-philosophy in the current literature, from naïve to messianic. The discussion is complicated by the fact that authors such as Nietzsche are often considered anti-philosophers without any direct evidence that they would agree. Some of the ideas of Boris Groys in English and Alain Badiou in French are even sufficiently congenial to our theses that we might even be tempted to refer to PIR, our Philosophy in Reality, as an anti-philosophy! We will not take this somewhat polemic approach, however, for the good and for us sufficient reason that the anti-philosophy we have found falls within the domain of *non*-Natural Philosophy, that is, it does not incorporate dialectic ties with a reality defined by physical properties. We thus allow the concept of anti-philosophy, following our general line, to coexist with that of the philosophy *tout court* defined in this book.

14.2 Some Current Views of Anti-philosophy

14.2.1 Antonio Palomo-Lamarca: "Philosophy is Dead"

Logic in Reality (LIR) must be considered as fundamentalist or foundationalist philosophy since it takes a physical principle, dynamic opposition expressed in matter-energy, as the basic ground or foundation of knowledge and inquiry about the world. In a short paper, a manifesto as he calls it himself, Palomo-Lamarca (APL) [20] defines an eliminativist position that makes anti-philosophy as an activity whose dynamics are closer to art than philosophy. For APL, the concepts of philosophy, including that of reality, and methods, such as logic, have been emptied of significance. In contrast, anti-philosophy is not a wisdom but a way of practical life, an art of feeling and of creativity.

While "philosophy is dead", APL relates anti-philosophy to Philosophy an idealization of what philosophy was (the old Lady) and perhaps still could be with a change in focus: "The aim of Anti-philosophy is not to destroy philosophy, but to call attention to its current state. In the other hand, science and the philosophy of science are "too tight and materialistic, controlled by "metaphysicists playing with language". Anti-philosophy in this view is most sensitive to what APL calls the misuse of philosophy, the exhaustion of Philosophy by incorrect usage and repetition. "The main purpose of anti-philosophy is to avoid the fundamentalism of analytical philosophy and science.

We will not spend more time on what is perhaps an isolated, marginal position. That fact that it exists is nonetheless an indication that some of the principles of philosophy are not as monolithic as they might appear, and a new critique may be justified.

14.2.2 Introduction to Antiphilosophy

A monograph by Boris Groys, with the indicated title [7], is a much more serious attempt to delineate anti-philosophy and ascribe some function to it. It summarizes the views of a number of authors, but it doing so does bring out the activity of 'anti-philolosophizing' as worthy of our attention, since it requires us to ask *ourselves* whether we are taking, as Husserl advocated a truly disinterested rejection of old doctrines in favor of a new discourse (a *metanoia*). Interestingly, Groys also uses the concept that creating traditional philosophy, at its best, is something like a poetic process, but anti-philosophy creates nothing new philosophical texts but only contextualizes and categorizes existing ones. Anti-philosophy from this perspective, as from that of APL, would be the end, the "final, absolute stage of philosophy", but only of philosophy (and standard metaphilosophy) in the ordinary sense. What anti-philosophers do generate is a new *form* universal evidence of reality. It is this capacity that is equivalent in part to the operation of Logic in Reality. According to Groys, "anti-philosophy—like anti-art—not only does not destroy philosophy as an institution but, rather, offers the only possible path for its survival." We might characterize our logic similarly, as an anti-logic, incapable of being accepted even within the Universal Logic of authors such as Béziau [2] and He [8]. Let leave this as an open question. In fact, for Philosophy in Reality, unlike Groys or Heidegger, or Sparrow for phenomenology [27] (Chap. 10), we do not need to claim the 'end' of anything, either because we believe it or for effect. Antagonistic views co-exist and interact, without entirely disappearing as references to classical philosophy continues to demonstrate. Let now look more carefully at the material Groys adduces in support of his approach.

Groys sees the origin of anti-philosophy in the tension between philosophy and truth as a commodity, essentially, a reduced, standardized and static form of thought. Anti-philosophy can then be characterized as a reaction to this state-of-affairs but entailing new functionality, that of 'command'. For Groys the command is for change, or to change the world. If real, this property would make his thesis congenial to ours, provided it is not taken too literally. As Groys correctly asks, "how might one react to the new anti-philosophical command giving tone in philosophy today without then repeating all the critical errors of philosophy? Groys sees the answer in the phenomenological reduction of Husserl, which he describes as integrating the 'as if' into one's ego such that "decisions lose their urgency, and the *opposition* (italics ours) between carrying out and rejecting a command dissolves in the infinite play of life possibilities." One should ignore this misuse, for effect, of infinite and concentrate on the valuable insight into the role of the opposition and the probability of the emergence of a new attitude that this passage provides.

We mentioned above the repetition that appears in much modern philosophy, and, based on his reading of the work of Walter Benjamin, Groys raises reproduction of truth rather than innovation—claims that a new text is closer to the truth—as the principle characterizing current anti-philosophy. This implies our emancipation from discourse in a post-discursive society in a 'theology' of capitalistic mass culture. Benjamin's stance also, then, would tend to suggest the 'end' of philosophy.

Groys makes the essential point, however, that exact reproduction, whether of a doctrine or object is not possible in reality. A 'discourse of difference' offers philosophy a new opportunity to question the process of total reproduction and repetition. Non-perfect reproduction is, then, production and/or emergence of something new, an 'other'. In a few words, we see the outline here of a potential post-anti-philosophy which, again, is not cut off from its origins in its precursors. It is Lupascian in its avoidance of absolutes of identity, but adds little to our original understanding.

The difficulty with much of the remainder of Groys' thesis is that it hardly refers to significant issues as anti-philosophy. We do not need Groys to tell us about Nietzschean nihilism or that there are logical inconsistencies in Derrida. Fortunately, the subject of anti-philosophy is far from being exhausted, as the next Sections will show.

14.2.3 What is Anti-philosophy?

This question is asked by Charles Djordjevic in a recent article [5] that addresses fundamental cognitive issues. First of all, he considers that anti-philosophy is a new *meta*philosophical position, because it focuses on the semantic status of philosophical utterances. Without reproducing Djordjevic' entire argument, he claims that there is conceptual space for an antiphilosophical position "understood as the denial of the (a?) cognitive status to the semantics of both philosophy and anti-philosophy".

The reader will by now anticipate, correctly, our conclusion that there is a non-separable dialectic relation between the two concepts: each is a part of the other, and one need not be 'against' either of them. Djordjevic provides a succinct discussion of what constitutes cognitive discourse, concluding that the truth or falsity of assertions depends on how the world actually is, something we think is not asserted enough. Rather, the assertions are assigned value if their content is propositions. Since philosophy does not give a justification for this assignment, the anti-philosopher concludes that the cognitive status of philosophy either rests on a non-philosophical assumption or leads rapidly to a regress.

Djordjevic notes that real philosophical debate often involves the failure of one partner to understand the most basic points the other is making. Thus one anti-philosopher, far from seeing a non-cognitive construal of philosophical semantics as negative, would accept it and see it as explaining (or a reflection of) the empirical aspects of philosophical discourse. This, of course, is exactly the line we take, since it is the empirical aspects which are domesticated in our logic.

Denial of a cognitive status to both philosophical and anti-philosophical claims allows Djordjevic to define his own position: "anti-philosophy maintains that both its and philosophy's semantics are non-cognitive and philosophical debates are best understood as rhetorical games of persuasion rather than rational exchanges of beliefs." We prefer not to see any cognitive exchanges as 'games' since their subjects can also be matters of life and death. What better seems to characterize anti-philosophy is a greater degree of openness to novel concepts and constructions, in a systems sense and to aspects of intellectual activity usually considered 'irrational'. Since LIR does this, we can label LIR as being in part anti-philosophical but this, so to speak, we knew already. Anti-philosophy can be a locus for debate about speech acts and attitudes, such as Wu Kun's Informational Stance [31] referred to below, anti-factual conceptions and so on.

Absolute concepts such as clarity and precision, from this rhetorical viewpoint, lose the value assigned to them by analytic philosophy; in our approach which is accordingly, antiphilosophical; concentration on philosophical sentences is not an objective. We agree with Djordjevic that anti-philosophy should be seen as a form of dynamic critique, an on-going attempt to try to get people to rethink entrenched positions. Civilized debate, then, gains an even more important role in ethics than it already enjoys.

14.3 Badiou: From Mathematics to Truth

In contrast to the previous authors, Alain Badiou is a recognized major figure in contemporary thought. We have been attracted to Badiou's work for several reasons, one of them being his attempt to break out of the apparently unchanging fixation of contemporary philosophy on language, as if an ontological turn had never occurred. Also, our LIR finds very congenial Badiou's attempt to reconcile a notion of the subject with ontology as well as his non-standard use of mathematics, in particular set theory. Our major points of reference will thus be his *L'antiphilosophie de Wittgenstein* and the careful analysis of Smith [26] of Badiou's central manuscript

Being and Event [1]. Somewhat more problematical for us is the religious context of Badiou's work, resolutely Christian, but we will simply allow the reader to establish his or her own relations to 'Good' and 'Evil', in a universalist spirit.

14.3.1 Grounding a Notion of the Subject

Badiou's ethics is based on the capacity of individuals of distinguishing themselves from their "finite, animal nature" and to become immortal, that is a *subject.* Badiou's unique thesis, however, is that a human being becomes a subject by exercising his or her capacity to use mathematics, in particular the set-theoretic Axiom of Choice, or rather the experience of its operation provides the connection between the individual, the events in which he is involved and himself as subject. It defines the conditions under which the freedom of the individual is possible and subjectivity can emerge.

In order to understand Badiou's conception of mathematics as ontology, we must first look at his basis in the work of Paul Cohen for a proof of the independence of the Continuum Hypothesis. As Brenner discussed in (2008), the continuum hypothesis refers to a conception of the universe founded on geometry, the Cantor–Dedekind view, as discussed by Longo [11], which sees not only in mathematics, but everywhere, continuity as ontologically preceding the discrete: "The latter is merely an accident coming out of the continuum background." Points are derived concepts, even if 'non-dimensional'. In this view, geometry in some deep sense more fundamental than dynamics, that is, energy in the standard view. This hypothesis has the advantage of corresponding to our intuition and experience, integrated into and confirmed by mathematics, of continuity in our perception of 'time' and linear movement. Penrose, on the other hand, had the strong intuition that "physics and space–time structure should be based, at root, on *discreteness* (emphasis his), rather than continuity". This discreteness is evidenced in quantum mechanical spin, combined with a fundamental concept of expressing phenomena as relation between objects, rather than between an object and some background space [21].

In the construction of the mathematical continuum, objective realities are not found in the mathematical entities involved, but in the process of constituting these entities as conceptions. There can be a cognitive interaction, dynamic in LIR terms, between the applicable *mathematical* logic and intuitions about continuity. Stating this somewhat more strongly, since the establishment by Gödel of the reciprocal relation between consistency and completeness in formal mathematical systems, the situation is no longer absolute. One should not be forced to choose between geometry and discontinuous objects and their relations. It is a foundational principle in LIR that both continuity and discontinuity are fundamental and are dialectically related.

A more serious critique of the above conception of the continuum is that it is restatement of a conception of general relativity (GR) as a pure geometrization of the world, from which the subjective aspects of space and time involving observers have been eliminated as inessential ontologically. LIR states that other readings of the physics and mathematics of GR that are possible, which restore the balance between geometry and energy that exists in reality.

Cohen showed essentially that a model of set theory could be constructed (forcing) in which the Continuum Hypothesis fails. Forcing provides the formal basis for a) the freedom of the individual and his subjective response to an event—becoming a subject. He 'becomes' philosophically to function with a freedom, inherent in the Axiom of Choice, that is equivalent to "pure chance" (see also the discussion of forcing in Sect. 7.2).

This is not very satisfactory: we have eliminated one idealization—continuity—and replaced it with another—pure chance. In Peirce, we may say at least, the two are maintained so that they can support one another [25].

14.3.2 The Philosophy of Forcing

Briefly, Badiou demonstrates that every discernable or constructible set is dominated by the conditions which don't possess the property that makes it discernable as a set. Badiou refers to Cohen's forcing, again, to "think beyond the strictures of the relativistic constructible universe of language", and concludes that while ontology "can mark out a space for an inhabitant of the constructible situation to decide upon the indiscernible, it falls to the subject—about which the ontological situation cannot comment, to nominate this indiscernible"—the undecidable event. The mathematical theory of forcing, in its emendation of the Continuum Hypothesis, according to Smith, provides Badiou with a paradigmatic model for the subjective response of an individual to an event.

The use of the word 'model' is not trivial; without further analysis, we must conclude that we know no more now than before about the nature of subjective responses. Badiou thus goes back to the eternal argument about the One and the Many and opts for a systematic ontology, *contra* Heidegger, not based on the One. This One, however, is designated as a *process*, the result of a counting operation—count-as-one—on the "pure multiplicity" of being.

The reference to process is interesting from an LIR perspective, as is Badiou's concept of different *non*-relations in philosophy: one is between being and the One, which is a *no* relation and the other is a non-determinate relation resulting from the presentations of multiples (Many) generating number systems and other intensive properties of his ontology. As Smith puts it [26], "Badiou's philosophy stands or falls on whether set theory provides an ontology of the *pure* multiple that avoids the pitfalls of the One." Of course, this problem never arises in LIR, since neither the One nor multiples are 'pure', a word that we believe should be banished from philosophy.

Not surprisingly, a set is a pure multiple if it is defined by nothing more than the multiples it presents. If sub-sets exist by a Power Law Axiom, including the empty set and the set itself, the original set cannot belong to itself, "on pain of paradox and inconsistency". We are thus in a philosophical domain of pre-paraconsistency [24].

Badiou's demonstration so far, by its own principles, belongs in the domain of philosophy *tout court*, without dynamic properties,

14.3.3 The Axiom of Choice in Lupasco and Badiou: Time and the Event

As discussed in Chap. 7, Lupasco grounded the Axiom of Choice in physics in a way that implies a dynamic relation between a set and elements of that set [14]. (The argument is not changed if the elements are also sets. Most of the time they will be.) It is nevertheless also true that Badiou's *use* of the Axiom of Choice is quite different, in that the elements of a set say 'humanity' are defined in terms of what does *not* belong to the set. Being is the null set to which nothing belongs. Further, Badiou thinks that the Axiom of Foundation means that a set contain or belong to itself. For us, this is the set-theoretical equivalent of binary logic that requires absolute separation of premises and conclusion, subject and object and so on (Suppes, [28] gives a good orthodox view of set theory).

From his view of mathematical structure, Badiou moves to the idea of the event, something that occurs and is therefore in time, in its usual conception. He dissects mathematical proofs and procedures and proposes a temporal structure on them, perhaps in an attempt to naturalize them. The 'matheme' of an event is an attempt at description that is not a consistent ontological statement but carries some idea of its structure. It is thus analogous to what we call an epistemon in Chap. 5. For Badiou, a matheme is a carrier of *inconsistency*. "The matheme acts as an inconsistent supplement outside of ontology that lets the subject know that its task is never complete."

The Axiom of Choice enters by extending the scope of the Power Set Axiom and 'allowing' the creation and naming of 'new', non-constructible subsets as the basis for novelty or, as we prefer, emergence. "Freely chosen non-constructible sets belong to the situation" and "a new situation is posited in which they are constructible". The 'movement' here is analogous and to a certain extent supports a structure of emergence of an included third that retains properties of the entities from which it has emerged. Badiou has described a mathematical counterpart of the Pauli Exclusion Principle in physics which, as we have seen, is the basis for novelty, initially, at the microscopic level. New situations can arise with new rules of construction. The Axiom of Choice, no longer dependent on forcing, is essential if a subjective response is to be possible within a situation. The Axiom of Choice, and therefore the individual's freedom is maintained but is determined by an inconsistent principle.

It is this last statement is our justification for this exploration of Badiou's philosophy, despite our disagreement with many of its premises. Inconsistency can and must operate in tandem with consistency. For us, however, Badiou inverts the existential order of dependence: freedom and non-freedom of choice are both inherent in nature but the Axiom of Choice does not 'produce' either. 'Evil' is perhaps the consequence of a local excess of non-freedom, but the theory of ethics developed by Badiou as a whole fails due to the absence of any real process dynamics. Let us still look at Badiou's vision of anti-philosophy as such. Badiou thus appears to define an anti-philosophy to refute the relativism of post-structuralist thought, recasting ontology in a new light, but what we exclude from our theory is Badiou's (mis)use of the Axiom of Choice for this purpose. Leaving aside this further example of ascribing operational capacity to concepts, we consider that an ontology of processes can 'comment' on any questions of Leibnizian discernability. We welcome Badiou's attempt to use set theory to explicate philosophy, but wish to select what set theory and what philosophy.

14.4 Badiou and Wittgenstein

The above summaries provide the context for the part of Badiou's work that is also of interest to us here, namely, his demonstration that the philosophy of Wittgenstein is in fact an *anti-philosophy.* In continuing a relatively simple concept we have noted above, Wittgenstein's work can be considered a form of poetry, inimical to philosophy-as-such. We do not like the implications of this approach, as its finality appears to be to exit from the 'worldliness of the world', that is from reality. To repeat, introducing a cut between poetry and philosophy (or science) is neither necessary, since they are linked dialectically, nor desirable, since there is an implied devaluation of ordinary reality.

Among the most useful suggestions of Badiou is that anti-philosophy (from now on AP) emerges in the mutual critique of two philosophers, Heraclitus and Parmenides, Pascal and Descartes. Its practice involves three operations: (1) the elimination of truth as a category; (2) the non-reducibility of philosophy to its propositions; (3) a stance and a new form of act that is super- or *a*-philosophical. Attentive readers will have found positions superficially parallel to (1) and (2) in his emphasis on reality rather than propositional truth. Badiou finds (3) exemplified in the AP of Wittgenstein as a withdrawal from what remains of dialectics in standard philosophy in favor of active non-thought (*non-pensée active*) beyond any meaningful proposition, that is, all science.

At this point one may ask whether Wittgenstein is 'doing' AP or simply misusing the universal quantifier. However, if in Badiou's terms, the anti-philosophic act consists in showing what there "is" in a way that no true proposition can, then Wittgenstein and Logic in Reality have the same objective but come to opposite conclusions about its nature. Wittgenstein wants to introduce a new and 'superior' principle of clarity that "establishes the laws of the inexpressible (*indicible*) or unthinkable". From this point on, it will be clear that Wittgenstein's views are, if nothing else, anti-LIR. LIR shows 'what no proposition can' by referring to the underlying dynamics of any process, what people like to call the empirical, without giving it adequate status versus the ideal.

The philosopher dual to Wittgenstein is Nietzsche: for Badiou, they have similar views on the properties of the thinkable and the unthinkable and are totally opposed on the role of Christianity. We will not study this topic here; rather, we note the anti-scientific quasi-mysticism in Wittgenstein's back and forth between what 'is' and what is 'said'. For Wittgenstein but also Nietzsche and Lacan, 'something' emerges from this anti-philosophical 'movement' that makes standard philosophy incomplete, as it is a part of the real that it does not capture. Unfortunately for anti-philosophers, a Real that their system does not appear capable of capturing is the *feminine*. The anti-philosophy of Wittgenstein, Pascal, Kirkegaard, and of course Nietzsche leads them toward what Badiou calls, correctly, a striking anti-feminism. From our point of view, of course, the problem is badly posed. Only from real interactions, if they are mutually recognized, could any new real principle, such as a proper attitude to women, traditionally, an 'Other', emerge.

Further examples could be given in the anti-philosophy of Wittgenstein which run contrary to our basic thesis of non-separability and interaction in the real world. In his idealism, there is no meaning left in 'our' world or its events—what happens. One thing that 'happens' is that philosophers produce meaningless statements—absurd ones, but Badiou claims that Wittgenstein's reduced explanation that this is explainable in terms of "confusions" is totally inadequate. Although he does not say so explicitly, Badiou's rejection of Wittgenstein's reliance on chance or contingency as fundamental allows space for a more balanced view, at least, of chance *and* necessity.

A comment on Wittgenstein's view of mathematics as part of *his* anti-philosophy is a key part of Badiou's own conceptions of philosophy, and the further extensions of the latter possible in our logic. Badiou first states, *contra* Wittgenstein, that the essence of a mathematical proposition is not equality. They are the theorems of existence, decomposition and presentation at the basis of set theory. These are far from any (standard) logical reductionism principle of independence. We say standard here because it is again our Logic in Reality, grounded in physics that establishes the correct relation of logic and mathematics. As stated in Chap. 6, we can agree with the concept that any Boolean algebra can be presented as a set-theoretic algebra, without excluding that *non*-Boolean algebras are better representations of reality.

In contrast to Badiou's humanistic concept of the subject outlined above, that of Wittgenstein and Lacan is of an idealized global entity of existence, at its 'boundary' with non-existence. These authors refuse what we call the dialectics of and in the world, considering it a 'crooked meta-language'. Finally Wittgenstein's elimination of the Other, as woman or as subject, makes his anti-philosophy a-moral, a-political. He is lost in Salvation.

14.5 The Anti-philosophy of Stéphane Lupasco: Affect

Intellectual honesty requires that we state that a major portion of Lupasco's work, not referred to earlier in this book, is devoted to a principle which in our view is in total contradiction to the rest of his *oeuvre*. We place a summary of this work in this Section because it constitutes an anti-philosophy in what we view as the *worst* sense possible. Lupasco designated all real energetic processes as 'non-ontological', which meant everything that was becoming, experience, dialectics and logic. He used ontological to refer to being, which for him consisted only of affectivity (affect). In some incomprehensible way, affect 'accompanies' the evolution of real cognitive phenomena. In some even stranger way, affect is fated to decline and disappear, leaving behind what could perhaps best be described as zombies. Lupasco's 'Error' was pointed out most cogently by his contemporary Fondane [6].[1]

Lupasco's devoted a major portion of almost all of his publications to his concept of affectivity. As first laid out in his Thesis of 1935 [12, 13], Lupasco defines existence as consisting of energetic phenomena subject to the principle of dynamic opposition—the "dynamic logic of contradiction". What is not subject to this logic consists of what he calls quality or affectivity ("*affectivité*"). Affectivity is the only ontological given, external to logic and its existential dynamic opposition. It can be suggested, however, that his logic prepared the ground for an "onto-logic", a less restrictive combination of logic and ontology that emerges from the subsequent work of Nicolescu. Lupasco's designation of emotional and physical affectivity (pleasure-pain) as objects with a non-energetic, non-physical and non-relational ontological being requires amplification. Lupasco wrote in 1947 [15]:

"It would seem that all existence is a network of antagonistic and system-forming dynamic events, stretched on an ontological substance which manifests itself by affectivity, in which existence bathes from time to time. This ontological affectivity is like a kind of Kantian noumenon or Leibnizian monad, operating by an incomprehensible imperative or through a parallelism which escapes knowledge as such".

This is close to "dual matter" positions in the literature that seem to have been introduced to insure the survival of subjectivity and a degree of transcendence. A preferred reading of Lupasco, however, in more agreement with current cognitive science, would be to consider his affectivity as involving property but not substance dualism, real but essentially, a-logical, non-physical and a-causal. For Lupasco, logical existence is thus continually accompanied by, inhabited by or "bathed in ", quality or affectivity, the latter being ontologically "real", corresponding to being or "*être*"—what is, and the former to non-being, "*non-être*", what exists. Lupasco's equating of existence with non-being is, of course, perfectly orthodox. As Heidegger reminds us, being for the Greeks meant permanence in the sense of arising (*physis*) and as such enduring (*ousia*). "Non-being means accordingly to depart from such generated permanence: *existasthai*, "existence", "to exist", meant for the Greeks precisely non-being." Heidegger goes on to castigate today's "estrangement" from being, implied by the general misuse of existence as designating being, but this is not our concern here. Heidegger provides some perhaps unexpected (and certainly unintended) support of Lupasco when he describes his "being" as both wholly indeterminate and at the same time highly determinate [12, 13] "From the standpoint of the usual logic we have here an obvious contradiction: ...determinate, wholly

[1]Fondane and his sister died tragically in the Holocaust, despite efforts by Lupasco and Picasso, among other prominent non-Jews, to save them.

indeterminate being. *If we decline to delude ourselves* (emphasis ours), ... *we find ourselves standing in the very middle of this contradiction. And this stand of ours is more real than just about anything else we call real; it is more real than dogs and cats, automobiles and newspapers.*" For Lupasco and Nicolescu, being is a-logical, characterized by neither contradiction nor non-contradiction, an emergent included middle, at another level of reality. Heidegger, like Hegel before him, retained an essentially classical point of view of the logic and dynamics involved.

For Lupasco, logical existence was constantly trying to expel affectivity from itself, and move in away from pain and suffering toward its absence, a state of interior peace and lucidity. As noted above, Lupasco says explicitly that ontological affectivity is the *only* (emphasis ours) ontological given of which we have the experience [16]. He tries to show that affectivity is not relational, has no time nor space, is not contradictorial, does not contain a subject or object, is not a concept, has neither intension nor extension (as do concepts), is neither homogeneous nor heterogeneous. The relationship between the concrete aspects of logical becoming and affectivity remained problematical and paradoxical [23] (not only for Lupasco). He postulated a unique form of causality and finality of which affectivity is an indication, but to which it is exterior. Affectivity has an action on energetic phenomena by "infiltrating" the nervous system to appear in the conscious mind, manifesting itself in the content of the latter that, as we will see, consists primarily of potentializations. Francesco Varela [29], in one of his contributions to the volume we refer to frequently [22], in discussing the "specious present", points out that as shown by Damasio [4], affect and emotions can be associated with a relatively stable set of neural correlates, to which, however, they are not to be reduced. He considers emotions as an integral part of an ontological readiness with roots in "minute events in brain physiology". Subject to further analysis, we see these two views as compatible. Lupasco uses the term "salutatory causality", which suggests a jump from the neurobiological to the ontological level, a kind of "ontological bypass" of the logical.

In all his writings subsequent to the Thesis, Lupasco refers only to affectivity, not to quality. Further, he does not delve , in his discussion of the 'psychic universe [16], into the (ontological?) nature of individuality or the self, which can be considered the "hardest" problem. As discussed elsewhere, consciousness as such, as well as the consciousness of consciousness, remain part of logical existence. With this limitation in mind, however, Lupasco's detailed analysis of the way in which affectivity in fact accompanies phenomena described on the basis of dynamic opposition is still worth noting briefly. A painful affectivity becomes perceptible when there is inhibition of vital functions or parts of consciousness [12]. This painful affectivity is "located" in, or attributed to the body in the first case and to the "I" in the second, a truly psychic pain. In both cases, the elimination of the affectivity present by the resistance and the reactions of the corresponding systems generate an agreeable affectivity *attributed* to them, as they are the locations of its elimination.

14.5.1 Onto-Logic?

In view of the above, the term suggested by Nicolescu of "onto-logic" to describe the Lupasco system is clearly not consistent with Lupasco's own view of ontology. Our suggestion is that despite his denial, his insistence that what was logical related only to "non-being" and only affect was to be included in "being", Lupasco can be understood without giving too much emphasis to his "ontological cut". This does not mean that an "onto-logic" is neither necessary nor possible. In fact, a major contribution of Lupasco may be that he liberated logic from its limitation to an idealized meta-logic and thus created the conditions for an eventual "union" of ontology and logic, one that he simply did not wish to recognize in his own work. A proposal along these lines is that of Nicolescu [19], and his description of the *hidden* included middle, in addition to the logical and ontological included middle. The logical included middle applies to emergent phenomena at "lower" physical levels of reality; the ontological included middle to "higher" ones, and it accordingly implies the consideration of several levels of reality at the same time [18]. Writing in the same Lupasco Memorial Volume, Craciunescu [3] concurs that the logic of Lupasco should be termed an "onto-logic" in that it is an "extension-virtualization of Hegelian formalism in what one might call a T-cosmology." The hidden included middle is, as a philosophical concept, at a still higher level of reality, involving further processes of recursion. But it is still *in* reality. Is a part of natural philosophy about human beings and their changes, not about 'unseen worlds', and we believe it is perfectly possible to equate it to the *Tao*, for which a 'concrete' definition is neither possible nor desirable.

The monograph written by Fondane in 1944, referred to above, makes other serious criticisms of Lupasco's view of the special status of affectivity. While many of Fondane's questions were answered in Lupasco's later work—some are noted here—Lupasco maintained his position on affectivity until the end of his life, as can be seen in the trialogue with Basarab Nicolescu and Solange de Mailly-Nesle [17]: ontological affectivity transcends energy and logic, and it is the only given that does so.

Lupasco was fully aware of the problem posed since antiquity by the existence of affectivity. According to Fondane, what Lupasco accomplished was to include diversity in logic, and give affectivity its own autonomy, according it the status of a Kantian noumenon, outside all knowledge, and accordingly outside existence. However, Fondane is entirely correct when he states that Lupasco's position on affectivity is inconsistent and untenable, since the strange relations he posits between affectivity and existence are, nonetheless, relations.

The most serious criticism that can be made of Lupasco's view is as an implied and even stated *project* that would eliminate affectivity from knowledge and existence. In Lupasco, pain accompanies contradiction; pleasure, primarily as absence of pain, accompanies the removal of conflict and contradiction. (This already looks quite suspect to the modern reader: to consider pleasure as the "negative aspect of affectivity" might have some methodological value in the exploration of the concept, but Lupasco repeated it often enough to convince us that he meant it literally.) The entry of affectivity on the logical scene each time there is a crisis or conflict looks to Fondane like a real, dramatic opposition that Lupasco refused to recognize as such. Worse, Lupasco's view of the end of existence is almost "inhuman": "Existence, as crisis, will always contain affectivity, positive or painful, although more and more pale (going from anxiety to uneasiness, finally to ennui) but in potentiality proceeding towards action, there will always remain something of the pure concept of affectivity like the hope of a rigorous affective vacuum, of a definitive break with being that realizes, perhaps only asymptotically, the becoming emerging from existence itself.[2]" This is a dismal picture indeed, which none of Lupasco's later work did anything to dispel. In contrast, the absence of affectivity that he noted was associated with mental illness was correct, as was his idea that some people, well or ill, resort to combat or violence to try to recover the "ontological given of affectivity."

There seems to be no place in Lupasco's system for any liberty outside the oscillations of the immanence of logical, energetic existence that excludes Being. As Fondane put it "When the termites of knowledge have passed, the pulp of things has been gnawed away, the participation of existence in Being through an affective communion has been evacuated, there is nothing left of things, but the thin film of knowledge, which will be blown away by the first wind." One must not take Fondane's justified criticisms as applying to Lupasco's logic of existence as such and its basis for a theory of knowledge. These remain, in our view, major achievements, as we have attempted to show in this book.

14.6 Non-philosophy

As an interesting variation on the theme of anti-philosophy, we refer to that of non-philosophy was a discipline proposed by Laruelle [10] that has not received further attention but it relevant to our thesis. It is a curious document that links aspects of science and philosophy in an original manner essentially by ascribing operational properties to certain 'non-philosophical' perspectives on philosophy. As an example, a key problem, in his terms, is the ontological status of the 'One' within philosophy, which associates it, whether explicitly or not, to Being and to the Other whilst forbidding it any measure of radical autonomy. The other problem is that of philosophy's theoretical status, insofar as philosophy is practice, affect, existence, but lacking in a rigorous knowledge of itself, a field of objective phenomena not yet subject to theoretical overview. We have addressed this situation in our treatment of the application of the Principle of Dynamic Opposition to the ontological process of 'theory-making' as opposed, but not totally separated from, the epistemological content of a theory or a philosophy itself.

Non-philosophy typically operates in the following way: everything is processed through a duality (of problems), which does not constitute a Two or a pair, and through an identity (of problems, and hence of solution) which does not constitute

[2] At 65 words, this is one of Lupasco's 'shorter' sentences.

a Unity or synthesis but a 'Unilateral duality' with an ontological status equivalent to that of 'Identity'. Non-philosophy is a transcendental discipline, which makes use of philosophy's transcendental dimension in order to formulate itself. It is the determination-in-the-last-instance of a theory (of a knowledge that remains distinct from its object—a model taken from science, and also of a pragmatics (of a *usage* of philosophy 'with a view to' the non-philosophical subject—a model taken from philosophy. It is theoretical by virtue of one of its models: science. But it is neither a philosophical and self-positing theory, nor a philosophical and self-positing pragmatics. It is theoretico-pragmatic only by virtue of its aspects as non-philosophical operation, but real or practical by virtue of its cause. Thus, it is not a 'negative' theory-pragmatics either, but rather one requiring that any vision of the 'One' be validated by invariant scientific and philosophical models.

In Laruelle's system, non-philosophy 'assists' philosophy in the following manner: it cannot 'transform' the objects of philosophy or the entities of the world. But it can *transform* philosophy as a whole defined as a self-presenting hybrid of identity and difference. It does not intervene 'within' the specificity of experience, as philosophy often and mistakenly claims to do, nor does it even provide that specificity with meaning. It is not, generally speaking, an operation or activity to which the subject would remain external. The Subject *is* assistance in its very essence. If assistance is neither interpretation nor practical intervention, it is the bringing-forth, one that is practical only in-the-last-instance, of world thought. It is this being-brought-forth or being-given which transforms the latter's autonomy and liberates it, and thereby liberates the Subject (as transcendental identity (of) world-thought), from its entrapment by an illusory belief in its own sufficiency (see Priest on the 'philosophy' of belief revision [24]).

This transcendental identity, that which *philosophy* as such constitutes, remains incommensurable with 'philosophy' in the standard philosophical sense. Logic in Reality, however naturalizes and accepts the 'contradictory' immanent/transcendent aspects of reality in its cognitive dialectics. It is both philosophy *and* non-philosophy in coexistence.

14.7 Interim Conclusion

Following our discussion of metaphilosophy, the second significant dialectic relationship we have analyzed is that of philosophy to anti-philosophy and non-philosophy. Its simplest characterization is as a critique, questioning the assumptions made by 'philosophy' about essentially all aspects of knowledge and existence. Because of the clarity, here a positive characteristic, with which the opposition between philosophy and anti-philosophy can be expressed, it becomes a valuable methodological tool to be added to the armentarium of Philosophy in Reality.

In this Part II, we have presented the first synthesis of our views of a new natural philosophy and a critique of its 'opposite', anti-philosophy. We now turn to a second synthesis the process terms which we find congenial of a Philosophy in Reality as Process.

References

1. Badiou, A.: L'être et l'événement. Éditions du Seuil, Paris (1988)
2. Béziau, J.Y. (ed.): Universal Logic : An Anthology. From Paul Hertz to Dov Gabbay. Birkhäuser Springer, Basel (2012)
3. Craciunescu, P.: L'état T et la transcosmologie poétique. In: Badescu, H., Nicolescu, B. (eds.) Stéphane Lupasco; L'homme et l'œuvre. Monaco: Éditions du Rocher (1999)
4. Damasio, A.: Self Comes to Mind. Constructing the Conscious Brain. Vintage Books, New York NY (2012)
5. Djordjevic, C.P.: What is antiphilosophy? Metaphilosophy **50**, 1–2 (2019). https://doi.org/10.1111/meta.12342. Accessed 15 Oct 2019 (2019)
6. Fondane, B.: L'être et la connaissance; essai sur Lupasco. Éditions Paris-Méditerranée, Paris (1998)
7. Groys, B.: Introduction to Antiphilosophy. Verso, United States. First published as Einfuhrung in die Anti-Philsophie. Carl Hanser Verlag, Munich (2009)
8. He, H., et al.: Principle of Universal Logics. Science Press, Beijing (2005)
9. Heidegger, M.: An Introduction to Metaphysics. Yale University Press, New Haven (1987)
10. Laruelle, F.: A summary of non-philosophy. Pli **8**, 138–148 (1999)
11. Longo, G.: The mathematical continuum. In: Petitot, J., et al. (eds.) Naturalizing Phenomenology. Issues in Contemporary Phenomenology and Cognitive Science. Stanford University Press, Stanford (1999)
12. Lupasco, S. 1973a. Du devenir logique et de l'affectivité; vol. 2: Essai d'une nouvelle théorie de la connaissance. J. Vrin., Paris (Originally published in J. Vrin, Paris, 1935)
13. Lupasco, S. 1973b. Du devenir logique et de l'affectivité; vol. 1: Le dualisme antagoniste. Essai d'une nouvelle théorie de la connaissance. J. Vrin., Paris (Originally published in J. Vrin, Paris, 1935)
14. Lupasco, S.: Le principe d'antagonisme et la logique de l'énergie. Editions du Rocher, Paris. (Originally published in Éditions Hermann, Paris, 1951) (1987)
15. Lupasco, S.: Logique et Contradiction. Presses Universitaires de France, Paris, France (1947)
16. Lupasco, S.: L'Univers Psychique. Denoël-Gonthier, Paris (1979)
17. Lupasco, S.: L'homme et ses trois éthiques. Editions du Rocher, Paris (1986)
18. Nicolescu, B.: Le tiers inclus. De la physique quantique à l'ontologie. In: Lupasco, S., et l'œuvre, L. (eds.) Horia Badescu and Basarab Nicolescu. Éditions du Rocher, Monaco (1999)
19. Nicolescu (Dir), B.: Le Tiers Caché Le tiers cache dans les différents domaines de la connaissance, pp. 171–176. Éditions Le Bois d'Orion, Paris (2016)
20. Palomo-Lamarca, A.: Anti-philosophy and Its Manifest. https://serbal.pntic.mec.as/AParteRei. Accessed 30 Oct 2019 (2001)
21. Penrose, R.: The Road to Reality. Knopf, New York NY (2004)
22. Petitot, J., et al. (eds.): Naturalizing Phenomenology. Issues in Contemporary Phenomenology and Cognitive Science. Stanford University Press, Stanford (1999)
23. Priest, G.: Logic of paradox. J. Philos. Logic **8**(1), 219–241 (1979)
24. Priest, G.: Paraconsistent belief revision. Theoria **01**, 214 (2001)
25. Short, T.: Peirce's Theory of Signs. Cambridge University Press, New York NY (2007)
26. Smith, B.A.: The limits of the subject in Badiou's *Being and Event. Cosmos and History*. J. Nat. Soc. Philos. **2**(1–2), 134–158 (2006)
27. Sparrow, T.: The End of Phenomenology. Metaphysics and the New Realism. Edinburgh UniversityPress, Edinburgh UK (2014)
28. Suppes, P.: Axiomatic Set Theory. Dover Publications, Mineola NY (1972)
29. Varela, F.J.: The specious present. In: Petitot, J., et al. (eds.) Naturalizing Phenomenology. Issues in Contemporary Phenomenology and Cognitive Science. Stanford University Press, Stanford, CA (1999)
30. Wu, K.: The essence, classification and quality of the different grades of information. Information **3**, 403–419 (2012)
31. Wu, K., Brenner, J.E.: The informational stance: philosophy and logic. Part I: the basic theories. Logic Logical Philos. **20**, 453–493 (2013)

Chapter 15
Philosophy in Reality as Process

15.1 Introduction. From Kant to Levinas

Before proceeding to the final study, of the application of philosophy to systems, let us pause for a moment to review what our Philosophy in Reality is and is not. In Chap. 11 we referred to an *informational* 'revolution' in philosophy, based on the major impact of information on all fields of knowledge and the formulation by Wu Kun of the Philosophy of Information as a Metaphilosophy [42]. The title of a paper by Wu and Brenner [43] was " Philosophy of Information: Revolution in Philosophy." We ask the reader's indulgence for hyperbole in the titles of papers, but we do believe that if some of the precepts in this book were to be implemented, it would amount to a substantial change in the way philosophy is done.

To demonstrate this, in line with our overall interest in change, let us compare our theses with some changes in philosophy that were perceived as major, starting with Kant and continuing through to neo-Kantianism. This period in philosophy, roughly 1880 to World War I, saw the dissatisfaction with some of the basic Kantian principles of philosophy, including that of the a priori and the 'given'. It was thus a 'transitional' period before the consequences of quantum mechanics were appreciated in philosophy. To a certain extent, except by Lupasco and Sklar [38], they *still* have not been. This allows us to position the work of Lupasco and bring out its originality, but also in relation to Kant himself.

The period of Neo-Kantianism was also that of the precursors of Husserl and phenomenology, also a response to the difficulties of the perhaps too simple earlier picture of immanence and transcendence. Much of the debate about what it is necessary to replace in Kant is couched in the terms of a philosophy that has resulted in the exclusion of references to science and change but has integrated the truth-functionality of standard logic. As we have argued at length, such logic cannot be the basis for a scientifically sound natural philosophy. Kant himself, as a follower of Newton, was in fact perhaps more comfortable with natural science than many of his followers [10]. The objective of this chapter, then, is to explore notions in both Kant and neo-Kantianism, to bring into relief the bases of the natural philosophy we

J. E. Brenner and A. U. Igamberdiev, *Philosophy in Reality*, Studies in Applied Philosophy, Epistemology and Rational Ethics 60, https://doi.org/10.1007/978-3-030-62757-7_15

proposed in Chap. 15. Our methodology will be that suggested in Chap. 5, a Lupascian form of 'dialectomethodology'—seeking out the non-truth-functional elements in processes and doctrines which are in real opposition or contradiction.

It is possible to find statements about the relations between real phenomena and a philosophy in which science in embedded that appear to go beyond the relations of standard propositional logic. Even the simplest subject-predicate relation refers to entities that are more real than the propositions about them. To pursue these lines of thought, however, requires the further dimension provided by the incorporation of the dynamics of change following the rules of Logic in Reality. Accordingly, for its intrinsic interest and as a test of the validity of our approach, we will concentrate on a few subjects which have been and still are at the core of philosophy. These were addressed in a depth, in particular by Kant. Hegel and Heidegger that we do not pretend to try to match here. We have selected some subjects, however, originating in the work of Kant, to which Logic in Reality can bring a new appreciation and extension.

From the middle of the twentieth century, philosophy has been challenged to develop new insights into the eternal questions that life and existence pose for human beings capable of asking them. The background is the fundamental progress that has been as is being made in the understanding of our physical world of energy, of time and of space, that of itself has rendered obsolescent many philosophical considerations, starting with antiquity. At the same time, this progress enabled new interpretations of classic insights in both the West and East, especially from China, that offered the possibility of a new transcultural synthesis of science logic and philosophy. One of the closest observers of the terms in which philosophical questions should be posed today is the ethicist and information theorist Rafael Capurro. In Chap. 12 Communication, we called attention to his rethinking, with John Holgate, of messaging as a process. Brenner has proposed a naturalization of Capurro's philosophy of the evolving digital paradigm in knowledge and the society. We conclude this section with a discussion of a, for us, key twentieth century philosopher—Emmanuel Levinas.

15.1.1 Process and Change; Becoming and Being

The entities of reality are constituted by their properties and the processes of their formation and evolution. These processes are *change*, changes in the most general terms—becomings. The fundamental axioms of LIR also imply a major reformulation of the key aspects of intrinsic and extrinsic properties. No property of a system that is involved in some form of thermodynamic interaction, that is, above the quantum level and the biological and mental levels can be, according to Conditional Contradiction, separated from its opposite or negation. In the previous chapters, the interactions present in reality at the physico-chemical, biological and cognitive levels, where the latter include perception, the content of knowledge and social dynamics were looked at as processes. As described in the original Brenner description of Logic in Reality, processes constitute the world of experience—from nature to social reality to perception, cognition and the content of knowledge—they should not be represented in a reductive fashion—in terms of their results, input–output pairs, or by static computational or discontinuous, step-wise interpretations involving static identities, in which the underlying dynamics has been occulted. We thus see philosophy itself as process, involved the dynamic process of human existence as one of the possible systems of description of reality. At the same time, as we have seen in Chap. 15, it is impossible to separate becoming from being, from what we can call a natural transcendence, the 'something different' about existence in a dialectical relation to becoming. It is this conception of philosophy that is the basis for the critique below.

15.2 Toward a Non-truth-Functional Philosophy

15.2.1 Recovery

The Logic in Reality of Brenner [3], expounded in *Logic in Reality* and in several articles [4, 5] (2015), has not become a subject of widespread discussion or analysis. The on-going resistance to the broad acceptance of his logical system was predicted by Lupasco himself, and explained as due to the preference of people for the identities and tautologies of standard bivalent logic, rather than the uncharted, changeable seas of a new dialectical logic of processes. We should thus alert the reader to the fact that except for the paper by the authors of this book [7], there is no literature dealing with a Philosophy in Reality as a subject in its own right.

As pointed out in Chap. 3, the field sof paraconsistent logic was first established by Newton da Costa and then by Graham Priest and da Costa's Brazilian students in the late twentieth century. Priest's brief reference to paraconsistency in real processes [34] has not led, even recently, to new insights about *them.* Paraconsistent logic is flourishing today, in part due to the efforts of Jean-Yves Béziau and other students of da Costa including Otávio Bueno, Walter Carnielli, Itala D'Ottaviano and Dieter Krause. For us, however, the abrogation, in paraconsistent logic, of the Aristotelian Axiom of Non-Contradiction in semantics is not sufficient to serve as a description of contradiction, better, of the counteraction or countervalence instantiated in real processes. Paraconsistent logic remains bound by and to a concept of truth.

Logic in Reality, however, like the Lupasco logic from which it stems is not a 'logic of contradiction'. It accepts real contradictions in the sense of interacting opposing forces in/of a real process, from which new entities can emerge. In the terminology of the Brenner/Igamberdiev paper [7], we 'recover' logic from the limitations imposed on it to propositions and their mathematical and category-theoretical equivalents.

15.2.2 Discovery

The discovery that is presented here for the first time as such is the extent to which links can be made between Logic in Reality and our concept of Natural Philosophy set forth in Chap. 13 on the one hand, and on the other what has been referred to above as Non-Natural Philosophy or philosophy *tout court*. We are aware of the vast volume of work that has been done on the relation and interaction between the fields of Philosophy, Logic and Science, and of some of the sensitivities involved. Many philosophers are concerned about the encroachment of science on philosophy and the loss of its specificity. we will not enter into the substance of this debate, which is in part otiose. We have established in Chap. 13 that a further naturalization of what has been defined as Natural Philosophy can be accomplished by taking into consideration the new set of logical principles of our Logic in Reality. Naturalization in our acceptation of the term is a bringing-into-science of aspects of Natural Philosophy and an enhancement of its explanatory power. The part of philosophy to which our principles do *not* apply was referred to as non-Natural Philosophy or philosophy *tout court*. This judgment was not only '*tout court*' itself, it neglected the capacity of our non-standard logic of processes to be brought into or contribute to previously established philosophical insights which were 'orphans', without adequate theoretical support in their existing context. In this chapter, we will focus on a few of these areas as further constitutive of our proposed Philosophy in Reality.

We therefore propose to do for philosophy what we have proposed for logic, dialectics and semiotics: to eliminate as far as possible reductionist interpretations and include the remainder in a new synthetic paradigm centered on the science of real processes as well as meaning and its communication. Our concept unites science, our logic and philosophy in a common meta-thesis and could provide the contours of a new form of understanding of nature and civilization. The product of this exercise is thus neither an anti-philosophy nor a non-philosophy in the sense of the preceding Chap. 14. LIR insists on the absence of absolute separations between fields and doctrines (of the kind 'not this but that'), especially those which serve no purpose, philosophical or scientific, as an excuse for the absence of seeing the possible interactions present. The separations that one must make are in fact between a system that decrees that logic must be based either on subject-predicate dualism or on a propositional calculus and a synthetic non-semantic system like ours.

15.2.3 Method. Two Features of Logic in Reality

Each of the issues in philosophy we could address has its own large literature. Here, we have followed the sequence in several reference texts and made statements about the relevance of our logic and dialectics to them. There are thus two features in the dialectical structure of Logic in Reality (LIR) that we will apply in this way: F1: LIR implies that two opposing terms or processes will be present in a phenomenon, to a different and changing degree; F2: a single term, principle or process will be more or less instantiated to a changing degree in a phenomenon. The first area of research we address from this standpoint is the central role in philosophy given by Kant to a theory of judgment. We will then look in turn at other applications and implications of LIT for naturalism, neo-Kantianism and, briefly. Popper's Three Worlds. We will summarize the work of Emmanuel Levinas as being closest inspirit to the philosophical implications of LIR. We summarize this chapter by looking at our philosophy in terms of process, drawing on the pertinent critique by Petrov and Scarfe of the concept of Dynamic Being [30].

15.3 The Question of Judgment

If one takes an open-minded, dialectical stance, the reading of Kant is a fascinating experience. One moves, almost continuously, from concepts that are purely idealist to ones that qualify in our view as demonstrations of the validity of a logic and philosophy based on energy. When we say that we disagree with the former and agree with the latter, this should be taken to mean that we neglect either of Kant's two kinds of thesis. For us, they constitute a dialectics, albeit one for which, even recognized, no basis in reality has yet been provided for. We also recognize that our 'judgments' are not based on reading Kant in the original German [15], but we hope that some of them, at least, will be perceived by Kantian scholars and others as consistent with it. According to Kant, judgments are complex conscious cognitions that (i) refer to objects either directly (via intuitions) or indirectly (via concepts). They have a further series of properties that are related to those of Kant's concept of logic, follow rules related to the picking out of objects by intuition, express true and false propositions and so on. Kant takes the innate capacity for judgment to be the central cognitive faculty of the human mind, in the sense that judgment is the joint product of all of the other cognitive faculties operating coherently and systematically together under a single higher-order unity of rational self-consciousness.

Kantian judgments are not merely psychological objects and processes as in psychologistic theories of judgment, nor are they essentially mind-independent, abstract objects (as in platonistic theories of judgment), nor again are they inherently assertoric takings of propositions to be true (as in Frege's theory of judgment). Instead, Kantian judgments are intersubjectively shareable; rationally communicable; cognitively-generated mental-act structures or types whose logically-structured truth-apt semantic contents can be the targets of many different kinds of epistemic or non-epistemic propositional attitudes. As just noted, every judgment has an intrinsic logical *form* that is both syntactic and semantic in nature, centered on predication. Even more fundamentally however, every judgment also has an "intension" (*Inhalt*) or semantic *content*: the "proposition" (*Satz*). A propositional content need not be monolithic but can be a unified composite of individually meaningful proper parts. More specifically, a proposition is the logically well-formed and semantically well-composed, truth-valued, unified objective representational content of a judgment; and more generally it is "what is judged" in the act of putting forward any sort of rational claim about the world.

15.3.1 Transcendental Idealism and Transcendental Realism

In our first presentation of Logic in Reality, we stated two major claims: (1) space and time were not independent of objects, but produced by them; and (2) appearance and reality were non-separable, dialectically connected. Placing these claims in relation to Kant, as might be expected, is not easy, but the differences are instructive: (1) space and time *are* appearances, but since they are also real and cognizable, they are more than simply "the forms of our sensible intuition of objects"; (2) the objects we perceive do exist independently of our intuition, even if they are also appearances; (3) we can cognize objects, but, in agreement with Kant, we cannot cognize things-in-themselves; (4) but we can think about things-in-themselves outside the Kantian categories. Students of Kant will quickly realize that this paragraph is a combination of transcendental idealism and transcendental realism, the latter being the commonsense view that objects in space and time *are* things-in-themselves. What Lupasco added, to repeat for clarity, is that appearances are not just 'mere' representations but participate in the ontology of the world. Kant states that transcendental idealism is a kind of empirical realism because it entails that we have knowledge of objects merely through self-consciousness, whereas Lupasco, by accepting the reality of appearances, accepts perception as a source of knowledge. Let us see how our position plays out in relation to Kant's central doctrine of judgment.

15.3.2 Innate Capacity for Judgment. Propositional Content

The existence of human rationality implies an innate capacity for judgment that for Kant is the central cognitive faculty of the mind, and judgment is embedded in the metaphysics of transcendental idealism. What is central to a judgment and has priority over all other properties is its propositional content. From here on, Kant and we will often be at odds because we do not accept the logical, epistemic or ontic priority of propositions in reality. Thus, we can agree with Kant that judgments are complex conscious cognitions that refer to objects via intuitions or concepts, enter into inferences and are central in a sense of their involving all cognitive faculties in a highly unified rational self-consciousness. What we have left out of this description are the, for us if not for Kant, less essential properties of 'pure' concepts and laws and operation in the domains of belief and truth and falsity.

A most interesting characterization of judgment is as a faculty of judging. This transforms the activity into a process, like the faculty of thinking, and gives the basis for a "mind for itself that is entirely life", or in which the principles of life are instantiated. It is a, perhaps not entirely conscious, powerful expression of the "power of judgment". We thus are drawn to a concept of an immanent doctrine of judgment as possible addition to the debate. Kant was concerned primarily with transcendent concepts, and did not describe in detail the structure and properties of immanence. In Chap. 15, we referred to the duality of immanence and transcendence and immanence of Deleuze. This is perhaps the place to repeat that both are instantiated dialectically in the logic and philosophy of LIR, but Deleuze's 'plane of immanence' [12] is as abstract as a plane in Euclid. However the entire question of immanence versus transcendence is far more important and central to our central logic and philosophy than in relation to Kant. It will be explored below in connection with the work of Emmanuel Levinas.

15.3.3 Intuitions and Concepts

Concepts are at once (a) general representations having the logical form of universality; (b) discursive representations expressing pure logical forms and falling under pure logical laws; (c) complex intensions ranging over "comprehensions" that contain all actual and possible objects falling under those intensions, as well as other narrower comprehensions; (d) mediate or indirect (i.e., attributive or descriptive) representations of individual objects; (e) rules for classifying and organizing perceptions of objects; and (f) "reflected" representations expressing the higher-order unity of rational self-consciousness. Intuitions in contrast are conscious object-directed representations that are (1) singular, (2) sense-related (3) object dependent, (4) immediate, or directly referential and, above all, (5) non-conceptual.

We can accept that truth is nothing but the objective reality of the total propositional form-and-content of a judgment. Another expression of this is that true judgments are ways of rationally projecting ourselves onto actual truth-makers. Kant is at least not saying that truth is nothing but asserting the corresponding actual facts; our rational interest in good judgment or the good action that should result—the core moral concept of sincerity—take us into to the domain of empirical judgments and thus, a little by the back door, into reality in LIR terms.

15.3.4 Transcendental Logic

As discussed by Hanna [15], Kant's logic amounts to an intensional logic of existential commitments, primitive modalities, and fine-grained conceptual structures, his narrow conception of a classical predicate (monadic) logic of forms notwithstanding. The latter kind of logic is, in its modern forms, decidable, provable and complete. Kant's general logic can be considered its core and at the same time the precursor of the intensional logics and the non-classical logics of the twentieth century—paraconsistent and paracomplete logics—even if not explicitly of LIR. Kant's pure general logic is a formal science and a moral science, but not a natural science: *ought* is irreducible to *is*. As we knew already, Kant is not a natural philosopher. For Kant, every judgment has a logical form, collated into a science of pure general logic: *pure*, because it is a priori, necessary, hence strictly universally true, and also without any associated sensory content, *general*, because it is strictly underdetermined by all objectively valid representational contents and also abstracts away from all specific or particular differences between represented objects; and *logic* because it systematically provides categorically normative rules for the truth of judgments (i.e., the law of non-contradiction or logical consistency) and for valid inference (i.e., the law of logical consequence).

In contrast, *transcendental* logic, in that, although also pure and a priori, is "special" in that it is *ontically restricted*, i.e., objectually-committed, and thereby presupposes the existence of certain specific categories or kinds (including natural kinds) of objects. In this way, transcendental logic presupposes pure general logic, and is synthetic and a priori, not analytic. Perhaps the most one can say is that this logic is closer to reality, and to Logic in Reality, than Kant's general logic, but we are still very far from any conception of process, even of ways of thinking.

15.3.5 Analytic and Synthetic Judgments

The familiar and today less accepted analytic-synthetic distinction, which originated with Kant, is that (1) analyticity is truth by virtue of linguistic meaning alone, independent of empirical facts; (2) syntheticity is truth by virtue of empirical facts; and (3) the statement of necessity versus contingency is formally and materially equivalent to the analytic-synthetic distinction. Since the latter, since Quine, has been discredited, we consider that (1) and (2) must be modified also, and that a total disjunction cannot exist between the two. The further casting of judgments in terms of propositions makes comparison with our non-propositional theory difficult. For Kant, in a synthetic judgment it is the intuitional components that determine its meaning and truth, not its empirical-conceptual components. A synthetic judgment is an intuition based proposition. In LIR, judgments are processes of judging, and both intuitional and empirical components will play a role.

15.3.6 Non-theoretical Judgments

According to Kant, theoretical judgments have the purpose or function of being true propositions about the world, whether the world is phenomenal or noumenal. Non-theoretical judgments are used *non-truth-theoretically*. This definition does not, however, yield a non-truth-*functional* philosophy, since according to Hanna "what matters for a non-theoretical judgment is how things *seem* to the judger, not how they actually *are*. This is the unsatisfactory result of making only the propositional content of a judgment relevant to its non-theoretical uses. *Contra* Hanna, we state that non-theoretical judgments—teleological, of taste, and pragmatic/moral judgments—are individuated not only by their propositional contents but by their dynamic meaning that goes outside propositions alone. In our view, the objectivity of the objects of experience does support the assumption that human cognition is centrally judgmental, in Kant's phrase. But neither the categories nor the further assumption that judgmental cognition is essentially propositional add anything to this picture. The curious concept of 'rogue' objects, ones that do not fit the artificial constraints of a static system, is perhaps a demonstration of their value!

15.3.7 Conclusion: Immanent Judgment?

In the strongest version of transcendental idealism, Kant asserts that from transcendental affinity, an "empirical affinity" of the sensory manifold of intuitions also directly follows. Since empirical affinity is the complete application to actual empirical nature of the system of causal laws under transcendental principles, it follows that empirical affinity is the same as a *systematic unity of nature*. Kant is saying that the systematic unity of nature is a trivial consequence of transcendental affinity. Later, Kant treats the principle of the systematic unity of nature as stemming only from a regulative but not a constitutive use of judgment. But if the principle of systematic unity is only subjectively and not in fact *objectively* necessary, then Kant has *not* shown us that the system of causal laws of nature *must be* completely applied to sensory appearances or objects. Rather he has shown only that we *must epistemically believe it to be* completely applied to sensory appearances or objects. So there remains the real possibility of relatively or absolutely chaotic aggregates of sensory appearances or objects that are not subsumed or even in principle cannot be subsumed under the transcendental affinity of the manifold, i.e., the real possibility of the categorial anarchy of rogue objects. In other words, even assuming transcendental affinity there might still be no complete application of transcendental laws to nature. So the transcendental schematic of the pure concepts is insufficient to bridge the gap between categories and sensory appearances, and the transcendental doctrine of judgment fails.

For us, this is not the end of judgment: subjectivity and objectivity are not totally separated, and no philosophical property such as systematic unity is part of a system of causal laws under only *transcendent* principles. Our Principle of Dynamic Opposition, and this is perhaps our key finding, can be considered as an *immanent* principle or doctrine of judgment that, among other things, allows all or at least some of the dichotomies to represent reality.

15.4 Neo-Kantianism

The break effectuated by the essentially German neo-Kantian philosophers with prior philosophy was in some ways as great as the one we propose here. Let us see where the two approaches coincide and where our logic and dialectics go beyond the most visionary of the neo-Kantians. To begin with, they took as fundamental the existence of a priori concepts and principles, but these did not need to be fixed forever. The very brief summary that follows includes references in the *Stanford Encyclopedia of Philosophy* article by Heis [17] to both the Marburg and 'Southwest' schools of neo-Kantian philosophers.

15.4.1 Forwards and Backwards

In seeking ways to distance themselves from Kant, neo-Kantians seem to have adopted positions that in fact added nothing to the original insights of Kant about the complex structure of thought and existence. Their major step seems to have been the rejection of the dualism of appearances and things-in-themselves, related respectively to Kant's categories of sensibility and understanding. As we have seen in the previous section, the additional alternative of a 'dynamic', evolving dualism provides a more workable picture of reality than a total separation.

The neo-Kantians saw philosophy as independent of psychology, but a psychology which was, also, separate from logic. Psychology was the domain of the subjective and logic of the objective, Apart from the Lupascian position, such a dichotomous view is no longer tenable today. Neo-Kantians rejected naturalism, since they stated that it made philosophy indistinguishable from natural science, and the theory of knowledge was applied empirical psychology. However, they avoided speculative metaphysics or that philosophy could deliver facts of science from purely philosophical a priori principles, or that it could be done without taking principles of science into account. This is, if nothing else, the basis for the compatibility of neo-Kantianism, in particular that of the Marburg school, with our Principle of Dynamic Opposition. We will discuss just two aspects of this domain.

15.4.2 The Marburg School: The Transcendental Method

According to the neo-Kantians, The object of a transcendental philosophy was human 'culture', or the 'fact' of culture. The fact of culture is a highly complex entity, constituted in or by concrete facts of culture in somewhat the same way that the 'informosome' of Wu Kun I constituted by information, and the transcendental method begins with the fact of culture. Heis cites Hermann Cohen as follows: "The transcendental method does not research the principles of human reason but rather the foundation of science that conditions scientific validity." The facts of science and culture were to be found, in Cohen's day, in 'books'. Today, following Wu we can say that they are to be found in all sources of information.

Another concept to which we can relate is that the facts of culture described are stated to be not static but dynamic. If they change, they must change in response to some 'force'. Culture changes as one scientific theory follows another, in an evolutionary or revolutionary manner as discussed by Wu and Brenner. We thus totally agree with the idea that, as Natorp put it, "the fact is not a 'factum', the thing done, but a '*fieri*', a becoming, a *devenir* (Lupasco). We are not used to thinking of ourselves as idealists, but we are in the sense of the Marburg school if this means simply that the objects of the world should not be limited to those of the senses. Cohen used the physical reality of electromagnetic fields and Special Relativity to vindicate idealism. But he also stated that at our conception of the world can *only* be supplied by thought and not by the senses. Perhaps the simplest counterexample to this is the *gravitational* field.

15.4.3 The Marburg School: The Philosophy of Logic

The Philosophy of Logic in this school has some elements that are congenial to our Logic in Reality. We would not agree that logic is involved only with the content of thought and not the act of thinking, as if it were part of psychology, because this would make logic 'less general'. LIR is the "science of the laws of thinking", and does not require that the special sciences look to it to know what is true. The Marburg philosophers questioned the distinction between sensibility and understanding that Kant's distinction between formal and transcendental logic used. Transcendental logic identifies the a priori concepts and principles that make experience possible, while formal logic gives the rules of the operation of understanding, abstracting completely from sensibility. We believe this distinction, also, is not required in science, and we can also agree with the Marburg philosophers that the laws of all logics are found by reflecting on the content of the concrete sciences. Formal logic is in principle revisable, since it is an abstraction from transcendental logic, which must evolve as science does.

Still not forgetting that that we are talking about a formal logic that is constituted by propositions, the preferred Marburg view sets the stage for LIR: logic is grounded in the most general form of experience. Formalism *pace* Frege and Russell, does not explain why logic is *applicable*. Logic is applicable to experience because its laws are discovered in an analysis of empirical knowledge in science; it has content because its laws are part of the content of science, treating science non-skeptically, and it is true because it is a part of science, that is, of reality, which is primary. The further step, that logic, as a part of science, is non-semantic and non-truth-functional was left for Lupasco to make.

15.5 Naturalism

We have already, in Chap. 13, defined a Natural Philosophy coming from the side of science, based on the properties of real physical, biological and cognitive phenomena. Any philosopher would say that we espouse naturalism, as we have allied *ab origine* philosophy closely with science. For the more complete philosophical system we would like to describe, we should refer to the distinction, made primarily in mid-twentieth century American philosophy, between ontological and epistemological naturalism. The objective is to see, also, to what extent these designations correspond with our identification of ontic and epistemic 'units' of reality—ontolons and epistemons, defined in Chap. 9.

15.5.1 Ontological Naturalism

Logic in Reality supports is a physicalist as well as naturalist attitude to complex special subject matters which appear at first to be at least partly outside physics. When in our view the reality of potentiality is accepted, it is sufficient to explain how, for example, mental processes can causally influence physical systems such as our bodies, since they are not totally disjoint. While this chapter deals essentially with doctrines that are specific to philosophy, the remainder of this book has been and is directed to showing how they apply in other areas of study. We should emphasize that philosophy that discusses 'pre-quantum' alternatives is only of historical interest and will not be reviewed. Also, our demonstration of existence of quantum-*like* macroscopic processes should not be taken to imply a new set of *sui generis* non-physical causes. There is no need to return to the view, for example that mental states must be identical with physical states [3].

15.5.1.1 Non-reductive Physicalism

Our theory provides a basis for the absence of ontological separateness between mental, so-called 'special' causes and their physical effects. But strict identity of special properties and physical properties is not required. Causal efficacy of special causes can occur as long as the properties they involve are realized by physical properties, even if they are not reductively 'identified' with them in the sense of standard epistemic type identity. As long as one is talking about real processes, we do not need to postulate different kinds of physical properties. Accordingly, we can eliminate at the same time metaphysically independent non-physical causes, again, of a historical interest only. Modern physics states that all physical effects or their probabilities are fully determined by their physical antecedents (Principle of Physical Closure). No metaphysically independent non-physical factor can make a causal difference in the evolution of the physical world.

Epiphenomenalism is a theory that conscious states, clearly caused by physical processes, cannot have physical effects. Physicalist naturalism (our theory) offers in fact a novel way of integrating the mental realm with the causal development of the spatio-temporal world in a way that should become more familiar with the acceptance of LIR.

15.5.1.2 Moral and Mathematical Facts. Nominalism and Modality

In our opinion, part of the problem in defining a naturalist stance toward moral and mathematical issues, among others, is to talk in terms of facts, and whether moral facts are or not identical to natural facts. To cut through a lot of debate, our position is that a moral intuition exists as a *natural* faculty, due to our seamless embedding in nature. We can accept that moral judgments qualify as a 'species' of truth for people who prefer thinking in terms of truth, without it being essential in practice. We thus answer Moore's open question as we believe he might have wanted it to be, namely, that there is no principled barrier to inferring moral facts from apparently non-moral natural facts. Whether the judgments are a priori or a posteriori should be beginning to be less and less relevant, in line with Williamson's intuition regarding knowledge of metaphysical modality [14, 41],

In the case of mathematical facts—the 'nature' of the existence of numbers—we refer to our earlier statement that our natural capacities include one of imagining non-real or abstract entities. In this book, we have tended to ignore discussions revolving around the reality, or lack of it, of such entities described by the terms possible worlds, abstract objects, universals and particulars. We acknowledge the reality of such philosophical concepts with respect essentially to the process of their conceptualization. Does that make us nominalists or non-nominalists? There seems to be some agreement that "in and of themselves" numbers and other such objects of thought are not causally efficient [13], and with this we obviously agree. This is a form of the nominalism that asserts that everything is "concrete". This is a weak metaphor, however; we find it more fruitful to examine change and the evolution of dynamic entities, constituted by energy as real processes. Naturalist irrealism is not to be confused with an ontologically non-naturalist realism. It is the cognitive processes employed to imagine non-spatiotemporal mathematical or other fictional entities which are real, not the entities themselves *qua* content of cognition. Thus, Field stated explicitly that mathematics "can be viewed as a useful fiction which facilitates inferences between nominalistic scientific claims, but is not itself implicated in our most serious beliefs about the world [28]." Papineau suggests that the most popular contemporary alternative to fictionalism is a Fregean version of non-naturalist realism: abstract mathematical beliefs can be justified as analytic truths that follow from propositional logic. Our response here is a respectful 'no comment'.

Regarding modal claims about non-actual, possible worlds, about which so much has been written (from 'possible knowledge'?), we can remain equally non-committal. Our best scientific theories certainly do not *commit* us to possible worlds. Modal knowledge is analytically a priori, as in the previous paragraph. Thus for us it is not only not obvious that knowledge of possible worlds cannot be construed realistically, it is impossible.

15.5.2 Methodological Naturalism. Intuition

We have no difficulty in asserting, from an LIR standpoint that as Papineau puts it "at bottom philosophy and science are both concerned with establishing synthetic knowledge about the natural world by a posteriori investigation", pursuing similar ends and using similar methods with the exception that scientists gather empirical data in ways that philosophers do not. Also, methodological naturalist philosophers typically test their findings against *intuition* rather than observation and experiment. A debate then arises as whether a priori philosophical intuitions are analytic or synthetic. If the intuitions are analytic, then accordingly a priori, their significance is doubtful. If they are synthetic, they may be or probably are significant, but then not a priori.

We believe that Williamson is right when he claims that there may not be an absolute distinction between the two [40]. If we consider with Lupasco that intuition is not and does not have to be separate from knowledge-as-such. The mind recognizes the properties of an intuition as a mental in which less concrete, vaguer entities are actualized—uppermost in the mind, sometimes but not always as a sudden 'flash'. But knowledge and the path to *a posteriori* synthesis is always present, 'waiting its turn' so to speak. One way of expressing the difference between LIR and philosophy is that standard philosophy starts with synthetic theories endorsed by everyday thought and then looks to more fundamental theories of reality to see what makes these everyday theories true. LIR is a more fundamental theory whose truth is the truth of reality and that does not seek a further truth to validate it.

15.5.2.1 A Priori Synthetic Intuitions. 'Experimental Philosophy'

The argument of some philosophers for the existence of a priori synthetic intuitions seems to us to stem primarily from a desire to 'have it both ways'. The process of conceptual analysis involves the articulation of synthetic theories and assessing them against a posteriori evidence, an acceptable form of methodological naturalism. The idea that conceptual analysis generates a priori knowledge or intuition is in our view ungrounded. Williamson assigns to philosophers a 'special faculty' of deciding whether intuitive judgments are acceptable, but is unable to classify the judgments as one or the other. There is still no positive explanation of how synthetic philosophical knowledge might be established without a posteriori evidence.

The idea of experience-independent access to synthetic truths brings us into the realm of belief. Our biological heritage, and the cultural heritage which derives form it do not yield separate justifications which would somehow make the a priori analytic. Unreliability is intrinsic to the nature of philosophy, but not of nature! LIR thus supports what Papineau calls a revisionist attitude to philosophical methods. Certain scenarios and theses may not be ruled out by imagined possible, often 'catastrophically counterfactual' cases (the famous, or infamous mind to new body transfers), but this is not for us proper empirical theorizing about real intuitions which *do* have a relation to synthetic claims about the world, based on the process of their emergence as a posteriori intuitions. Supported in this way by their congruence with the general theory of LIR, such claims may be considered part of 'experimental philosophy'. If some of the above considerations are found to be applicable at least to areas such as the philosophy of mind and epistemology, further work here, rather than in the modal realm, may justify the effort.

As methodological naturalists, we simply insist that most primary philosophical concerns are synthetic and a posteriori, independently of whether modal claims, which are not, are implied. In fact, LIR offers a basis for saying that this would be a natural consequence of the PDO operating in theories. We thus agree with Papineau when he says that science is interested in synthetic a posteriori claims as such, rather than their modal implications. In our concluding Chap. 20, we will return to naturalism in connection with Husserl's critique of both the natural and human sciences,

15.6 Popper's Three World Ontology

Karl Popper is among the twentieth century philosophers for whom we have greatest respect, in particular for the social implications of his work. His triadic system or 'Three Worlds Ontology' [33] also bears some resemblance to the triadic system of actuality, potentiality and emergent T-state of Lupasco. The part of his doctrine that discredits the concept of knowledge as justified true belief is consistent with our approach to a non-truth-functional logic and science. In other respects, the resemblance is, in our view, superficial, and differences with our approach bring out the relatively greater scientific aspects of the latter. Briefly, Popper's First World (W1) is the 'world' (scare quotes ours) of material objects, events and processes, including living systems, the Second World (W2) is the world of mental events, processes and predispositions; and the Third World (W3) is that of products of the human mind. Popper imagined real dynamic interactions between his worlds, with aspects of W3 impacting W1, working through W2. In our opinion, in the absence of a discussion of the origin of cognition is science, Popper's distinctions and approach to processes fall rather into the domain of epistemology rather than ontology as the real. The term process is used for W1 and W2 but then why not W3? What is included in the genetic code, for example, is better termed information rather than knowledge, since all would agree that there is no human knowing involved.

The triadic aspects of Popper are a selection, *in fine*, no more grounded in science than that of the categories of Peirce. Our program thus differs from that of Popper in this essential way: Popper emphasizes the features of one of his worlds which separate it from another world and from the objects of that world: subjective and objective knowledge. Lupasco and Logic in Reality are intended to describe our world in its dynamic unity, and to bring out what unifies the 'worlds' rather than what separates them. To put it simply, there is no post-Hegelian dialectics in Popper in the energetic sense that we have proposed in Chap. 5. Perhaps what is most attractive in Popper is his *intuition* of a complexity of structure in society and human relations which he was not able to characterize in the necessary terms. The concept of falsifiability, as discussed in Brenner [3] in respect of scientific theories is good as far as it goes, in an epistemic truth-functional context. Building from this, Brenner proposed a new criterion of the validity of a theory in terms of whether or not it functionally included its opposite or contradiction. It demands an additional modicum of intellectual honesty to argue carefully against one's own opinion, but we cannot conceive of a true Philosophy in Reality without it.

15.7 Existence. The Concept in Lupasco

In this book, in discussing the work of Stéphane Lupasco, we have concentrated on his proposal of a new logic in and of reality, in particular of a logic of energy and real, energetic processes. This concept of a non-propositional, non-truth-functional logic, despite its roots in as well as applications to science has not (yet) been widely adopted. This situation notwithstanding, we have demonstrated in this book how Lupasco's basic Principle of Dynamic Opposition is instantiated at all levels of reality, especially in biological and cognitive phenomena, the theory of knowledge and, finally, in philosophy via the implications of scientific concepts. His language and form differed that typical of standard 20th philosophy, but Lupasco did address many specifically philosophical problems—but some of them only *once.* As mentioned earlier, Lupasco's earliest major work was his French State Thesis of 1935 entitled *Essai d'une nouvelle Théorie de la Connaissance* (Essay on a New Theory of Knowledge), of which Volume I was *Le dualisme antagoniste.* (Antagonistic Dualism) and Volume II was *Du Devenir Logique et de l'Affectivité*, (On Logical Becoming and Affectivity) [23]. It was then only after the end of World War II that Lupasco set out his logical, scientific and philosophical system in some fifteen books. This section will concentrate on Lupasco's views of existence and the concept as background for the reader to compare with more familiar philosophical texts.

In the dialectical vision of Lupasco, the essence of existence is logic-as-process, a logic of reality; existence is a duality of potential antagonistic dynamisms capable of actualization simply because any constitutive antagonism implies potentiality and actuality. The outcome of this process of actualization is knowledge and/or knowing. One is no longer conscious of the original logical conflict which is transcended by consciousness as knowing, deductive or inductive (translation Brenner).

For Lupasco, existence is not to be confused with the cognitive existential, whose content is the nature and properties of existence. Lupasco goes so far as to say that the more we know or are conscious of, the less we exist. The existential includes its own set of antagonisms, namely, a subconscious or unconscious of actualization and a consciousness of potentialization. This implies that existence itself, as the locus of maximum antagonism our counteraction (contradiction in the dynamic sense) contains the basis for neither consciousness nor unconsciousness. Existence neither is nor is not conscious, it is *a*-conscious. This is the defining property of the concept, among other things.

Lupasco first defines a sensation as a conflict involving the contradictory potential terms immanent to logic and constituting it (see previous section), the concept is the conflict of the logical becomings generated by the dynamism of this antagonistic duality. This can best be understood by first noting that it is difficult to be consciously aware of a concept, one can only be aware of the duality which constitutes its entire nature and structure, that of its (the concept's) comprehension and extension. The mind moves continually from one to the other, while one tries to grasp the two together. One has, nevertheless the certainty that the concept exists, even if it is neither entirely conscious nor unconscious, neither known nor unknown. In a very complicated argument, Lupasco arrives at the conclusion that existence, as sensation is potential non-being, as potential concept is the potential of non-being and finally that a actualized concept and pure sensation at the same time, existence is the absoluteness of non-being, that is, its potentiality and actuality at the same time, the totality of itself as conflict.

This section has been presented to allow the reader to make his/her own comparison with the 'concept of concept' in Kant, starting in Sect. 15.3.3. It is clearly a compromise between excluding the Lupasco picture as irrelevant to Kant and much longer discussion of both. We believe this is a domain for further study. From a similar perspective, we discuss the relation of immanence and transcendence, culminating in the work of Emmanuel Levinas [2] and his concept of Transcendence-in-Immanence.

15.8 Transcendence-In-Immanence

In our discussion of information in Chap. 13, we mentioned that its complex properties justified being referred to as a "unity-in-diversity". This expression has a long and familiar history going back to the expression coined by the fifteenth century German philosopher Nicolas of Cusa [27] of the coincidence of opposites (*coincidentia oppositorum*). The reference to opposites in functional manner suggests a relation to a logic of opposites such as that of Lupasco in its original formulation [24]. We do not agree with this attribution in the case of Logic in Reality: the term "coincidence of opposites" is inadequate to express a dynamic relation of 'opposites'. What 'happens' in the coincidence, do the opposites get wiped out? Or do they co-exist comfortably in the pleroma? Nicolescu has referred to de Cues as a precursor of Lupasco, but this is true only in a very general way. We do not. There is no 'physics' in de Cues' doctrine.

Unity-in-diversity, on the other hand, implies some very real relations in both the physical and cognitive domains, especially when the term is taken to include a *changing* degree of unity and diversity. A question to which we have not found an answer is why the opposite term of diversity-in-unity has not received attention, 'equal time' in philosophy? In fact it has received only anecdotal attention in the titles of conferences and a of a book series in psychology. This would be simply another instance of the fact noted by Lupasco to which referred earlier, of the lower ontological value assigned in general to what is diverse, vague, contradictory and so on.

In our general interpretation of critical opposing concepts, in addition to unity and diversity, we include transcendence and immanence. Immanence and transcendence in the original logical philosophy of Lupasco are defined in relation to causation and determinism, also in a proposed a contradictorial relation (mutual actualization and potentialization). The causalities of non-contradiction are the cause of the causality of contradiction, and the latter is the cause of the former. Then, the causalities of non-contradiction, of rationality and irrationality, can be called transcendent to the extent they transcend contradiction. However, these causalities are the cause of the causality of contradiction, which can be called the causality of immanence or immanent causality. Logical values that imply immanence and transcendence thus 'cause' themselves reciprocally. An essential corollary of this point is that there are no 'pure' immanent and transcendent phenomena. One cannot, therefore, separate completely immanent 'real' events and transcendent 'abstract' facts (statements, propositions, categories, etc.) *qua* their existence but only *qua* their meaning by abstraction and elimination of any dynamics, that is, as non-spatio-temporal entities. This conclusion justifies the terms of immanence-in-transcendence and transcendence-in-immanence, of which we will give two relatively recent examples.

15.8.1 Gilles Deleuze. Immanence and Life

Deleuze has probed deeply into the relation between real events and philosophical concepts and 'events', especially, immanence, transcendence, life and meaning. His work is significant for this study because of the way in which he rejected dialectics [19], although he accepted a reciprocal relation between his most important terms—Immanence and Life. The best way of summarizing his system is to see it as a structure of abstract relations between terms which define two domains—'philosophy' and 'science'. Deleuze's philosophy includes transcendental structures of several kinds: two levels of idealized structures: a pre-philosophic chaos and a plane of immanence (in which language games operate), which 'cuts' through the chaos, the transcendental field; the metaphysical surface and the plane of immanence. Examples of separability in 'philosophy' are to be found in Deleuze's construction of meaning [12] as a metaphysical surface, or a 'line', a middle between extremes [1], that are the loci of the separation of different aspects of phenomena, propositions and things [11].

In this theory, Life is simply a more affirmative and better-specified concept than immanence. Life rather than immanence opposes transcendence not only as a general concept, but as the form in which (or by which) transcendence is specified, namely dialectics. Term-to-term oppositions remain an essential part of philosophy—Negation-Transcendence versus Affirmation—Immanence, and Dialectics, placed in opposition to Life, in opposition to an integral 'immanentism'. Transcendence is thus specified by dialectics, but its relation to immanence is not dialectics. In our view, these are not dynamic relations in the LIR sense, in which the real relations between non-absolute, non-abstract elements constitute the 'dialectics', and there is no difference between opposition and dialectics.

Deleuze's plane of immanence provides a field in which concepts and meanings are supposed to be produced, circulate collide, etc. Life is transcendental, but the plane of immanence is a life; "it is not immanence in life, but immanence which is not in anything." If it were immanence in life it would lose its character as being which possesses in itself the reason for its being, as opposed to a being whose existence depends on that of another. A critical task for this philosophy is to retain a property—the infinite—that is allegedly 'lost' in science. The objective of philosophy is not to recognize objects; it is the task of science to convert the objects of the plane of immanence into determined states-of-affairs.

The philosophy of Deleuze illustrates the results of an application of concepts of immanence and transcendence that does not define or include any dynamic dialectical relation between them. The domain of Deleuze's philosophy is a realm, governed by binary logic, of undetermined, idealized entities, Humean in its lack of effective interactions. In the domain of reality to which LIR applies, the existence of all beings depends and is defined by that of others. Infinities and infinitesimals do not exist, but are replaced by transfinite values, and immanent and transcendent aspects of phenomena are alternately actualized and potentialized.

Our interim conclusion is again that LIR can discuss philosophical problems in physical, dynamical terms that do not require recourse to any imaginary, abstract structures to separate or define aspects of reality. Any such aspects that are considered 'virtual' or 'possible' in Deleuze are so 'in philosophy' but 'in reality' are instantiated as potentialities.

15.8.2 Emmanuel Levinas

There are several reasons for including in a book, which by definition cannot include references to all significant philosophers of the twentieth century, the next sections on the work of the Lithuanian-Jewish philosopher Emmanuel Levinas [2]. Levinas differs in one important aspect from that of the other philosophers we have studied as precursors, including, to a limited extent, Kant himself. Levinas does not direct his attention either to science or to logic in a way that would allow comparison with our Logic in Reality. However, he states that the concept of a scientific principle is also open to interpretation. If causation can be viewed as a physical process, as in LIR, then it belongs as in Cassirer's conception to a new type of physical statement in which both measurements and laws or principles are interwoven. Laudisa [19] and Massimi [26] proposed that a scientific principle is best understood in the context of Cassirer's reinterpretations of the Kantian a priori and principle of systematicity in regulative terms. A scientific principle fulfills a regulative task of systematizing and conferring order on empirical knowledge, *while being an integral part of that knowledge* (emphasis ours). This could serve as a definition of the function of Logic in Reality. In our view, it is otiose to try to argue whether entities bearing properties and in relation with other entities are ontologically prior to laws of nature or not. As quoted by Laudisa, Cassirer talked about an "ultimate common element of all possible forms of scientific knowledge, never perfectly achieved." That the PDO might be such an element we leave as an open-ended possibility in the spirit of Cassirer's inquiry.

15.8.3 Levinas, Husserl and Heidegger; Transcendence-In-Immanence

The work of Lévinas does not belong to a single recent school or trend. Our interest in it is based on the *dramatic structure* he gives to transcendence, better to the transcendent states that our existence implies. Unlike Heidegger, Lévinas situates his very concrete transcendence in the *interruption* of the first temporal levels of occurrence of *Da-sein*. Heidegger gradually translated his conception of the silent call of being (to *Da-sein*) into the notion of the event (*Ereignis*) in the 1930s, but Lévinas makes *his i*nterruption intersubjective, the basis of ethical intersubjectivity. Being in Lévinas thus entails both dynamic forces and a conception of natural processes and causality.

As Bergo has pointed out [2], the novelty of *Otherwise than Being* lies in its three innovations: (1) the analyses and tropes for transcendence-in-immanence (recurrence, proximity, obsession, persecution, and substitution), (2) the deconstruction of language understood as the site in which existence is said and explained, and (3) Lévinas' stepping out of philosophical reasoning into a context that both "says" and "unsays" itself by turns, in a cycle of alternating actualization and potentialization. Lévinas' philosophical project is constant: *to rethink the meaning of existence in terms of the ethical transcendence of the other*. To that end, he consistently revisited Husserl's phenomenological method. He reconceived Heidegger's ontological difference as an irreducible separation between being and the good we enact.

For Lévinas, any good apparently beyond being is not radically separated from existence. "This affair is a human one." Any philosophical translation of concrete embodied life must therefore approach the human subject as it emerges through its relations with others, even though the intersubjective situation entails a loss of "egoistic mastery". Lévinas argues that the lapse of time *between* lived immediacy and its reflective representation is *never* fully gathered by the *logos*. The rethinking of the lapse along with transcendence-in-immanence is Lévinas' ultimate modification of Heidegger's project. Together, the lapse and this new conception of transcendence ultimately pass into language as words addressed.

Transcendence is one of Lévinas words for the spontaneity of responsibility for another person. It is experienced in concrete life and is variously expressed, from words like "here I am" (as in the Islamic statement of one's being available to the deity: *Labaykh, allahumma, labaykh*) through to apologies and self-accounting.

In Lévinas later work, transcendence, understood as the other "outside", became transcendence-in-immanence. Merleau-Ponty's account of intercorporeity similarly urged that philosophies of embodiment should never *oppose* (purpose of opposition) immanence and transcendence. However, in LIR, opposition is not to be understood negatively. In his major *Otherwise than Being*, Lévinas approaches transcendence in sensuous and temporal terms, arguing for a past that eludes thematization. In that sense, transcendence for an embodied being is always transcendence-in-immanence. The affective "experience" of our relations with others is thus preserved as a trace or pre-thematic reminiscence of the flesh, as "a relationship (but not as an abstract relationship) with a singularity, without the mediation of any principle, any ideality". Invoking *Otherwise than Being*, Llewelyn has called this affective dimension of invested selfhood, or "ipseity", a "'deep' saying" that "bears witness to what is neither recollected nor forgotten *in the epistemic sense* of these terms when (or once) it is represented as a sign". For a relevant discussion of remembering and forgetting as processes following the principles of LIR (see, [8, 18]).

Proximity and vulnerability as properties are the *loci* of transcendence-in-immanence and the birth of signification (whether words are actually uttered or not). For Lévinas, there is more in living affectivity than Heidegger's conception of being speaking through language captured. The moment we understand signification originally as an affective proto-intentionality, in our terms potentialization and not as some thought, already formulated, that the I thereupon chooses to communicate to another. Lévinas thus conceives language as more than denotation and description but part of his felt concern. There is nothing like this in the way that Lupasco approached existence. Lupasco's role was to confirm the physicality of the acts of responding, of trusting and inspiring trust. When Lupasco talks about affect, *affectivité*, it is always in terms of its absolute separation from real processes, thus more like Heidegger. Key terms in Lévinas' system are *Fürsorge*: what one must make available to the other, namely, both solicitude *and active* assistance. Another key term is *Vertrautheit* which we read as trust *and* confidence. Both are processes, trusting and having confidence. One can perhaps say that Lévinas reconnected the philosophical thought with its necessary component of feeling that Heidegger had separated.

15.9 Philosophy in Reality as Process

We find ourselves driven, by our own logic of processes, to a description of our Philosophy in Reality (PIR) as a process. We thus include ourselves among philosophers of process, described as precursors of Lupasco in Chap. 3. In PIR, two major components can be perceived. One is philosophy as philosophizing, the active creation of philosophical thought. At the beginning of the process of its creation (or recreation during, reading, studying, etc.), concepts move conceptually between more or less actualized or potentialized states, second, an 'activated complex' of changing and stable concepts, to use our term from chemistry, or T-state emerges at a higher level of reality or complexity.

In addition to its dialectics as a directed cognitive activity, PIR as a body of knowledge is something very mobile. Additions are being made to it in real time, so to speak, very much in the spirit of the evolution of science in which there is flow of concepts, which is not and does not have to be constant. Elements enter and leave as they become trivial or inapplicable. This is no different from the pattern of development of any text or document, edition of a book into which *changes* are incorporated. In general, most changes are undervalued and discarded, the exception, of course, being the original *Book of Changes* itself.

We thus arrive at a principle like that of Relational Quantum Mechanics (RQM). RQM [21] is an interpretation of quantum phenomena that discards the notions of the absolute state of a system, the absolute value of its physical quantities, or an absolute event. The theory describes only the way systems affect one another in the course of physical interactions. State and physical quantities always refer to the interaction, or the relation, between two systems. If we translate this theory into the philosophical domain, the value or preferred principle of a philosophy should not depend on any actualized published composite text but its cognition and re-cognition by philosophers who add to and subtract from it. For a more recent discussion of RQM see Laudisa [20] This does not mean the end of Book Series such as the one of which this book is a part. It is rather a concept of the further naturalization of natural philosophy discussed in Chap. 13. It is a reassessment of the role of a book of philosophy within philosophy. If it too, is nothing but change, then its existence becomes one of a living system and of works of art as they are cognized.

15.10 Logic in Reality and Dynamic Being

We need to include, in our Process Philosophy in Reality, a treatment of issues in existence at which we have only briefly looked it, viz., the nature of being and becoming. Being is often thought of as a kind of One of Parmenides, eternal and immobile, an entity within 'pure' philosophy and therefore outside the main thrust of this book. The term in the heading of this section, Dynamic Being is not one commonly used, but it is of immediate interest because it appears to imply that an absolute cut between being and becoming must and can be avoided. As with certain other areas that we have identified, for example, the Philosophy of Complex Processes (cf. Chap. 16), the recent serious discussion of Dynamic Being seems to be limited to a single Compendium of articles, edited by Vesselin Petrov and Adam Scarfe in 2015 [35]. The sub-title of the volume is *Essays in Process Relational Ontology*. We have been able, using our logical and dialectical approaches to some of the articles, propose significant advances in this domain, as we feel the point of intersection between our dynamic logic and metaphysics of processes and standard analytical approaches to dynamics is a fundamental one in philosophy: it is the notion of being as dynamic.

15.10.1 The Characteristics and Categories of Dynamic Being

Dynamic Being is of particular interest to us because of the number of critical issues raised by the juxtaposition of different approaches and understandings of these terms. From the perspective of our logic and dialectics, we find a large number of definitions of being, ontologies and process which we consider neither dynamic nor ontological. One major difference, which we must emphasize at the start, is that when we talk about dynamic ontology and Seibt (Sect. 15.10.2) about the ontology of dynamicity, we are talking about different things. She (and others in the volume) describes her work as a "formal theory of dynamicity and dynamic entities". We suggest that it is, rather, "an analytical theory of formal dynamicity and formally dynamic entities". Some of the contributions of Vesselin Petrov include statements that are in remarkable conjunction with our views about the dialectical nature of dynamicity 'in reality'.

The historical flavor of the Petrov's title, "Dynamic Aspects of the *Development* of Process Ontology" should not obscure the fact that his objective is meta-ontological, being about both the principles of theory formation but also about the underlying (ontological) characteristics of potential process ontologies. Petrov shows, among other things, that these are more dynamic in character that Whitehead's own process-relational ontology, discussed in Sect. 15.10.4.

The most important point is that Petrov is driven to the position that "in a truly dynamic ontology, such as process ontology, the basic philosophical categories should themselves be *variable* (italics his), so that they can express the dynamic nature of reality". In other words, the process ontology of the future that has variable categories ought not to be dialectical in the strict sense of the word. It will be a dynamic, process ontology. For Petrov, the concept of "dynamical systems" is broader than the notion of process ontology, but he admits that it may be "too early to specify what meaning the word "dynamic" would take on in the process ontology of the future". This objective, namely, of 'opening up' the static structure of the categories of category theory was explicitly addressed in [3]: "LIR (is) an applied categorial ontology, based a view of the fundamental dualities of energy, and hence of reality, as inherent in the effective quantum field description of quantum phenomena. Categories of a New Energy Ontology (NEO) are established, including the essential categories of Dynamic Opposition, Process and T-state (emergent included middle) and Subject-Object that are both formal and physically meaningful. The categories are shown to fit the LIR axioms and the need suggested to view most important, interactive physical, biological and cognitive phenomena as instantiating the category of Non-Separability, related but not identical to the existence of non-separability at the quantum level of reality." In addition, however, the author saw no need to retain the requirement of standard category theories that any particular category have the properties of exclusivity and exhaustivity. In his chapter, Petrov recognized that Brenner did point to a further development of process ontology—one that involves, in some sense, a "variability" of categories, but his concept was open to a criticism that, with its emphasis on dialectics, it embraces a neo-Hegelian standpoint. The discussion of dialectics in Chap. 5 of the present book should go a long way in the direction of further distinguishing our interpretation of the term "dialectics" from Hegel.

The critical point for us is that variable categories are suitable for standard analytic ontologies and bring significant changes to them. However, variable categories are the *natural* categories of our views of process and change, hence of dynamicity as ontological in the physical sense of what is irreducibly real, namely energy itself. Thus process views, if correct, would make putatively causal claims by scientists subject to a critical test. Those that pick out real processes could be causal; those that don't can't." Bueno says [9]: "..., the partialness modeled by the partial structures approach is not understood as an intrinsic, onto-logical 'partialness' in the world (as an aspect about which an empiricist will be glad to remain agnostic. We are concerned here with an 'epistemic', not an 'onto-logical' partialness." LIR adds the critical detail of the operation of the PDO on the logical elements of real processes, better, of the processes involving real spatiotemporal entities, which naturalizes this position, respecting the principles of both the primacy of physics constraint and naturalistic or physical closure.

From the point of view of formal ontology, Brenner suggested that the sole material category is energy. The opportunity that results from the LIR approach is that from a *metaphysical* standpoint, for real systems or phenomena or processes in which these physical and metaphysical dualities of the world are instantiated, their terms are *not* separated or separable. This argument, however, suggests that the most useful categorial division that can be made is exactly this: (1) phenomena that show non-separability of the terms of the dualities as an essential aspect of their existence at their level of reality (NSC) and (2) those that instantiate separability, SC. One may ask at this point to what extent these sub-categories, SC and NSC, are simply restatements of the logical operators of disjunction and conjunction respectively, as they are understood in standard logic. This would be valid, however, only for the classical view of conjunction and disjunction as non-spatio-temporal, quasi-abstract concepts, whereas in LIR they also are considered to be dynamic process entities, instantiated in phenomena and subject to the predicates of actuality and potentiality in transfinite sequences.

Petrov suggests [31] that the characteristic features of the development of process ontological systems up to now, have not been necessary or inevitable but features of realized possibilities, namely, as logical features. The development of process ontology after Whitehead has its own logic, and Petrov outlines the key elements—the basic features—of that logic. He defines three kinds of change respecting the development of ontology: (1) theory revision by way of partial replacement of elements of an existing category system and/or local alterations to a categorical network,(2) radical ontology theory revision by way of wholesale replacement of the existing category system; and (3) the expansion of theory by way of specialization. Petrov points out that there have been many different interpretations of the category of "substance" and there have been many attempts "to develop a dynamic conception of being." Some of them begin with the claim that "there is a dynamic ontological framework" implied by Aristotle's own original notion of "substance." In the current work, there is the substitution of "process" for the old category of "substance." According to Rescher, there are weak and strong versions of process philosophy. The weak version posits the priority, but not the primacy, of "processes" over static "substances," while the strong version holds to the primacy of the former over the latter. Rescher states, pertaining to a weak version, "in a dynamic world, *processes* are more fundamental than *things*. Since substantial things emerge in and from the world's course of changes, processes have priority over things." Rescher's further claim that "the process approach provides an effective framework for better understanding both the conduct and product of rational inquiry" does suggest a substitution of a new category of "process" for the old category of "substance."

This characterization of process is close to our preferred notion, *provided* it is understood to be part of a logic and ontology of energy (cf. Chap. 6). To philosophers who might claim that we have simply substituted 'energy' for 'substance' and that our ontology has gone backwards toward discredited substance theories. Our response is that our notion of process is not separable from any notion of an underlying physical substrate to which, however, following Lupasco, critical additional properties based on the Principle of Dynamic Opposition have been assigned.

In concluding his chapter, Petrov suggests that Brenner's Logic in Reality has a function as an ontological system, based in a new kind of dialectics. Another possibility is that it will rather be a new kind of "dynamic" process ontology, and that he feels that this potentiality is the one more likely to be realized. We quote again from Petrov: "All of the tendencies that have been considered in this essay have been based on the present situation surrounding the development of process ontology. In this sense, they represent next *logical* steps in the course of development of process ontology. This means that not all process ontological systems of the future will see to the realization of these tendencies, but that inevitably there will be new process ontologies that will realize them and they will be the most successful and influential ones."

At the present time, it is the formal analytical approach that sets the tone for philosophy. We strongly oppose this approach. For us, formal process theories, like the formal logics on which they are based, pick out just enough of the characteristics of real processes to appear dynamic but their fatal scientific and philosophical weakness is expressed in a key passage from Seibt [36]: in her General Process Theory (GPT), "a central role is played by "subjectless activities. … General processes are concrete dynamics that are best understood on the model of subjectless activities…. Subjectless activities are *dynamic* but they are not changes. Constitutive 'phases' of their overall dynamicity (for example: the change of place of every single flake constituting the dynamicity of the snowing) contribute to the functionality of the activity but not *as* temporal stages or phases. In contrast to developments, activities have no internal temporal differentiation. …[The logical properties of subjectless activities like 'it is snowing' can illustrate] the main category features of general processes: they are *independent, concrete, spatiotemporally extended, non-particular, non-countable, determinable*, and *dynamic* entities…".

There is certainly place in philosophy, at least, in a philosophy *tout court*, in which the dynamic referents are falling snowflakes. However, this doctrine can also be read as a refusal to address the dynamics of, for example, people writing articles about snowflakes, when the world is "falling apart" around them. In the next section, we will look further at Seibt's positions and those of other contributors to the *Dynamic Being* Compendium.

15.10.2 The Dynamicity of Johanna Seibt

The chapter in the Compendium by the process philosopher Johanna Seibt, entitled "Aristotle's 'completeness test' as heuristics for an account of dynamicity" [37] introduces a concept that one might imagine would be central to our study, in view of the importance of Aristotle as a precursor of Lupasco. The reasons why this is *not* the case should help the reader understand why the work of the latter represents an unrecognized but essential contribution to philosophical thought. Seibt begins by recognizing that analytical ontology has little to offer for an ontology of 'dynamicity', so she tries to go back to the difference between *dunamis* and *energeia* and between *kinesis* and *energeia*. She starts by using aspectual rather than tense inferences to develop a test for the completeness of an action, which she relates to standard mereology. This leads her further to two modes of dynamicity: 1) a developmental mode of coming-about or coming-to-pass (kinesis) and of going-on as self-propagation (*energeia*). However, and this is crucial, Seibt cannot think about or represent symbolically an *energeia*. "This creates a curious interdependence: the second mode of dynamicity, going-on or self-expression (autotelicity), is the way in which the first mode of dynamicity is there—coming—about is there by (the virtue of) the going on of some *dunamis* to become A." However, for Seibt, as apparently for Bergson, what can be conceptualized of a coming-about is the pair of initial and end states,the transition between these states, however, the 'ongoingness' of coming about, entirely eludes any effort of descriptive conceptualization. To anticipate our conclusion, this is clearly no problem for a real, physically dynamic conceptualization such as ours. Seibt then introduces the concept of automerity (self-containment) and says that it dovetails with autotelicity and asks if automerity is conceptually coherent if it requires that a whole, A, contains itself as a part? The LIR answer is, of course, partly, but let us go on. Seibt's solution is to go from a transitive to a non-transitive part-whole relation where the relation "is part of" is defined relative to partition levels, parts of parts of parts, provided the partition has a self-similar structure. If **a** and **d** are 1-parts of **b**, then **a** can be a 2-part of itself. A moment's reflection will show that this solution is vacuous except for mathematical entities for which strict self-similarity holds, *e.g.* fractals such as the Mandelbrot Set. For all real 'dynamic beings', some additional principle is required and that is, of course the role that we propose for the Principle of Dynamic Opposition.

The difficulties that Seibt has with becoming and dynamicity and temporal flow reflect the problems with her assumptions. Thus "temporal passage does not seem to be either a necessary or sufficient condition for dynamicity (LIR: because it does not exist)—temporal passage may be only one special manifestation of dynamicity (LIR: LIR!). Then "only if one could show that (1) change is the only type of dynamicity (LIR: then dynamicity is superfluous); (2) that we *must* conceive of change in terms of states with contradictory features (LIR: no problem) and (3) temporal passage *must* be conceptualized in terms of states with contradictory temporal features (LIR: no problem), such new implications could come into view". Generalizing further, however, if analytical philosophy can only come up with a picture of process like the above, can any application of it to reality be valid? To keep the attention of readers active in the field of analytical philosophy, we ask their permission to leave this question open for the time being.

15.10.3 Dynamic Ontology in Whitehead

The relationship between the concepts of reality and being of Whitehead and those of Logic in Reality has always involved a great deal of tension. His logic, to begin with, developed with Russell, is at the basis of modern first and second order predicate logics, but involves strictly mathematical and linguistic counters. It bears no relation to a logic of real processes such as LIR, grounded in physics. In [3], an attempt was made to find interpretive bridges between Whitehead's idealism and the realism of Lupasco, despite its partial reliance on idealized, abstract objects of reality.

The discussion of Whitehead's metaphysics by Maassen in *Dynamic Being* [25] was read in the same spirit, but with somewhat different results. Maassen describes the basic components of Whitehead's metaphysics that appear to reflect its dynamic character: the ontological status of 'actual entities', centers of activity, subjects which prehend or involve each other by reason of their prehensions of each other. Maassen emphasizes Whitehead's view that an actual entity never is, but is always in the process of becoming. Why never? We will return to this point in a moment. Whitehead advocated a new cosmological scheme, in which the dominant scientific materialism was to be replaced with a modified organic or holistic theory of reality. He introduced an ontological principle according to which "there is nothing which floats into the world from nowhere." Whitehead writes that "this (causal) ontological principle means that actual entities are the only reasons, so that to search for a reason is to search for one or more actual entities.

Three statements in Whitehead found by Maassen look at first as if they belong in *our* system:

1. Metaphysics has to refer to the here and now, namely, to the actual and not to generalities, abstraction, or to something which is beyond the world.
2. The metaphysics of immanence is based in experience. The subject "grasps"—No transcendence—at most into the immanence.

3. The 'essence' is not hidden behind the phenomena. The phenomena [themselves] are the 'essence'."

In Whitehead [25], the "concrete facts of relatedness" are classed as prehensions among the categories of existence, while the "world as process" is the first category of explanation. One does not need to argue about the hierarchy here; what is important is the existence of a conceptual relation between relation and process.

There also seemed to be a relation between thee LIR potentiality and actuality and Whitehead's potentiality and givenness as ontological predicates. There is a 'correlation' between them, and the "completion of givenness in actual fact converts the 'not-given' for a fact into 'impossibility' for that fact." More significantly, potentiality and givenness are meaningless apart from the entities to which they refer, and they are requirements for, in a nexus of actual things, the "process of supersession by novel actual things". This is a description of the metaphysical basis of what is designated in current terms as emergence.

From another standpoint, however, if actualization and potentialization are process entities in their own right, as well as predicates, they belong *themselves* to the category of dynamic opposition. In this case, they can no longer be characterized as Whiteheadian entities.

According to Strawson and Simons, cited by Maassen, "descriptive metaphysics" describes the actual structure of our thought about the world, while revisionary metaphysics is an attempt concerned to produce a better structure. Simons considers Whitehead's metaphysics as revisionary, but at the same time, he emphasizes that there is also an empirical, descriptive, basis to it. One could conclude that Whitehead tries to overcome these distinctions in his own form of speculative philosophy. Similar to Plato's philosophy, the eternal objects—the forms—are the potentials for each process of becoming. Whitehead summarizes eight categories of existence which will not be repeated here. The use of the word 'pure' in several of them, despite reference to matters of fact cuts the ties to phenomena. The Category of the Ultimate shows what underlies all of these processes of becoming. "Creativity," "the many," and "the one," are the notions pertaining to this category. Creativity is the principle of novelty. Each process is never just a repetition of what has been or what will be, but more or less something new.

Whitehead's generic notion of actual entities must be taken as that and nothing else, that is, as a metaphysical description of the nature of real things, and not as involving any claim that actual entities are real things, or the real constituents of things. In Whitehead's philosophy, real things must be described as actual entities, but this does not mean that things are really actual entities or that actual entities are real. The point of Whitehead's cosmology—as for any cosmology—is that its generic concept [the concept of an actual entity] is the true descriptive model of the world, and is in that sense actual or real; Bradley summarizes Whitehead's relation to Kant: whereas medieval transcendentalism thinks being in terms of its representable features, and Kantian transcendentalism thinks being in terms of the conditions of its representation, Whitehead's transcendentalism thinks being in terms of its immanent, serial conditions of self-actualization.

In this new kind of transcendentalism, an actual entity therefore never is (an object), but is in a process of becoming; it is the subject. It is an object only for those actual entities which prehend it in their own processes of becoming. The opportunity for self-determination lies in the free self-creating process of concrescence; the goal of this process leads the actual entity either "upward" or "downward." This is where moral considerations occur. Whitehead: "the actual entity, in a state of process during which it is not fully definite, determines its own ultimate definiteness. This is the whole point of moral responsibility. Such responsibility is conditioned by the limits of the data, and by the categorical conditions of concrescence."

What follows for morality from this organismic perspective? Whitehead says that "the whole concept of absolute individuals with absolute rights, and with a contractual power of forming fully defined external relations, has broken down." For Whitehead, "the notion of the unquantified stability of particular laws of nature and of particular moral codes is a primary illusion which has vitiated much philosophy." In his view, "moral codes are relevant to presuppositions respecting the systematic character of the relevant universe," and "the details of these codes are limited to the social circumstances of the immediate environment." As such, from the perspective of the organismic, process philosophy, "morality is always the aim at that union of harmony, intensity, and vividness which involves the perfection of importance for that occasion." What is to be avoided is the exaggerated claim that there is only one behavior system belonging to the essential character of the universe, as the universal moral ideal. What is universal is the spirit which should permeate any behavior system in the circumstances of its adoption. Thus morality does not indicate what you are to do in mythological abstractions. It does concern the general idea which should be the justification for any particulate objective. Rather, for Whitehead, "morality consists in the control of process so as to maximize process. It is that aim at greatness of experience in the various dimensions belonging to it."

In the first critical years of the twenty-first century, we cannot accept a regression to cultural relativism, no matter what its philosophical origins, coming through a philosophical 'back door'. *Dynamic Being* is first and foremost about the being/becoming of human beings, which might be better called that to start with.

15.10.4 The Difference Between Dynamical Systems and Process Theories

In several places, particularly in our discussion of continuity and discontinuity in our Logic in Reality, we have suggested that standard dynamic systems theories (DSTs) do not fully merit the designation of dynamic if the only change they mirror is that modelled by differential calculus. We met the sociologist and philosopher Roberto Poli in our chapter on anticipation in communication. In Chap. 2 of *Dynamic Being* [32], Poli makes a careful comparison of characteristics of dynamical systems and process theories in relation to Dynamic Being. His answers stem from the obvious fact that dynamical systems theory is a theory of a class of formal models. It is a scientific theory, while process theory is ontological (metaphysical), i.e., it is a philosophical theory. We do not entirely agree with the cut made, here or elsewhere, between science and philosophy, but let us accept it for the sake of argument.

Poli outlines two differences between the two theories: process theories require some type of "glue" to keep the states of the underlying temporal continuum together, and they accept causation as providing such "glue." Classic dynamical systems do not need any "glue" to keep the points of the continuum together. The second main difference concerns time: dynamical systems theories consider time only as order, while for process theories need concepts such as unidimensionality, the present moment, parallelism, and actuality in their understanding of time. In that sense, dynamical systems theory is more general than process theory and does not necessarily have a process theory as its underlying ontological framework (see our discussion of the Continuum Hypothesis in 4.3.2 and subsequently,)

Differences with our perspective on what constitutes process and consequently a theory of process appear immediately. Their source is the standard concept of an independent time and space in which 'process' is alleged to operate. Poli concludes by stressing that in contrast to dynamical systems theory, process theories explicitly require causation as the link that keeps together the states of the underlying temporal continuum; dynamical systems theories see time only as order. The perspective on time that is embedded in the latter is an external one, while process theories add a vision of time from "inside," due to the apparent structural features of the moment-now. According to Poli, for DSTs, time, as an external parameter, has, at most, the structure of the set of the real numbers and is therefore composed of non-denumerable "atoms" ordered in the usual way. In contrast, the temporal continuum for processes seems to be composed of units that fuse or merge together. Apparently, its continuum consists of "elastic" units, or units endowed with some sort of limited "duration." Not surprisingly, these units are close to Leibniz's infinitesimals.

In the concept of process theory of Poli, time is different and remarkably more complex than in DSTs (state-transitions systems). Understanding the differing importance of time for state-transition systems and process systems is of crucial importance if a thoroughly developed process theory is to be achieved. As stated earlier, for state-transition systems, time is a parameter. In the case of processes, time is instead a constitutive (ontological) dimension. To be clear, process theory does not deny that time is order; what it claims is that time is "more than" order.

The source of our disagreement with Poli is now obvious: the view in Logic in Reality of the continuum. The continuum hypothesis refers to a conception of the universe founded on geometry, the Cantor–Dedekind view, as discussed by Longo [22], which sees not only in mathematics, but everywhere, continuity as ontologically preceding the discrete: "The latter is merely an accident coming out of the continuum background." Points are derived concepts, even if 'non-dimensional'. In this view, geometry (statism) is in some deep sense more fundamental than dynamics, that is, energy in the standard view. This hypothesis has the advantage of corresponding to our intuition and experience, integrated into and confirmed by mathematics, of continuity in our perception of 'time' and linear movement.

Penrose, on the other hand, had the strong intuition that "physics and space–time structure should be based on discreteness (emphasis his), rather than continuity". This discreteness is evidenced in quantum mechanical spin, combined, however, with a fundamental notion of expressing phenomena in terms of a relation between objects, rather than between an object and some back-ground space [29]. For us, in the construction of the mathematical continuum, objective realities are not found in the mathematical entities involved, but in the process of constituting these entities as conceptions. There can be an interaction, dynamic in LIR terms, between the applicable mathematical logic and intuitions about continuity. Stating this somewhat more strongly, since the establishment by Gödel of the reciprocal relation between consistency and completeness in formal mathematical systems, the situation is no longer absolute. One should not be forced to choose between geometry and discontinuous objects and their relations. This opens the door to a different foundational principle, using the principles of LIR, in which both continuity and discontinuity are fundamental and are dialectically related.

A more serious critique of the above conception of the continuum is that it is restatement of a conception of general relativity (GR) as a pure geometrization of the world, from which the subjective aspects of space and time involving observers have been eliminated as inessential ontologically. It is one of the major conclusions of this book which other readings of the physics and mathematics of GR that are possible, that restore the balance between geometry and energy that exists in reality and are supported by LIR.

We therefore conclude that our approach is orthogonal to that of Poli, with regard to time, space and being. If the continuum does not exist, it is otiose to discuss what keeps the spatial points of a continuum together. The law of the Excluded Middle fails as it does in all intuitionist mathematics and logic which depend on 'real' infinitesimals. It is only in some loose, abstract sense that it is acceptable to say that processes extend in time, which is not a real category in LIR. Time is not "the general dimension of the many real processes unfolding in reality". We can agree with Poli however that past, present, and future are different temporal characterizations of real processes. This implies that past and future processes are both real. The difference between them is that past processes have been actual in some temporal present, while future processes have never enjoyed actuality. Time is both the dimension and the direction of what becomes. What exists does so in a temporal way. It lasts, but the existence of a temporal mode does not establish the independence of time.

15.10.4.1 Actuality and Parallelism

In our view, the approach of Poli is epistemological. For Poli, parallelism means that all temporal processes run at the same velocity. No temporal process overtakes any other temporal process or falls behind it. Events are chained (JEB/AUI: by whom or what?) to their temporal positions. They march together with the entire flow of time. All the dimensions that traverse time—such as spatial or causal dimensions—are such that there can be movements in them. The only dimension without any internal movement is time. This is an accepted possible view the in etermalism concept of time, but there is no evidence for its reality or lack of it.

Poli continues by saying that the moment-now and parallelism characterize actuality. Actuality is the feature of the present as it advances in the flow of time. As the constantly actual, the present is the window through which events continually enter. Actuality is a constantly new moment-now. It maintains itself in the flow of time; it is not fleeting. In this sense, the actual is an "eternal present." The series of the fleeting moments of the now constitutes the constancy of actuality, or of the now-in-movement. What happens to an entity while it lasts? The lastingness of the entity is its continuance in its actuality. While it lasts, it proceeds together with the now-in-movement; it maintains itself in the flowing-now. The duration of a process means that the process proceeds together with the now; it maintains itself in the now-in-movement. This applies to all durations, be they things, living beings, psychological acts, or social phenomena. Their durations mean that they keep themselves in the now-in-movement. Duration does not imply, however, that all the states of a process are actual at any given point of time. If they were so, there would be no process at all. This is a correct conclusion, but its refusal to introduce potentiality in conjunction with actuality is what we unfortunately have come to expect. Poli approaches the problem by stating that what will happen after the actual moment-now is not unreal. It is as real as the present moment-now. The difference is that it is non-actual. The main difference between past and future is that what is now past has enjoyed its moments-now, while what is future has still to enjoy its moments-now. By denying being to the past and/or to the future, one severs the roots of the now.

Poli brings out the well-known concepts of continuum versus point, in other words, the idealistic constructions that it possible to avoid with LIR. "What, then, is a point? A point is something that depends on a surrounding whole. The point exists as a boundary, as a limit.

How should one understand the interplay between what is actually present and the continuum to which it pertains? One further quotation will be of help here [16]: "perhaps we can say that the concept of being in the proper sense coincides with the concept of that which is now or present"—that is, actuality according to Hartmann's categories "but everything that is now or present is a temporally extended thing which is now or present with respect to one moment after another." It follows that what exists in the proper sense can be taken as what is actually present, as long as we do not forget that everything that is present also pertains to a temporally extended whole. The ontological difficulty of time lies entirely in the double form of dependence of actuality from a previous stretch of its continuum, and of the continuum itself from the actuality of the moving present. Hartmann's distinction between "being" and "actuality" clarifies the issue by distinguishing between the being of the extended thing or process and the actual existence of each of its boundaries when they enter successive moments-now.

Categorially, causation is a species of the more general category of determination. What distinguishes causation from the other species of determination is that causation has the form of a temporal series. Furthermore, causation includes a productive moment. This is precisely the moment that distinguishes causation from temporality. Time itself is not productive; only causation can be.

The character of series of causation implies that each effect is the cause of another effect and each cause the effect of a previous cause. Their ordered collection is a process. The production of new states is what keeps the process in movement, and this is the work of causation. Therefore, beyond being an ordered series of states, causation has, as its most characteristic moment, the production of new states. The temporal series, on the contrary, can be void. Causation is what produces the states that constitute a process unfolding in time, moments-now, while what is future still has to enjoy its moments-now. By denying being to the past and/or to the future, one severs the roots of the now. This is not completely satisfactory; it is quite easy to establish epistemological boundaries when one is not required to give them any physical meaning. We are thus, again, up against a real boundary between a theory that can at least be discussed in real dynamic terms and an analytical simulacrum.

15.10.5 A Chinese View of Dynamic Being. Harmony

Our intuitions on the relation of Chinese thought and its dynamics to Logic in Reality, suggested in Chap. 2 and in Brenner's discussion of LIR and the Tao in [6], find confirmation in the chapter by Chenyang Li, "Being as a Process of Harmonization: A Chinese View of Dynamic Being" (2015). In this one title we find all four of the themes of process, harmony, being and dynamics of interest to us. In (2015), Brenner discussed the Chinese concept of the harmony throughout nature as being present implicitly or explicitly all of the Chinese classics. It is clearly one of homeostasis, an integrated dynamic equilibrium with the potential for the emergence of new phenomena. The complete title of Hansen's version of the Tao Te Ching includes the critical 'On the Art of Harmony'. Throughout Jullien's treatise on art one finds references to the harmonic dimension in music and elsewhere, harmonic coherence, images with a regulatory effect that are generators of order. Harmony is that of image and phenomenon, also, for which Chinese uses the same word. Harmony appears also in the work of the Western ecologist Ulanowicz [39]. In his semi-classical conception, under some non-equilibrium circumstances, the production of order becomes an inevitable feature of increasing entropy. In particular, perpetual harmonies can emerge from the collapse of non-equilibrium configurations. Thus in Ulanowicz' view, "an appropriate metaphor for the course of nature seems to be less the headlong drive towards an endpoint than a more nuanced interplay between agonistic tendencies, as in our dialectics and the Yin and Yang alternance. Brenner noted that a society which is respectful of the environment is in a relation to and appreciation of it which is also a harmony. This is the ground of ecology, discussed in another Brenner study and briefly in this book.

One of points that has to be emphasized is that unlike the mathematical Pythagorean conception of harmony, Confucian harmony is defined in qualitative terms; it can not be measured quantitatively or precisely, but nevertheless purports to describe the dynamic reality of the world. It is thus 'in harmony' with the philosophy defended in this book.

15.10.6 Process-Relational Ontology in Evolutionary Biology

The work of Igamberdiev in the area of biology and its philosophy, to be reviewed in the next Chap. 16, will be looked at here for points of overlap with study of process in biology by Scarfe [35]. Scarfe develops a process-relational perspective that stands in contrast to radical "biological kind anti-essentialism". The latter aims at destroying genetic determinism without remainder, reducing species to individuals and/or over-emphasizing difference in a manner that renders biological research untenable. The process-relational interpretation of conceptions of "biological kinds," "ethotype," "behavior," "mentality," and "species" that is more consistent with the notions that the genetic, physiological, morphological, hormonal, epigenetic, and behavioral characters of organisms are complex and subject to transformative change rather than being static. This developmental complexity and plasticity is best understood as a process-relational metaphysics or dynamic ontology. The term "ethotype" is construed in light of Whiteheadian process-relational ontology and/or of other dynamic ontologies, rather than by way of static substance ontology. We call attention here to the extensive work on the plasticity of biological and social systems of Mark-Williams Debono, which Brenner has placed in the context of Logic in Reality.

The potential ramifications of emphasizing the causal role of behavior and the notion of "ethotype" in explaining evolutionary processes extend into the philosophical domains of evolutionary epistemology and ethics. In relation to evolutionary epistemology, an emphasis on the behavioral habits and the structures of activity that are engaged in by the organisms in question raises several issues. By placing emphasis on the role of both the organism's common, enduring, and changing patterns of behavior in evolutionary processes, the field of evolutionary ethics stands to become revolutionized. We see a parallel here to the 'revolution' in the field of information, proposed by Wu Kun, that is the consequence of, again, the change from the static reductionist concept of information as data to information as process.

Scarfe's chapter points to the notion that behavior can be, and often is, a causal instigator of evolutionary change, and that, in abstraction from morphological and/or genetic transformation over time, the concept of evolution should be thought of as involving behavioral change, of itself, over time. In so doing, it has sought to advance the scarcely employed concept of ethotype, taken as the behavioral analogue to the widely used terms, "genotype" and "phenotype," and interpreted along the lines of Whiteheadian process-relational ontology and/or other dynamic ontologies, as an indispensable tool for biological research. He sides neither with anthropocentric humanism or transhumanism of figures such as Julian Huxley nor with biocentric anti-humanism of radical environmentalism. Rather, critical pan-selectionism, embracing biological- and ecological wisdom, occupies the middle path beyond these extreme positions, members of the human species being finite, compositional parts, among many, of the complex, interdependent whole that is Nature, yet utilizing their cognitive and rational capacities creatively in adapting themselves to, and integrating themselves in, the environment.

Scarfe states that behavior can be, and often is, a causal instigator of evolutionary change, and that, in abstraction from morphological and/or genetic transformation over time, the concept of evolution should be thought of as involving behavioral change, of itself, over time. In so doing, it has sought to advance the scarcely employed concept of ethotype, taken as the behavioral analogue to the widely used terms, "genotype" and "phenotype," and interpreted along the lines of Whiteheadian process-relational ontology and/or other dynamic ontologies, as an indispensable tool for biological research. These issues will reappear in our Chap. 17 on Living Systems.

This chapter completes the Part II of our discussion focused one the different forms of philosophy. We now turn our attention to systems and the construction of a natural philosophy that integrates, without conflation, systems science and philosophy with our previous theses and their logic, dialectics and science.

References

1. Badiou, A.: De la vie comme nom de l'être. Philosophie **89**, 78–93 (2006)
2. Bergo, B.: Emmanuel Levinas (2019). https://plato.stanford.edu/archives/fall2019/entries/levinas/.
3. Brenner, J.E.: Logic in Reality. Springer, Dordrecht (2008)
4. Brenner, J.E.: The philosophical logic of Stéphane Lupasco. Logic Logic. Philos. **19**, 243–285 (2010)
5. Brenner, J.E.: Information in reality: logic and metaphysics. Triple-C **9**, 332–341 (2011)
6. Brenner, J.E.: Linking the Tao, biomathics and information through the logic of energy. Prog. Biophys. Mol. Biol. **131**, 15–33 (2017)
7. Brenner, J.E., Igamberdiev, A.U.: Philosophy in reality: Scientific discovery and logical recovery. Philosophies **4**, 22 (2019)
8. Brenner, J.E.: The logic of forgetting. Plastir **20**(3), 1–25 (2011)
9. Bueno, O.: What is structural empiricism? Scientific change in an empiricist setting. Erkenntnis **50**, 59–85 (1999)
10. Capurro, R.: Personal communication to J. E. Brenner, 6 January 2019
11. Deleuze, G.: Logique du Sens. Les Éditions de Minuit, Paris (1969)
12. Deleuze, G., Guattari, F.: Qu'est-ce que la Philosophie? Les Éditions de Minuit, Paris, France (1991)
13. Field, H.: Science Without Numbers. Blackwell, Oxford UK (1980)
14. Field, H.: Realism, Mathematics and Modality. Blackwell, Oxford UK (1989)
15. Hanna, R.: Kant's Theory of Judgment (2018). https://plato.stanford.edu/archives/win2018/entries/kant-judgment/
16. Hartmann, N.: Philosophie der Natur: Abriss der speziellen Kategorienlehre. De Gruyter, Berlin (1950)
17. Heis, J.: Neo-kantianism. In: Stanford Encyclopaedia of Philosophy (Winter 2018 Edition) (2018). https://plato.stanford.edu/archives/sum2018/entries/neo-kantianism/
18. Heller-Roazen, D.: Echolalias: On the Forgetting of Language. Zone Books, New York, NY (2008)

19. Lardreau, G.: L'histoire de la philosophie comme exercice différé. In: Gilles Deleuze; Immanence et Vie. Presses Universitaires de France, Quadrige, Paris (2006)
20. Laudisa, F.: Open problems in relational quantum mechanics. J. Gen. Philos. Sci. **50**(2), 215–230 (2019)
21. Laudisa, F., Rovelli, C.: Relational quantum mechanics. In: Zalta, E.N. (ed.) The Stanford Encyclopedia of Philosophy (Spring 2002 Edition). Stanford University: Stanford, CA (2002). https://plato.stanford.edu/archives/spr2002/entries/qm-relational/
22. Longo, G.: The Mathematical Continuum. In: Naturalizing Phenomenology. Issues in Contemporary Phenomenology and Cognitive Science, J. Petitot et al. (Eds.). Stanford University Press, Stanford CA (1999)
23. Lupasco, S.: Du devenir logique et de l'affectivité; Vol. 1: Le dualisme antagoniste. Essai d'une nouvelle théorie de la connaissance. Vol. II : Du devenir logique et de l'affectivité. J. Vrin, Paris (1973). (Originally published in Paris: J.Vrin, 1935)
24. Lupasco, S.: Logique et Contradiction. Presses Universitaires de France: Paris, France (1947)
25. Maassen, H.: Aspects of dynamic ontology in whitehead's process and reality. In: Petrov, V., Scarfe, A. (eds.) Dynamic Being: Essays in Process-Relational Ontology, pp. 68–85. Cambridge Scholars, Newcastle-upon-Tyne (2015)
26. Massimi, M.: Pauli's Exclusion Principle. The Origin and Validation of a Scientific Principle. Cambridge University Press E-book, Cambridge, UK (2005). https://www.cambridge.org/978 0521839112,
27. Miller, C.L.: Nicholas of Cusa. In: Zalta, E.N. (ed.) The Stanford Encyclopaedia of Philosophy (Spring 2017 Edition) (2017); https://plato.stanford.edu/archives/sum2017/entries/cusanus/
28. Papineau, D.: Naturalism (2016). https://plato.stanford.edu/archives/win2016/entries/naturalism/
29. Penrose, R.: The Road to Reality. Knopf, New York NY (2004)
30. Petrov, V., Scarfe, A. (eds.): Dynamic Being: Essays in Process-Relational Ontology. Cambridge Scholars, Newcastle-upon-Tyne (2015)
31. Petrov, V.: Dynamic aspects of the development of process ontology. In: Petrov, V., Scarfe, A. (eds.) Dynamic Being: Essays in Process-Relational Ontology, pp. 44–67. Cambridge Scholars, Newcastle-upon-Tyne (2015)
32. Poli, R.: The difference between dynamical systems and process theories. In: Petrov, V., Scarfe, A. (eds.) Dynamic Being: Essays in Process-Relational Ontology, pp. 28–43. Cambridge Scholars, Newcastle-upon-Tyne (2015)
33. Popper, K.: Three Worlds. The Tanner Lecture On Human Values Delivered at The University of Michigan, 7 Apr 1978
34. Priest, G.: In Contradiction. Martinus Nijhoff, Dordrecht (1987)
35. Scarfe, A.C.: On ethotype, epigenetics and organic selection: process-relational ontology and behavior in evolutionary biology. In: Petrov, V., Scarfe, A. (eds.) Dynamic Being: Essays in Process-Relational Ontology, pp. 230–261. Cambridge Scholars, Newcastle-upon-Tyne (2015)
36. Seibt, J.: Free process theory: towards a typology of occurrings. In: Seibt, J (ed.) Process Theories: Cross-disciplinary Studies in Dynamic Categories. Kluwer Academic Publishers, Dordrecht (2003)
37. Seibt, J.: Aristotle's 'completeness test' as heuristics for an account of dynamicity. In: Petrov, V., Scarfe, A. (eds.) Dynamic Being: Essays in Process-Relational Ontology, pp. 2–27. Cambridge Scholars, Newcastle-upon-Tyne (2015)
38. Sklar, L.: Philosophy of Physics. Westview Press, Boulder and San Francisco (1992)
39. Ulanowicz, R.E.: Increasing entropy: heat death or perpetual harmonies? Int. J. Des. Ecodyn. **4**, 83–96 (2009)
40. Williamson, T.: The Philosophy of Philosophy. Blackwell Publishing, Malden, MA (2007)
41. Williamson, T.: Modal Logic as Metaphysics. Oxford University Press, Oxford, UK (2013)
42. Wu, K., Brenner, J.E.: An informational ontology and epistemology of cognition. Found. Sci. **20**, 249–279 (2015)
43. Wu, K., Brenner, J.E.: Philosophy of information: revolution in philosophy. Towards an informational metaphilosophy of science. Philosophies **2**, 20 (2017)

Part III
The Philosophy of Structures and Systems

Chapter 16
Structures and Complex Systems

16.1 Introduction: Structures and Systems

The term system has already been used in many connections in this book. We have focused our discussion on real, ontological as opposed to epistemological systems, at the same time as we have insisted on the absence of total separation or distinction in reality. In relation to living systems, references to systems and complex systems are so common that their underlying meaning is usually ignored. Complexity has received the denomination of a science, as in the complexity science of Laszlo, which refers to systems, and complexity philosophy in the work of Morin which does the same. Another term that is used routinely without any necessary prior definition is *structure*. The emphasis in the description of structures is generally on their identity and stability.

The English terms of system and structure are derived Greek and Latin roots: system comes from the Greek verb *sunistanai*, to stand, to be or to have been placed together, leading to stance and attitude. Structure is based firmly (*sic*) on the Latin root *struere* to pile up or build, hence what is built.

In our view, systems and structures are among the concepts that should be defined philosophically to insure that extraneous meanings and interpretations are not introduced by default. Following the original discussions by Lupasco [46,47], we define systems and structures with the understanding that in reality, the terms correspond to the dynamic processes that underlie them, that is, how a system or a structure is first pro-duced, 'led-forth' from some substrate, and how it evolves and changes without losing its essential characteristics or 'standing'.

We would like to ask for the reader's indulgence regarding the organization of this chapter: he or she may feel that there is a substantial tension, perhaps indicative of something quite fundamental, between the concepts of structure and system. Is either more fundamental than the other? This makes it at first difficult to decide which one to discuss before the other; perhaps this is an indication that they are in fact connected dynamically in the sense of the Principle of Dynamic Opposition, each sharing characteristics of the other.

J. E. Brenner and A. U. Igamberdiev, *Philosophy in Reality*, Studies in Applied Philosophy, Epistemology and Rational Ethics 60, https://doi.org/10.1007/978-3-030-62757-7_16

We have elected to begin with a discussion of structures, recognizing that the concept of system will surface and be actualized from the beginning of the discussion. Not all structures are systems, but some are. The critical difference is that systems are readily identifiable instances of change and process, but not all structures are. A pile of bricks and even a building is a structure, but apart from the inevitable and irrelevant microscopic changes, there are no systemic interactions in progress in it. Microscopic deterministic changes can indeed lead to macroscopic ones, the collapse of a bridge for example, but these forms of change have only a binary logic associated with them. Despite the potential real consequences for individual human beings, the minimum set of interactions necessary to form a system is not present in such structures.

16.2 What Is a Structure?

A further characterization of structures is thus necessary because structures *are* processes when looked from the point of view of their existence, their 'coming-into-being' on an appropriate time scale. This concept emerges from the logic in/of reality, and is the consequence of the inability of any real process to go totally to completion. Especially in the case of man-made structures, the constructed object cannot be separated from the dynamics of its *con*struction, which includes as a component the intentionality of the builder. It is a truism, but one which we believe that has philosophical purport, that "some buildings fall apart before they are completed".

Lupasco thus approached the question of dynamic structure from two angles: structure-as-process as we have begun to outline above and structure-*in*-process in complex emerging systems. The question that is the title of this section was originally posed by Lupasco [46] "What is a structure?" The answer he gave was that structures are also dynamisms, not to be objectified and reified. Regardless of what rules and materials of construction are used, "for these rules to generate a veritable structure, they must obey these logical laws or conditions necessary for its existence." All structures are the result of trade-offs between actualizations and potentializations and are constituted to different extents by primarily bonding forces and homogenization, by primarily heterogenizing forces and anti-bonding and a third at a T-state between the two, emerging at the point of equivalence of the two. Any individual structure is never rigorously actual, that is, absolute in any sense, given the nature and logic of energy. It is a dynamic structuring that is always functionally associated with an antagonistic and contradictory potential structuring, a force of change.

Lupasco made the following link between structure and form: the energetic dynamisms that constitute all matter and all existence, and the systems they generate, are "pure structural forms, containers of containers, structures of structures, subject to an essential and ineluctable chain of transformation. There is therefore no such thing as a full and static form; devoid of a present, going always from past to future, or even inversely, temporality is immanent to form". Every form, every system, to which as well will see the same criteria apply, all matter, in a word, is thus in LIR terms a real process. Lupasco proposed the term structuring (Fr: *structuration*) to describe the emergence of structures, a term used some twenty years later by Giddens [25] to describe the same phenomenon in a social context.

In the second aspect of Lupasco's theory, the structure of real processes involves the change of an energetic entity and its opposite or antagonist from a state of subjective actuality to one of objective potentiality or T-state. As for physical structures, these more epistemological entities are to be placed in the category of Non-Separability. The values (degrees) of actualization and potentialization or T-state are logical in that they depend on this syntactical structure as well as the context.

What does this mean for a structural model or explanation? Does it make sense to consider them, also, as dynamic forms, subject to potentialization and the actualization of their contradictions? The answer is yes and no. As formal objects *qua* their meaning, the structural descriptions of LIR as such, like all semantic entities, belong in the category of Separability, in this case that of non-spatio-temporal entities, the data of standard inference. However, considered as processes capable of change, they are from this point of view in the natural philosophy of LIR. We would include in the list of structural models those displaying a sequence of argumentation in which the advantage oscillates from one antagonist to the other.

The structures of all elements or entities in this category gain their explanations from LIR as a metaphysical but also physical theory. The elements are process structures and their deterministic dynamics is that described using the non-Kolmogorovian probability language proposed in Chap. 3. The criteria for applying this concept of structure-as-process, given a process of two elements, are those indicated above: a 'structuring' seen externally is a kind of form; looked at internally, it consists of the processes themselves. Metaphysical structural explanation picks out the dialectically opposed elements and shows what is involved in the operation of the PDO with respect to them.

16.2.1 Structural Realism and Relations

There is a similarity between the Lupasco view of the fundamental nature of relations and that of Ladyman and Ross of the primacy of relations (Chap. 10) and [47]. These authors maintain a metaphysics of structural realism (Ontic Structural Realism, OSR) according to which there is, primitively, structure in the sense of concrete, physical relations, with objects derived from relations. LIR gives a picture of this 'derivation' in terms of a pause in the ortho-dialectic concatenation of processes. The ontological commitment to relations in LIR and OSR, as well as Relational Quantum Mechanics (RQM) places objects and their relations on the same level within a 'holistic metaphysics'. Esfeld and Lam [21] sees our world as one of holism, tied together by relations that do not supervene on (whose source is not) intrinsic properties. There is no ontological priority, but rather a mutual ontological dependence between space–time relations and the objects that stand in the relations, considered as space–time 'points' or point-events as developed in the theory of Lusanna and Pauri [44]. These point-events also define the process entities which LIR sees as populating the universe, instantiating both identity and diversity.

16.2.2 Scientific Realism, Structural Realism and Structuralism

These further concepts, definitions and attempted descriptions of reality individually capture correct but limited aspects of a dynamic world. Several closely related terms apply to the positioning of structure in science and philosophy. A comprehensive survey is in [49] and related *SEP* articles.

- Scientific realism is the view that we ought to accept as real the unobservable entities posited by successful scientific theories. A powerful argument in favor of scientific realism is the no-miracles argument, according to which the success of science would be miraculous if scientific theories were not at least approximately true descriptions of the world.
- LIR: Logic in Reality is realist in this sense.
- Structural realism is the view that scientific theories tell us only about the form or structure of the unobservable world and not about its nature. This leaves open the question as to whether the natures of things are posited to be unknowable for some reason or eliminated altogether.
- LIR: From the outset, our theory refutes the assumption of any cut between the structure and nature of the part of the world that is and is not observable.
- Structuralism in its simplest form focuses on empirical structure as a defense of the on-going validity of science against Kuhnian 'revolutions and 'minimal structuralism', cf. Otávio Bueno [12] and Ladyman [38].
- Epistemic Structural Realism (ESR) is the claim that all we know is the structure of the relations between things and not the things themselves.
- Ontic Structural Realism (OSR), crudely, is the claim that there are no 'things' and that structure is all there is. This is called 'radical structuralism' by van Fraassen [24].

16.2.2.1 Scientific Realism

Scientific realism is the conception that, subject to the recognition that scientific methods are fallible as suggested above and that most scientific knowledge is accordingly approximate, one is justified in accepting the findings of scientists, validated by consensus, as representations of reality, that is, that the unobservable entities postulated by the theories in fact exist. Constructive empiricism argues that the best current scientific theories do not require such belief, and the success of modern science can be understood without it. It is such philosophical challenges to scientific realism that convert it to a philosophical position, as well summarized in Boyd [7]. We can discuss only a few of these challenges, and the realist response to them here, but one notion stands out as clearly supported by and supporting the logic of/in reality, that of approximate truth. As we saw in Chap. 3, truth in LIR is the truth of reality, which cannot be absolute. It is accordingly *unscientific* as well as metaphysically false to require that science generate absolute truth. LIR describes, in a way that makes it appear less accidental, the relation between the actual experimental methodology used to obtain knowledge of unobservable phenomena and the theory involving prior knowledge of other unobservables upon which the methodology depends. The non-actual entity that is intended as the consequence of the experiment has a potential existence, not yet proven but present as a non-localized process in the mind of the experimenter. Entity realism (ER) is another form of scientific realism. ER consists of the thesis that science does provide knowledge of a mind-independent reality, but it does not accept the strongest scientific realist claim that science provides, or can provide, *complete* knowledge of unobservables and their properties. This is, obviously, not a claim that LIR makes either.

16.2.2.2 Structural Realism

More kinds of structuralism abound in contemporary *analytic* philosophy. These include causal structuralism concerning the individuation of properties, mathematical structuralism concerning the nature of mathematical objects, and structuralism about laws and dispositions. These will not concern us here.

Typically, ontic structural realists will argue that what we have learned from contemporary physics is that the nature of space, time and matter are not compatible with standard metaphysical views about the ontological relationship between individuals, intrinsic properties and relations. We have already defined Logic in Reality broadly as an Ontic Structural Realism OSR as it is a form of structural realism based on an ontological and metaphysical thesis that gives ontological priority to structure and relations.

As quoted by Laudisa [41], Cassirer held that the possibility of talking of 'objects' in a context is the possibility of individuating invariants and Max Born said: "Invariants are the concepts of which science speaks in the same way as ordinary language speaks of 'things', and which it provides with names as if they were ordinary things" and: "The feature which suggests reality is always some kind of invariance of a structure independent of the aspect, the projection". "I think the idea of invariant is the clue to a relational concept of reality, not only in physics but in every aspect of the world." Eddington says [20]: "What sort of thing is it that I know? The answer is structure. To be quite precise, it is structure of the kind defined and investigated in the mathematical theory of groups." Poincaré understands group structure in Kantian terms as a pure form of the understanding. It would seem that without positing knowledge of individual objects we cannot explain why certain properties and relations tend to cohere.

Van Fraassen argues that the heart of the problem with OSR is that it must imply that what has looked like the structure of something with unknown qualitative features is actually all there is to nature. But with this, the contrast between structure and what is not structure has disappeared. Thus, from the point of view of one who adopts this position, any difference between it and 'ordinary' scientific realism also disappears. It seems then that, once adopted, it is not to be called structuralism at all! For if there is no non-structure, there is no structure either. But for those who do not adopt the view, it remains startling: from an external or prior point of view, it seems to tell us that nature needs to be *entirely re-conceived* [24].

This is in essence the position defended in this book. The reconception we propose, following Lupasco, is to place the Principle of Dynamic Opposition at center of our notion of both structures and systems in a process of change. The essence of van Fraassen's objection here is that the difference between mathematical, uninstantiated/abstract structure and physical, instantiated/concrete structure cannot itself be explained in purely structural terms. The notion of the object is open. The PDO operates in time such that an object cannot be understood in isolation from its process of being formed and forming (*cf* [39]).

Saunders [60] points out that there is no reason to think that ontic structural realists are committed to the idea that the structure of the world is mathematical. Ladyman and Ross [37] argue that no account can be given of what makes the world-structure physical and not mathematical, but this position obviously depends on a form of mathematical structuralism: mathematical structures exist independently of their concrete instantiations. Another view is eliminativist: statements about mathematical structures are disguised generalizations about their instantiations that exemplify them.

Informational structural realism in the context of the foundations of computer science is defended by Floridi [23]. Structuralism has also become popular in metaphysics recently in the form of causal essentialism. This is the doctrine that the causal relations that properties bear to other properties exhaust their natures. This doctrine was refuted by Brenner [8] and above. There are two versions of mathematical structuralism: a realist view according to which mathematical structures exist independently of their concrete instantiations, and an eliminativist position according to which statements about mathematical structures are disguised generalizations about their instantiations that exemplify them. We subscribe to the latter view, taking the occasion to remind the reader that the term 'an instantiation' must be taken in a dynamic sense as part of a process.

16.3 Structural Reality

Structural Reality (SR) is the term used by the mathematician and scientist of information Mark Burgin to describe his proposed General Theory of Structures (GST) [14] as the 'furniture' of the world. Basically, Burgin sees the world as constituted by sets of triads of entities in functional relations with one another, recognizing that the relations refer as in Logic in Reality to entities from all the different levels of reality. As Burgin accomplished also in his *General Theory of Information*, [15] he provides a very complete taxonomy of these entities as 'units' in the sense of our own exploration of the units of reality. It is worth while to list some of the most important triads, Once the existence of a triad has been established, anyone can then 'mine' from his personal perspective, background and skills and explore the purport of the relations in it. Some triads will mean more for some people rather than others, but the properties of all them are grounded in the respective science.

Burgin was led toward mathematics as the basis for developing a GST, but this theory goes quickly beyond pure mathematics. This theory transparently demonstrates that the concept of structure used in contemporary physics and mathematics is essentially incomplete. The argument is similar to the one we have used for logic: standard theories of structure do not encompass all structures that mathematicians, physicists and other scientists *actually* use in their work.

16.3.1 Some Key Structural Triads

Let us, to follow Burgin for a moment, consider some of the basic triads he considers:

1. The dialectic triad of Hegel, thesis, anti-thesis, synthesis.
2. The *co-dialectic* triad of the Chinese, Yin, Yang and their conjunction, the Tao.
3. The triad in mathematics of name, mapping, domain.
4. The Freudian triad of ego, id superego.
5. Existential/philosophical triads: Plato, Spenser, Popper (cf. Chap. 15)
6. Hartmann-Poli triad: material—psychological—social.

It is clear that these and other triads to which Burgin refers and which correspond closely to everyday experience are useful tools for the organization and classification of knowledge. They are thus comparable in form with the systems of philosophers such as Pierce and Spencer-Brown, but differ in that, as Burgin brings out, they have a foundation in science.

16.3.2 Structuralism and Structural Realism

Burgin also provides a review of these subjects, covering much of the same ground and many of the same authors that we have. He states that his primary concern is *structure* rather than the essence and nature of the phenomena studied, and that to investigate a phenomenon means to find and study its structure. As a scientific methodology, *structuralism* is based on the structural vision of reality and on mathematics as the science of structures. It explores the relationships between fundamental principal elements in language, literature, and other fields upon which some higher mental, linguistic, social, or cultural structures and structural networks are built. Through these networks meaning is produced within a particular person, organization, social system, community or culture.

The reader who has been attentive to our method will note another instance of the insertion of a dichotomy between structure and nature or essence for which there is no scientific or philosophical justification. As to structuralism, Burgin provides an adequate hermeneutic description, but, as discussed in Chap. 9, we cannot agree with the origin of meaning in the indicated epistemic fields. Meaning, structure and process cannot be separated in our conception of philosophy.

16.3.3 Other Differences with the LIR Concept of Structure and Reality

There are other significant differences between Burgin's GST and our own view of both of structure and of the reality of which it is a part (see also Sect. 11.3.1). Burgin refers to structures as primarily static entities, even if it is clear that they are subject to *change.* We have emphasized the dynamic, changing aspects of structure, structure-as-process. There is thus a strict parallel between our respective approaches to structure and information. Burgin's General Theory of Information (GTI) is, rather, a Theory of General Information (TGI), epistemological rather than ontological. His General Theory of Structures is a Theory of General Structures (TGS).

Relations in Burgin are the standard ones of equality and inequality, and implications are defined as 2nd order relations, but they apply only between static entities. Change and process are present but by implication. Regarding reality, Burgin gives the following, also triadic outline of three *pure forms* of reality:

1. *Actual reality*, consisting of natural objects and processes, directly or indirectly perceived by the senses and reflected in human nervous system:
2. *Virtual reality*, or *virtuality*, created or simulated by some technological system;
3. *Imaginary or fictional reality*, the product of human mentality.

Burgin then uses, for the conception of potentiality *vs.* actuality, the view of Aristotle: substance is the *potentiality* of things, (which do not exist without form), and form (structure) is the *actuality* of things, where the substance of a thing is "that out of which" a thing is made or as the "constituents" of that thing. Substance in the sense of Aristotle is not necessarily physical. Letters are the substance of words, and Aristotle even calls the parts of a geometrical shape, such as segments, rectangles or triangles, by the name *intelligible substance.* A thing can both *have* substance and also *be* substance. What is missing here, of course, is the sense of dialectic interaction between actuality and potentiality and the concept of emergence from that interaction. It is an expression, in modern terms, of a mathematics based on the Pythagorean Theorem outlined in Chap. 7. In summary, Burgin's book can be seen as a useful introduction to our more active conceptions of both structure and mathematics in reality.

16.4 What Is a System?

The notion of a system, an entity constituted by a multiplicity of other entities acting in concert, is probably deeply embedded in the human psyche, as immediately given from the properties of living individuals and groups. The roles of the entities of a system are different in a no more mysterious way than arms and legs, mouth and stomach are different. It is less easy, however, to define the functional relations between the parts of a system that captures their structure and evolution scientifically and philosophically.

16.4.1 "To Be and not to Be; that is the System"

As an introduction to such a discussion of systems, we refer to the statement by Sir Stafford Beer, the founder of the field of synergetics: "To Be *and* Not to Be, that is the System" [22]. This statement recognizes that a system is characterized by a fundamental duality, and this implies that a way must exist for the system to express this duality. It requires an understanding of *how* a process and its negation can exist at the same time in the same place. Usually, this aspect of the problem is not discussed, but in Logic in Reality the problem is resolved by assuming that one applicable ontological property—being—is predominantly actualized or present and the other—non-being—is predominantly potentialized or absent. But this absence and this presence, which are by the basic principle of dynamic opposition never actualized or potentialized to the extent of 100%, characterize a system, carry information and are causally efficient.

16.4.2 The Lupasco Concept of System

In his "Notions of General Systemology", first outlined in [47], Lupasco set forth the principles underlying all of his work in terms of systems. These principles constitute an alternate expression of the non-propositional, non-truth-functional logic we have outlined that enables an interpretation of the dynamics of cohesion and interaction one observes in real processes. We will first list Lupasco's Axiomatic Statements of systems' principles, first published in English in [8] and then provide further details of their operation as systems theory.

Our ontological approach to reality starts from its simplest aspects, the classic 'one', but where the one is one real object. To this we add a concept of the existence of more than one thing, the 'many' and their subsequent collection into more or less stable groups or systems. Systems are constituted of elements linked by either some internal property, or by the forces or operations that they express of which they are or transmit. The positive side, so to speak, of negation in Lupasco, can be seen in his approach to systems. Lupasco replaces the tautology of standard logic, (1) A c A, implies A by adding (2) A c Â, A implies non-A and (3) A c-bar A, A non-implies A. In other words, the existence of A implies the existence of non-A and the non-implication of itself.

16.4.3 The Lupasco Systemology as a Philosophy of Systems

The concept of a system developed by Lupasco has not previously been designated as a 'philosophy' but this is what we propose here. It certainly would appear first to satisfy the requirements defined by von Bertalanffy for inclusion in a General Systems Theory. Logic in Reality is constituted not by it reference to any particular system or body of knowledge. It rather presents "principles that apply to generalized systems or their subclasses, irrespective of their particular kind, the nature of their component elements and the relationships or 'forces' between them". LIR fits this description; in fact, it is not only a 'universal' principle valid for systems in general, it provides additional (scientific) insights into the relationships and forces involved.

The systems philosopher Rousseau [58], following Laszlo [40] stated that GST provides a formal model of the nature of Nature, but that the *understanding* of the nature of Nature requires "an interpretation of GST involving concrete commitments that systems *philosophy* (italics ours) aims to provide". Rousseau has suggested, many years after Lupasco but almost certainly independently, the introduction of the new term of "General Systemology" to better describe the situation.

Lupasco thus embedded his concepts of both the *requirements* for the formation of systems and their *properties* in an overall vision of origin and development of the universe, a cosmogenesis. The logical laws that all energetic processes at lower levels could follow, leading to systems—systemogenesis—could not be in conflict with laws, at the appropriate level of reality, operating in the cosmos. In Chap. 3, we discussed the grounding of the Logic in Reality in physics and its expression in the Principle of Dynamic Opposition. This Principle 'guarantees', ontologically speaking, the incompleteness of real processes, their alternation between predominantly actuality and predominantly potentiality of their elements, and the emergence of new states or entities from the point of maximum contradiction between the two major sets of elements in opposition. If we take the statement of Laszlo that "there is no separation between forming and being formed", it sounds quite profound, but the reality is better expressed by pointing to the dialectic which avoids any absolute separation and points to the process of exchange between the processes.

16.4.3.1 Aspects of Systems in Morin

Edgar Morin has given his own, highly personal and humanistic readings of systems theory since its codification by Ludwig [2]. He has developed his own logical framework, dialogic, and showed how it can apply to complex phenomena, leading to his fundamental principle of complexity—the ecology of action—in a new epistemology of complexity. The relation between complexity and dialogic is that the latter is one of the principles of the former: the dialogic principle allows us to maintain duality at the heart of unity. It associates two terms that are at the same time complementary and antagonistic.

Another expression linking systems and complexity is that "extremely complex systems (are those) where the part is in the whole and the whole is in the part". One in beyond holism and reductionism in a recursive relational circuit in which parts and wholes "explain" one another, neither term being reducible to the other (Morin's "holographic" principle of complexity). Three terms, for example species, individual and society, also can refer to one another in a circuit that itself is the true system: its three terms are at the same time concurrent and antagonistic. Morin collaborated with Lupasco, and Nicolescu, in the foundation of the International Center for Transdisciplinary Research in 1984, and we may take this date as the beginning of the major effort by Nicolescu, after Lupasco's death in 1988, to develop the critical notions of theoretical transdisciplinarity [55]. Unfortunately, neither Nicolescu nor Morin have reviewed the notions of Lupasco of the real dynamics of the origin and evolution of systems, and the present study is the most detailed one available.

From the standpoint of Wu Kun's Philosophy of Information discussed in Chap. 13, the Morin notion of complexity suffers from being, like the views of systems outlined above, one-dimensional and not sufficiently complex. In our view, the lack of grounding of all of the systems approaches or "ways of thinking" has blocked them as ways of gaining further insights into nature. Thus the "systems thinking" in this case assumes *ab origine* a mathematical structure of reality which it may not have, or have only in the case of simple processes that take place "spontaneously", that is are highly linear.

Morin's system of logic, dialogic, which is often referred to in systems theory bears some relation to that of Lupasco, with whom he was loosely associated. Morin [54] defines a "dialogical principle that allows us to maintain duality at the heart of unity. It associates two terms that are at the same time complementary and antagonistic." However, neither this principle nor the basis for its operation is grounded in physics. The Lupasco Principle of Dynamic Opposition describes not the abstract elements or concepts of complexity, philosophical, political, etc., but the instantiation of the complex elements in reality. Our working hypothesis is, therefore, that Systems Thinking, even enhanced by the concept of complexity, neither further defines information or how it can be both a constituent and a display and a representation of reality. We will therefore look more closely at the concept of Informational Thinking as an alternate paradigm, its dynamics and the LIR logic of its functioning.

16.5 The Lupasco Theory of Systems

Given these assumptions, Lupasco's theory [47] of what constitutes the operative 'law' or linkage resulting in a system is composed of two key relations and an axiomatic determination, as follows:

16.5.1 The Relation of Antagonism

The natural property that in our view determines the possibility of all systems, is the following: systems are not possible if there is no force of repulsion or exclusion between elements which prevents their "agglomeration" into an undifferentiated mass, and not possible if nothing attracts or associates two or more elements; they all fly apart, so to speak. (For us, repulsion, exclusion and dissociation are equivalent terms.) Accordingly, for a system to form and exist, its constituents must be able, at the same time, to both attract and repel one another, associate and dissociate, to integrate and disintegrate. Every system is therefore a function of two antagonistic forces, linked to one another, constituting a relation of antagonism [47]. In Lupasco's notation, a system s is a function of the relation of antagonism R that is a function of two antagonistic dynamisms that can in principle be measured and quantified. The quantity of antagonism is a maximum when the two are equal. One can also say that this quantity measures the dynamics of the system. Every system, be it nuclear, atomic, molecular or at the level of the macroscopic objects of our senses is always, in this view, a function of, in its constitution, this relation of antagonistic or opposing forces. Systems which tend towards an equality of tension, such as the nucleus of an atom, will be more stable and resistant to disintegration than those in which one of the dynamisms is heavily favored over the other.

16.5.2 The Relation of Contradiction

In the dynamic view of reality discussed here, the appearance of a principle of contradiction (or its equivalent) in this line of argument, that is similar to the one derived from considerations of the dualistic aspects of energy as such, supports the idea that there is something of fundamental theoretical importance about contradiction or antagonism, physical dynamic opposition, throughout nature. This consideration leads to a form of argument similar to the first: a system would not be possible if all its constituents or elements were strictly identical, strictly also meaning with relation to their location and configuration in space–time. They would be "confounded" in the same continuity or homogeneity. No system would be possible, either, if all elements were totally heterogeneous, without some degree of homogeneity that would prevent this diversity being neither a system, class or set.

Every system thus implies at the same time homogeneity and heterogeneity. The relation of contradiction is maximal as identity and diversity approach equality. The relation of non-contradiction thus also depends on the levels of both antagonism and identity. As antagonism decreases, non-contradiction increases and contradiction decreases.

16.5.3 The Principle of Antagonism Applied to Energy

The axiomatic determination, which places the previous notions in the systems context, is based on the concept that every system requires the energy involved in its dynamic relations in order to exist. All its constituents and elements, according to the equivalence of mass, energy and information, must consist of energy. (Exceptions are the elements of classical mathematics and all other abstract elements, including semantic elements, such as paradoxical sentences, in which the quantity of energetic interaction is nil.)

Lupasco developed a "logical algebra of energy", applicable to all processes, as a chain of implications expressing the above. The key concept is that every energetic element (or phenomenon) passing from a potential state to an actual state finds itself necessarily, at a certain moment in an intermediate state T, called the T-state, where it conflicts with the antagonistic energy passing from a state of actualization A to one of potentialization P, which can be summarized as follows (cf. also Chap. 3): each of these three elements is an antagonistic energetic duality or alternatively an antagonistic conjunction.

Each is a system, and all more complex systems are generated by such antagonistic dynamisms. One can therefore replace, in the systems formulas, the d values by the values of any element e, adding, to its coefficients of homogeneity and heterogeneity, those of the states of A, P and T. Simple substitution shows that two of these three types of system imply a progressive actualization of non-contradiction, and the third an actualization of contradiction and potentialization of non-contradiction. These logical systems of energy constitute all phenomena or aspects of experience, from microscopic to macroscopic; antagonism and contradictory values are irreducibly constitutive of all real events. Their validity is experimental, and their logic is a logic of antagonism. These systems are the basis for the generation of systems of systems, formally, by the extension of the concept of actualization, potentialization and T-state in two ways: first to the operation of implication itself as used above, and second to the system and its members, operating jointly. In the real world, three forms of system can be identified: the first of these systems of implication corresponds to progressive homogenization, *i.e.*, the 2nd Law of Thermodynamics; the second to the progressive actualization of heterogeneity, a "systemogenesis" of living matter or life, and the third to the situation of maximum contradiction applying to both the quantum and mental levels of reality. Finally, all these chains of systems and systems of systems can be considered as constituting dialectics and dialectics of dialectics, generated by the contradictory nature of energy and its principles as outlined above.

In these principles, we see a basis for von Bertalanffy's concept of continuous multivariable interactions as well as their modern formulation by Heylighen, Hofkirchner and others of metasystem transitions [29]. This was a notion developed by Turchin [62] to describe the process by which control emerges in evolutionary systems. Already in 1962, Lupasco related the concepts of feedback and a non-Shannon type of information. Lupasco noted that any cybernetic system has the capacity for *feedback* or "counter-action", initiated by some perturbation and he describes a control level as follows:.

These principles exemplify the Logic of Transdisciplinarity and point toward the need to consider its implications in the understanding of the evolution of real complex systems. In particular, in my view, it is the above systems view that insures the foundations in reality of the recursive aspects of complex processes described as loops [32].

Pouvreau [57] has studied the importance given by von Bertalanffy to mathematics in the GST. In my view, although standard mathematics is necessary, it is not sufficient to capture all of the non-Boolean, non-Markovian aspects of systems. This is where the Lupasco approach may be useful. However, to make a satisfactory comparison, a mathematization is required, in more familiar terms, of Lupasco's calculus of chains of chains of implications as describing the evolution of real processes.

16.6 Contemporary Systems Theory

16.6.1 von Bertalanffy and Laszlo

The analysis and discussion of complex scientific and social phenomena from a systems standpoint, starting with the work of Ludwig von Bertalanffy in the mid-twentieth Century, [63] has proven to be a major addition to knowledge and the understanding of the world. As Brenner noted, the authority for many systems is limited to a grounding in forms of complexity theory. These in turn fail, in our view since they themselves embodied principles of standard bivalent or multi-valent truth-functional logics inappropriate to real processes and change. We will begin by briefly reviewing the basic General Systems Theory (GST) of von Bertalanffy [2] and the derivation from it by Laszlo of a first Philosophy of Systems [39]. In his foundational work, von Bertalanffy defined systems simply as "complexes or sets of elements standing in interactions or interrelations," but his GST was supposed to be capable of giving exact definitions of and even quantifying complex concepts. von Bertalanffy stated himself, however, that no axiomatic basis was provided for his theory of systems. He failed to see beyond differential calculus, as does current so-called Dynamic Systems Theory, as its basis.

In one of his last articles, "The History and Status of General Systems Theory" [3], von Bertalanffy wrote that ultimately all the boundaries of real objects are dynamic rather than spatial. "Hence an object (and in particular a system) is definable only by its cohesion in a broad sense, that is, the interactions of its component elements".

This is an extraordinary statement, which goes beyond the concept of information as 'constraints' in the theory of Kauffmann mentioned in Chap. 13. It means that a system is characterized by *change*, in particular, but limited to, change in and of its boundaries. The Principle of Dynamic Opposition offers a way of conceptualizing the progression of the movements of systems from more or less actual to more or less potential and vice versa. Of course, objects and simple systems do exhibit uniform, Newtonian motion, but as a limiting case of emergent simplicity. von Bertalanffy also stated that development of GST could lead to the unification of science, a science of the future that could play a role similar to Aristotelian logic in the science of antiquity. At a Symposium on Lupasco, Brenner showed that his extension of logic to encompass reality effected a metalogical rejunction, restoring logic to its original role, in antiquity, of a science of nature [10]. A major conclusion of this study is thus that the logical approach of Lupasco, which we have designated as Logic in Reality, might provide a unique framework for accomplishing the task that von Bertalanffy set for his General Systems Theory. Logic, in the extended Lupasco sense, could be an integral part of systems thinking.

16.6.2 Systems Theory and the Role of Logic in Reality

By taking a step away from debates about systems, emergence and complexity, it becomes fairly obvious that these are not independent concepts but that their definitions are closely intertwined, a term frequently used by Hooker and others. Another major problem is that much of systems science and complexity theory, like that of communication, is cast in epistemological terms, referring to more or less abstract observers and static models. There is one concept of philosophy that could be considered a 'competitor' to or of Philosophy in Reality. It is the Philosophy of Complex Systems defined by, among others, a transdisciplinary group of major contributors to theories of computational philosophy, model-based reasoning, system dynamics and paraconsistent logic including Cliff Hooker, Dov Gabbay, Paul Thagard and John Woods.

There is a clear antagonism, it is perhaps justified to use the term non-metaphorically between the work of this thinkers and ourselves. Although the former is based on and includes the most recent scientific references, for example to non-linearity, it is cast in a framework of classical concepts of logic and interaction separability. As we have done in the case of the philosophers considered as precursors to Lupasco, we review briefly the work of the above four authors. All that we have 'asked' of them is a more or less clear definition of what their views of philosophy, complex and systems amount to. In particular, what is the relation of the Philosophy of Complex Systems to the Science of Complex Systems *tout court*?

If we go back to the criteria enunciated by Lupasco for the formation and on-going existence of systems, summarized above, it should be clear that they are necessary but by no means sufficient. Systems, by which we will mean essentially complex systems, and ones which have critical properties and whose evolution follows dynamic laws and principles of which the Principle of Dynamic Opposition is essential. However, we find that other essential laws and principles are compatible with LIR and the PDO and, used together, add to the explanatory power and hence the philosophical weight of the original philosophy of complex systems.

In his discussion of what he prefers to call, following von Bertalanffy, a *general* systems theory, [39] reemphasizes the role of complexity—"the complex theoretical enterprise of systems philosophy". He sees all the 'new disciplines', including GST, cybernetics, information, decision and game theories as parts of a new natural philosophy, as mentioned in Chap. 13. But this new natural philosophy, for Laszlo must include new concepts and models "adequate to deal with the biological, behavioral and social universes … isomorphisms which in their formal structure applied to diverse realms of phenomena, function in an interdisciplinary and integrative manner." We see Logic in Reality as such a new concept, "the world as organization", but with the additional principle of dynamic opposition operative throughout nature, its science and its philosophy.

Let us reiterate the PDO in this context: it is first of all a *scientific* principle that all complex systems move from states of predominant actuality to predominant potentiality and vice versa, alternately and reciprocally. The basis for a philosophy of change is grounded in this principle. Like Lupasco, Laszlo sought a physical ground for the existence of the most complex, that is, living systems and found it, as echoed by Kaufmann and his colleagues, in the sets of constraints which produce "relatively stable, non-probabilistic sets of energy-flows". In a most interesting way, also reminiscent of the energy-based approach of Lupasco, Laszlo ties the concept of natural-cognitive systems to a normative ethics defined as a state of sufficient information input. "The physical aspect of natural-cognitive systems which correlates with optimum value (for society) is a state of sufficient energy import.

It is nevertheless necessary to repeat that at the level of a specific process, a feeling for that which is absent, partially or completely, as in the definition of information of [19], is essential for a grasp of its probable evolution. Let is therefore see where else in the philosophy of systems one may find what everyone would agree is a desirable synergy between scientific and philosophical perspectives.

16.7 Systems Science and Complex Systems

Systems science developed after the appearance of von Bertalanffy's General Systems Theory from the interaction of standard information theory and cybernetics. One contemporary definition of systems *science* was the following[1]: "A new discipline that combines theoretical, practical and methodological approaches relative

[1]French Association of the Science of Cybernetic, Cognitive and Technical Systems (*AFSCET*), 1994.

to research topics that are recognized as being too complex to be accessed in a reductionist fashion, and that pose problems of (1) boundaries, internal and external relations, structure and laws or emergent properties characterizing the system as such and (2) modes of observation, representation and model building or simulation of a complex totality." The reader will recognize in this definition issues that have been addressed in a logical fashion that hopefully is non-reductionist.

Systems science overlaps with complexity science, in that the latter is based on a definition of the complex systems that are the objects of systems science study, albeit from a less computational standpoint. A complex system is loosely defined as constructed by a large number of simple, mutually interacting parts, capable of exchanging stimuli with its environment and of adapting its internal structure as a consequence of such interaction. The non-linear interactions involved can give rise to coherent, emergent complex behavior with a rich structure. Key concepts in complexity science are, for example, the coexistence of diversity and stability, for which LIR provides an interpretation. Complexity science also looks at the dynamics of systems in transition regions of self-organized criticality. Schematic systems are used to investigate self-organization, but without the grounding in dynamic opposition and potentiality that I have proposed as necessary to explain the functioning of such organization, as well as the ambiguity in the term 'self'-.

As stated at a Congress in 2005,[2] the major objective of systems science today is to provide a consensual, transdisciplinary approach to the increasingly complex problems faced by workers in all areas of society, with the laudable intention of 'placing man at the center of its preoccupations'. Models and strategies are designed to develop effective operational tools as well as conceptual and philosophical ones.

Systems science includes aspects of such a diversity of sciences and disciplines that makes it difficult to capture in a few words. One example is the science of ago-antagonist systems (SAAS), developed by Bernard-Weil [1], which bears as superficial resemblance to LIR. SAAS purports to identify and take into account, in concrete systems, pairs of elements that are both conflicting and cooperative, either at the same time or alternatively. As we show here, it is necessary to specify more completely what is meant by 'at the same time' or 'alternatively' and to look for the origins of both conflict and cooperation in the potentialities of the systems' elements. This theory, like many others in systems science, has practical applications as a step in understanding the role of pairs of antagonists in living cells, the human body, business enterprises, etc.. These applications are most interesting but will not be developed further here.

The views of Wu and LIR of real-world processes as emergent are not epistemological but ontological, similar to that of Hofkirchner. As Hofkirchner stresses in his discussion of information and computation, "only if computation is meant as a self-organizing process involving emergence in a *non*-epistemological sense can it do justice to the generation of information [30]."

---

[2] 6th European Systems Science Congress, Paris, September 19–22, 2005.

16.7.1 Transdisciplinarity and Systems Thinking

Systems Thinking is been defined primarily as an approach to problem solving, by viewing problems as features of an overall system which are best understood in the context of relationships with each other and with other systems, rather than in isolation. In principle, systems thinking techniques may be used to study any kind of system—natural, scientific, engineered, human or conceptual. The difference and advantages of systems thinking *versus* traditional forms standard analysis is clear.

In practice, two things are missing that are necessary, in our opinion, to give systems thinking the necessary philosophical depth: one is an adequately grounded definition of a system in the first place, in which contradictorial interactions are present constitutively and second a proper conception of how qualitative properties of systems may be expressed.

Hammond [28] summarized developments of systems thinking since the establishment of its categories of application—technology, science and philosophy—by von Bertalanffy. We can all agree today with his conception of General Systems Theory that emphasizes a more holistic and humanistic approach to knowledge and practice, while deploring the fact that such an approach has not materialized.

We consider this article a very accurate reflection of the "state-of-art" of systems thinking. Starting with von Bertalanffy, the author points to many significant contributions to a systems view, which she defines: "The systems view reinforces a constructivist orientation to knowledge as a dialectical, pluralistic and participatory process that emphasizes the importance of mutual understanding, meaning and values." All of the well-known difficulties in achieving such goals are indicated, the fragmentation of knowledge, the use of systems thinking for social control and that indeed "we have yet to discover the appropriate approach to systems". We, of course, believe that the approach outlined in this book offers one.

The approach of one of the most influential systems thinkers, Senge [61], amounts to not much more than an exhortation to look at the "whole", an organization as a holistic, dynamic process and to balance short-term and longer-term cost–benefit parameters. This is fine as far as it goes, but no one can say today that it goes far enough.

Hammond's statement of objective nevertheless merits repetition here: "Perhaps the primary challenge for systems thinkers in the twenty-first Century is to find ways of integrating the insights emerging out of the various branches of systems thinking over the past fifty or sixty years." We propose that it is Informational Thinking, which must include something like our notion of information-as-process (Chap. 13) that suggests new ways of accomplishing this integration, despite the difficult *de*-fragmentation of knowledge that must take place.

System dynamics is an approach to understanding the behavior of complex systems over time. It deals with internal feedback loops and time delays that affect the behavior of the entire system. While the approach is in principle applicable to ecosystems and political systems, in fact it can only be used for the most mechanical, quantitative features of such systems, capable of being modeled in causal loop diagrams. Accordingly, system dynamics adds nothing fundamental, as a form of systems "thinking" to the understanding of information or other complex phenomena as such.

With hindsight, the notion of applying systems theory to the solution of practical problems, for example, those of organizations, is neither more nor less than common sense. The unfortunate state of the world, however, is a demonstration that such solutions have been limited in scope. As a systems as well as information scientist, in his major book on the relation of systems, semiotics and information, [11], clearly understood the limitations of a systems theory such as that of Niklas Luhmann, in which the subject is lost in a functionalism that is not adequately grounded in external reality and a proper philosophical framework.

16.7.2 Complexity

It is often suggested that notions of complexity provide substantial additional insight into the nature of (systems and real processes) information. it turns out to be easy to show that current relatively rigorous notions of complexity are all tied back to computer science, specifically, algorithmic information theory, as in the Kolmogorov complexity of an informational object [18]. We believe, however, that none of the existing approaches based on systems or standard computational notions of complexity are adequate to define the unique ontological status of information.

While the lack of formalism in the complex systems approach serves to differentiate it from strictly computational ones, the lack of foundations diminishes the value of its humanistic and ethical characteristics. The only complex systems studied in detail seem to be those simple enough to be computationally tractable. If the essence of complexity is *non*-computability, then the right of such systems to be called complex is open to question.

A far more appropriate methodology is to relate complexity to simplicity, dialectically, as situations in which simplicity can emerge from complexity and vice versa. This is, of course, where concepts from Logic in Reality are useful in insuring some coherence in the discussion of the dynamics of the changes involved. No reason is given as to why such things as complex processes can and do emerge in nature, but it might be essential for the understanding of how they operate what the basis for the emergence might be.

16.7.2.1 The Philosophy of Complexity

In an article devoted explicitly to this subject [45], Chris Lucas attempts to define the congruence between organic or holistic ways of thinking—systems thinking, organic thinking and connectionist thinking relating to non-reductionism: the system is defined by its connections and not by its parts. All of the approaches that these terms imply are useful, more so if they can refer to non-static processes. This philosophy includes the concepts of autopoiesis, which we have criticized as well as limiting the discussion to *Boolean* networks to the exclusion of non-Boolean structures. Interactions are the source of new emergent 'modules' in the system via autocatalysis or self-production. For us, these last terms beg the question as to the forces involved in these operations. It is not adequate to say that "two-way structural interplay is common in complex systems", without defining the process of the changes in progress as they move from potential to actual.

We see complexity in operation not only far from equilibrium but close to it, *pace* Prigogine. Some of Lucas' formulations however, are felicitous and deserve analysis since they refer to at least some underlying dynamics. Complexity certainly is involved as "social systems replicate to create additional systems", although there is no need to limit the origin of their emergence to "copying errors", However, we agree with Lucas that "this discards the fixed-in-time assumption of most science." And further, "the co-evolutionary or epistatic nature of interrelated systems requires us to take a contextual approach, studying the dynamics of interactions rather than the static makeup of parts studied in more conventional science." This is of course the scientific side of our Philosophy in Reality.

16.7.3 Gianfranco Minati: Logical Openness

In his most recent work, the systems scientist and philosopher Gianfranco Minati and his school at the University of Milan has made what we consider a unique critique of his own theory. While not referring directly to the logical system we have proposed here and elsewhere, Minati's approach includes many concepts which both support and are supported by our logical and philosophical stance. In their systems approach to logical openness [53] and emergence [53], Gianfranco Minati and his associates show that the usual picture of systems is too limited to deal with logically open systems, in which the internal state of the system, as well as of the environment, needs to be taken into account. These are the systems which are the main subjects of Logic in Reality.

The major strategy of Minati is to establish a principled role for the observer that defines epistemological levels of logical openness. However, for a description of the dynamics of complex phenomena, especially change, an account in which the observer has an epistemological role needs to be supplemented by an ontological non-conceptual account in which the rules governing the real interactions between entities, including the observer, are also applied. In LIR, the observer is in an *ontologically prior* dynamic relation with the observed of which he is a part. One such relation is that between the scientist and his experimental configuration.

The sequence originally proposed by Minati et al.—REAL SYSTEM → REDUCTION BY OBSERVER → MODEL 1,2,3 … → EPISTEMIC CHANGE IN OBSERVER—is an epistemological construction in "epistemic" space; nothing has "happened" to the system. This sequence will now be related to emergent change in a broader sense that includes both epistemic and ontological change.

16.7.4 Toward a General Theory of Emergence

Minati discussed an approach to a General Theory of Emergence (GTE) in the context of the concept of generalization itself [52], but in our view this early theory does not deal *directly* with the physical, cognitive processes of change. Minati sees emergence generally as the emergence of new properties and operators whose validity can be extended from one domain to another. GTE focuses on collective phenomena that establish systems and is better described as a meta-theory that deals with.

- the establishment of systems through emergence (which is also a process) but where the dynamics do not relate only to processes with respect to time but also to multi-modeling, epistemological hierarchies of logical openness and the acquisition of new properties, *e.g.*, those emerging from lower-level structures and systemic properties having theories as particular cases;.
- correspondences between models and representations of phenomena considered emergent;
- identification and classification of possible non-equivalent kinds of emergence, such as biological and physical.

The nature of the differences between Minati's GTE and LIR are clear. First of all, a system for Minati as discussed in a companion paper [52] is the *model* of a phenomenon, not the phenomenon itself and its evolution. The effective modeling of such systems is not based on separate models, but rather on integrated models able to represent interactions between levels and processes of emergence, in which Minati's conception of the dynamic usage of models (DYSAM) plays a key role. This process of acquisition of emergent properties and its codification in a GTE may offer a wide new variety of epistemological approaches (Minati's term) applicable to life, mind and development in general. In contrast, the above LIR view of real-world processes as emergent is not epistemological but an ontological one, similar to that of Hofkirchner. As Hofkirchner stresses in his discussion of information and computation [30], "only if computation is meant as a self-organizing process involving emergence in a *non*-epistemological sense can it do justice to the generation of information."

Minati said "we have phenomenological approaches, but not a theory able to model embedded processes of the emergence of properties". LIR, from Minati's perspective, is like theories of quantum mechanics, dissipative structures and phase transformations which are not themselves GTE's. He asked what is missing to generalize *them* (emphasis ours) in such a way as to extend them and make them more general. We considered LIR is a candidate for the "theory able to model processes embedded in reality". Further, the process of development of a GTE is itself a process of *change* in LIR terms. To a certain extent, the two approaches informed and complemented one another. The LIR Theory of Emergence interacts with GTE, and with similar theories, with either ontological or epistemological aspects of the processes studied predominating at any times. However, although the ontological priority lies in LIR, LIR is no more able to completely define the domain of systems than it can of physics.

In their most recent book, *From Collective Beings to Quasi-Systems* (2016), Minati and Pessa present a new theory that is a major contribution to Systems Theory that from our standpoint facilitates positioning Logic in Reality in current thought, as we will now see.

16.8 Logic in Reality and the New Systems Theory of Minati

16.8.1 Beyond von Bertalanffy

With the above background, let us compare in more detail our three systems thinkers, Laszlo, Minati and Hofkirchner, mentioned previously in connection with his approach to the science and philosophy of information. We will begin, however, by again reviewing the basic General Systems Theory (GST) of von Bertalanffy and the derivation from it by Erwin Laszlo of the first Philosophy of Systems.

In his foundational work on General Systems Theory, von Bertalanffy defined systems simply as "complexes or sets of elements standing in interactions or interrelations," but his General Systems Theory (GST) was supposed to be capable of giving exact definitions of and even quantifying complex concepts. Von Bertalanffy stated himself, however, that no axiomatic basis was provided for his theory of systems. He failed to see beyond differential calculus, just as current so-called Dynamic Systems Theory, as a basis for systems theory.

In their initial approach to systems and emergence, Minati and his colleagues [53] show that the usual picture of systems is too limited to deal with logically open systems, in which the internal state of a system, as well as its environment, need to be taken into account. As noted above, the strategy of Minati was to establish a principled role for the observer that defines the epistemological levels of logical openness mentioned.

However, for many complex phenomena whose description and overall dynamics have not been captured by current theories, such as information, change, intentionality, etc., an account in which the observer has an epistemological role needs to be supplemented by an ontological non-conceptual account in which the rules governing the real interactions between entities, including the observer, are also applied. In LIR, the observer is in an *ontologically prior* dynamic relation with the observed of which he is a part. One such relation is that between the scientist and his experimental configuration. In their earlier *Collective Beings* [50], Minati and his colleagues analyzed in great depth the characteristics of the formation and evolution of entities constituted by multiple individuals from the quantum to macroscopic levels of reality, from bosons to flocks of birds and swarms of fish and insects.

From our LIR standpoint, these 'collective beings', real entities are within natural science and hence discussable from a natural philosophical standpoint as they instantiate a dynamism of movement, *e.g.* change of position. However, the philosophy of such entities is limited to their epistemological designation. There is no additional meaning to the interactions and changes in interactions of the members of the being as a consequence of those changes. That one of small fish at the edge of the swarm can be and generally is eaten by a shark or other predator has meaning for that fish, but not for the rest of us.

The situation with crowds of human beings capable of self-reference is more complex, as might be expected. The structure of the interaction between the group or crowd and the individuals constituting it also requires attention. Brenner saw the group-individual interaction as a process to which Logic in Reality is applicable.

16.8.2 Toward a Post-Good Old Fashioned Systems Theory

As discussed in Chap. 3, Lupasco effected an ascent from a static, binary logic of propositions to a dynamic logic of processes, which we refer to as the logic in and of reality (LIR). In his *From-Collective-Beings-to-Quasi-Systems (FCBQS)*, Minati has now effected a similar ascent from a binary static conception of systems, designated by him as GOFS, 'Good Old Fashioned Systems' (GOFS) to define the contours of a new post-Bertalanffy theory of systems, designated as post-GOFS. Minati writes that an ontological approach might be intended as the search for "*fundamental systems*, if not *the* system, to be then considered in different non-equivalent *actualizations* into *real* systems. Is such an approach still viable? *Can we look for the general network?*".

We join Minati in this effort in the following way: such an approach may now be considered a more appropriate one for collective systems with structural dynamics and where coherence(s) and related properties are *invariants*. In our view however, this holds also for *non*-collective systems, even those with only a few members but involved in complex relationships.

In order to establish the maximum useful synergy between the two approaches, one needs to differentiate between Minati's use of the term ontology and ours. Minati notes Brenner's definition as the philosophical study of the nature of existence, of being and that it is intended in philosophy as the science of what is *currently existent*, "the study of the kinds, structures and properties of objects, events, processes and their relationships in every area of reality. We do not agree, however, that ontology is "a part of the branch of philosophy known as *metaphysics*." Ontology for us deals with the existence of entities but not their categorization, grouping within hierarchies or according to similarities or differences related to different kinds of applications. The domain of ontology for us is the extant domain and not a part of a discipline that studies structures of concepts and meanings pertaining to that discipline.

Having stated that, we still conclude that a notion of the dynamic logical structure of processes and their evolution, that is, change, requires coupling with a notion of systems, as was recognized by Lupasco himself. The post-GOFS theory of Minati completes LIR by adding a formalism that, in the LIR spirit, includes a reference to its own limitations. While LIR focuses on establishing the operation of the principle of dynamic opposition between two entities in terms of three parameters of degrees of actualization, potentialization and emergent entities (T-states), Minati's theory permits the consideration of many systems at the same time interacting in different ways, both simple and complex, even if any individual interaction follows the Lupascian principle of dynamic opposition (PDO). Let us now look at some of the specific ways in which the Minati and Lupasco approaches relate to and support one another.

In the previous section, we began to show that standard concepts of what a system is and how it can change are not sufficiently complex to characterize it scientifically, logically and philosophically. In particular, they fail to distinguish between epistemological and ontological aspects of systems and how the latter may interact. Our first suggestion was to propose a description of a system as constituted in accordance with the principle of dynamic opposition of Lupasco. To repeat, we consider that the Lupasco principle of dynamic opposition founds a systems theory that is an essential aspect of any philosophy in/of reality. Everything of interest and importance to knowledge is a dynamic system with both epistemological and ontological properties and characteristics.

The systems theory of Lupasco, however, did not develop the principles of his systems approach to nature and reality to account for the variety of entities and forms, static and dynamic, emerging and disappearing all at the same time. There are other quite extraordinary correspondences of Minati's with our current approach to logic and philosophy in reality. In (2016), Minati showed that one must go far beyond a simple systems/non-systems dichotomy. This statement implies the existence of entities that are not, not yet, or even no longer systems, and it requires maintaining an active relation between the all of these concepts to place the dynamics of systems in proper perspective.

It should be obvious that we can address only a small fraction of the topics subsumed under the title of systems theory or systems science in the literature in general. We have therefore concentrated in this chapter on the properties of systems in the post-GOFS theory of Minati. In general, these will be properties of *changing* systems, a new window on the eternal subject of change in reality.

16.8.3 From a Reductionist to a Post-Reductionist Systems Theory

In previous studies, Minati and his colleagues introduced the concepts of Multiple Systems, Collective Beings and the DYnamic uSAge of Models (DYSAM) to deal with these systems phenomena, focusing, in particular, on collective behavior in swarms, flocks, herds, traffic, crowds and industrial districts. A wide variety of approaches on these themes have been published, but they all fail to describe completely the complex operators driving changes in the systems. Minati has stated explicitly that "the question about the existence of some simple underlying laws of nature (such as the principles of thermodynamics) that produce the whole variety of the observed phenomena we discussed is still to be uncovered." This comment attracted our attention because the Principle of Dynamic Opposition (PDO) is in our view part of these underlying laws that has not been recognized as such but could be related to the outstanding problems of systems theory. These include:

- modeling and recognizing a phenomenon as emergent;
- inducing the emergence of collective behavior;
- acting on collective emergent phenomena to affect their properties;
- merging different collective emergent phenomena.

As defined by Minati, typical properties of systems include self-regulation, feedback and complex functionalities. Such properties can also be studied in a non-systemic conceptual framework and adapted to describe new states reached by systems. They are considered when dealing with reproducible phenomena obeying classical evolutionary dynamics, intended as a description of changes occurring in entities conserving their structures in space and time (a typical case is, e.g., motion in mechanics and even in organizations).

This reduced application of systemics is unable to deal with complex phenomena such as the properties acquired by collective systems with changing structures. In these cases new conceptual problems, such as that of coherence, arise, and it becomes impossible to deal with them using GOFS, except in particular cases. However, GOFS is suitable to deal with processes of acquiring and maintaining the same or only a few, fixed systemic properties. On the contrary complex systems continuously acquire new (often delocalized) and sometimes *in*coherent sequences of properties.

Simple extensions of GOFS are not sufficient, primarily because of the novel nature of problems often formulated in classical or in GOFS terms, but which require different and more suitable representations and approaches. These problems include those related to dynamical multiplicity and transformation, the latter being viewed as a continuous acquisition of even simultaneous but non-equivalent properties.

It should be clear from this first brief introduction to Minati's work that Logic in Reality, including its concept of the basis of emergence at a point of maximum interaction between actual and potential states, does not address all these problems. It is not in and of itself a systems theory. It is, however, as will attempt to show, an indispensable component of the description of the evolution of complex systems.

16.8.4 Properties of a New Systemics. Non-separability

Between, Pre- and Quasi- are for us three of the most important functional concepts which ground Minati's post-Bertalanffy's Systems Theory, introduced in (Minati 2016). These terms will be capitalized without scare quotes in what follows. First of all, Between does not refer to binary entities such as static equilibria, scalar averages or halves. Appropriate entities are emergent third terms, as in Logic in Reality, creative social and economic compromises with potential for further development, and Minati's models of dynamic equilibria. The meaning of Quasi is not limited to almost or approximately. All three terms for Minati, refer to properties of open systems in a Logic of Openness.[3]

The terms also refer to Incompleteness; lip service is also paid to this term without comprehension of its implications for and in philosophy and science. We discussed in Chap. 8 the characteristics of some inorganic and simple organic chemical reactions which 'go to completion' to all intents and purposes. The backwards process from products to reactants has a high activation energy. Hence its rate constant under normal conditions of temperature and pressure, and in the absence of high inputs of energy, that is vanishingly small. This is however, not the case for many of the chemical reactions in biological and especially cognitive systems in which the forward reaction can be reversed by small inputs of energy.

For Minati, a "*new systemics* should be the place where the *objects* under study are the systemic properties themselves, their emergence, the dynamics of their dynamics, their correspondence, their coherence and their possibly non-homogeneous occurrence." Changes towards a new systemics will involve a conceptual shift from Multiple Systems (MS) and Collective Beings (CB) to what Minati now defines as *quasi-systems* (QSs). The latter are entities in which systemic properties are partial, sometimes regular, but in other cases partially lost. He introduces the concept of *pre-property*, which can be related to the potentialities present in complex systems in our logic. QSs are useful for studying changes in, and the recurrence of, properties at various levels of emergence and inter-level dynamics. He revisits the classical systemic concepts of inter- and trans-disciplinarity, by resorting to specific disciplinary examples.

Many aspects of such a new systemics involve explicit reference to contradictorial relations for which LIR provides the logical framework, in particular non-separability and separability. The discussion we develop identifies dualisms that are epistemic in nature, such as true and false, and the ontological ones of greater interest mentioned by Minati the *world* between macro and micro, open and closed; uncertainty; and the nature of phenomena occurring *within* degrees of freedom.

16.8.5 Quantum Field Theory

From the perspective of LIR, The effective quantum field (EQF) is an adequate description of the structure of the quantum world (Chap. 6). It can be seen as a self-sufficient, self-included middle that does not require a further independent underlying substrate. Indeed, it has been referred to as instantiating 'self-duality' [26], but it is not easy to see what this means physically. One proposal defines that self-duality as the structure of the effective field, and it would appear to involve fundamental dualities between, for example, electricity and magnetism. Its dualities can be related to those of string theories (T-duality), developed in the 'second super-string revolution' by

[3]Bueno and Costa [13] propose a concept of quasi-truth which to a certain extent 'discovers' truth as a non-purely epistemic property.

Witten [64]. This involved the introduction of new symmetry operations and sets of dualities, e.g., the strong–weak duality, where strong and weak refer to the strong and weak forces operating between heavy (nuclear) and light particles respectively. It is still impossible to say today what the physical status of string theory is, but from a mathematical standpoint, the dualities seem to be extremely deep and indicate types of structure that point towards the dynamic opposition of LIR. A two-level description of Nature was soon recognized in Quantum Field Theory (QFT) as the duality between field and particles' [6], and as indicated in Chap. 7, led to the basis for LIR.

In a manner reminiscent of our derivation of the Principle of Dynamic Opposition and its *non*-Boolean logic from quantum mechanics, better quantum field theory (QFT), Minati views QFT as a theoretical framework incorporating systemic principles, dealing with the quantum aspects of a new systemics or even a quantum systemics *tout court*. LIR does not deal with the mathematical physics of quantum mechanical concepts such as entanglement, non-classical interactions from the properties of the vacuum, long-range correlations and quantum decoherence, collapse, and other microscopic quantum effects. LIR does deal with the non-classical interactions operative at macroscopic levels (Chap. 6), emergence and information. Similarly LIR joins the new systemics with regard to topics such as modeling phenomena occurring *between* levels of emergence (are they classical or non-classical?) and the coexistence of classical and non-classical representations recalling wave-particle duality.

In this dual perspective, properties of systems are studied with regard to their partiality, instability, uncertainty and incompleteness as being of equal importance as their 'positive' counterparts. Post-GOFS considers phenomena of multiple emergences where the dynamics relate to changes of levels or of the kind of emergence.

Minati's adds additional levels of complexity to the Nicolescu presentation of trans-disciplinarity [56] in relation to the topological behavior of networks. Network properties can be reasonably considered as trans-disciplinary properties. Minati stresses the importance of a new inter- and trans-disciplinarity approach as fundamental to establish knowledge, or better, meta-knowledge, for the knowledge society by acting upon language and adopting constructivism as a general framework.

The relation of LIR to Minati's new approach to a post-Bertalanffy or post-Good Old Fashioned Systems (GOFS) is our major interest, we should not forget that GOFS had already made substantial contributions to knowledge, for example by seeing the limitations of von Bertalanffy's use of the language of ordinary differential equations and its replacement by Poincaré's so-called 'qualitative theory' of differential equations, the basis of modern dynamical systems theory. GOFS theory incorporated the need to take pairs of interacting entities into consideration; relevant aspects of statistical mechanics as a new way for dealing with multiplicities; and, by relying on probabilistic methods, showed the inadequacy of absolute determinism, characteristic of traditional classical mechanics. On the issue of multiplicity, GOFS adopted a phenomenological approach, inspired by the methods of Mathematical Ecology and of economic dynamics.

In a general way, we can say that Minati responded to the obstacles facing advances in systems theory in a way similar to Lupasco in his philosophical logic of science. These include giving adequate importance to problems in which information is incomplete or imprecise; avoiding approaches which depend on assumptions of absolute precision, lack of ambiguity, time independence and observer independence; including views of macroscopic structures as counterparts of suitable microscopic dynamics The approach of looking for defining with absolute precision, independently of the observer, is replaced instead by examination of the properties of reality as presented.

Thus, in the cases with which we are concerned possessing qualities in opposition such as inaccuracy, imprecision, ambiguity and non-exhaustiveness, Minati and we agree that different assumptions and strategies are often required. These qualities may be understood as aspects of processes possibly converging towards different levels of accuracy, precision, exhaustiveness and lack of ambiguity, which characterize arrival points, attractors or temporary points of stability. Moreover, in the systems supporting these processes, the dynamics of their levels may be symptoms of the establishment of novel equilibriums and pre-properties (see below). Minati shows that further interesting symptoms are the ways in which the degrees of freedom are respected and used during the process. Violations of the degrees of freedom may be precious sources of information about the ongoing process in order to discover whether it might be induced and not just forced towards a unique, i.e. 'correct', evolution. Within these contexts it is convenient to make use of diversity, rather than of homogeneity, by resorting to suitable strategies. This is one of the few documents (another is by Hofkirchner) that make explicit reference to the *need* and not merely the tolerance for diversity in science and philosophy.

16.8.6 Change and Non-change. Symmetry

Let us repeat that our objective is to give the proper ontological value to both change and non-change. In the context of symmetry, the latter includes the beautiful forms of inorganic and organic crystals, including those of their X-ray pictures. This kind of non-change, of course, has nothing to do the brutal, inhuman principle of *ne varietur* (this must not be changed), that exists only as a support of totalitarianism.

A reality constituted by changing processes includes those possessing states that are temporarily stable and unchanging to all intents and purposes—objects. A rock is such an object, although in its history, or world-line, it was at some time a liquid, molten, without definite contours. One of the most important classes of properties of objects is that of symmetry, defined most commonly in terms of *non*-change, that is, invariance under the operation of a relatively small group of geometrical transformations. The most familiar members of this group are reflection or mirror-symmetry; rotation (rotational symmetry) and repetition (translational symmetry). As explained on the *Symmetrion* site of the International Symmetry Association, a less familiar but much more general meaning of symmetry exists (https://symmetry.hu): "In this generalized meaning, one can speak about symmetry if

- under any (not certainly geometric) kind of transformation (operation)
- at least one (not certainly geometric) property
- of the (not certainly geometric) object is left invariant (intact)."

This generalized meaning of symmetry makes it applicable to material objects in nature, including the products of mental operations. In addition to geometric symmetries of shape (morphological), one can discuss symmetries and asymmetries of function, *e.g.* in human beings; the gauge symmetries of fundamental physical phenomena, and the complex properties of cognitive constructions like those of artistic creations such as painting, sculpture and literary-poetic form.

In order to see more clearly the operation of symmetry in change, we restate the generalized meaning in more positive terms: "One can speak about symmetry if

- under an energetic transformation (change) of the elements of a process
- at least one physical scalar, vector or tensor property (magnitude, direction)
- of the elements is left invariant *for a finite time.*"

The purpose of this generalization is to extend the concept of symmetry to such complex cognitive phenomena such as human behavior, creative acts and informational processes. Readers who have followed our discussion of harmony in Chinese thought will also find it in symmetry. Symmetry as so defined comes into a closer relation to its opposite or countervalence, non-symmetry, for which it is also necessary to define three key forms; these are not totally disjoint, but co-exist or accompany symmetry. György Darvas, head of the Symmetry Association, has formulated them as follows on the reference site:

Asymmetry	The lack or absence of symmetry.
Dissymmetry	An object is symmetric in its main features, but this symmetry is slightly distorted, as in some Islamic ornaments and designs.
Antisymmetry	The object is symmetric in one of its properties, but another property changes to its opposite (*e.g.*, a chess board).

It is thus possible to identify two forms of change of symmetry: (1) going from one form of symmetry to another, *e.g.* rotational to translational symmetry; (2) going from symmetry to one of its three negations, that is, forms of non-symmetry. The existence of the potentiality and actuality of mental (creative) and/or physical movement from one to the other this implies that a principle of *incompleteness* applies since only formal (epistemic) changes are instantaneous, better timeless. Formal here means theoretically defined and does not refer to real forms capable of undergoing physical change. We cannot go further here into this most important subject at the interface of science, philosophy and art, and we refer the reader to the *Symmetrion* site.

16.8.7 The Question of Computability

The modern systems theory, 'soft computing' tools had already been introduced to deal with complex systems characterized by imprecision, uncertainty and stochasticity. As Minati points out, "the description of these systems is often approximate and allows only a partial knowledge of what actually occurs. These soft computing tools include neural networks, cellular automata, genetic algorithms, artificial life models, ensemble-learning algorithms, multi-agent modeling, swarm intelligence models, fuzzy systems and quantum computing. Within the context of hard computing, the models can describe only very abstract and idealized situations, whereas soft computing tools, able to deal with ambiguous and noisy data by using multi-valued or fuzzy logics, are more suited for realistic cases. That is, they are characterized by low computational cost, self-evolving software, easy tractability and high tolerance for imprecision. In a sense, their operation is more akin to that used by the human mind. Unfortunately, in a number of cases, such novel approaches are used only as technical tools without replacing the old approaches. Thus the general reasoning is still based on the combination and use of the new tools in an antiquated conceptual framework. This is a kind of *second-order reductionism* characterizing GOFS.

In our opinion, this correct critique does not go far enough. To the extent that phenomena cannot be captured by algorithms and the multi-valued and fuzzy logics used are still truth-functional, propositional logics, any resulting systems theory would be limited in its capacity of describing the evolution of real systems. Cellular automata, in particular, apply to only the simplest repetitive properties of living systems. Even within computability theory the existence of undecidable problems, *i.e.* problems that, in principle, do not admit an algorithmic solution, has been proven. These results of undecidability (which constitute the true core of computation theory) do not seem to have been taken into consideration within GOFS. The identification of computation with the existence of an algorithm has been combined with other assumptions such as that everything is measurable, quantifiable and numerable, producing another second-order level of reductionism embedded into GOFS as a disciplinary heritage.

In Minati's post-GOFS, as in LIR, the focus is no longer on the states but rather on the dynamics of processes and on their nature, *i.e.* classical or non-classical, even or rather especially when occurring in non-separated ways. This aspect can be taken as characterizing the new systemics. Thus recent advances in theoretical physics have now renewed interest in transient phenomena, considered the most important ones for complex open systems in non-equilibrium situations. In Minati's view, unfortunately, these developments are, so far, not being utilized in the building of a new form of systemics. The latter is, of course, necessary as the study of complex systems forces one to consider multiple dynamics, transients, non-equilibrium states and indefinable or moving boundaries.

Within these domains, the study of 'Between' requires models based on various combinations of both classical and non-classical approaches. That is, the ability to differentiate 'between' is conceptually substituted by an undefined number of possible approaches where the ability to distinguish is a particular case to be used for macroscopic and mono-structural regimes of processes. Regarding the separability of systems from the environment, a simple example of the inapplicability of this assumption is given by ecosystems where the differentiation between external and internal is unsuitable. In these cases the environment pervades the elements which produce, in their turn, a further active environment. This dynamic environment is active and not an amorphous, abstract space hosting processes, but has the holistic, complex structure suggested in Chap. 9.

This view impacts on the concept of identity. The new assumption is not to have single identities but, rather, different, eventually coherent, dynamics, intended as identities. Such dynamics may be multiple, superimposed and possessing multiple, hanging coherences. Coherence and the related acquired property is identity, just as flocks and biological living bodies keep their identity while their structures or materiality changes over time. This is, in other words, the essential feature of existence of identity-within-diversity. (It is of possible interest to consider eventual conceptual correspondences with the quantum vacuum pervading everything, but the latter is not a good model in the absence of the physical bases for interaction.)

Minati expresses exactly our view on how binary logic has been an unsuitable basis for describing complex phenomena with out reduction. In his words, application of the dichotomous representation of events, processes, questions and answers should be limited to the use of binary digital devices, such as those operating in real computers. It is assumed that this representation is induced by the logic used in making deductions. For Minati, however, the terms 'true' and 'false' are nothing but technical expressions used in a specific part of logic, that is, *model theory*. Outside the specific domain of model theory, formal logic theories never deal with the problem of truth or falsity but with other problems, such as the proof of deducibility of a formal symbolic expression from the axioms of the theory taken into consideration. Given the history of logic, it is not surprising that it has often been used, sometimes in an incorrect way, to support the conceptual frameworks sustaining many different disciplines.

In any case, it is evident that resorting to binary logic is a kind, perhaps the most common kind of simplification. This approach is suitable *tout court*, *i.e.* without any simplifications, for problems such as those of elementary games theory. There are multiple solutions are possible as in the case of the so-called prisoner's dilemma, but they are all based on linguistic dichotomies. GOFS has been compatible with such a view and has had difficulties in adopting a generalized alternative approach. Minati reflects our view in the following passage. "Instead, in the case of processes such as emergence, the logical aspects should be modeled in a suitable way and, in their turn, not even be explicitly formalizable, possibly non-causal and dynamical, i.e. with classical and non-classical aspects and with diverse structural regimes of validity. The new systemics should not be fully compatible with a binary framework but, at most, must be considered as useful only in particular and non-generalizable cases.

The new conceptual approach we propose could be very useful in understanding the different complementarities existing in the study of an open system, such as those between observed and observer, between process and structure, between rate-dependent and rate- independent and between part and whole. A similar conceptual approach could be introduced for other famous dualities such as language and thought, mind and body and entities and environment.

16.8.8 Coherence and Emergence

In GOFS, *coherence* is defined as some sort of dynamical regularity. However, the definition of dynamical regularity must be tailored to suit the kind of system considered, as the traditional definition of dynamical regularity, borrowed from dynamical systems theory is too limited. This leads to the idea that perhaps it would be possible to introduce a *universal theory of coherence*, based entirely on quantum-like arguments. In this case, QFT would become a fundamental basis for a new general systems theory (see above). This is a critical philosophical point, raised also in LIR, where, however, macroscopic processes are considered to be in part only isomorphic to processes defined by QFT.

Beyond coherence, its acquisition and preservation, Minati sees are processes of *transition* between coherence and incoherence which may be considered, on the one hand, as degenerative and, on the other, as seeding, an *occasion* to establish new coherences, i.e. to make coherent what is incoherent at a given level of representation. This in our view is a further instance of instantiation of the PDO, in the language we use throughout this book, of semi-actualization and semi-potentialization at a T-state being the locus of emergence.

Minati, as we, underlines the richness and creative potential of indeterminacy when systems with equivalent choices are driven by fluctuations implying irreversibility. There are a number of references to properties intended as still missing or inaccessible such as *un*defined, *in*accurate, *in*comprehensible and *un*determined, all in contrast with categories favored in GOFS such as precision, accuracy, completeness, uniqueness and optimality. The invariant common to older frameworks is the imprecise, negative qualification of uncertainty. However, uncertainty and undecidability do not always refer to deficiencies but, rather, should also be intended as specifying *spaces of equivalent configurations* explored and created by the system to choose a unique and irreversible configuration.

The overall activity of scientific research consists in a continuous and complicated interaction between the two activities of decomposing and composing, and, despite the objections raised against these procedures, they contribute to the enlargement of our knowledge. In this regard, GOFS, while stating the uselessness of decomposition ('the whole is different from the sum of its parts'), rarely tried to prove the validity of this statement. Moreover, GOFS inspired a lot of research, within systems engineering, devoted mainly to formalizing decomposition procedures. Admittedly, the problem of assessing the usefulness of decomposition and composition procedures is very difficult. This difficulty stems from the fact that the existence of general procedures of both kinds is lacking, and upon the circumstance that the number of particular procedures so far introduced for specific classes of systems is very small. After all, a complete theory of decomposition and composition procedures would be fully equivalent to a complete theory of emergence.

LIR starts from accepting the equivalent value of the 'negative' elements. It is easy from this point to develop a picture of the interaction between processes of composition and decomposition and a new entity emerging from it. We thus consider that the need for new conceptual categories for discussion of systems has been partially met by LIR. LIR does not ignore the real separability of properties and representations and their *dynamical* relationships, interdependence and coherence. The new conceptual environment we describe here as part of *Philosophy in Reality* relates to strategies which have been developed and used, for instance, by quantum physics and biology, and based on representation of one phenomenon in terms of another so that they cannot be represented as being *separate.*

Separability is a matter of simplification, allowing the conceptual possibility of distinction and independence. In contrast, an *unavoidable* space of properties is the place where single and separable properties, elements and interactions may be active *only* at a specific level of representation superimposed upon the entangled ground, ignored for convenience by the researcher. This space may be conceptually intended in all respects as the quantum vacuum, but this is not required in our approach, as long as the principle ultimately derivable from it is accepted as a basis for *change* and the time evolution of state variables is not governed, as in Poincaré, by a system of *ordinary differential equations.* In Minati's post-GOFS, it is possible to introduce various new approaches, for considering multiple probability features and coherent states rather than sequences of *single* properties of the same system. Among these approaches, he quotes neural networks; cellular automata; genetic algorithms; artificial life systems; multi-agent modeling; swarm intelligence models; neuro-fuzzy systems; crowd computing and quantum computing. Within these contexts, the function *f* conceptually, and continually, *emerges* from the individual cases when considering clusters of variables and their properties as well as properties of sequences of rules as in the case of meta-structures. This corresponds in GOFS, to some principles of classical logic being retained to deal with simple cases, but requiring something like LIR for the rest.

The idea that the current state of nature follows deterministically from its state at the previous instant is attributed to Laplace (1749–1627). The problem is that the world *emerges* rather than *functions.* The previous way of thinking has proved to be ineffective because of the systemic and complex nature of the world where processes are characterized by non-linearities of various kinds. LIR suggests some new basic features of reality that help to define this emergence, but it does not pretend, by its own philosophy, to do so exhaustively or exclusively.

According to classical physics, the reason why deterministic models are not applicable is an *incomplete knowledge* of all physical features of the components of the given system. At the microscopic level, it is possible to deal with *single* elements—*multiplicity* is not microscopic—such as molecules, cells and particles, even though they may be indistinguishable because of our limits and inabilities, rather than obeying some intrinsic theoretical principle. On the other hand, macroscopic features should be intended as indices, related to the general properties of various types of *collections* of microscopic entities considered for their *aggregations* and related properties such as pressure and temperature. These two levels, microscopic and macroscopic, may be considered as two temporary *extreme* hierarchical levels. Salthe has studied the properties of hierarchies divided into two major classes, subsumptive and compositional [59]. Respectively, they define domains of (1) diachronic taxonomic relations of general to specific 'inheritance' and (2) synchronic mereological relations of scale. Salthe claims that this hierarchy provides an 'instantaneous' modeling of systems, focusing on "dynamic energy transactions and on-going processes". This is, of course, the domain of LIR, but our view, as his, is that these hierarchies must be recognized primarily as classificatory, epistemological tools. In any case, the assumption of the existence of only the microscopic or the macroscopic levels is a matter of simplification. The point is not only to consider the various hierarchies and their possible extremes but their dynamics as a means of characterizing the systems under study. *We may consider various windows for such dynamics adopted as balance between cognitive strategies and physical effects, i.e. philosophically speaking, between constructivism and objectivism.* The study of such windows is an ongoing systemic trans-disciplinary project, which needs to take into consideration various aspects including neurological, technological, physical (related to classical and quantum physics) and epistemological ones.

Non-symbolic computation is one interesting tool since the processing is not *explicit*, i.e. the result has an emergent nature [43] since it cannot be recognized or anticipated stepwise. Steps cannot be suitably understood as microscopic computational steps, all sequential parts of a general algorithm playing here the conceptual role of a macroscopic level looking for explicit solutions.

16.8.9 Mesoscopic Variables and Meta-Structures

Mesoscopic variables, when not relating to the quantum level, relate, in several disciplines such as physics, chemistry and biology, to an *intermediate* level between micro and macro where the micro is not completely neglected as usually happens when adopting the *summary statistical* macroscopic levels ignoring details. They have been discussed by Brenner in [5]. Minati quotes the approach is based on a *philosophy of the 'middle way'* [51] considering the mesoscopic level of description as an *area of continuous negotiations between micro and macro* and the definition of families of possible observables as a *research strategy*.

The meta-structure approach for modeling complex collective behavior is also based on mesoscopic variables adopted to represent structural dynamics, i.e. changes in the structures of interaction by considering, for instance, suitable clusterings, their single and cross-correlated properties, the mesoscopic vector and the usage of degrees of freedom. Meta-structures are considered as coherent sequences of multiple dynamical structures represented by properties of suitable sets of mesoscopic variables and related clusters intended to *transversally intercept* and represent values adopted by aggregates of microscopic variables. Values of mesoscopic variables and related clusters are then considered to suitably represent the effective application of rules of interaction. Suitable properties of sets of such values represent coherence in sequences of configurations, i.e. collective behavior. All the cases considered above are intended as windows between the micro and the macro.

For example, lasers have been intensively studied using mesoscopic variables, later generalized by the Synergetics of Haken and Stafford Beer [27] using the *slaving principle.* In this context, the amplitude of fluctuations of the unstable mode is called an *order parameter*, as it drives the dynamics of pattern formation. The meta-structure approach for modeling complex collective behavior is also based on mesoscopic variables adopted to represent structural dynamics, *i.e.* changes in the structures of interaction considering, for instance, suitable clusterings, their single and cross-correlated properties, a mesoscopic vector and the degrees of freedom present. Furthermore, the properties of clusters corresponding to mesoscopic variables, including metrical and topological, can also be studied. For Minati, a *structure* is an organization with specified parameters (layers, *weighted* interconnections as in neural networks or electronic circuits). In the simplest case, it is possible to consider a population of elements all interacting through the *same* structure given, for instance, by the *same* rules of interaction as in Brownian-like motion or simple cellular automata. There are also cases of *dynamical structures* where, for instance, elements may interact by using simultaneously or sequentially different rules as for Multiple Systems. Rules of interaction are contextually decided from time to time on the basis, for instance, of cognitive processing performed by agents. The term *meta-structure* denotes multiple simultaneous (of the same, in this case *superimposed* or different elements) or sequential structures of interactions. Such dynamical structures are suitable for studying and modeling collective behavior. The concept of meta-structure is intended as a dynamical set of simultaneous, superimposed and possibly *interfering* structures of interactions. In short, a meta-structure is the *result* of processes of any, explicit or non-explicit, linear or non-linear, dynamic combinations of structures of interaction. What for us is striking about this passage is the reference to the structures of interaction. This notion points to the potential formalization of the interactions present in any dynamic, evolving process. It is thus a basic notion underlying the properties of processes in reality as well as in all disciplines.

16.8.10 Conceptualizations, Descriptions and Representations in Post-GOFS

We focus here on the rules, models and representations suitable to represent the dynamics, levels and coherences of and in systems. We are concerned with the eventual, that is, anticipated properties of populations of processes of emergence rather than on phenomenological subsequent results of such processes. The task is to focus on processes as operators rather than on their performance and results. Logic in Reality reconciles these two alternatives. Formally, Minati's approach is based on conceptualization as being a formal, symbolic representation of entities, such as objects and concepts, assumed to be *existent*. *Ontology is then intended as an explicit specification of such conceptualization.* In *computer science*, for instance, the term is used to denote a file containing the formal definition of terms and relationships. An ontology should be built by analyzing the domain to be represented and by conceptualizing it *explicitly*, that is, symbolically. We claim, however, that conceptualizations can be made that refer non-symbolically to the cognitive processes in progress.

Minati asks the following further questions of interest to philosophy: if the uncertainty of coherence represents its *open valence*. Can we *generalize* the study of uncertainty when related to such phenomena of collective interaction? Can the *structural* quasiness of Quasi-Collective Beings be the suitable place to study such eventual forms of generalization? Can properties of levels of coherences of sequences of collective interactions be described by a single general theory able to deal, for instance, with super-coherences and super-analogies? It looks as if a doctrine of post-GOFS theory could benefit from the maintenance of several of these principles at the same time. The concepts introduced about regimes of validity may be considered not only suitable to model, but even as eventual representations of the intrinsic dynamics of nature. This is a fantastic challenge for the post-GOFS and its philosophy. In particular the quasiness at a single level of emergence can be considered as an *open valence* suitable to *combine* with other simultaneous processes of emergence occurring at the same level. This calls for formalizations able to deal not only with well-defined cases as for classical approaches.

Post-GOFS needs formalizations (can we still use this term?) representing such processes where the connection is not given by rules but by coherence of multiple dynamical rules. This relates to the concept of *super-coherence*, to the second-order dynamics of/and coherences of processes of emergence. These are tools to act on such processes to keep, change or avoid, for instance, coherences. We think with Minati that a post-GOFS theory will use representations and models suitable to deal with the global, emergent rather than summative or averaged effects or properties of systems. As Minati proposes, properties of network or meta-structural representations will define a new kind of macroscopic level of emergence when actions will be considered on such properties rather than on their *causes. The new problems will be, for instance, how to vary the topology of networks, their scale invariance, meta-structural properties and regimes of validity.* The challenge for LIR is to show the validity of a non-statistical approach that emphasizes the uniqueness of systems in their evolution, as part of a universe of change.

16.8.10.1 Softness and Emergence

Softness in Minati's theory [51] refers to the effectiveness and importance of non-explicit, non-*analytical* actions—for complex systems as discussed above in its various aspects like changing levels or weights for neural networks. The *soft multiplicity* of complex phenomena should be respected not intended as a limit to approaches, but as aspects of any strategy assumed to be *effective* with complex systems that cannot *suitably* process analytical interventions as *orders* and *prescriptions*. We to this list the alternation between actual and potential parts of phenomena defined by LIR. It is will serve to characterize the possible dynamical evolution of some kinds of collective systems but also to forecast in advance their survival ability in the presence of damage and their usefulness as supporters of information transmission.

Regular or potential persistence of possibly different, multiple, superimposed emergent properties, coherences and non-equivalent representations should be considered as the dynamical identity of the system as well as being the property of such persistence. The presumed effective separability of properties and representations ignores their *dynamical* relationships, interdependence and higher coherence as discussed in Chap. 7 which deals with *systems of levels of emergence*. The new conceptual environment relates to strategies which have been developed and used, for instance, by quantum physics and biology, and based on representation of one phenomenon in terms of another so that they cannot be represented as being *separate*. Variables, properties and interactions should be formulated by *generalizing* the traditional *uncertainty principle* of quantum physics. This generalization includes, *e.g.* (1) the Heisenberg principle, concerning the position and momentum of a particle, as well as time and frequency of signals (from Fourier analysis), and (2) the fact that the observer and the observed are a model-dependent dipolar entity, closely linked to what and how we want to observe (theory of cognitive operators based on theoretical principles introduced by von Foerster).

16.8.11 The Philosophy of the Middle Way

It is for us highly significant that just as Lupasco found it necessary to ground his new logic of reality in the quantum mechanics of his day, Minati has found it necessary to ground his new systems theory in the most recent findings of quantum field theory (QFT). Even in the Brenner book of 2008, we read that QFT has properties that support non-classical interactions, dualities and self-dualities and in general a picture of the quantum field as consisting of operators associated with an ensemble of probabilities. Also, these operators represent not the values of physical quantities but those quantities themselves. This is the 'directness of representation' that we see in all instances of strong interaction (contradiction) leading to emergence: there is no need for intermediary epistemological structures, such as signs, or ontological structures, especially in the areas of causation (and mental phenomena). Cao says that the theory of quantum electrodynamics (QED) within quantum field theory (QFT) has an ontology underlying the mechanism of interaction that is essentially that of the field rather than of a particle. Further, as locally quantized fields, they have to a great extent (not completely!) lost their continuity [17]. Therefore, in LIR terms, quantum fields instantiate both continuity and discontinuity. Emergent processes at quantum critical points, unlike simple macrophysical changes of state, have both particle and field aspects.

Minati proposes a 'Philosophy of the Middle Way' which we believe leads to descriptions of phenomena which are equivalent to those of the Lupasco philosophical 'Logic of the Included Middle' described in Chap. 3. The Logic in Reality (LIR) is our formulation of this logic. In relation to energy, QFT appears to be not only more fundamental, but provide a much clearer concept of the origin of duality, opposition and an included middle, despite the absence of full agreement among quantum physicists as to what the field is.

Against this background, the Philosophy of the Middle Way has the following descriptors (the term as quoted in Minati is from [42]: mesoscopic variables relate, in all the supra-quantum level disciplines such as physics, chemistry and biology, to an *intermediate* level between micro and macro where the micro is not completely neglected. This Philosophy considers the mesoscopic level of description as an *area of continuous negotiations between micro and macro* and the definition of families of possible observables as a *research strategy*.

While this passage is useful on its own, it also points to toward "continuous negotiations" as describing the interactions at the same level, between *macro* and *macro* to model complex but *non*-collective behavior. From this point of view, the actuality-potentiality dualism also implies an unstable mode from which patterns, as new entities do, may emerge. In the opinion of Minati, such a philosophy is necessary to deal with the problem of complexity and *its* philosophy. As discussed further below, the Philosophy of Complexity or Complex Systems, in contrast to the two fields themselves, has undergone a very irregular development and seems dormant for the time being (2020).

16.8.11.1 The Between as a Middle Way. True and False

The Between mentioned above, represented by states, is of a 'middle way', The *between*, again, is considered as the place where important kinds of phenomena, usually not considered by classical approaches and GOFS, occur, such as emergence and *negotiations* between macro and micro above. Interfacing may be usefully discussed by considering logical openness (as in Sect. 2.7). Interfacing is often an inescapable or even unwilled process due to spatial and temporal *adjacency*. These makes interfacing inescapable and its opposite *total* isolation and separation, is only partial or ideal. Interfacing can also be considered to occur even in temporal processes of *transition* between social era, scientific or philosophical theories and approaches. In these cases, interfacing is initially bi-directional and then changes into unidirectional thus adapting the system to the new situation.

Minati then asks us to *consider such interfacing occurring between what is considered to be true or false. Can we consider as interfaces processes where the truth is becoming false and the false is becoming truth?* We consider this an unusually pertinent example of a new philosophy that shows the advantages of a dynamic dualist perspective. Taking this to the next level, Minati envisions *degenerative* processes when truth *decays* into falsehood and consider *correcting* the reverse. Considering the first case, this decay is usually combined with, if not implied by, the emergence of a new truth. On the other hand, in general processes of the emergence of new truths, *the truth comes first*. Truth may thus be considered to come after the decay of another truth rather than as the transience from false.

This process can be considered as the construction of second-order dynamic truth and is of great interest when non-trivially reduced to *relativism*. The *between* is given by processes of understanding, discovery, falsification, demonstrating, testing, conjecturing, proposing and validating. Regarding science, *truth* may be understood as demonstrated, robust, consistent, and supported by experimental evidence. Minati sees as a process of *liberating thought.* We are beyond standard objective or non-objective constructivism with only its network of dual possibilities but almost in a *fuzzy network* where *entities*, like concepts, theories, approaches, experiments and conjectures, have several levels or degrees of truthiness and links among them clustered within specific theories and disciplines.

The entities will be again, our epistemons in a systems context. Further and most importantly, in such a network, truth could be considered as an emergent property, at an interface, or interfacing between epistemology and ontology. We literally find it liberating not have to maintain artificial barriers where none really exist. This view further supports the Lupasco concept, to which we refer several times in this book, that "truth is the truth of reality", something which precedes and grounds standard notions of epistemic truth.

16.9 Hooker and the Philosophy of Complex Systems

At the beginning of this chapter, we discussed existing concepts of complex systems in the literature and how they do or do not follow the rules of Logic in Reality. We were obviously interested, of course, in the philosophy of complex systems (PCS) and how it might be positioned in relation to LIR. We found ourselves confronted with the following curious situation: after the publication of a major Compendium, edited by Hooker [33–35], we could find almost no references in the Internet to the philosophy of complex systems as such.

We therefore analyzed some of the articles in the Compendium to try to determine what the most accepted ideas were of the content of the PCS and the origins of why it seemed to have disappeared. What we found was that the emphasis in this book on reality and dialectics can represent an important turn in current philosophical thought about complex systems and accordingly about philosophy in general. In fact, precisely because our Logic in Reality explicates the dialectics of the underlying physics, we can contribute to clarifying applicable notions in *both* natural and *non*-natural philosophy as defined in Chap. 15. In this section, we will review the three contributions by Cliff Hooker to this Compendium, starting with those of Hofkirchner and Bishop.

16.9.1 Hofkirchner and the General Systems Theory of Bertalanffy

The restatement by Wolfgang Hofkirchner of the objectives of von Bertalanffy's General Systems Theory shows their parallelism with those of this book: GST was to be a theory of homologies or isomorphisms characteristic of the organization of "wholes"—systems appearing throughout the various scientific domains. GST was a theory of general principles of systems, a metatheory offering a new world view and consequently a new paradigm for all sciences and humanities, including philosophy. According to Hofkirchner, Pouvreau and Drack suggested 'General Systemology' might be more appropriate than GST, and Systemology was a term also proposed by Lupasco.

Some of the goals defined by Bertalanffy for a GST, mentioned by Hofkirchner, include defining a theory focussed not on systems of a more or less special kind, but on universal principles applicable to physical, biological and social systems in general.

Another goal of GST was to "bring together the analytical and organismic, as well as the descriptive and normative, to general a new type of unity of science, one based on an unworkable reduction to a single 'science' but a formal unity based on the generality and ubiquity of the system concept and on the isomorphisms it induces between sciences of which the logical and methodological autonomy is guaranteed." The term 'organismic' implies the 'recovery' (see [9]) of the properties of 'wholeness' and relative autonomy of living beings. In a phrase, GST was intended to be anti-reductionist, anti-mechanist and anti-vitalist. The first, epistemological objective involves postulating isomorphisms between disciplines; the second, ontological, between real world systems—the major domain of Logic in Reality; and the third of using the theory as a tool to support humane, humanistic interventions in the real world of facts and processes, that is, an ethics.

We have referred above to the concept of transdisciplinarity (TD). In the acceptation of Basarab Nicolescu, as noted, the friend of Lupasco and continuator of his work, TD is what lies within, between and beyond all disciplines. Rigor and tolerance are major requirements for transdisciplinary research [56]. As Hofkirchner confirms [31], operationally, TD comes into play when each discipline is engaged in constructing a common bases of methods and concepts, grasping for a bigger picture. "In fact, GST and systems science constitute transdisciplinary science per se."

In contradistinction to LIR, GST did attempt to specify systems in terms of their mathematical structure, despite accepting its limitations in the representation of complex phenomena. GST accepts analogous comparisons in a qualitative form, but it 'made life difficult for itself by maintaining analytical distinctions such as that between substance and form. In the concept of Rapoport, quoted by Hofkirchner, the appropriate mathematics for systems is one of organized complexity, the notion of a system as "an interwoven network of relations beyond linearity and additivity". As we have tried to show throughout this book, the term 'interwoven' must be given the ontological, physical meaning of being subject to the principle of dynamic opposition of the elements which are designated as interwoven. It is only in this way, we claim, that there is an ontological meaning to the difference, introduced by von Bertalanffy, of the difference between a "dynamic interaction of parts and relations of organization". In organization, organization and systems all of which involve levels of parts and wholes, bottom-up processes are ones in which interaction on the level of the parts results in relations at the level of the whole, and top-down processes in which relations at the level of the whole manifest themselves at the level of the parts, that is, in their behavior.

In the mereology of Logic in Reality the picture is simplified by making the key postulate that that parts and wholes are not totally independent; each shares properties of the other, This provides a grounding of both intra- and inter-system hierarchies and their 'superposition', which we designate as partial. We thus propose, here as in our discussion of holons and ontolons, the ontological units applicable to dynamic, especially living systems. LIR is thus in this sense a candidate for the missing theoretical model for the structure and evolution of higher-level systems. In Chap. 19 dealing with social systems more explicitly, we will look further at Hofkirchner's critical social systems perspective on the commons, that is, the common good.

16.9.2 Robert C. Bishop. Against Ontological Indeterminism

The contribution by Robert C. Bishop to the Compendium, entitled "Metaphysical and Epistemological Issues in Complex Systems" [4] allows further relevant parallels to be made with the core dialectical concepts of Logic in Reality and the Natural Philosophy proposed in this book. Bishop begins by stating that the loss of the capacity for linear superposition is one of the crucial features of complex systems. However, complex behavior is not limited to large multi-component systems but can arise in fairly simple systems as well. This idea fits well with the general applicability of the LIR Principle of Dynamic Opposition to all systems. For us, a systems theory is not complete if it does not capture, at least in part, the dynamics of the three-body system constituted by the nuclear family of husband, wife and child.

One major difference with LIR is, however, that, unlike Bishop, we do not restrict the concept of empirically accessible features of physical systems to only epistemic states and properties. The operation of the PDO implies that ontic states and properties are also accessible [5]. Better, the accessible states have both ontic and epistemic properties on one or the other of which one may concentrate at a particular time.

It is encouraging to see that Bishop characterizes some complex systems phenomenologically, despite the difficulty in translating such characterizations to measures. Of the eleven criteria proposed, three are of particular interest from our standpoint:

- *Relations*. System constituents are coupled to each other via some kinds of relations, so are not mere aggregates like sand grain piles.

- *Integration.* Various forms of structural/functional relations, such as feedback loops, couple the components contributing crucially to maintaining system integrity.
- *Intricate behavior.* System behavior lies somewhere between simple order and total disorder such that it is difficult to describer and does not merely exhibit randomly produced structures,

As we have seen in the concepts of Minati, it is the specification of the conditions for the formation of systems defined by Lupasco that offers a reasonable origin for the existence and operation of such properties.

Bishop makes a useful distinction between *resultants*—the outcomes of the simple decomposition of linear systems and the linear superposition of sub-systems and *emergents* in the case of non-linear systems where the tight coupling between components prevents the separation of the interactions and their Hamiltonian. This is the origin of the Collective Behavior studied by Minati. "The interplay among hierarchical levels in nonlinear systems blurs distinctions like part-whole, system –environment, constituent-level and so on." For us, 'blur' is a not very scientific or philosophically valid term which we replace by the lawful evolution of entities and processes between degrees of actuality and potentiality. The restatement of the situation is hardly better: "distinctions between parts and wholes, hierarchies and the like are pragmatic rather than absolute. The best answer to the question of whether a complex system can be individuated identified as a distinct individual from its environment is "yes and no". Logic in Reality underwrites an objective ontology that does not require or imply absolute ontic/epistemic cuts, even between the classical observer and observed, by focussing rather on the mutual changes effected during the process of observing.

The availability of the LIR view of dynamics explicates the problems regarding determinism as enabling predictability. Bishop states that predictability of (the evolution of) systems "has much to do with epistemic states while determinism has to do with ontic states and while the characteristics of ontic states should have implications for the nature of epistemic states, it is difficult at best to draw any conclusions about the ontic states of a system, or a 'faithful model' of the system based on our access to epistemic states. But we do not need to infer a determinism of the system from one of the model. As discussed above, the model and the system are not independent entities. The determinism we infer is not from a mathematical or deterministic *model* but from the system as a dynamic process.

Bishop's development confirms our approach by showing the difficulties in supporting ontological indeterminism, allowing the LIR interpretation of probability distributions of interacting actualities and potentialities. Determinism does not fail in macroscopic systems because quantum effects *might* influence them; they live quite comfortably with their isomorphic version of quantum superposition due to the PDO.

The sense of Bishop's reasoning that we wish to convey is its openness, like that of Hooker and Minati, to the basic tenets of the science and philosophy of Logic in Reality which can accept and respond to the challenges the existence of complex systems make to "our philosophical and scientific lore about the world". Nonlinear modelling and complexity, among other things, require seeing how the principle of non-separability may apply to model-system duality. A prime example is the difficulty that theories of self-organization, from Kant to Hofkirchner have had in relation to standard concepts of causality. As Bishop puts it succinctly: "While process and structural accounts (of causation) each appear to have some pieces of the puzzle for understanding self-organization, process theories lack an adequate account of the structural constraints of wholes on parts, while structural theories lack an adequate account of processes." We believe it was the as unrecognized genius of Lupasco to have eliminated the artificial separation of the two theories by showing the process nature of structure as summarized above.

In complex systems, when higher hierarchical levels in a systems add necessary conditions for behavior as a top-down constraint, it looks very much, as Igamberdiev stated in [36] like an Aristotelian notion of formal cause. Bishop criticizes analytical philosophers who have left formal causes unanalyzed, focussing on logical and formal relationships among efficient causes in bottom-up phenomena. Lupasco deserves the historical credit, outlined in [48], for establishing the dialectical relation between formal and efficient cause and the operational capacity of the former.

Bishop places the dialectic between reduction and emergence in the same way: reductionists argue that the properties and behavior of systems as a whole and completely determined by the states and properties of their parts (ontic claim) or are at least explainable by them (epistemic claim). Emergentists deny one or both of these claims. In our theory, in which system components are *ab origine* not independent of one another, interplay, better interaction between parts and wholes in the a priori contingent contexts provide the principles basis for emergence. In Bishop's terms, relevant information about lower-level properties and behavior of constituents or elements *plus* the specification of an appropriate contingent context allows in principle for the prediction of higher-level properties and behavior *in many cases*. We ask no more, but also no less of our theory.

Similarly, LIR adds the context for the debate about the priority of fundamental causal laws *vs.* structuring laws to explain the relationships between lower- and higher-level dynamics. Again, the common assumption that models can be used to make the connections between laws and physical systems (the 'faithful model' assumption) is flawed because it assigns operational power to the model which it does not possess. Models, mathematical or other, often simply mimic the behavior of the system in question. In contrast, application of the principles of LIR, which is 'model-free', offer at least a minimum of, in Bishop's expression, "genuine correspondence to the actual (we: and potential) properties of the target system."

16.10 The Contributions of Hooker

Hooker [33] defined the concept of Naturalistic Realism (NR) and explored the relationship between a science, its philosophy and an applicable metaphilosophy.

Hooker's approach was congenial to us in its rejection of the distinctions of standard binary logic and an acceptance of a dialectical interaction between the physical and cognitive domains. Wu's concepts of the complex structural role of information in the extant domain, in which the purely physical-energetic and cognitive-energetic components are not totally separated or separable is an informational version of NR.

Hooker further defined a Philosophy of Science as follows: "The content of a philosophy of science comprises a theory. The theory is of the same kind as those of science in every philosophically relevant respect except logical role, insofar as the subject matter of the philosophy is science (or a science) itself". A scientific aspect of Hooker's work is its essential humanism, including (1) the idea that the goal of rational thought, the rational acceptance of beliefs and means in both science and life in general, is the maximization of human potential as well as of the depth and scope of understanding and (2) the rehabilitation of intuition as part of rational function and hence of science and (3) maintaining awareness that philosophical work does not take place in a social and ideological vacuum. These ideas, in particular that intuition and knowledge are dialectically interacting cognitive processes, are central to the non-standard logic and epistemology of Stéphane Lupasco, outlined in Chaps. 3 and 4 and discussed in detail in by Brenner in [8].

Further in this spirit, the Philosophy of Information Science is to be distinguished from Information Science by its focus on issues similar to those that serve to distinguish the philosophy of physics from physics, as discussed in a compendium (Butterfield, Earman, [16]. These include questions on the one hand of the interpretation of physical theories as such and how these inform philosophical discussions, and on the other underlying "substrate" of the theories expressed as statements about reality, probability, space–time and so on. In the case of information, the Philosophy of Information provides a framework for inclusion of the simplest as well as the most complex "philosophical" features of Information Science, especially, meaning and value. In both cases, we feel that the emphasis should be placed on the absence of absolute separation between a science and its philosophy. This should be, in our view, a key doctrine of the Philosophy of Science. Not only are they not separated but their non-separability is their most essential categorial feature, as it is in the Logic of Reality which the cognitive dynamics of the relationship follow. Our approach cuts through the debate as to whether a Philosophy of Science follows the rules and logic of philosophy, science or both.

16.10.1 A Scientist Turned Philosopher

Hooker identifies himself as an ex-scientist turned philosopher working to provide an intellectual framework for understanding the foundational, *i.e.* scientific issues raised by the complex systems revolution. Both authors could describe themselves in the same way, the time lapse of the 'ex-' being somewhat longer in the case of Brenner. Hooker's goal was to help develop mature foundations/philosophy of complex systems. Our goal is of course more general—a philosophy of nature which is *ipso facto* complex, but the discussion which follows will show how unnecessary it is to work as if there were hard boundaries between the approaches.

Two brief comments on the above: as shown by Wu and Brenner, the term revolution also applies to the field of information, and a revolution in philosophy may be one of its specific consequences (Wu, Brenner, [65]. We propose that the reader meditate on Hooker's phrase that "complex winds of change" are blowing through all disciplines, including Chinese medicine!

We understand Hooker's preference for dynamical analysis over logical and semantic analyses when it comes to fundamental concepts and principles in view of what logical and semantic analysis has been largely limited to. Hopefully, the 'recovery' of the second two that we propose in this book will enable better integration of insights from both domains. Hooker provides a perspicacious account of the co-evolution of the scientific paradigm and the philosophy of science which we will not reproduce here in full. A key idea of Hooker, however, is that of the dynamics of science, or science as a dynamic system. He states that "this perspective is wholly foreign to the traditional epistemological conception, *the logic-centred conception of science* (italics ours)". In agreeing, we propose our logic as exactly one of the new concepts, methods and inference patterns that Hooker says is needed for a transformative dynamics. To paraphrase Hooker, we consider ourselves as actors in a new metascience process that regrounds and naturalizes transcendent normative epistemology. We ally ourselves with Rescher "who shifted the focus of *change* from theory to method in defiance of the eternal logic model" which we parse as standard truth-functional logic".

It is noteworthy that from a systems perspective, we can avoid separation between biological and cognitive evolutionary processes. One views knowledge "as simply one aspect of biological life that emerges as phenotypic cognitively-based capacities increase, and view science as its concrete institutional communal/species form." We include our Logic and Mathematics in Reality as additional such aspects, allowing the epistemic process of science to emerge from the material process in the broadest sense of material. Epistemic normativity can be extracted from non-normative processes because they are not *totally* non-normative, not because non-normative processes are dynamic ones.

Logic also enters Hooker's study of cognition, as something still normatively determined by its (standard) logical structure. But the processes underlying cognitive capacities define a more complex logical structure in the sense of LIR. This reflects the non-standard logical categories of inputs and outputs, leading to the kind of shift we look for throughout knowledge, here from node-focused neural net functions to a focus on the dynamical states themselves. Hooker cites Churchland as leading a philosophical revolution—yet another—from computational functionalism to a more empirically connected philosophy of mind. Very much in the spirit of LIR is Hooker's statement that assuming, as we do, dual interrelations of emergence and reduction, there need not be any absolute opposition between the two approaches to neuropsychological relations.

16.10.2 Conceptualizing Reduction, Emergence and Self-Organization

The themes touched above are elaborated further in the 2nd and 3rd Hooker contributions to the Compendium. In the second, entitled "Conceptualizing Reduction, Emergence and Self-organization in Complex Dynamical Systems", the focus is on their interrelationships. Organization itself is defined as a relational condition of systems where the components play distinct roles which, however, are so interrelated as to produce a coherent global outcome. We have discussed the question of whether the 'self-' is justified or should be replaced by 'hetero-' above. What is important here is that the coherent global outcome is an irreducible new, emergent dynamic entity. In Hooker's view, contrary to the standard one, reduction and emergence are not totally opposed, but emergence 'underpins' functional reduction and reduction, both compositional and functional allows specification of the processes and the initial and boundary conditions that underpin emergence. "The two are thus intricately interwoven and mutually supportive". As we have tried to show on several other occasions in this book, the existence of mutually defining capacities of actualization and potentialization, also at the conceptual level, provide a more natural description of the relation than interwoven.

Attempting to solve the dynamics of complex systems by calling attention to their ontic properties as processes does and will not make it analytically tractable, but do they offer in principle a "new, more generous foundation for them"? Since the variables of the complex systems involved (their degrees of actuality and potentiality) are *not* mutually independent D'Alembert's criterion for constraint holonomicity fails. As Hooker brings out clearly, a coherent Lagrangian mechanics may not be available—'scandal'! Hooker concludes—we quote exactly—"the logic machine is not dead". Much can still be done with standard dynamical laws and representations but the domain of complex systems 'confronts' their assumptions and their monolithicity. Here Hooker again refers to the "delicately entwined dance of emergence and reduction providing constraints on compression modular (the consequence of the operation of condition dependent laws) that permit its expansion. Stated in LIR terms expansion had been potentialized or repressed, then actualized and expressed.

16.10.3 Advances in the Philosophy of Complex Systems

In Part B of his Introduction to the Philosophy of Complex Systems [13], Hooker describes a scientific paradigm and philosophy of science as they appeared in 2009, now over ten years ago! Does the absence of discussion and literature of these formulations imply that no major advances over them have been made since what Hooker called his first presentation? Might the present book be considered, at least in part as the missing second presentation?

Logic in Reality does not provide a general account of the modularties and hierarchies that occur in complex systems let alone a complete set of general systems principle. Hooker cites Cartwright in describing what at first sight spears to be an unavoidable trade-off between generality and explanatory power and phenomenological empirical adequacy. In his section on discriminating model structure, Hooker states that a valid model must provide reasonable descriptions of interaction relations, variable and parameters.

The way forward may lie in a comment, not further developed, to the effect that "it proves necessary to express threshold location values a probabilistically distributed, making explicit their uncertainty. Hooker then proposes shifting to a higher order methodological focus, using a functional notion of adaptive resilience, essentially "the capacity to survive functionally intact the perturbing relation of uncertainty."

Subject to much further work, we believe that the probabilistic values of the actuality and potentiality of a system, as described in LIR, may constitute the threshold location values as basis for the emergence of new states—change. Hooker's lapidary statement: "*Change is the normal condition and stasis now requires explanation* (italics ours)". We are clearly in a domain of process dynamics rather than statics as the new scientific paradigm requires. The additional critical contribution of LIR is not only to have the focus on difference, disorder, singularity and 'aberrant' behavior in both science and society; it is also their uninterrupted dynamic interaction with their positive counterparts.

In LIR terms, the correlative philosophy of science for complex systems that emerges from and must accompany them requires *synthetic* solutions which must nevertheless be embedded in science. The principle of dynamic opposition (PDO) in reality is a condition dependent law, which can provide condition dependent explanations. Confirmations are a still more difficult problem. In the theory developed by Hooker up to this point, condition-dependence addresses only dynamically-based interrelations, physical in the usual sense, and context refers to features which are non-dynamical in the same sense. Application of PDO, however, is the way we propose to unify the two without conflation in the sense that both sets of relations cannot be totally separated or separable. This is our tentative contribution to the science of complex systems that is, or at least was, still being invented.

16.10.4 What Has Been Missing from LIR?

As one consequence of this study, we have a better idea of what has been missing in presentations to date of the philosophy of Lupasco. It is a demonstration of how it can be used effectively in a unified, or partly unified, systems approach to phenomena, without slipping into dogma. Thus, Logic in Reality and its Principle of Dynamic Opposition is a *reduction* in the sense Hooker uses the term, roughly equivalent to a synthesis using factual (dynamic) characteristics of systems. What have been missing from the LIR picture to-date are definitions of the formal constraints under which the interactions operate within and between levels of reality. A new method is required for the representation not only of entities and processes and their movement form actuality to potentiality but of the complexity of the systems they become.

Appropriately, Hooker uses the metaphor of a "dual revolution in science and philosophy as the most important large-scale development in science and cognition in a century". His Compendium was designed to "pull together the complex strands of this revolution and focus attention on its foundational/philosophical problems and challenges". We believe that the revolution in philosophy that derives from Wu Kun's conception of the role of information is an integral part of and/or companion to Hooker's. Exploration of this synergy is one of our major goals for further work. With this final Philosophy of Systems component of Philosophy in Reality in hand, let us now conclude this study by seeing how both may be applied to living systems, to society, and to concepts of social systems and their philosophy.

References

1. Bernard-Weil, É.: La théorie ago-antagoniste. Le Débat **106**, 106–120 (1999)
2. von Bertalanffy, L.: General System Theory. George Braziller, New York, NY (1969)
3. von Bertalanffy, L.: The history and status of general systems theory. Acad. Manag. J. **15**(4), 407–426 (1972)
4. Bishop, R.C.: Metaphysical and epistemological issues in complex systems. In: Change Hooker, C. (ed.) *Philosophy of Complex Systems. Handbook of the Philosophy of Science*, vol. 10. Amsterdam, Elsevier, pp. 105–136 (2011)
5. Bishop, R., Brenner, J.E.: Potentiality, Actuality and Non-Separability in Quantum and Classical Physics: Heisenberg's *Res Potentiae* in the Macroscopic World (2018). arXiv:1801.01471.
6. Blasone, M., Jizba, P., Vitiello, G.: Quantum Field Theory and its Macroscopic Manifestations. Boson Condensation, Ordered Patterns and Topological Defects. Imperial College Press, London, UK (2011)
7. Boyd, R.: Scientific Realism. The Stanford Encyclopedia of Philosophy (Summer 2002 Edition), E.N. Zalta (Ed.). http://plato.stanford.edu/archives/spr2002/entries/scientific-realism/ (2002)
8. Brenner, J.E.: Logic in Reality. Springer, Dordrecht (2008)
9. Brenner, J.E., Igamberdiev, A.U.: Philosophy in Reality: Scientific Discovery and Logical Recovery. Philosophies **4,** 22 (2019)
10. Brenner, J.E.: Stéphane Lupasco et la Rejonction Métalogique. In: A la confluence de deux cultures. Lupasco aujourd'hui. Proceedings of the International UNESCO Colloquium, Paris, 24 Mar 2010. Editions Oxus, Paris, pp. 250–285 (2010)
11. Brier, S.: Cybersemiotics. Why Information is not Enough. University of Toronto Press, Toronto (2008)
12. Bueno, O.: What is structural empiricism? Scientific change in an empiricist setting. Erkenntnis **50**, 59–85 (1999)
13. Bueno, O., da Costa, N.C.A.: Quasi-truth, paraconsistency, and the foundations of science. Synthese **154**, 383–399 (2007)
14. Burgin, M.: Structural Reality. Nova Science Publishers, New York, NY (2012)
15. Burgin, M.: Theory of Information. Fundamentality, Diversity and Unification. World Scientific, Singapore (2010)
16. Butterfield, J., Earman, J. (eds.) Handbook of the Philosophy of Science. The Philosophy of Physics. Elsevier, Amsterdam (2007)

17. Cao, T.: Conceptual Developments of 20th Century Field Theories. Cambridge University Press, Cambridge (1997)
18. Chedid, F.B.: Kolmogorov Complexity and Information Content. *arXiv-1710*, 17 Sept 2017
19. Deacon, T.: Incomplete Nature: How Mind Evolved from Matter. W.W. Norton & Co, New York, NY (2012)
20. Eddington, E.: The Philosophy of Physical Science. Cambridge University Press, Cambridge, UK (1939)
21. Esfeld, M., Lam, V.: Holism and structural realism. In: Vanderbeeken, R., Weber, E., (eds.) Worldviews, Science and Us: Studies of Analytical Metaphysics. A Selection of Topics from a Methodological Perspective. World Scientific, Singapore (2010)
22. Espejo, R., Schwaninger, M.: To Be and Not to Be; That is the System: A Tribute to Stafford Beer. Carl Auer Systeme-Verlag, Wiesbaden (1996)
23. Floridi, L.: A defence of informational structural realism. Synthese **161**(2), 1–51 (2008)
24. Van Fraassen, B.C.: Structure: its shadow and substance. Br. J. Philos. Sci. **57**, 275–307 (2006)
25. Giddens, A.: The Constitution of Society. Polity Press, Cambridge, UK (1984)
26. Gomez, C.: Electric-Magnetic Duality and Effective Field Theories. ArXiv: hep-th/9510023 (1995)
27. Haken, H.: Synergetics: Introduction and Advanced Topics. Springer, Berlin, Germany (2012)
28. Hammond, D.: Philosophical and ethical foundations of systems thinking. triple-C **3**(2), 20–27 (2005)
29. Heylighen, F.: Evolutionary transitions: how do levels of complexity emerge? Complexity **6**(1), 53–57 (2000)
30. Hofkirchner, W.: Does computing embrace self-organisation? In: Dodig-Crnkovic, G., Burgin, M. (eds.) Information and Computation. World Scientific Publishing (2010). https://www.idt.mdh.se/ECAP-2005/INFOCOMPBOOK/
31. Hofkirchner, W., Schafranek, M.: General system theory. In: Hooker, C. (ed.) Philosophy of Complex Systems. Handbook of the Philosophy of Science, vol. 10. Elsevier, Amsterdam, pp. 177–194 (2011)
32. Hofstadter, D.R.: Gödel, Escher Bach; an Eternal Golden Braid. Basic Books Inc., New York, NY (1979)
33. Hooker, C.: Introduction to philosophy of complex systems: A. In: Hooker, C. (ed.) Philosophy of Complex Systems. Handbook of the Philosophy of Science, vol. 10. Elsevier, Amsterdam, pp. 3–90 (2011)
34. Hooker, C.: Conceptualising reduction, emergence and self-organization in complex dynamical systems. In: Hooker, C. (ed.) Philosophy of Complex Systems. Handbook of the Philosophy of Science, vol. 10. Elsevier, Amsterdam, pp. 195–202 (2011)
35. Hooker, C.: Introduction to philosophy of complex systems: B. In: Hooker, C. (ed.) Philosophy of Complex Systems. Handbook of the Philosophy of Science, vol. 10. Elsevier, Amsterdam, pp. 841–910 (2011)
36. Igamberdiev, A.U.: Time and life in the relational universe: prolegomena to an integral paradigm of natural philosophy. Philosophies **3**, 30 (2018)
37. Ladyman, J., Ross, D.: Every Thing Must Go. Metaphysics Naturalized. Oxford University Press, Oxford, UK (2007)
38. Ladyman, J.: Structural realism. In: Zalta, E.N. (ed.) The Stanford Encyclopedia of Philosophy (Summer 2020 Edition) (2020). https://plato.stanford.edu/archives/spr2020/entires/structural-realism/
39. Laszlo, E.: Introduction to Systems Philosophy, Towards a New Paradigm of Contemporary Thought. Harper & Row, Publishers, New York, NY (1972)
40. Laszlo, E.: The Systems View of the World. A Holistic Vision for Our Time. Hampton Press, Cresskill, NJ (1996)
41. Laudisa, F.: Le principe de causalité entre empirisme logique et néokantisme. Philosophie **89**, 78–93 (2006)
42. Laughlin, R.B., et al.: The middle way. Proc. Natl. Acad. Sci. U.S.A. **97**(1), 32–37 (2000)

43. Licata I., Minati, G.: Emergence, Computation and the Freedom Degree Loss Information Principle in Complex Systems. Found. Sci. **22,** 863–881 (2017)
44. Luca, L., Pauri, M.: Dynamical emergence of instantaneous 3-spaces in a class of models of general relativity. In: van der Merwe, A. (ed.) Relativity and the Dimensionality of the World. Springer, Dordrecht (2006). arXiv:gr-qc/0611045v17
45. Lucas, C.: The philosophy of complexity. In: Prinicipia Cybernetica Web Page version 4.83, Original 2000 (2005). Accessed 13 Apr 2019. www.calresco.org/lucas/philos.htm
46. Lupasco, S.: Qu'est-ce qu'une structure? Christian Bourgois, Paris (1967)
47. Lupasco, S.: L'énergie et la matière vivante. Editions du Rocher, Monaco. (Originally published 1962, Julliard, Paris) (1986)
48. Lupasco, S.: Du devenir logique et de l'affectivité; vol. 1: Le dualisme antagoniste. Essai d'une nouvelle théorie de la connaissance. J. Vrin, Paris (Originally published in Paris: J. Vrin, 1935) (1973).
49. Miller, A.: Realism (2019). https://plato.stanford.edu/archives/win2019/entries/realism/ (Winter 2019 Edition)
50. Minati, G., Pessa, E.: Collective Beings. Springer, New York, NY (2006)
51. Minati, G., Pessa, E.: From Collective Beings to Quasi-Systems. Springer Science+Business Media, New York, NY (2018)
52. Minati, G.: General theory of emergence. Beyond systemic generalization. In: Processes of Emergence of Systems and Systemic Properties. World Scientific, Singapore (2009)
53. Minati, G., Penna, M., Pessa, E.: Thermodynamical and logical openness in general systems. Syst. Res. Behav. Sci. Syst. Res. **15**, 131–145 (1998)
54. Morin, E.: La méthode 3. La connaissance de la connaissance. Éditions le Seuil, Paris (1984)
55. Nicolescu, B.: Manifesto of Transdisciplinarity. State University of New York Press, Albany, NY (2002)
56. Nicolescu, B.: Nous, la particule et le monde. Editions du Rocher, Paris. (Originally published in Paris: Éditions Le Mail, 1985) (2002)
57. Pouvreau, D.: Une histoire de la systémologie générale de Ludwig von Bertalanffy – Généalogie, genèse, actualisation et postérité d'un projet herméneutique. Thèse de doctorat, 2013. Research (2015)
58. Rousseau, D.: General Systemology. Transdisciplinarity for Discovery, Insight and Innovation. Springer, Singapore (2018)
59. Salthe, S.: Hierarchical Structures. Axiomathes **22**, 355–383 (2012)
60. Saunders, S.: Structural realism again. Synthese **136**, 127–133 (2003)
61. Senge, P.: The Art and Practice of Learning. Doubleday, New York NY (1990)
62. Turchin, V.: The Phenomenon of Science. A Cybernetic Approach to Human Evolution. Columbia University Press, New York NY (1977)
63. Valentinov, V.: Ludwig von Bertalanffy and his enduring relevance: celebrating 50 years of general systems theory. Syst. Res. Behav. Sci. **36**, 1–4 (2019)
64. Witten, E.: String theory dynamics in various dimensions. Nuclear Physics B 443, 85–126 (1995)
65. Wu, K., Brenner, J.E.: Philosophy of information: revolution in philosophy. Towards an informational metaphilosophy of science. Philosophies **2**, 20 (2017)

Chapter 17
Living Systems: The Epistemic Relation to Reality

17.1 Introduction: Life as Internally Determined Activity According to Aristotle

Aristotle was the first to formulate in philosophical terms the concept of coding in living systems. This concept was not recognized until the invention of genetics [51]. In his tractate *De Anima,* Book II, Aristotle introduced the fundamental triadic structure that underlies all phenomena related to the internal activity of living systems [1]. It includes (1) Matter (potentiality), Entelechy as knowledge, which is the first or pre-actualized actuality; and (2) Entelechy as an exercise of knowledge, which is activity of recognition or actualized actuality. This profound structure is related to the biological level of reality. It can be associated directly with genetic organization where Entelechy as knowledge corresponds to the embedded description, i.e. to genetic structures. This means that Aristotle was the first who in general terms introduced the idea of *code* as the main characteristics of living systems similarly to the reference frame in physics. In this respect, he can be considered as a founder of biology as a natural science. In physics, the reference frame was not correctly defined before Galileo despite of the book of Aristotle known as "Physics". Only after the rediscovery of Mendel's of inheritance, the idea of code was firmly reinstated in biology. The idea of code is directly connected to such main property of living system as the "closure to efficient causation" [46]which also refers back to Aristotle.

Aristotle's concept of life is expressed in his ideas on the relations between a part and the whole: if the parts are present the whole may be absent, but the whole is definitively expressed in its parts, being present as "Entelechy as a possession of knowledge", i.e., as a possibility of the realization of the whole. In Lupascian terms, one would say that the whole is present as potentiality, which in LIR is not limited to "matter". The wholeness of living organisms can thus be rationally described and logically analyzed. It determines the characteristic features of biosystems, realized at the level of the internal structure of an organism as well as at the level of interactions between organisms.

J. E. Brenner and A. U. Igamberdiev, *Philosophy in Reality*, Studies in Applied Philosophy, Epistemology and Rational Ethics 60, https://doi.org/10.1007/978-3-030-62757-7_17

A profound distinction between physical motion and biological transformation is made in terms of Aristotelian causes, and the interrelation between matter and the two types (or levels) of entelechy is not simply reduced in the Aristotelian framework to the logic that he developed in his Organon (the collection of six works on logic). In the philosophy of Aristotle, the logic of life includes a profound non-classical component that appears when Aristotle considers temporal transformations. While Plato analyzed the forms 'beyond time', Aristotle observed them also in the temporal world, and for this observation he developed a complex conceptual apparatus for his philosophy that dialectically describes the phenomenon of actualization.

17.2 The Problem of Complexification. Change

Living systems are complex and this complexity is irreducible to their elements as realized early in philosophical thought. Complexification in the course of individual development is a mostly ordered and complicated process taking place in the natural world. The dynamic oppositions involved in ordering it were the opposite concepts of preformationism and epigenesis, outlined already in ancient philosophy. They became scientifically defined with the development of science in the seventeenth century. Despite all their shortcomings, these approaches introduced new visions for understanding the generation of biological forms, and in the concept of embryology, as formulated by von Baer [2], the opposite approaches to understanding morphogenesis were fairly balanced. Baer's laws of embryology contain important holistic ideas. In particular, they refer to the fact that general characteristics of the group to which an embryo belongs develop before special characteristics and the general structural relations are formed before the most specific appear. Modern understanding of the individual development of an organism, being far from final resolution, includes both preformationist and epigenetic elements in their complementarity and mutual interaction.

While the development of individual organisms was comprehended as a visible biological phenomenon, the ideas of evolution and evolutionary complexification were not fully comprehended even in major systems of philosophical thought throughout the centuries. Evolutionary views can be traced to the Ancient Greek philosophy, even to Anaximander who claimed that humans evolved from fishes (which is generally correct), and then to the ideas similar to natural selection in the concept of Empedocles and then of Epicurus as presented by Titus Lucretius Carus in his poem "*De rerum natura*". In the modern era, the development of paleontological research and comparative anatomy led to the necessity of an explanation of the complexification of forms during the natural history of the Earth. The most important were the concepts developed by Georges Cuvier, Jean Baptist Lamarck and Étienne Geoffroy Saint-Hilaire in early seventeenth century. They all aimed at solving the problem of the *change* of biological forms in the course of natural history as we return here to one of the major themes of this book. Cuvier as a "father of paleontology" considered that these changes were triggered by the geological catastrophes, although he rejected any further explanation except the creationistic interpretation. Lamarck introduced the principle that he defined as a complexifying force (or constraint) to explain transitions from one level of organization (gradation) to another, while the concept of direct inheritable adaptation (adaptive force or constraint) that is usually attributed to him was considered only within one level of organization. Geoffroy Saint-Hilaire, by establishing the principle of the unity of composition of biological organisms, considered the changes at early embryogenic stages as crucial in generation of new forms and became a predecessor of the theories of evo-devo and of the punctuated equilibrium in evolution. Evolutionary developmental biology (evo-devo) compares the developmental processes of different organisms to infer the ancestral relationships between them with the emphasis of how developmental processes evolved.

17.3 Natural Selection: Can It Be the Most Fundamental Biological Principle?

Three great French biologists outlined different aspects of the changes and transformations in the organic world in order to explain complexification in natural history. However, the next development of biology resulted in the claim that not only the creationistic approach of Cuvier but also the introduction of the complexifying and adapting forces proposed by Lamarck are non-scientific and should be removed from the paradigm of natural science. Instead, the principle of natural selection, that avoids any explanation of complexification as such, was introduced by Charles Darwin and Alfred Russell Wallace for the ultimate explanation of the evolutionary process. A complexifying transaction was eliminated from the explanation of change whose cause was reduced to random variation that takes place without direction and apparent determinism.

The notion of natural selection appears in Empedocles and Lucretius who adapted the doctrine of Epicurus in his poem "*De rerum natura*". However, the main challenge is to explain why natural selection appears to be *generic* as claimed in the Darwinian interpretation. The genericity in this concept, in our view, is due to (1) *random variation* in Darwin's initial view or (2) to random mutations in genetics, i.e. to a broader variability from which the useful traits can be selected. Another important question is why natural selection results in complexification as originally perceived. These aspects are usually considered as given, without requiring detailed analysis. The problem of transition to a new level of organization is not resolved in the Darwinian concept of evolution. Although it may refer to natural selection, it requires the analysis of the internal structural and dynamic constraints that could make possible such a transition.

Natural selection as the basis for evolution looks like a notion that embodies a vitalistic antagonism but on closer inspection, there is nothing to distinguish it from a purely physico-chemical concept of life. It can be placed together with other reductionist notions of hierarchy and progress. Some sort of efficient cause seems to be the only basis proposed for natural selection to operate, whereas a fundamental theoretical role for the antagonisms found at the physical and chemical levels of reality and consequently the real causes of the origin of life and evolution at the biological level remained undeveloped. Both Lupasco [32] and Emmeche [11] have castigated the account of evolution in the neo-Darwinian paradigm of natural selection as algorithmic and mechanist, adequate at best as a theory for insentient zombies. The establishment of the categories of Emergence, Closure and Downward Causation is necessary but not sufficient as an approach to a theory of the origin of life and evolution which will go beyond the strictly formulated natural selection paradigm. The commonly used examples of replacing white by black butterflies under the pressure of environmental constraints as well as the examples of differently colored daisies in the Daisyworld concept may be perfect for modelling the operation of natural selection but they do not explain the main challenge of evolutionary biology which is complexification. To overcome this, Barbieri [3] has suggested that natural selection operates within one level of evolutionary complexity, while the transition to a new level of complexity is based on the development of new coding systems.

In fact, both the Darwinian natural selection and the so-called Lamarckian adaptive force act at the same level of organization; the question is to what extent these changes are genetic or epigenetic, and can epigenetic modifications become genetic changes. The concept of a Read–Write genome [49] states that the epigenetic-genetic transition is common in nature, other authors consider it improbable although there is substantial evidence for such a transition. Shapiro has reconsidered our understanding of the genome as a Read-Only Memory, changing via copying of casual errors, to a sophisticatedly formatted system of data storage Read-Write (RW), continuously changing according to cellular modifications and inscriptions. These inscriptions occur at three temporal levels: of cellular reproduction, of multicellular development, and of evolutionary transformations, and include the diversity of processes characteristic for each level such as the formation of nucleoprotein complexes, epigenetic formatting, and structural transformations of nucleotide sequences. The writing function of the genome directly depends on the generation of new "words" and loading them by meaning. Therefore, it is not a direct writing and not a directly memorized adaptation in the sense of Lamarck, but it is the process mediated logically by the generative properties of the system. Nevertheless, neither the Darwinian nor the Lamarckian approach to adaptation resolve the problem of complexification which Lamarck attributed to a complexification force.

Emmeche considered biosemiotics, as noted above, as a promising perspective, but was concerned that its concept of code-duality also might imply a hidden prototype fallacy, the genotype–phenotype duality of classical genetics [11]. He later described biosemiotics as a "corrective theoretical enterprise" that enables investigations of questions to be made that have been dismissed due to the materialist and reductionist assumptions of much Neo-Darwinism. As Emmeche remarked, "the real challenge is not just to consider life as semiotic processes rather than as organized molecular systems, but to investigate the relation between the molecular and semiotic aspects of life processes".

LIR may provide another form of 'interactionist' explication in the evolution-developmental debate that does not suffer from the absence of physical grounding. Actualization arises in evolution as a process of constituting new boundary conditions. These enable constraints that create new, typically unprestatable, adjacent possible opportunities for further evolution. New actual states emerge, in a persistent becoming, so evolution flows into a typically unprestatable succession of adjacent possibles [25]. Contradictorial or conditional dualism and its related conception of contradictorial cause and effect offer a non-traditional, non-mechanistic metaphysical and methodological approach. We expect that this approach, when applied to extended evolutionary synthesis, will provide the background for a new paradigm. This paradigm could transform the current evolutionary views from the messy and random set of different approaches into a strong concept incorporating different approaches into a single powerful theory. The search for a unified concept in theoretical biology is a major challenge of modern science (see the extensive review of [9].

17.4 Back to Plato: Nomogenesis

Relative independence of biological form from underlying molecular processes that include gene regulation and expression can be considered as one of the major biological principles. It allows studying transitions between forms independently of their generation and considering only the basic principles of transformation of geometry. This philosophical approach, whose origin is in Plato, was successfully applied by D'Arcy Wentworth Thompson in his famous book "On Growth and Form" [53] where he showed that transformations of coordinate systems correspond to generation of forms specific to biological species. However, he mentioned that "morphology is not only a study of material things and of the forms of material things, but has its dynamical aspect, under which we deal with the interpretation, in terms of force, of the operations of Energy".

The basic concept of a space of biological forms is that it is essentially curvilinear, non-Euclidean and more diverse than any physical space–time of the general theory of relativity. The ideas of D'Arcy Thompson were developed by several followers of his approach. Here we would like to note Van Valen [55], who emphasized the informational continuity of repetitively homologous structures, which is neither genetic nor environmentally imposed, and Petukhov [42] who described basic principles of biological transformations grounded in non-Euclidean geometry.

In other evolutionary theories, it was suggested by Berg [6] in his nomogenesis theory and further developed by many authors including Lima-de-Faria [31], that the laws of evolution are based on objective rules of transformations of forms independent of adaptability and natural selection. Although this thesis can support the idea of analyzing forms and their transformations without relation to molecular and submolecular processes, in fact, to understand a relation of transformations of forms in ontogenesis and evolution to the underlying phenomena, we need to analyze fundamental principles of biological organization based on percolation between different levels of physical reality. This would correspond (philosophically) to actualization of *potentia* and represent the Aristotelian approach to understanding the biological form. The nomogenetic approach, although truly scientific, is incomplete in abandoning the energetic transition between potentiality and actuality and concentrating on geometry. The same incompleteness we criticize in physics when it is limited to the geometry of background space–time and does not include quantum mechanical transformations (transactions) that generate space–time itself. The existence of metrics, according to basic ideas of Riemann, is a consequence of certain external physical forces of connection applied to individual objects. These forces of connection, as suggested by Riemann, are related to the discreteness of space, and form its actual observed structure. The challenge is to define these forces and describe the mechanism of their action. There is no mathematical preference for any particular curvature. The basic question here is "how geometry is formed", i.e. what is the basis for the observed curvilinearity of the space of biological forms? It may be resolved via the action of the observer, which is a living system in the biological realm. This corresponds to the idea of Lupasco that objects generate their own space–time and its "geometry".

D'Arcy Thompson was a pioneer in claiming that transformations of a coordinate system correspond to generation of forms specific to concrete biological species. However, his association with nomogenetic theories of evolution is usually exaggerated. He considered not only purely geometrical but also the dynamical aspect of transformations as arising from the percolation between different levels of organization of biological systems, i.e. he viewed the process of transformation as epigenetic. This understanding can be traced to the founder of embryology von Baer [2], mentioned above. Geometrical transformations emerging in biological evolution represent the universal laws of pattern formation appearing as the limiting conditions (constraints) on emergent phenomena. In addition to the transformations described by D'Arcy Thompson, there are macroevolutionary phenomena, which refer to the arbitrary, small genomic changes that result in non-intertransformable phenotypes. Rosen [46] finds their cause outside the entailments in the system but rather in the environment where everything is unentailed. The externality of the environment provides the new 'extra information' that thus appears in evolution. Further development of the ideas of D'Arcy Thompson should be concentrated on the limits of structural stability and similarity transformations in identifying the points of bifurcations where the continuity is lost and new topological forms arise. These transformations were described by Thom [52] in his introduction of catastrophe theory, they suggest further investigation with an approach that incorporates epigenetic mechanisms for major transformations. The framework of D'Arcy Thompson can be a starting point for analyzing such transformations, and he can be considered as a predecessor of the emerging concept of evolution defined as Extended Evolutionary Synthesis (see next section). However, the morphology changes described by D'Arcy Thompson do not include the major bifurcations that characterize macroevolution and appear beyond the similarity transformations between phenotypes. They can be regarded as major perturbations having their source outside the entailments in the system where nothing is entailed [46]. The extra information appearing in such macro-evolutionary transformations is assigned to an acausal environment that determines the emerging set of bifurcation points and a major complexification of organization. Entailing the outside corresponds in reality to major changes in evolution such as the Cambrian 'explosion'. This was based on the formation of hard skeletons into biological structures by incorporation of calcium and other inorganic ions from sea water. A special case of macro-evolutionary transformation is the incorporation of other organisms in the emerging complex structure which is defined as symbiogenesis.

17.5 Extended Evolutionary Synthesis: More Challenges and Expectations

In the first half of the twentieth century, the concept of modern evolutionary synthesis unified genetics: Darwinism replaced mutational variability by random variability. This version of Darwinism still dominates evolutionary theory despite significant limitations which have become evident. In response to new challenges, the ideas of extended evolutionary synthesis have been developed, although this is not yet shaped into a single powerful concept, being an agglomeration of several approaches that may contradict to each other. The paradigm of the extended evolutionary synthesis was announced in the 1950s by Conrad H. Waddington, developed in the concept of punctuated equilibrium by Stephen Jay Gould and Niles Eldredge in the 1980s, and was reconceptualized recently by Pigliucci and Müller [43].

In fact, there are the following challenges that the extended evolutionary synthesis has to deal with:

1. The adaptive constraint of the evolution without major complexification: genetic *versus* epigenetic changes and to what extent epigenetic changes can be transformed to genetic changes (the capacity of genome for writing *versus* reading).
2. The complexifying constraint of progressive evolution: the role of horizontal gene transfer, gene duplication, "language games" of various genetic systems including the role of RNA in generating new combinations, symbiogenesis, punctuated equilibrium, evo-devo, other possible "mechanisms" in the transaction to a new level of organization.
3. The nomothetic constraint in the transitions between forms based on the objective rules of transformation of geometry.

All these three constraints operate as a unity. In this new paradigm, natural selection will not be the sole or even major factor of evolution. It will be only one special case of the dynamics of opposites which includes many other forms of complementary interaction from conflicting to mutually beneficial. The main challenge is still to explain complexification, which, in fact, started to significantly expand with the origin of eukaryotic cell and then of multicellularity, i.e. only at the later stage of evolutionary time scale. We will therefore now analyze some approaches that appear in modern evolutionary theory to explain the gradual increase of evolutionary complexity.

17.6 RNA as an Intermediate Component of the Molecular Biology Triad Representing an Actualized T-State

The genotype–phenotype interaction is not one of determination but of interpretation. These ideas, if translated to modern language, can be traced to the founder of embryology K. E. von Baer. Genes are, in fact, the subsets of metabolites in the autopoietic structure of biological systems [19] stably reflecting the structure of proteins and their functions, but there is a further form that appears beyond this reflection. It is a projection in 3D of an entire structure consisting of metabolites, enzymes and genes bootstrapping each other. This means that this complex dynamic form is not "encoded" in genes. Although the rules of the genetic code are strictly fixed, even within the code certain variations in interpretation exist, and the common phenomenon is editing of the transcribed text, especially in mitochondria. The basic structure of the genetic language is more rigid than the human language and corresponds more to the unambiguous rules of formal logic. A limited area of freedom is expanded at the level of hypertexts, which provides the flexibility in the individual development and adaptation of organisms and opens potentially infinite possibilities for evolution. However, the main role in generativity of genetic language belongs to RNA. Different forms of RNA can be considered as an actualized representation of the T-state in the framework of Lupasco, included thirds between X (DNA) and Y (pool of proteins). DNA and proteins interact in a hypercyclic (see below) living structure, in which the dualisms are partly actualized and partly potentialized: DNA represents an actualized potential (via encoding) for proteins and protein molecules via their enzymatic activity provide a potential for transcription and translation of the coding system of DNA [17]. RNA, in turn, generates a high flexibility in this dual relationship possessing both the coding and the enzymatic properties. Thus, the RNA pool represents a powerful source of the "formal superposition" which represents the mechanism driving evolutionary adaptation.

Processes which oscillate between actuality and potentiality, but do not return to the identical point from which they started, actually describe a spiral in configuration space. We have concluded, and this point will hopefully be developed, that the elements of a recursive process, for example in communication as described by Leydesdorff, also evolve in a spiral form. It is essential to avoid the term cyclic for such paths. The concept of 'hypercycle' applies to such processes, and also refers to the presence of several entangled spirals at different stages of development.

It is true that generativity in the system "gene–protein" is limited to evolution by means of mutations and their selection. However, in the system with an intermediate component (RNA), a huge flexibility appears and its evolution can take place as a complex linguistic process, in which RNA becomes a part of the temporal component, mediating incomplete identification and increasing the plasticity between genes and their realized functions.

In the classical scheme of the central dogma of molecular biology, the general role of all forms of RNA is not clearly defined. In our view, the most general theoretical role of RNA consists in the generation of possibilities of new Gödelian (hypertextual) statements which can be memorized in genome. Different types of RNA fulfill a generative function that consists in providing dynamical flexibility in the system and enabling potentially infinite unfolding of the genetic language. In this anticipatory language game, RNA moderates an incomplete identification between genes and functions and increases the plasticity of this interaction [22]. Witzany [57] considers different ways of the appearance of new evolutionary features on the basis of genome editing with participation of different types of RNA molecules. The total pool of RNA is divided in the translational and combinatorial pools, the latter is formed through generative dynamics and can be memorized in DNA via reverse transcription [24]. Thus, the field of the set of RNA molecules, expanded by the number of possible combinatorial rearrangements, generates a potential field of possible realizations and represents the partly-actualized, partly-potentialized T-state in the sense of Lupasco's Logic in Reality.

17.7 Relational Biology as the Basis for Theoretical Biology

17.7.1 The Relational Biology of Rashevsky and Rosen

Relational biology was introduced by Rashevsky [44] and further developed by Rosen [46]. It describes life as ontologically independent generic phenomenon. The generic property of living systems can be analyzed as possessing self, which is related to internal determination (Aristotle's "*di aytoy*") or "closure to efficient causation" [46]. Life corresponds to a certain relation between elements, and its physical structure can possibly be substitutable. In the system closed to efficient causation. 158, the internal determinant, defined by Lefebvre [28] as "eidos-navigator", ultimately appears. A living system has at least one point of determination by the eidos-navigator, so any living body has a point referred to its "self" in which its movement is not determined physically. This movement unfolds into a structure that reflects the internal choice.

For understanding the nature of living beings, the problem of a self internally determining their activity, has to be analyzed in detail. The "self" is characterized by a spontaneous activity that introduces computation into the real world, which itself represents a non-computable decision attributed to the living system. Schrödinger [48] was the first who suggested that the nature of self is quantum mechanical and placed it beyond quantum reduction. It corresponds to the internal quantum state (IQS, the term introduced by [20, 21] that is the potentiality that directs possible actualizations, is delocalized and pre-programs the a priori forms of space and time, generating the spatiotemporal frame in which the world is observed.

When there is no "active" observer concerned, the modeling of space–time results in Newtonian geometry with a single unitary Euclidean coordinate system. When we consider the observer "actively" measuring space–time by signals with finite velocity, we come to Einstein's relational space–time but in which all coordinate systems are equal. When the effect of measurement is taken into account, we come to the choice (generation) of coordinate system and its transformation toward a stable (or homeostatic) state as an attractor. This generates various structures whose generation can be described by non-Euclidean geometry. For biological systems, we introduce a virtual notion of an endo-observer realizing the internal measurement and establishing space–time relations.

When we consider an individual living system embedded in physical space–time, we assume that the spatiotemporal structure of the living system itself should have certain optimality parameters that allow coordination of systems' processes inside this structure and result in optimal fitness in their embedding into the external space. This principle of optimality is an evident basis for understanding space–time generation by living organisms. Robert Rosen in his earlier book *Optimality Principles in Biology* [45] showed how morphogenetic structures correspond to most optimal functional solutions. Rosen considered some empirical relations that govern evolution, growth, and the transformation of biological forms. He analyzed feedback control principles to explain optimality in different, in particular related to morphogenesis, processes. In fact, living system is involved in a continuous internal measurement to maintain its structure and optimally respond to the challenges of environment.

17.7.2 Physics and the Relational Foundations of Biology

The development of a concept for explanation of the emergence of new actualizations is a difficult philosophical problem. It had different ways of solution in biology, and recently the same challenge appeared regarding the credibility of the multiverse idea in physics. A principle that is beyond standard logic was introduced in biology as the principle of natural selection. This principle, which was formulated by Darwin but appears in Lucretius and earlier in Empedocles, has been viewed from Darwin's time as the most fundamental principle in biology. It was generalized in the mathematical concepts describing evolution, and then was applied to the origin of life as the formation of hypercycles [10]. It still dominates as the basis of explanation in biology today. The application of this principle to non-living matter was initially abandoned until it was applied in cosmology as a 'natural selection' of universes [50]. It is always easy to use this principle as the basis of a final explanation of everything that is observed in the Universe including the actualization in quantum measurement in physics. However, we need to keep in mind the limitation of this principle for the explanation of actual existence and realize its philosophical deficiency despite the apparent transparency of such explanations. In the changing world, i.e. in Reality, there is no fixed "environment" in which the actual is selected, and this diminishes its role as a "frame of reference", not only in biology. If space–time were absolute in the Newtonian sense, such a principle could work, but in a relational Universe its applicability is limited.

In physics, the objective of a principle of "natural selection" is to bring the final cause for observability, i.e., to replace the anthropic principle, which refers to the fundamental constants in our universe being of the right size to permit life. The principle in biology is based on the concept of the survival of the fittest, which is also teleonomic in the Aristotelian sense. This principle, introduced explicitly by Lucretius, who borrowed it from Epicurus and probably from earlier philosophers (Empedocles), can be viewed in a more general paradigm as the consequence of spatiotemporality generated by the multiplicity of actualizations in reality. It, however, cannot directly explain complexification and the necessity of evolutionary growth unless it is viewed as a consequence of the more general principle of dynamic opposition and the actualizations and potentializations resulting from it.

Relational space–time as a condition for observability [23] should have an internal reference frame to meet the criteria of universality for all observers defined by the set of fundamental physical constants. These constants correspond to observability of the world and represent the natural limits of computation that generate the observable physical Universe [20]. (It is possible that their values themselves may evolve in a kind of meta-evolutionary process [38]. The physical world can be viewed as a set of active units that perform quantum measurements. In the original conception of Leibniz, as above, we can call them monads. An a posteriori monadic description of phenomena can be made, for example, under certain conditions, when the measurements are held for prolonged times with precise outputs within organized structures (bodies) where a higher monad rules other simpler monads, cognitive phenomena and consciousness arise. None of the monads act on any other directly but the patterns of their spatiotemporal representation physically interact in the external actualized world.

As far as we are concerned, and as discussed in Chap. 10 and elsewhere in this book, such monadic descriptions of the world are superannuated in philosophy and science. They are superseded by our logic of energy and our most recent casting of it in terms of ontolons.

Despite the formulation of the theory of relativity in 1905 (special) and 1915 (general), and of the universal acceptance of this theory as the basic fundamental paradigm, the physics of the twentieth century largely maintained a substance understanding of space–time. It incorporated such concepts as the age of the universe, its inflation and expansion, and raised the question what was before the Big Bang. The alternative to the general theory of relativity model suggested by Milne [37] excluded gravitational interaction from the model of an expanding universe and was not widely supported. In fact, quantum measurement occurs only in the gravitational field. All other thermodynamic fields have entropy, but the complexity of structures in the physical universe is related to the basic field, which is gravitational.

Complexity appears to be relational being not the property of the system but the property of its observation [46]. Relational complexity implies also relational entropy which acquires exact value only in relation to the primary non-entropic field appearing to us as the gravitational field. This actually means that the expanding property of the universe follows from the development and complexification of patterns generated in individual quantum measurements performed AS monads. Kineman proposed the sketch of theory that he called the "relational self-similar space-time cosmology" [27] where individual substances form the spatial relations to each other, which logically refers to the fundamental features of windowless monads in Leibniz philosophy that coexist forming the spatial order. The temporal relations in a similar way logically are the origin of arise to the timeless characteristics of individual monads.

This understanding can be seen in the statement of Heraclitus: "An invisible harmony is better than a visible one". The realization of computation could be possible only given certain preconditions expressed in the basic symmetries of transformations of physical objects and corresponding to fundamental physical laws [41]. However, we may not be able to formulate these preconditions in a deductive way and can only judge them from the observed structure of the visible universe. It seems that there is no way of substantiating fundamental constants outside of the anthropic principle because the parameter of the stable universe in which consciousness can appear remains the final cause of the existence of these parameters. Outside this final cause we cannot evaluate a possibility of actualizations with different sets of fundamental constants. We are locked in the world of the given fundamental constants and physical laws. The possibility of reflexive loops representing life and consciousness in a world with different physics may not be intelligible for the beings inhabiting this particular physical world. This may be the main reason that we can only follow reality as it appears to us and the statement of Wittgenstein "Everything we see could also be otherwise" cannot be assessed in practical activity. This statement belongs to a class of anti-realist positions which to some extent reflect a *wish* that things were different. The possibility of formulating this question, even if no answer is available, confers a feeling of desire for omniscience, omnipotence and immortality, attributes of an entity it is not necessary to name. Our reality is one of change, including our eventual disappearance, and it verges on irresponsibility to not devote our energy to understanding *it*, and acknowledging our real limitations.

Existing physical parameters may strictly conform to the observability of the world and represent a unique solution to the problem of free will and consciousness, as suggested by the anthropic principle. The free will theorem of Conway and Kochen [8] states that, if we have a certain amount of "free will", then, subject to certain assumptions, so must some elementary particles. The minimum action defined by the Planck's constant already implies certain freedom of will. It generates the loop of space and can be the basis of retrocausality, so a particle can be multiplied via the reversal of time and, *e.g.*, the multiplicity of electrons occurs via the movement of one particle forward and backward in time forming a complex world line as in the one-electron universe of Wheeler and Feynman (see [12]). Retrocausality is observed at the level of elementary particles, while the downward causality is present macroscopically and refers mainly to complex systems.

The existing values of fundamental constants may represent the only solution for the shielded from thermal noise coherent states of living beings. This unique solution may appear beyond mathematics; however, it *can* be substantiated by the sets of empirical data which show that it perfectly fits to the observability of real world. Like the reduction of uncertainty during decision-making occurs in the unconscious prior to its awareness [30], the proof of validity of the fundamental constants comes in a way as Diogenes 'proved' the existence of movement by walking, i.e., via establishing the limits of real processes, computable or not that shape the physical world. The Philosophy in Reality is thus the point of view of 'walking-Diogenes'.

17.7.3 Biological Transactions: Relational Biology and Internal Measurement

In modern theoretical biology, the internal measurement concept is suggested as part of the basis for an explanation of the phenomenon of life [18, 35, 39]. Quantum measurement occurs during the action of a measuring device on the measured system. When the measuring device is a part of the measured system, the measurement proceeds internally in relation to the whole system and the latter becomes more complex as a result of the measurement itself. This leads to an iterative recursive process which appears as the development and evolution of the system. The quantum measuring device can be organized in such way that it "encodes" the system in the course of the transactional interaction with the measured object and makes it possible for the measurement to proceed in a regular way. In this case, a system can memorize the evolutionary complication of organization and evolve further, in other words it contains an internal description memorizing the measurement result (the result of transaction). The evolutionary increase of complexity becomes possible when the genotype appears as a system distinct from the phenotype and embedded into it. This separates energy-degenerate rate-independent genetic symbols from the rate-dependent dynamics of construction that they control [40].

The internal measuring device measures, among other elements of the measured system, itself in the state of measuring, which limits the possible accuracy of quantum measurement—a manifestation of the Heisenberg uncertainty principle [34]. The increase in complexity as a result of quantum measurement corresponds to the increase in complexity of living organisms in the course of evolution. Indeed, evolution can be viewed as a potentially transfinite recursive embedding process, where life evolves as measurement of environment, and, through living organisms, becomes embedded into this environment, which affects its further development by its own presence. According to Magdalef [33], a measurement (observation) of the environment becomes a source of evolution of the living system itself. The complexity of environment increases as a result of life, which, in turn, produces more complexity in life as reflection of this fact in the course of measurement at the next level of recursion. Uncertainty comes about as a necessary consequence of such an embedding measurement. Thus the increase in complexity occurs simply as a result of quantum measurement, viewed as a biological transaction. Biological evolution, viewed as adaptation to the fitness landscape changing in the course of evolution, becomes partly its own cause, a universal property of the living world.

To describe the fundamental structure of a living system, Rosen introduced the concept of (M, R) system where M refers to metabolism and R—to repair (or replacement as in subsequent interpretations). The system is internally closed because metabolism uses its own components for replacement and is dependent on the external influx of matter and energy only as a material, but not efficient, cause. The level of complexity corresponding to (M, R) systems is homologous to a self-maintained internal model and has advanced generational (or "generic" in Rosen's terms) properties. The structure of (M, R) system is an example how the internal logic generates the topological structure and the abstracting capacity of the system. At the very high level of complexity, double homunculus structures appear having an internal reflexive model of themselves and corresponding to conscious beings [47].

17.8 Ontolons and Semiosis in Living Systems. Second-Order Non-linearity and Poincaré Oscillators

17.8.1 Non-linear Interactions in Living Systems

The principle of "closure to efficient causation" introduced by Rosen for living systems, can be seen as the application of Leibniz's "no windows" principle in biology. In the quantum mechanical concept of self, the growth of complexity results from living activity as a necessary consequence of the embedding measurement in which the reduction of uncertainty takes place. Evolution is a process that aims to overcome the physical limits of computability in which the incomplete identification appearing as an uncertainty in the measurement process is read and interpreted as a basis for new realizations. In such a process, the environment continuously changes in the course of adaptation, and evolution becomes a *generic* phenomenon having its own cause. The Universe, according to Kineman [26], consists of the units called "holons" that possess self and correspond to Leibniz's monads. The wave function superposition is limited by the single domain and does not expand outside it which corresponds to the statement of Leibniz that monads have no windows. As we show in this book, the notions of "holons", "monads", or "things-in-themselves" can be substituted by a more relevant term "ontolon".

We consider a cyclic structure with internal semiosis of the *kykeon*-type (cf. Chap. 9) in the sense of Heraclitus. More accurately, it should be described as hypercyclic as in Sect. 17.6. Generally, one tries to use a non-linear system of dynamical equations to describe a complex system. A complex system is loosely defined as constructed by a large number of simple, mutually interacting parts, capable of exchanging stimuli with its environment and of adapting its internal structure as a consequence of such interaction. The non-linear interactions involved can give rise to coherent, emergent complex behavior with a rich structure. Key concepts in complexity science are, for example, the coexistence of diversity and identity, for which LIR provides an interpretation. Complexity science also looks at the dynamics of systems in transition regions of self-organized criticality, but as these are difficult to solve, they are approximated by linear ones, but some phenomena may be hidden by this strategy such as chaos, singularities, and solitons or, we now would add, ontolons. Thus, the dynamic behavior of a non-linear system may appear to be counterintuitive and unpredictable.

17.8.2 Limitations of the Concept of Non-linearity

For us, this picture is quite primitive, since the interactions at the heart of a process and their movement from actuality to potentiality is obscured. We see the term non-linear as describing the complex path between prior and subsequent states of a process from both mathematical and physical perspectives and, we now add, logical ones as well based on Logic in Reality. Thus, any movement from cause to effect can be seen as a highly non-linear, multi-dimensional process. But in what exactly does the non-linearity of non-linear dynamics consist? Many physical phenomena are described as emergent: tornadoes certainly arise from complex temperature and humidity gradients, and such systems are considered to be the consequence of non-linear dynamic interactions. From the LIR standpoint however, they are (almost) pure, actualized macrophysical processes with no form of internal representation or semantics. Other examples are the dissipative, far-from-equilibrium systems described by Prigogine, and the intrinsically simple structures such as the convection cells in heated liquids or certain oscillating chemical systems, as discussed below. Bauer [5] was probably the first who considered far-from-equilibrium states as internally determined by living systems and not externally imposed on them.

Our conclusion from this brief overview of the standard notion of non-linearity is that it is not a domain to which the principles of LIR apply. We might call the phenomena referred to, to try to characterize the relation, as *first-order* non-linear processes. If we look at phenomena instantiating the properties of recursion and self-reference, and, consequently, of the Principle of Dynamic Opposition, we start to see an underlying unity which justifies referring to them as instantiating *second-order* non-linearity. We thus approach, from another perspective, the description of processes in terms of a *hypercyclical* evolution (see above).

Further parallels can be drawn to concepts of Synchronic Downward Causation. We propose that this semiotic system first described by von Uexküll [54] in relation to living systems represents the basic structure of an ontolon as the unit of actualized substance bearing not only spatiotemporal but also cognitive attributes. In a different context, this structure was introduced by Barham [4] in his model of dynamical informational semiosis. In his theory of the meaning of information, he identifies meaningful functions with generalized non-linear oscillators and their associated phase space attractors. By postulating the existence within such oscillators of a component capable of coordinating low-energy interactions with the correct environmental conditions supporting the dynamical stability of the oscillator, the meaning of information is interpreted as the prediction of successful functional action. This can be considered as the elementary structure of an ontolon, extending Barham's definition of 'epistemon'. The epistemic function should be considered as a secondary to the ontic nature of such a structure: it is the basic unit of nature that can be seen as constituting a complex cognitive system. Robert Rosen's (M, R) systems are other representations of such ontolons pointing their closure to efficient causation and relative independence of external factors, but not total separability from them.

17.8.3 Expansion of Non-linearity: Poincaré Oscillators

In fact, the original invention of such structures should be attributed to Henri Poincaré. He introduced non-linear dynamics and differential topology in physics in his concept of the Poincaré map, which describes the simplest non-linear system called the Duffing oscillator. Glass and Mackey [13] suggested that all systems exhibiting limit-cycle behavior should be termed Poincaré oscillators. This structure can be viewed as the basic structure of the ontolon that is realized at different levels of organization. The periodic orbit of the continuous dynamical system is stable if and only if the fixed point of the discrete dynamical system is stable. The trajectory of such oscillators can be called a periodical attractor or limit cycle. The simplest Poincaré oscillator does not assume a developed internal structure, but a complex internal structure can potentially arise as a result of its evolution towards the expansion of its external stability supported by more sophisticated internal dynamics attributed to its "agency". What is important to point out here is that the interaction of Poincaré oscillators having internal information dynamics can be described as a discourse, whose result may not be deterministically anticipated. The oscillator itself has its internal rule that appears as its internal logic, and for its understanding in our description as a meta-logic, it is necessary to describe its behavior as a dialectical process.

While classical mechanical movement can be described by bivalent formal logic, the description of the movement that involves control of potential realizations involves another logical scheme that incorporates potential-actual states as prerequisites of transition (LIR). This meta-logic is the logic of non-mechanical causality and it can be defined as a dialectical logic in its broad sense. It treats causality as the achievement of a solution established in the discourse of communicating ontolons. In this communication, ontolons can oppose each other or unite in larger ontolons in which the unified system of control is formed. We will further consider the internal structure of ontolons corresponding to living systems and to conscious systems. But if we consider all hierarchies of ontolons, we should briefly return to non-living reality and emphasize that even at the lowest quantum level reality is or can be seen as having two attributes, one spatial and the other dynamic. This is reflected in the characteristics of quantum as a "wave-pilot" in the de Broglie interpretation.

In relation to the phenomenon of life, we assign a presumed ontolon structure, based on the internal constraints that support the uniqueness of each biological system and its self-maintenance. In this representation, it is possible to understand how biological systems enter into a dialectical discourse that corresponds to the operation of the LIR Principle of Dynamic Opposition. It is thus an oversimplification to view biosystems as simply separable into two functional parts, corresponding to a hardware and software. This static approach is still popular in biosemiotics and other rather reductionist concepts. Such concepts ignore the dynamic aspect of semiosis which consists in the fact that the significant system of the genome is internally reproduced within the system and is repaired through sets of internal constraints representing the feedback from the elements that it encodes (which is described in Rosen's relational biology rather statically). Such a system represents a whole that incorporates a locally stable point attractor of the Poincaréan type.

The apparent independence of symbol systems from physical laws follows from the analysis of Poincaréan type oscillators representing the basis of the semiotic dynamical cycles described by von Uexküll and Barham. The view that genetic symbol systems have evolved so far from the origin of life and that semiotics does not appear to have any necessary relation whatsoever to physical laws is also true, but it occults the fact that the processes involved instantiate the same categories of Dynamic Opposition and Non-Separability. Pattee states that information is not in the category of universal and inexorable physical laws, but refers to initial and boundary conditions [40].

17.8.4 Constraints in Dynamic Semiotic Systems

Boundary conditions formed by local structures are called constraints, while informational structures such as symbol vehicles are a special type of constraint. Pattee claims that life and evolvability require the complex interaction of rate-independent symbol constraints and rate-dependent physical dynamics [40]. However, the concepts of initial conditions and constraints in physics make sense only in the context of the law-based physical dynamics to which they apply. This is also the case for the concept of information. We agree with Pattee that the illusion of isolation of symbols from matter can arise from the arbitrariness of the apparent epistemic cut. Further, the apparent isolation of symbolic expression from physics seems due to an epistemic necessity, but ontologically it is still an illusion, making a clear distinction is not the same as isolation from all relations. In general, one clearly separates the genotype from the phenotype, but from an operative standpoint, one certainly does not think of them as isolated or independent entities.

Further elaboration of the matter-symbol problem is possible using the two-level framework of analysis implied by LIR. If the illusion of isolation is an epistemic illusion, whose reality is accepted, this means that symbolic expression is not metaphysically isolated from physics. Consequently, their relation or interaction is real, and it can be considered to have an appropriate dynamics. The remaining question concerns the use of antagonism or constraints to characterize these dynamics. This can be resolved by a view of symbolic memory constraints as dynamic processes in themselves, co-evolving with the other components of biological systems. This is similar to the statement of Hegel [16] about memory as a sign-creating activity.

17.9 Rosen's Model and an Endoperspective

Besides its error-correcting cybernetic controls, a living system contains anticipatory mechanisms to pre-empt possible errors, as discussed by Rosen [46] and others. These controls are realized through the agency of a predictive model, converting present information into predicted future consequences. These are *ipso facto* essentially potential and we propose that their evolution follows the rules of LIR.

As noted above, Rosen suggested an alternative to a classical dualistic genetic model of the biological system. He called it the (M, R) system where M is metabolism and R is repair or, as later suggested, replacement [29]. Elements of the metabolic system are continuously replaced and the elements that replace them are also replaced, and this can go to the infinite regression. However, Rosen stated that the system can be "closed to efficient causation" and contain the internal principle of organizational invariance [46]. This results in avoiding infinite regression and placing the system in a stable non-equilibrium state in which the system, remaining open to material flows, becomes selective for them and closed to the efficient causes that are locked inside of it. This structure is based on the principle of "organizational invariance" [29] that escapes the infinite regression for the internal system's time T, during which the system stably performs its function and avoids "global system failure".

Rosen's theory formulates the basic ontolon structure of living systems. Other approaches include Eigen's theory of hypercycles [10] and the autopoietic theory of Maturana [36] and Varela [56]. The latter have common features to Rosen's model but remain less developed in relation to their formalizable and hence logical and ultimately computational value. Rosen's (M, R) system includes the operation which designates the system as a whole and acts as a generator of the complete enclosed structure of such systems. There is no rigid algorithm to apply as this operation has its own ambiguity. When it is applied, the ambiguity becomes frozen and internal computation becomes possible (cf. the 'frozen dialectics' of Lupasco [32]). This is basically a recognition of the real-life phenomenon that dynamic systems can and do slow down and even stop, before restarting. Later, Gunji et al. [14] came to the idea of describing the organizational invariance of a biological system from the point of view of "heterarchy" which involves self-reference naturally through the inherited logical inconsistencies between levels. One description of a heterarchy is as a dualism of the property of self-reference and the property of a frame problem, interacting energetically with each other. The research program of Gunji represents an important current development of Rosen's ideas in which a successful explanation of internally constrained principles of animal behavior and human consciousness has been obtained [15].

Rosen's understanding of organizational invariance is similar to Gödel's encoding of statements about the system within it, applied to biological systems. The parameter representing organizational invariance is equivalent to an agent establishing the set of Gödel numbers generated within the system. The whole biological system can be viewed as consisting of: (a) Metabolism—sets and relations; (b) Replacement—relations on relations; and (c) Organizational invariance—Gödel statements about the sets actually in place and their dynamic relations. It is important to note that Gödel statements are not sets or relations; they are meta-mathematical statements with the dual function of both sets and relations. Logic in Reality is the first metalogical system in which such statements can be embedded and given a real interpretation.

Thus, the triadic structure of life includes sets, relations, and meta-mathematical as well as metalogical statements (encoded within the system) that are as real and as causally efficient as any physical element of the system. Translation of these statements into the system occurs via a sophisticated 'machinery' of transcription, translation and recombination, *e.g.*, splicing events. We remind the reader that any such event is a consequence of the operation of the real dualities or polarities in chemical and biological systems and the movement within them between potentiality and actuality and vice versa.

17.10 The Epistemic Cut and Living Processes

The concept of an epistemic cut, mentioned briefly above, was originally formulated by von Neumann in his demonstration that the function of measurement of some physical variable is irreducible to the dynamics of the measuring device [40]. The logic here is related to the necessary separation of symbolic memory and the dynamic laws required for the self-replication of a biological system. It has been considered as a special case of a general epistemic problem: how to bridge the separation between the observer and observed, the controller and the controlled, the subject and the object.

An epistemic cut appears in a view of dynamical laws which requires that such laws and the initial conditions of a system are sharply separated, the initial conditions are capable of being measured, and measurement and laws have no reciprocal influence. This intellectual distinction between initial conditions and laws allegedly has its origin and embodiment in living organisms. In this conception, our perceptions as well as our natural languages support a deterministic, either-or logical syntax and causal semantics that conform to a classical dynamics. This happens to be true. We would say science is burdened with this concept of state-determined behavior as a modern form of Laplacean determinism, but it does not validate these considerations as the basis for a theory of biology in particular or reality in general. Such a view of syntax and dynamics is suspect, since it fails in many areas in addition to quantum mechanics. For example, natural language cannot be described even by categorial extensions of classical logic.

Non-integrable conditions, or constraints, first noted in the limits imposed to on Epicurean *clinamina*, can be proposed for bridging the epistemic cut. One constraint states that, in order to provide configurational space for hereditary processes, there must be more degrees of freedom available for the description of the total system than for following its actual motion. As stated by Pattee, since law-based dynamics are based on energy, in addition to non-integrable memory reading, memory storage requires alternative states of energy. Constraints are formally equivalent to laws, and the evolution of systems depends on both. The complementarity of dynamic laws and the measurement function is irreducible, based on a demonstration by von Neumann that the contrary would lead to an infinite regress of measurement devices operating on the systems of systems plus measuring devices and so on. However, Pattee makes the assumption that epistemic irreducibility does not imply any ontological dualism and that it arises whenever a distinction must be made between a subject and an object, or in semiotic terms, between a symbol and its referent. An ontological (read metaphysical) dualism is exactly what results, and the consequences are subject to the Leibnizian analysis (of similarity and difference, etc.). If the terms are different, they cannot communicate or interact; if they are the same there is no cut. The only possibility of a bridge is that they are the same and different, and for this to have meaning requires a Lupascian framework.

The classical view of logical disjunction is of something is totally different from something else. Is the epistemic cut, then, essentially equivalent to the classical disjunction classical? It is, as the terms are only epistemologically but not functionally connected. No one would think of 'separating' conjunction and disjunction. However, this does not confer any additional reality to the cut, but demonstrates its limitations. Possessing "self", i.e. being a subject means primarily instantiating actualization (efficient cause) and being an object instantiating potentialization (final cause). One can easily associate potentialization with symbolic control and actualization with measurement, following the approach implied by the categorial features of Subject and Object, without the standard categorial features of exclusivity and exhaustivity have been eliminated in LIR.

17.11 Conclusion: A Systems Picture and a Philosophical Perspective

The concepts of LIR explicate the systemic-historic perspective of developmental and evolutionary biology. Its chief tenet is that an epigenetic structural drift that is not solely genetically determined constitutes the ontogeny of an organism. Biological epigenesis implies that although the development of the phenotype is made possible by an initial structure including, but not limited to, the genome, it is not totally determined by it [7]. The genotype–phenotype relation (phenotype as cell or complete organism) contributes to the expression of new structural features, but only by participating in a process that takes place in a structural context that is distinct and operationally complementary to the genotype. The LIR picture is very similar, but provides in addition a description of the lower level, contradictorial processes that combine or couple to result in this complementarity.

In the gene-centered view, genes establish and determine the direction that the structural change of the whole undergoes, independent of the prior dynamics of the whole. The systems-historical view stresses the reciprocal relationship between the whole and the parts and the organism/environment relationship for ontogenetic changes during development of the phenotype as well as changes in the genotype or in any other component in the evolving lineage. The phenotype is the result of development understood as the ontogenic history of the individual. In other words, an organism is a unique organized whole of mutually correspondent parts that exist only in realizing a particular mode of relationship with their environment. It is not a consequence of design ('intelligent design'), acting an Aristotelian final cause, and not a result of an internal component, the gene, as an efficient cause, acting as a plan or program of construction.

Cells as biomacromolecular systems must have the capacity for continuous structural change and be at the same time discrete, as noted above, with a self-generated boundary as a condition of existence. Biological macromolecules are ontologically related to cells in the same way that organs are related to organisms. They do not exist nor can they be formed in nature outside their structural context or a laboratory environment. In the latter case, it is the cellular structural context that has proven even more difficult to duplicate. Both biomacromolecules and the cells that they compose are assumed to have arisen together in a historical process of origin and evolution of cells as multi-structural entities.

The preferred answer to the question of the origin during evolution of processes with global system functionality is that (1) a global system itself contains the relevant structural aspects, and individual organisms form and exist within it, but (2) there are also mechanisms for the global co-evolution of the traits necessary to propagate the global system. Both an ecosystem, for example, and the individual members of it themselves contain, as potentialities, some of the relevant structures necessary for such co-evolution. This view is consistent with the idea that natural selection has a major functional role in the potential production of new combinations of phenotypic character traits, but that the effects of mutations of the genome are constrained by the interactions with the environment of the organism's existing systems resulting from the non-mutated genes already present.

17.11.1 A Philosophical Reflection

Most readers will not have the familiarity with the foundational concepts in biology and chemistry that underlie the preceding section. We recall that its intent was to give the reader here, as in other chapters, some feeling for the principles underlying "life itself", to use the pregnant title of one of Rosen's books. By emphasizing our common biology (and physics and chemistry) as earlier in this book, we like to think that its will be easier for the reader to give, literally, substance to the principle of the common good. In the next two chapters, therefore, we will look at the principles of a 'philosophy of society', in which we as individuals are embedded, from two systems perspectives.

References

1. Aristotle: De Anima (On the Soul). In Barnes, J. (Ed) The Complete Works of Aristotle, pp. 641–692, transl. by Smith, J.A. Princeton University Press, Princeton NJ (1995)
2. von Baer, K.E.: Redesn gehalten in wissenschaftlichen Versammlungen und kleinere Aufsätze vermischten Inhalts. H. Schmitzdorff, St. Petersburg (1864)
3. Barbieri, M.: Code Biology. A New Science of Life. Springer, Dordrecht (2015)
4. Barham, J.A.: Poincaréan approach to evolutionary epistemology. J. Soc. Bio. Struct. **13**, 193–258 (1990)
5. Bauer, E.S.: Theoretical Biology. Akadémiai Kiadó, Budapest (Originally published 1935) (1982)
6. Berg, L.S.: Nomogenesis: Or, Evolution Determined by Law. M.I.T. Press, Cambridge, UK (1922)
7. Cecchi, C., Vargas, A.O., Villagrá, C., Mpodozis, J.: Answering cuvier: notes on the systemic/historic nature of living beings. Cybern. Hum. Know. **11**, 11–19 (2004)
8. Conway, J., Kochen, S.: The free will theorem. Found. Phys. **36**, 1441–1473 (2006)
9. Cornish-Bowden, A., Cárdenas, M.L.: Contrasting theories of life: historical context, current theories. In search of an ideal theory. BioSystems **188**, 104063 (2020)
10. Eigen, M., Schuster, P.: The Hypercycle: A Principle of Natural Self-Organization. Springer, Berlin (1979)
11. Emmeche, C.: Closure, function, emergence, semiosis and life: the same idea? Closure; emergent organizations and their dynamics. Ann. N. Y. Acad. Sci. **901**, 187–197 (2000)
12. Feynman, R.P.: The theory of positrons. Phys. Rev. **76**, 749–759 (1949)
13. Glass, L., Mackey, M.C.: From Clocks to Chaos: The Rhythms of Life. Princeton University Press, Princeton, NJ (1988)
14. Gunji, Y.P., Sasai, K., Wakisaka, S.: Abstract heterarchy: time/state-scale re-entrant form. BioSystems **91**, 13–33 (2008)
15. Gunji, Y.P., Shinohara, S., Haruna, T., Basios, V.: Inverse Bayesian inference as a key of consciousness featuring a macroscopic quantum logical structure. BioSystems **152**, 44–65 (2017)
16. Hegel, G.W.F.: Philosophy of Mind, trans. by William Wallace. Oxford University Press, Oxford, UK (Originally published 1830) (2010)
17. Hu, Z., Petoukhov, S.V., Petukhova, E.S.: I Ching, dyadic groups of binary numbers and the geno-logic coding in living bodies. Prog. Biophys. Mol. Biol. **131**, 354–368 (2017)
18. Igamberdiev, A.U.: Quantum mechanical properties of biosystems: a framework for complexity, structural stability and transformations. BioSystems **31**, 65–73 (1993)
19. Igamberdiev, A.U.: Foundations of metabolic organization: coherence as a basis of computational properties in metabolic networks. BioSystems **50**, 1–16 (1999)
20. Igamberdiev, A.U.: Quantum computation, non-demolition measurements, and reflective control in living systems. BioSystems **77**, 47–56 (2004)
21. Igamberdiev, A.U.: Physical limits of computation and emergence of life. BioSystems **90**, 340–349 (2007)

22. Igamberdiev, A.U.: Time, Life, and Civilization. Nova Science Publishers, Hauppauge, NY (2015)
23. Igamberdiev, A.U.: Time and life in the relational universe: prolegomena to an integral paradigm of natural philosophy. Philosophies **3**, 30 (2018)
24. Igamberdiev, A.U., Shklovskiy-Kordi, N.E.: Computational power and generative capacity of genetic systems. BioSystems **142–143**, 1–8 (2016)
25. Kauffman, S.A.: Prolegomenon to patterns in evolution. BioSystems **123**, 3–8 (2014)
26. Kineman, J.J.: R-Theory: a synthesis of Robert Rosen's relational complexity. Syst. Res. Behav. Sci. **29**, 527–538 (2012)
27. Kineman, J.J.: Relational self-similar space-time cosmology revisited. In: Proceedings of the 54th Meeting of the International Society for System Sciences. Waterloo, Canada (2010)
28. Lefebvre, V.A.: What Are Consciousness, Animacy, Mental Activity and the Like? Leaf & Oaks Publishers, Los Angeles, CA (2016)
29. Letelier, J.C., Soto-Andrade, J., Guíñez Abarzúa, F., Cornish-Bowden, A., Cárdenas, M.L.: Organizational invariance and metabolic closure: analysis in terms of (M, R) systems. J. Theor. Biol. **238**, 949–961 (2006)
30. Libet, B.: Unconscious cerebral initiative and the role of conscious will in voluntary action. Behav. Brain Sci. **8**, 529–566 (1985)
31. Lima-de-Faria, A.: The atomic basis of biological symmetry and periodicity. BioSystems **43**, 115–135 (1997)
32. Lupasco, S.: L'énergie et la matière vivante. Julliard, Paris (1962)
33. Magdalef, R.: Information and uncertainty in living systems, a view from ecology. BioSystems **38**, 141–146 (1996)
34. Matsuno, K.: The uncertainty principle as an evolutionary engine. BioSystems **27**, 63–76 (1992)
35. Matsuno, K.: Quantum and biological computation. BioSystems **35**, 209–212 (1995)
36. Maturana, H.R.: Autopoesis, structural coupling and cognition. Cybern. Hum. Know. **9**, 5–34 (2003)
37. Milne, E.A.: Relativity, Gravitation and World-Structure. Clarendon Press, Oxford, UK (1935)
38. Nakagomi, T.: Mathematical formulation of Leibnizian world: a theory of individual-whole or interior-exterior reflective systems. BioSystems **69**, 15–26 (2003)
39. Pattee, H.H.: The limitations of formal models of measurement, control and cognition. Appl. Math. Comput. **56**, 111–130 (1993)
40. Pattee, H.H.: The physics of symbols: bridging the epistemic cut. BioSystems **60**, 5–12 (2001)
41. Penrose, R.: Emperor's New Mind. Concerning Computer, Minds, and the Laws of Physics. Oxford University Press, London, UK (1989)
42. Petukhov, S.V.: Non-Euclidean geometries and algorithms of living bodies. Comput. Math. Appl. **17**, 505–534 (1989)
43. Pigliucci, M., Müller, G.B. (eds.): Evolution, The Extended Synthesis. MIT Press, Cambridge, MA (2010)
44. Rashevsky, N.: Topology and life: In search of general mathematical principles in biology and sociology. Bull. Mathe. Biophys. **16**, 317–348 (1954)
45. Rosen, R.: Optimality Principles in Biology. Springer, Berlin (1967)
46. Rosen, R.: Life Itself. Columbia University Press, New York, NY (1991)
47. Sawa, K., Igamberdiev, A.U.: The double homunculus model of self-reflective systems. BioSystems **144**, 1–7 (2016)
48. Schrödinger, E.: What is Life? Cambridge University Press, Cambridge UK (1944)
49. Shapiro, J.A.: The basic concept of the read-write genome: Mini-review on cell-mediated DNA modification. BioSystems **140**, 35–37 (2016)
50. Smolin, L.: The Life of Cosmos. Oxford University Press, Oxford UK (1997)
51. Stent, G.S.: Molecular Genetics; An Introductory Narrative. W.H. Freeman, San Francisco, CA (1971)
52. Thom, R.: Structural Stability and Morphogenesis. W.A. Benjamin Inc., Reading, MA (1975)
53. Thompson D'Arcy, W.: On Growth and Form. Cambridge University Press, London UK (1917)
54. Uexküll, J. von.: Theoretische Biologie. Frankfurt-am-Main, Suhrkamp (1973)

55. Van Valen, L.M.: Homology and causes. J. Morphol. **173**, 305–312 (1982)
56. Varela, F.J.: The Specious Present. In: Petitot, J. et al., (eds.) Naturalizing Phenomenology. Issues in Contemporary Phenomenology and Cognitive Science. Stanford University Press, Stanford, CA (1999)
57. Witzany, G.: Crucial steps to life: from chemical reactions to code using agents. BioSystems **140**, 49–57 (2016)

Chapter 18
Social Systems: Transformation of External Reality by Communicating, Reflexive Subjects

18.1 Introduction. Self-Reference and Social Reality

The emergence of social from biological systems is consistent with the origin of a new level of cognitive self-reference which involves an active anticipatory transformation of externality. While self-reference is the main property of life, constituting the foundation of the logic of biological structures [61, 62], the *internal* significance of externality is related directly to the origin of human civilization. It is mediated by the appearance of a reflexive image of the external world that arises through communication between the human agents constituting social systems. In human reflection, the image of objective world is internalized via a specific type of reflexive loop. This structure represents a logical pattern that describes the interrelation between the self and the external world and can be modeled on the basis of referral of the self to its symbolic image. This is recognized as an external agent that defines basic modes of the subject's behavior. When biological systems acquire consciousness and become social beings, essentially the world as a whole can be cognized. This perpetual activity corresponds to the growth of knowledge in which externality becomes embedded in the cognitive system, and its elements acquire properties of utilitarian tools possessing pragmatic value. The agents inhabiting the external world perceive it via its internalization: external objects become recognized in terms of their affordance, defined by Gibson [17] as a relationship in which the environment and the agent inhabiting it operate in conjunction.

Fundamental aspects of human actions like playing and experimenting or exploring the environment and becoming skilled in 'managing' it appeared previously in animals. These were the basis for the ability to acquire more specific knowledge of using objects as external tools and to the development of social organization. At the level of the nervous system, the idea of externality is reflected in the structures that can generate similar activity, both in the case of the subject's action and in the case of the subject's substitution by the other, i.e., in the observation of the same action as external (e.g., performed by another subject). Mirror neurons are such structures, stimulated by the observation of behavior of the other, as though the observer were

J. E. Brenner and A. U. Igamberdiev, *Philosophy in Reality*, Studies in Applied Philosophy, Epistemology and Rational Ethics 60, https://doi.org/10.1007/978-3-030-62757-7_18

itself acting. In fact, the consistent use of tools becomes possible when they are defined in the language as signs. Externality, being internalized in the structure of human language, becomes the starting point of continuous social evolution. Fundamental reflexive operations, being memorized and repeated in language, establish the basis for creating and maintaining the complex structures of human social institutions [68].

18.2 Objective Patterns of Reflexive Consciousness

18.2.1 Anticipation of Reflexive Structures in Psychoanalysis

The process of reflection, taken not only within the logical framework of consciousness (*res cogitans*) but also in relation to the observable spatiotemporal physical world (*res extensa*), establishes the structure in which the physical event and its sense representation appear as dual constituents of the wide-ranging physical reality perceived as a whole unity of *res potentia* by executed by the *cogito* (cf. Chap. 12 on Communication). This fundamental structure acquires its particular shape in human reflexion, being generated in the process of establishment of the "language-ready brain" [1]. Basic patterns generated in conscious events are associated with the certain types of reflexive loops. The fundamental reflexive structure for human social behavior, metaphorically described by Freud [16] as the Oedipus complex, represents a logical pattern establishing interrelations between consciousness and the external world and thus determining a projection of someone's image into the other, which appears as a possibility to substitute the other [27]. The structuralist interpretation given by Jacques Lacan [42] discovers in the basic subject structure a dynamic flow of the signifier and the signified, united by a "common denominator" through which "the meaningful stops the shift in meaning" and thereby specifies the object of desire. This "denominator" is, in fact, a generalized Cartesian *cogito* including the structure of the subject in its totality.

Incorporation of the image of the entire world into the dynamic process of acquiring knowledge can be viewed as a starting point of social evolution [27]. (Rightly or wrongly, it may be the origin of the notion of an absolute). Lacan [42] used the torus and the Möbius strip to demonstrate how the image of the subject is formed through internal exclusions and external inclusions. Steven Rosen [63] used similar illustrations to substantiate basic principles of shaping the real world from potential reality, the Principle of Dynamic Opposition in operation, so to speak. The biological level in the social reflexive structure relates to the unconscious (*Id* in Freudian terms), while the two other components of Freud's scheme (*Ego* and *Super-Ego*) represent the psychological reality which raises biological reality to a new level of structuration (cf. Chap. 16). Here, the *Super-Ego* appears as a socially determined external codification of the subject's dynamics. The entity of conscious subject has its biological foundation in the background, while its awareness as a self (*Ego*) is formed through a social symbolic codification (*Super-Ego*). While the reflection of *Ego* is efficiently computable, the Super-*Ego*, in fact, is a result of an inverse Bayesian inference [19] and presumes a "robust computation". This operates as a metaphorical representation of a new kind of determination that corresponds to the structure of conscious subject.

18.2.2 Formalized Reflexive Structures in the Model of Lefebvre

Topological depiction of the subject by Lacan, although more generalized than the initial Freudian mytho-semantic representation, remains merely illustrative than strictly formalizable. It therefore requires further analysis to attempt to match the formal framework of mathematical logic. A new development was undertaken by Vladimir Lefebvre [43, 45] who suggested a powerful model of human reflection via mapping of the components of the structure of subject into simple equations of a commutative algebra. Such equations can successfully handle the dynamic oppositions appearing in human reflection within a cognitive limit beyond which these algebras do not commute. We believe this development has been largely and incorrectly ignored in recent versions of social philosophy.

The model of human reflexion, suggested by Lefebvre and based on dynamic oppositions, assumes, in agreement with the ideas of psychoanalysis, that the subject's self-esteem and the feeling of this self-esteem as negative or positive, take place largely unconsciously. This assumption makes it possible to reduce the basic structure of psychoanalysis to recursive Boolean schemes and to formulate specific formal laws of reflexion governing this part of human behavior. Non-Boolean aspects of behavior are not captured in this model. Although in Lefebvre's model the principles of human reflexion are not derived from psychoanalysis, the subject's own image, which converts the initial intention into action—behavior, corresponds to the imaginative activity of the Ego, i.e., how the subject represents (estimates) own intention. The latter is perceived as being imposed by moral law and corresponds to the Super-Ego symbolism. In an internal process of making choices, a reflexive system maximizes the pragmatic status of its image of itself. Reflexive models generated internally by the subject introduce a natural form of computation [12] in the operation with the externality. At the biological level, living systems generate and maintain internal predictive models of themselves and of their environments [61]. Conscious beings are able to expand this capacity by incorporating the elements of external reality into their reflexive models and computing them. This certainly does not imply any validity of computational models of consciousness, which are not sufficient by themselves for the explanation of conscious phenomena in agreement with the ideas of Searle [67]. Consciousness is not a computation but generates the computable models of externality. The connection between the phenomenon of subjective reality and the corresponding brain process is considered as a relationship between information and its carrier. According to such views, mental causation is a kind of information causality that differs from ordinary physical causation because information is invariant with respect to the physical properties of its carrier and can be coded in different ways.

In this non-processual model, each component of reflexion attains one of the two values that are positive or negative (1 or 0), and the triadic combination of these two values forms a specific structure of the subject. The subject's state a_1 appears as a composition of the statements a_2 and a_3, all having either positive or negative values. The structures consisting of two subjects A and B characterize the situation in which reflexion is realized via inclusion of the second subject (B). Gibson [17] emphasized that our behavior depends on what we perceive the other person intentions to be. The construction of formal models of reflexion led Lefebvre to the conclusion about the existence of two principally different reflexive systems, the 'Western' (W) and the 'Eastern' (E), which he originally attributed to the ethical constituent of human behavior. In one of them the combination of events characterized by different perspectives is estimated as negative whereas their separation is estimated as positive, and in the other the combination of 'the bad' and 'the good' is estimated positively, whereas their separation carries a negative value [43]. These systems appear as complementary solutions of the main reflexive equations, and represent parts of the universal oppositions that have been generated in human behavior throughout the history of mankind. In the process of social dynamics, a unique system of dichotomous constructs serves as a special axis for projecting (mapping) of other person's intentions. This model explains the experimentally tested values of binary choice, the patterns of musical intervals in the European culture, and the appearance of golden section in the creations of architecture and art [43, 46].

In the E system, according to Lefebvre, a positive valuation of the combination of 'the bad' and 'the good' in referral to the same system generates a binary opposition to a neighboring system considered as an enemy, while in the W system, a negative estimation of the same combination generates a negotiable relation to a neighboring system. The negotiable relation incorporates the *res potentia* in which a ternary relation is established, where the opposites are in principle balanced by the medium (T) value. The evolutionary dynamics of ternary structures is able to avoid self-destruction and permeate multiple layers of the social continuum at different speeds and different intervals, so the valuable meanings of the opposite system can be incorporated within the culture, being transported from their periphery to center, and the ternary (or generally many-valued) system strives to adapt itself [50]. The "binary" and "ternary" systems defined by Lotman generally correspond to the E and W systems defined by Lefebvre. Interpretations that seek a simple resolution of the paradox of two reflexive systems can be introduced; however, it seems that both types of solutions (based on either W or E) will appear in any future development of the social order. However, if and when a principally new and advanced social organization arises, they may be incorporated in it as resolved constituents of the basic social structure. Paradoxically, Lefebvre's analysis of the W and E systems itself has all characteristics of the W type: it reduces all reflexive choices to bivalent schemes. Strictly speaking, the more advanced reflexive Eastern systems based on continuous logic are simply *not* isomorphic to the binary E systems [48]. A radical binary form of the E system, which is characteristic of totalitarian societies, appears when continuity is distorted by the intervention of binary oppositions into a continuous model of reality. In fact, binary outputs are commonly generated by nervous system in the course of perception of external stimuli [33], and this basic property of the actualization process determines the appearance of simple binary solutions in the course of reflexive actions. We insist, however that these binary solutions represent only a percentage of the dynamic tertiary solutions available to the human subject, beyond the AND/OR dichotomy.

The way to overcome radical models is to transform them to a new many-valued structure that can emerge from the binary representation of the world. In fact, Lefebvre emphasized the difference between AND and OR to distinguish the discrete referential systems W and E. The W system in his model is based on the statement "me (a) seen by you (b) AND me (a)" and expressed logically as a ∧ (a AND b), and the E system is based on the statement "me seen by you OR me", expressed as a ∧ (a OR b). By developing the advanced models based on many-valued and continuous logics, in particular Logic in Reality, it should be possible to assimilate both AND with OR and at the same time distinguish AND from OR, thus smoothing out the contradiction between Western and Eastern cultures. In this framework, the E model represents a ternary relation of another kind, based on other logics (continuous, or continuous/discontinuous). A necessity of the logic such as LIR which refers directly to the state and evolution of processes rather to the statements about such processes may provide a better framework for discussion of psychological dynamics if we assume that these processes are driven by choices made by the endo-observer incorporated into the evolving reality. Because the Eastern thought is less-language dependent *ab origine*, it reflects the latter logic. A continuous logic with the para-consistent interpretation of the real number line allows discontinuities that remain at the ontological level higher than numbers and corresponds to the Logic in Reality picture formulated by Lupasco [51] and Brenner [7].

In the formal model of reflexion, the structure of the subject is modeled by triads of binary oppositions (intentions of the subject), and the total quantity of formed structures is divided into four (it is eight in the simple structure including one subject A). In a certain sense, these combinations are similar to the structure of the genetic code, in which apparent emergence through sets of reflexive actions may be analyzable. The basic representation of such structures was introduced in the Chinese *I Ching* [25]. These combinations infer multiple types of arrangements of interactions between subjects, representing alternative combinations of intentions and resulting in various dynamical patterns of conscious behavior. Their detailed analysis can be a subject of a separate study (see also Chap. 2).

The structure of the subject represents a new level of self-identity of a system based on a more developed abstracting capacity [55] arising from a new set of relations generated by the fundamental reflexive capacity forming the self-conscious subject. This structure generates a new measurement system that potentially can expand the changing border between the internal and external to limits that cannot be reached within the biological realm. This cognitive system also expands the limits of retro-causality that is crucial in shaping biological systems and directing their evolution [56]. The potential of social systems to directly operate on the outside world [19], and thus creatively modify and internalize external reality, exceeds in by many orders of magnitude the corresponding capacity of biological systems.

18.3 The Logic of Consciousness and Lupasco

Applicability of an emergent bivalent Boolean logic of internal choice is related to the classicality of observable space–time [70]. While the law of the excluded third is generally not applicable to dynamic complex systems, the observer, acting inside classically framed objective space–time, is nevertheless permanently involved in a Boolean experiment, in which stable collections of experiences are distinguished, being represented as dynamical systems of the objective world 'roaming' through their possible worlds. Any observation of a dynamical system perturbs its evolution resulting occasionally in the creation of a new observable property of the system with respect to its actual world [70]. In this paradigm, classicality is originally attributed to the logic of the observer, which allows making a selective choice from the field of potentialities, i.e., it has a psycho-biological rather than a physical nature, and by this means the agent/observer chooses the most adequate "habitable" model of the Universe [32, 33].

Operation of Boolean logic in classical systems is reflected in the two major methodological principles defined by Bohr [6] as complementarity and correspondence (cf. also Chap. 4 in relation to LIR). The complementarity of oppositions, when they are actualized from the potential field, means that both opposite concepts can work and interact in the course of realization determining the dynamics of the whole system. The duality of oppositions appears as a complementary realization of potentiality. It forms a new potentiality and possesses the energy as a measure of difference (opposition) that can be converted into work. This applies not only to the physical world but also to human societies operating with the dualities of reflexive choice [44].

The principle of correspondence, taken in its broad sense, means not only the correspondence of the quantum theory to classical physics within certain limits but also a necessity of description of measuring devices in terms of classicality of the time–space of their operation. It means also that biological systems operate in the space which is well approximated to the conditions described by Newtonian mechanics and Euclidean geometry [33]. Sawa and Gunji [64] suggest that a transitive law appears in the model society whose members have knowledge about the causal relations obtaining between subjects. A particular pattern of opinions generates a dialogue in which the local logic is connected with the global one, and thus incorporates a particular member of society embedded into the external world. This external world has parameters in which the spatial orientation and possible forecast of consequences of particular dynamics generated by the subject can be realized. Sawa and Igamberdiev

[65] expanded the concept of reflexive choice by the inclusion of observed objects into reflexive structures where the dialogue of agents generates the knowledge about these objects and assigns meanings to them, thus transforming them into utilitarian tools. The knowledge acquired about the external objects perceived by reflexive agents raises the question of their objectivity as a pre-supposition of using them in pragmatic social activity. Actual physical objectivity has its origin in the objectivity of the ultimate logical structures of the world. These structures are actualized via the unique sets of physical parameters making the world observable and intelligible [30]. Self-maintained reflexive systems exhibit themselves via introduction of the universal computable and non-computable laws harmonizing their communication and interaction with external objects. Operation of reflexive systems using bivalent (two-valued) logic can be further expanded when we consider them using the many-valued (multivalent) logic, and then the logic operating with continuous values [66]. This complicates the initial Lefebvre's model of reflexive consciousness but significantly increases the number of options that arise in the dynamics of human societies, when various complementary types of local values are transmitted globally and become the acceptable alternate rules for human behavior and for world vision. The difference, logically and philosophically, between this approach and that of LIR is that the latter takes explicitly into account the co-operation of discontinuous and continuous values.

Temporal dynamics follows from complementary actualizations of external objects as being measured by conscious reflexive agents. Instead of being permanently actualized, the alternate actualizations are separated by a time interval [29, 31]. The incompleteness of a single actualization in Gödel's sense entails actualization of the opposite and thus prevents the dynamical system from collapsing. As discussed in detail in Chaps. 3 and 4, the 'logic in reality' of this dynamical process that contains "the axiom of included middle", determines the state of three dynamic elements (opposites A and non-A, and the middle state T) and describes dynamic processes in physical reality rather than formal change. It was introduced by Lupasco in 1930 [51], see also [7, 8,9], one precursor was Nicolai Vasiliev [73]. In fact, Aristotle (~350 BC) in his description of the process of actualization, e.g., in the tractate *De Anima* developed a non-classical logical scheme, which is different from the well-known Aristotelian logic operating in the actualized world [2].

While in the classical concept of the world the middle state T is excluded, it is always present in the dynamic process involving the energy of actualization either of primarily A or non-A at a particular moment of time and space, and turning from A to non-A at a later moment of time and space. As a result, A and non-A are not only separated by a time interval and diverge spatially but also form new entities in the course of their dynamical movement, generating new events through the separation and synthesis of the particulars of oppositions, incorporating new spatiotemporal solutions achieved in the course of this dynamics. Time separates contradictory statements, potentiates the opposite and forms a horizon of being in terms of Heidegger [23]. While Husserl [26] developed the model of consciousness by making a "phenomenological reduction" of consciousness from the temporal materialized world, Heidegger aimed to return to the question of how consciousness operates in the finite world shaped by apparent time as a horizon of being. The Being-in-the-World (*Dasein*) emerges from the abstract Being (*Sein*) through this time and is realized in history by an agent ("hero") who transforms the actualized world into a new evolutionary stage.

Moving from the perception of the physical world to the psychology of reflexive choice, we find that the reflexive choice in itself is approximated by the Boolean logic of opposites, while the dynamics generated by the reflexive choice needs the apparatus of the non-classical, non-Boolean logic for its description. The dynamics of reflexive oppositions may resemble the dialectical process described in the philosophical system of Hegel [22] represented there as a movement of the abstract idea, while that of Marx and Engels (see [15]) is considered as being determined on a rather narrowly defined material basis. Contrary to Hegel and in a much broader sense than Marx, Lupasco defined a physical basis for the description of reality. This introduces a causal necessity for thesis to generate anti-thesis that Hegel claimed to be logically deductive. The Lupasco logic corresponds to a real dynamics, but remains a true logic because it permits inferences to be made provided that it includes the endo-observer in a dynamic emergent *middle state*. The Logic in Reality of Lupasco and Brenner, in combination with the psychology of reflexive choice, describes the true contradictions that which govern the dynamics of human societies and, in turn, irreversibly actualize our knowledge about the external world into its utilitarian application. Accordingly, this dynamics does not simply oscillate but generates social progress in the course of social evolution. This 'progress' is by no means linear, but includes stagnation and regression. This is not new; what is, we believe new is that these movements are also logical in the sense of LIR, scientific and philosophical.

18.4 Reflexive Choice as a Filter Operating via the Res Potentia

The discrepancy in the opinion of two agents about external objects generates a possibility of different types of their respective cognitive designations. While the communication process assumes that the word designating a particular object in the language is single, the meanings of this word for different agents can be different. This implies that words acquire their meanings within specific contexts, and this process appears as a selection from a potential superposition of meanings which takes place contextually in the human mind. In other words, consciousness operates in the real world as a filter, and language application is in fact a choice of meanings. The idea of language as a tool that enables the operation of consciousness as a filter belongs to Nalimov [57], who introduced a probabilistic model of language. Following the Logic in Reality of Lupasco as described in Chap. 3 [8], in the course of use of language, the reduction of T to A or anti-A takes place resulting in the reduction of spectrum of meanings that reflects the structure of reflexive choice. The fundamental property of human consciousness, its operation as a filter during its action with meanings of words, distinguishes it from mono-semantic formal languages. This property can be recognized in the frames of an inverse Bayesian inference, which is the basis of a concept of free will underlying the verification of free inverse choice [19]. Language refers to a semantic activity that gives meaning to sounds by codes and by processes of interpretation [5].

A conceptual approach explaining the selection of meanings in the brain was developed by Matsuno [54]. He suggested that the human brain is capable of realizing "non-programmable computation" in which "natural language processor" disambiguates words thus making semantic commitments, as well as revising previous semantic commitments. This continuous process of revising semantic commitments generates new information. Any common language word is characterized by a probabilistic pattern of meanings that can be expressed by the distributive function having a characteristic curve [57]. In the course of using context-dependent words, the selection of meanings takes place from the large domain in which certain meanings are more probable and others less. The process of understanding, according to Nalimov [58], occurs spontaneously and generates a self-organizing structure of the text of human language. Its dynamics is based on the context-dependent spontaneous appearance of a filter $p(y/m)$ interacting with the initial distributive function $p\ (m)$ and generating the function $p(m/y)$ which determines semantics of a new text:

$$p(m/y) = k \times p(m) \times p(y/m)$$

This formula, that was first introduced by Thomas Bayes as a foundation of probability theory, represents here a syllogism of the two premises $p\ (m)$ and $p(y/m)$ that paradoxically generates a new semantics, while k is a normalization constant of this process. Bayesian logic probabilistically selects meanings via application of a filter to the spectrum of all potential meanings. This logical operation has a numerical value because it contains the multiplication sign giving it an arithmetical character. It determines the inapplicability of the law of the excluded third to the operation of human language, i.e., it corresponds to a non-classical multivalent logic of the type developed by Nicolai Vasiliev and Stéphane Lupasco. The Bayesian approach of Nalimov [58] can be considered as a version of a non-classical multivalent logic applicable to the particular aspects of operation of human language during the selection of meanings.

18.4.1 The Problem of Spontaneity in a Bayesian Picture

The philosophical question that remains is the nature of spontaneity of appearance and modulation of the filter $p(y/m)$. It can be related to the issue of the existence of apparent free will of individual cognitive systems in the case of sudden deviations in the process of decision-making. In fact, defining meanings through the application of the Bayesian filter leads not only to the reduction in the probability space by limiting available data but also to the expansion of the probability space for generating new testable hypotheses that in turn generate reductions of the number of probabilities. Thus, the processes of contraction and relaxation of the probability space form the basis of reflexive consciousness. The open quantum systems approach of Asano et al. [3] takes Bayesian probabilities as the personal judgments of individual cognitive agents. The process of reflexive choice acts to elicit a response from the world and to recover it back for that agent; non-quantum language is the only means by which different agents can compare their internal experiences. In line with this paradigm, Gunji et al. [18, 19] propose a measurement-oriented inference system comprising Bayesian and inverse Bayesian inferences. In this model, Bayesian inference contracts the *epistemic* probability space, while the inverse inference relaxes the constraints on this space and permits apparent spontaneous deviations from the classical Bayesian scheme. These two inferences allow an agent to make a reflexive choice that changes the initial probability pattern. This results in the dynamics that underlies any development of entities that include reflexive choice as the basis of their structure. The reader should refer to our 'Bayesian' picture of probability indicated in Sect. 3.3.3. We call attention to our concept of relating probability to the Lupasco logic by defining it as the extent to which a given process will tend to continue on the same path, stop or reverse itself.

18.5 The Transactional Reflexive Action of "Challenge and Response"

Reflexive choice in a relation to a transactional model is always related to the challenge-response relationship. Mimicking the situation at the lower biological level of reality, a formal scheme of binary choice is unfolded into the temporal dynamical process of response to internal or external challenges in the social system. In this process, challenge is viewed as a cause and reflexive choice as a consequence-effect becoming a new cause in the social dynamics. The law of challenge and response, according to Toynbee [69], determines the development of human civilization. The challenge can be related to a particular historical situation or to natural factors. These factors by themselves cannot drive the change unless they introduce a contradiction into the system, which appears in it as a real antagonism leading to counter-actions. The contradiction, viewed, according to Gunji et al. [19], as a binary set of the direct and inverse Bayesian inferences can spontaneously generate an adequate response. The choice of response possesses a certain apparent freedom without any predetermination. This choice, according to Toynbee, comes from a "creative minority" which promotes and realizes the ideas and enthralls the rest of the society, but the existence of that minority does not have to be randomly caused.

The reflexive aspect of challenge and response can be analyzed in the case of the well-known semantic paradox of Epimenides [27, 28], see also Sects. 9.7 and 12.2.1). The contradictory statement of the Cretan "pre"- philosopher of the Seventh Century BC Epimenides "All Cretans are liars" in the social reality is preceded by an unsatisfactory feeling by the members of the Cretan society caused by a particular historical or natural situation. Otherwise, this statement would not be recognized in the society as a challenge. In a certain historical background, this statement will be reflected in the society with the consequences that depend, in particular, on the type of reflexive system of the Cretan majority. Epimenides represents the "creative minority" that assigns to the whole system—all Cretans—the property of signification—liars—finding himself in a unique state within the whole society [27]. In this "meta"-state, Epimenides acquires the position where he is different from being simply one of the Cretans, simultaneously belonging and non- belonging to the set of Cretans. The contradiction in the paradox is an initial causal point of the system's shift to a new state. The society of Cretans through its reflexive action acquires its own dynamics via obtaining the property signified by Epimenides. The process of social reformation is a responsive consequence of this reflection: the Cretans confess to being liars and to want to change for the better, which theoretically would result in reformation of the whole system. This contradictorial construction becomes imprinted in the society, causing its evolution, and to which physical space–time is only the usual precondition.

This model example does not correspond to a specific historical reality but illustrates the structure of many cases of social transformation. It has the same features as the quantum transaction described by Ruth Kastner [39] for the explanation of the phenomenon of quantum measurement. In this "social transaction" the new social spatiotemporality emerges by the way of the transactional reflexive action. In this action, the two potential "waves" that correspond to the challenge and response, one coming from the system itself and the other from an agent, form a "handshake"-type interaction resulting in social transformation.

The transactional model of challenge and response assumes the existence of complex interactions between the opposite types of reflection in the set of fictional characters, whether or not they have an historical exemplar. Generally, a 'hero' (literary definition) or an 'agent' (scientifically neutral definition) who transforms a system belongs to the opposite reflexive system as compared to the majority of the system's representatives. He/she possesses a capacity to formulate a statement as a challenge that potentially can transform the system into the opposite one. A prerequisite of such transformation can be the development of a new concept or technology that cannot be implemented in the current reflexive structure of the system with "outdated" relations of its members, or it can be the pressure from other, e.g., neighboring, social systems. As a result, different spatiotemporal solutions can be realized that include the strengthening of the system in a newly generated structure or its dissolution into the opposite subsystems corresponding more to the pure W and E types (see above) according to the traditions prevailing in local parts of the original system. At the scale of global civilization, these processes can be viewed as components of the general globalization process (see [76]).

Social transaction represents the case of metasystem transition as it was defined by Valentin Turchin [72] and Francis Heylighen [24]. In the course of a metasystem transition, the integration of initially independent or loosely connected components is achieved via the emergence of control of the interactions of these components through the action of a goal-directed agent. The transaction that results in the development of a more complex and flexible metasystem is an example of a metasystem transition that is preceded by various quantitative changes generating the precursors or 'reactants' of the transaction. In this regard, the principle of metasystem transition replaces a statement of the passage of quantitative changes into qualitative changes. The principle of *metasystem transition* should be considered as the second principle of a system's development after the first principle of *dynamic opposition in reality.* It transforms an apparent cyclic dynamics into a spiral spatiotemporal pattern that results from memorization of new information in the reflexive loops that incorporate externality [34]: [35]. The principle of *spiral dynamics* can be regarded as the third principle of a system's development. Hegel's principle of "negation of negation" represents its reductionist version that was transformed to the "third law of dialectics" by Engels [15] , but remained at an epistemic level in the absence of any clear description of the underlying physics.

18.6 Transformation of Eidoi into Technoi as a Basis of Social Evolution

18.6.1 Reflectiveness and Invention of Tools to Transform External Reality

In the concept of Gunji et al. [19], the conscious action of subject corresponds to the inverse Bayesian inference that cannot be described computationally in a strict sense. In this concept, while Bayesian inference contracts probability space, the newly defined inverse inference relaxes the space. It possesses a potential to be the basis for pragmatic use in the process that implies "robust computation". This can be referred to such fundamental practices as the use of tools, agriculture, and early industries. Human history can be viewed as a process of the utilization of the symbolic representation for practical purposes. In this transition, ideal symbolic forms (*eidoi*) become materialized and transformed into tools ("*technoi*") [53]. The same process occurs in modern times: the fundamental concepts of science appear as abstract new possibilities for meaning and then exhibit an exceptional power for the utilitarian use. As a result of this knowledge, an external object can be incorporated in knowledge as a *technos*: it becomes not only memorized as an object but fixed in the system through its function. It can then serve as a tool for converting the repetitive process of antagonism of oppositions into the non-linear, recursive process of social evolution that uses new *technoi* in its development. This process also affects the biological transformation of the human body in the course of social evolution, in accordance with the ideas of co-evolution of brain and hand initially formulated by Engels [14] in his work "The part played by labor in the transition from ape to man". These ideas have been confirmed in recent studies and expanded in the concept of co-evolution of language and consciousness [11].

Creation of meaning (signification) becomes possible within social systems as a result of incomplete knowledge of the object of external reality, followed by a dialogue between the agents as members of social system to obtain this knowledge [65]. This basic principle of social development originates as described by Vygotsky [75]: he considered conscious regulation of human behavior as the establishment of additional connections to the external world through the process of signification mediated by higher psychic functions. These signification patterns (*eidoi*) are not only formed inside the self through communication of social agents ("through others we become ourselves", as Vygotsky mentioned, cf. Chap. 12 on communication), but they mediate the appearance of tools (*technoi*) that are directed outside for the transformation of the external world. The concept of double self-reflexive consciousness [65] transforms the ideas of Vygotsky into verifiable models and explains why humans have better abilities to perceive the outside world than animals. The single reflection can generate only memory while the double reflexive entity, in which the subject represents (estimates) its own intention, possesses two other standpoints, one connected with whole of the world, and one related to the logistic map.

Together with the two other standpoints, memory defines (1) the dynamic internal structure of the human perception of externality and (2) the mode of the subject's action on and within the external structure, making it accessible for utilitarian purposes. These standpoints can be used consistently and also can be applied to the development of more advanced tools using the logistic maps acquired.

18.6.2 Communication and Connection

It is the communicating set ("*omnium*") of "non-equivalent observers" [30] which forms the relational spatiotemporal environment with commonly observable external objects. The question is how these objects can be incorporated into the active social medium and be used as the tools of transformation of this environment into a comfortable habitable space. In the reflexive theory developed by Sawa and Igamberdiev [65], a dialogue model between two agents is considered, which refers to the development of understanding by the agents regarding what kind of object is in front of the agent. It is not only a memorization about what to do with the object as observed in the instincts and the acquired behavior of animals, but it is a kind of investigation of the object based on the interaction between agents, i.e., on social activity. In the dialogue model of Sawa and Igamberdiev [65], an agent makes a choice between the observation of an object and the acceptance of the opinion of another agent. In this case not only the observation and memorization of an object takes place but also the appearance of the induced argument via acceptance/rejection of the other agent's observation. This means that the dialogue induces a logical map of the object with a possibility of choice of the meaning attributed to this object. In this logical map, the object represents a chosen version of multiple possible meanings ("affordances" according to [17], see above) that represent a whole range of possibilities, the reduced part of which an object is.

This also induces the idea of connectivity of the perceived object with another object and potentially with the whole world, a part of which the object is, and an ability of the object to signify the whole world, e.g., as a fetish in early religions. This double self-reflexive perception possesses the ability to map the outside world that results in the generation of knowledge. Double self-reflection is an inherent capacity of conscious human beings while the single one is inherent to all perceiving subjects. Double self-reflexiveness generates a tripartite structure of externality which includes an element of usual memory, the whole of the world, and the logistic map. These three constituents form the innate hierarchical relation to each other, and through this relation they define the structure of perception of the inhabited world and the mode of subject's incorporation in it. The social activity of double self-reflexive agents assumes flexible establishment of the balance between the observation of an object and the acceptance of an opinion of other agent. This picture corresponds exactly to that of the double contingency in the communication theory of Leydesdorff (Chap. 12). The logical map of this establishment includes the paradoxical statements about the same reality and generates a search of their resolution during which new statements are generated to resolve logical oppositions. This leads to the growth of knowledge about the object and its connections with other objects.

18.6.3 Transformation of the Industrial Basis of Societies

Vavilov [74] established that settled civilizations coincide with the centers of origin of cultivated plants and thus he formulated the main principles of the Neolithic revolution. It has been suggested that the genetic variability of plants in the center of origin of agriculture was used initially for sacral purposes [21] and only later transformed for utilitarian use. The origin of agriculture overturned totemic religious use of animals and allowed their generalized domestication; thus cultivated plants and domesticated animals were incorporated into the society as *eidos*-signs which later acquired the property of *technoi*.

A transformation of sacral meaning into utilitarian application and its reverse relate to the opposite representations of the structure of the reflexive subject established in the opposite reflexive systems referred to as Western (W) and Eastern (E) [43, 45], as noted above. The contradiction of challenge and response appears in the structures of opposite systems in a way such that they generate opposite answers to the same questions about the internal evaluation of the mode of behavior and consider the opposite false. Due to this contradiction, which becomes in the framework of Logic in Reality [7] a real antagonism leading to counter-actions, new spatial and temporal structures can be generated and unfolded in the process of social evolution. While the movement from E to W involved the initial separation and individuation in the society to transform it into a market-oriented and industrialized type, the movement from W to E could lead to the formation of global empires. The transition of E to W corresponds to a desacralization of the objects as signs, while the transition from W to E can generate new or forgotten sacral meanings.

The transition of transcendent to utilitarian reality takes place in many areas of human social activity. One example is science in which initially irrational events find their rational explanation, albeit sometimes transient, as in the return to ignorance and superstition we are witnessing as this book is being written. To make something be utilitarian, it should first be treated as rational. This basic principle clarifying human social evolution works for the explanation of the origin of agriculture, industry, and language. Such activity can be viewed as transcendental, philosophically ascending beyond some concept or limit, and constituting the basis of social activity and social evolution. Ordinary knowledge is the knowledge of objects; transcendental knowledge, according to Kant [38], is the knowledge of how it is possible for us, as subjects, to experience those objects as objects. Kant considered self-consciousness as a prerequisite of the ability to experience the world of objects, in which the self-conscious mind substantiates both the structure of objects and its own unity (cf. Chap. 15).

Karl Jaspers [37] defined four major periods in the development of human civilization. They include prehistory, ancient cultures, the axis time, and the scientific revolution. Jaspers included in prehistory the origin of languages, of tools, the first use of fire, the origin of myths, etc. According to Jaspers, human history as a history of ancient civilizations starts in V-III millennia BC with the origin of settled civilizations and the centralized state system, which appeared mainly for solving irrigation problems. It led to the origin of script and spiritual aristocracy of scribes, and resulted in the origin of ethnical groups united by common language, culture and myths. On the time axis, which corresponds approximately to the period of 800–200 BC, the spiritual basis of mankind was formed independently in different civilizations which resulted in the formation of the modern social typology. From this stage, history acquires a global dimension as compared to local histories of the previous stage. The formation of the scientific stage belongs to the Seventeenth Century. The original scheme of Jaspers can now be modified by the incorporation of additional stages of globalization, which include the Neolithic stages. The industrial society is based on the W type of reflexion and promotes the formation of legal and juridical system which defines the laws prohibiting illegal actions violating rules of civil and economical society. The formed system is dynamic, flexible and self-stabilizing via powerful error-correcting mechanisms based on elections and other democratic procedures [52].

18.7 Reflexive Choice and Cliodynamics

18.7.1 Cliodynamics as a Modelling of Reflexive Social Processes

Opposite reflexive patterns correspond to certain types of societies that interact dynamically at different levels of social organization. This is analyzed in the doctrine of cliodynamics [41, 52, 71] which takes history as a science with its conceptual basis based on the theory of business and social cycles formulated by [40]. We briefly mentioned cliodynamics in the Sect. 13.7.1. Cliodynamics is an attempt to develop the analysis of history as a mode of Western thought via building mathematical models and databases and forming our digital ontology [10]. The cliodynamic approach is based on the assumption of a non-random nature of social processes which opens the possibility of mathematical modeling. The power of Western approach is applied to analyze the dynamical processes in which the opposite modes interact. Such formal analysis, however, included a Gödelian incompleteness in its basic nature; this implies the probability of events that escape purely mathematical modeling and result in its continuous complexification.

Cliodynamics can also be considered as a temporal-scale representation of world-systems that are defined in the concept of Immanuel [76]. The linear oscillatory dynamics of social evolution, in accordance with Kondratieff's ideas, is related to the inventions of new technologies followed by the decline caused by exhaustion of their effects, which, in turn, triggers the transition to a new rise. Its connection to reflexive psychology can clarify important aspects of social development. Lefebvre [43, 45] succeeded in distinguishing between the two opposite types of reflection, and explained their interaction first in societies, and finally at the scale of global civilization. This interaction exhibits interference between the opposite reflexive types, in particular, in establishing the relations between the center and periphery of the society. Each reflexive system has a tendency to generate the opposite trend within itself. In reality, there are different modes that can be realized in particular social systems and conditions. The processes can also stop or move in no particular direction and sometimes backwards. As noted above, [51] defined systems generating such alternative outcomes as *paradeductive*. They possess multiple combinations of linkages and ruptures based on the subjective deductive inferences related to the context-dependent choice of the endo-observer.

Kondratieff waves in the dynamics of social systems correspond to the economic cycles that include expansion, stagnation and recession. They can be accompanied by periodic shifts in the preference for reflexive structures approximating to the Western and Eastern types described by Lefebvre [48]. Expressed as changes in the probability pattern of distribution of reflexive types within a society, they can be traced in the course of the history of human civilization.

According to Malkov [52], the life-supporting resource in early societies (fertile soil) was limited and caused a struggle for its possession. This resulted in the development of early agrarian households as closed systems that induced the first forms of collectivism inside and aggression outside. It led to the strengthening of authoritarian power of the agrarian elites and facilitation of non-economical methods of social management. In this authoritarian E system, further development of economic and political independence of individual social subjects resulted in appearance of the prerequisites of the W system generating instability in the E system, but for prolonged times led to its restoration during new political-demographical cycles [59]. Religious and political power was merged in the same individuals, and this stage of civilization could be referred as "the sacred settlements" that appeared initially in the areas of

Levant and Anatolia starting from 7000 BCE or even earlier (Jericho, Çatalhüyük, then Ur).

18.7.2 Imperiogenesis

The W-type society emerged in history in the course of transformation of sacred settlements into free merchant cities such as Uruk and Ebla. It appeared in the Near East in 4000–3000 BCE, in particular, due to the development of the system of individual property. The objects of sacral value acquired utilitarian use. The IV millennium BCE was the time of merchant cities in the Near East which had some features of the society of W-type and ruled by people's assemblies and the elected ruler. In the third millennium BCE these cities were conquered and included in the extended empires. The extended empires were of the E-type and their appearance corresponded to the use of a new technology (bronze). Thus, the oscillations between the E and W system took place from the very birth of civilization despite of the dominance of the E system over large historical epochs. These oscillations in certain periods coincided with the climate changes during the Holocene.

The invention of new technologies drove the transformations between the dynamic oppositions over human history. Stealing of technologies (such as domestication of animals and production of warfare) by nomadic people from settled civilizations strengthened the E-type of reflection and resulted in the formation of large empires. Imperiogenesis, according to [71], was based on antagonistic interactions, very much in the Lupascian sense, between settled agriculturalists and nomadic pastoralists, which resulted in an autocatalytic process of the type well approximated by non-equilibrium thermodynamics, as well as by the Principle of Dynamic Opposition. The interaction between the two types of societies, which belong initially in some essential constituents to the alternative types of reflexive view on the external world, united spatially separated parts of human civilization belonging to the center and periphery via an autocatalytic process, resulting in feedback loops between the settled and nomadic societies. By 700–500 BCE, the territory covered by empires significantly increased, which corresponded to the epoch of Axial Age defined by [37]. He considered great religions as philosophical breakthroughs that appeared in response to the political and social instability brought on by the increased pressure from nomadic steppe dwellers. The new mega-empires were unified by supranational ideologies such as Zoroastrianism, Buddhism and Confucianism. The types of meditation in Hinduism and Buddhism were based on a free reflexive choice leading, for an individual, either to absolute wholeness or nothingness as analyzed by Lefebvre [48]. While the E-type of reflexion was dominating, the "islands" of W-type were present in certain areas, e.g., in Ancient Greece. The social system of Chinawas dynamically determined by major oscillations between Confucianism that possesses certain freedom in symbolization within the structure of subject and implements social relations that can be compatible to the W-type, and Legalism which is purely the E-system. It is hardly necessary to repeat that the original *Book of Changes* faithfully reflects this evolution. The development of social ideas of Plato was characterized by the shift from the flexible system of E-type that incorporates economic relations of the W-type (in "*The Republic*") to the rigid totalitarian E- type system in "*The Laws*".

The feudal system established in Christian Europe supported the fragmented structure of individual feudal households and formed a new political culture of contractual relations at different levels of social hierarchy, which resulted in development of the culture of political balance and coalition preventing strengthening of one political subject over another. This all promoted the formation of the W-system. Gurevich [20] considered feudalism as a Western European phenomenon developed when free tillers in search of defense accepted the patronage of their barons and thus exchanged freedom for safety which reinstated the W system in parts of Europe. The transition from small feudal states to an industrial civilization was facilitated through unification movements in Europe in the Nineteenth Century. This was despite the tendency for the formation of large empires in the Twentieth Century which lasted for a limited period of time and resulted in the most devastating wars in the history of mankind.

18.7.3 The Formation of Global Civilization

The two modes of social consciousness interact at the levels of one individual, of a small group, of large society, and ultimately at the level of global civilization. According to Grigory Pomerants [60], different dynamical patterns in social history correspond to relevant spatiotemporal scales. These are chaotic ("Brownian") at the level of single individuals, sinusoidal at the level of separate cities, cyclic at the level of individual civilizations, and linear at the level of global civilization. Globalization trends are dependent on the invention of new technologies, which are the main determinants of the ultimate progressive development of the whole human society possessing a major linear constituent. Pomerants also defined several stages of globalization. The picture of the two opposite W and E types of social organization is extreme, but even more important is the interaction of the opposite types of reflection based on the presence of the components of both types in one society and even within one individual. This is the kind of onto-epistemic event that is ultimately the objective of our work to characterize philosophically and scientifically.

The two systems correspond to the two types of social time. In the extreme case, the E system generates a static *durée* with catastrophic transitions while the W system generates a dynamical temporal organization with smooth transitions. Interaction between the static and the dynamic modes of organization in turn generates social evolution. Further development of reflexive psychology should include the transition from the two-valued logic to the many-valued logic and then to the logic operating with continuous values of Sawa and Igamberdiev [66]. While the distinction between the W and E systems will remain valid, this transition will substantiate the existence of more rigid and more flexible patterns within both types of reflexive structures.

18.8 Conclusion. Progress in Social Evolution

Social progress was defined by Durkheim [13] as the development of an increased differentiation of properties and functions in the social organism. As we have shown, differentiation is based on the interaction of several dynamic oppositions represented in the social organism as opposite reflexive systems. The cyclic dynamics of civilizations acquires historical linearity due to the memorization of technical achievements; technical progress introduces linearity and, besides the utilitarian goals for improvement of the well-being of the society, it also fixes the progress of knowledge in human history. New ideas, when codified only in manuscripts, may be lost, but when they are established also in advanced technologies, their loss becomes very unlikely. Only few major technologies in human history were forgotten and then rediscovered. This fundamental aspect of social progress is based on cumulative knowledge while ethical knowledge based on reflexive choice is not cumulative, but arises at a more fundamental level in the cognitive dynamics of the individual.

Although the linear development of the global world-system is determined by technical progress and social memory, the basis of civilization contains the pattern of formation of the world image in accordance with its language and culture. Language, defined as the instrument of formation of the objective pattern of culture, generates by itself the alternative realizations possible within a single culture. There is a polyphony of interactions between cultures and also in the inner structure of a particular culture [4]. In this process, two systems of reflexive coding participate which cannot be fully translated into one another, and this generates a non-predictable creativity of transformation [49]. In the dialogue of oppositions, the dynamic relation between the East and West types of reflexive systems is formed generating novel texts. The dialogic (generally polyphonic) principle of culture developed by Mikhail Bakhtin and then by Yuri Lotman, underlies the structure of the internal energy of the system which opens the possibilities of its evolution. It can be related also to the principle of existential communication of Jaspers [36].

The duality of dynamic oppositions in reflexive choice is a real source of energy that produces the actual work of social evolution. It can be either constructive or destructive but in both cases it can reach very high levels of capacity for introducing social change. Social systems develop, experiencing rise and decline, and disappear as a result of both internal processes and/or external influence. Today, we face a process in which the interaction and dialogue between cultures turns into internal self-dialogue of the global culture, corresponding to a new type of social development. In the course of social progress, opposite reflexive types can transform into the resolved constituents of new social structures and become the elements of management of societies which instantiate many-valued, continuous and process logics such as LIR. Increasing the number of options where Eastern practices of reflexion are based on continuous logical models may acquire a new significance complementing the bivalent Western schemes [9]. This development may finally open into a new dialogue between civilizations at what Lefebvre imagined at 'cosmic' scales [47]. Simply, the basic reflexive principles that constitute social evolution of human societies from their first appearance in reality will hopefully continue to operate in the direction of the common good.

References

1. Arbib, M.A.: Toward the language-ready brain: biological evolution and primate comparisons. Psychon. Bull. Rev. **24**, 142–150 (2017)
2. Aristotle: De Anima (On the Soul). In *Aristotelis: the Complete Works of Aristotle*. Princeton, NJ: Princeton University Press (1984)
3. Asano, M., Basieva, I., Khrennikov, A., Ohya, M., Tanaka, Y.: Quantum-like generalization of the Bayesian updating scheme for objective and subjective mental uncertainties. J. Math. Psychol. **56**, 166–175 (2012)
4. Bakhtin, M.M.: Problems of Dostoevsky's Poetics, p. 1984. University of Minnesota Press, Minneapolis, MN (1929)
5. Barbieri, M.: The semantic theory of language. BioSystems **190**, 104100 (2020)
6. Bohr, N.: Can quantum-mechanical description of physical reality be considered complete? Phys. Rev. **48**, 696–702 (1935)
7. Brenner, J.E.: Logic in Reality. Springer, Dordrecht (2008)
8. Brenner, J.E.: The philosophical logic of Stéphane Lupasco (1900–1988). Logic and Logical Philosophy **19**, 243–285 (2010)
9. Brenner, J.E.: Linking the Tao, biomathics and information through the logic of energy. Prog. Biophys. Mol. Biol. **131**, 15–33 (2017)
10. Capurro, R.: Towards an ontological foundation of information ethics. Ethics Inf. Technol. **8**, 175–186 (2006)
11. Deacon, T.W.: The Symbolic Species: the Co-evolution of Language and the Brain. W.W. Norton, New York, NY (1998)
12. Dodig-Crnkovic, G.: Significance of models of computation, from Turing model to natural computation. Mind. Mach. **21**, 301–320 (2011)
13. Durkheim, E.: The Division of Labor in Society. Free Press, New York, NY, USA (Originally published 1893) (1997)
14. Engels, F.: The part played by labor in the transition from ape to man. In *Dialectics of Nature*. Wellred Publications, London UK (Written 1976, Originally published 1896) (2012)
15. Engels, F.: Dialectics of Nature. Wellred Publications, London, UK (Written 1883, Originally published 1925) (2012)
16. Freud, S.: Introductory Lectures on Psychoanalysis. Liveright, New York, NY (Originally published 1917) (1989)
17. Gibson, J.J.: The theory of affordances. In: Shaw, R., Bransford, J. (eds.) Perceiving, Acting, and Knowing: toward an Ecological Psychology, pp. 67–82. Lawrence Erlbaum, Hillsdale, NJ (1977)
18. Gunji, Y.P., Sonoda, K., Basios, V.: Quantum cognition based on an ambiguous representation derived from a rough set approximation. BioSystems **141**, 55–66 (2016)
19. Gunji, Y.P., Minoura, M., Kojima, K., Horry, Y.: Free will in endo- consciousness driven by Bayesian and inverse Bayesian inference. Prog. Biophys. Mol. Biol. **131**, 312–324 (2017)
20. Gurevich, A.J.: Categories of Medieval Culture. Routledge and Kegan Paul, London, UK (1985)
21. Haas, J., Creamer, W., Mesia, L.H., Goldstein, D., Reinhard, K., Rodriguez, C.V.: Evidence for maize (*Zea mays*) in the late archaic (3000–1800 BC) in the norte chico region of Peru. Proc. Natl. Acad. Sci. U.S.A. **110**, 4945–4949 (2013)
22. Hegel, G.W.F.: Wissenschaft der Logik (Science of Logic). Cambridge University Press, Cambridge UK (Originally published 1812) (2010)
23. Heidegger, M.: Being and Time. SUNY Press, Albany, NY, USA (Originally published 1927) (2008)

24. Heylighen, F.: (Meta)systems as Constraints on Variation: a classification and natural history of metasystem transitions. World Futures J. General Evol. **45**, 59–85 (1995)
25. Hu, Z., Petoukhov, S.V., Petukhova, E.S.: I Ching, dyadic groups of binary numbers and the geno-logic coding in living bodies. Prog. Biophys. Mol. Biol. **131**, 354–368 (2017)
26. Husserl, E.: On the Phenomenology of the Consciousness of Internal Time. Dordrecht: Kluwer (Originally published 1928) (1990)
27. Igamberdiev, A.U.: Semiosis and reflectivity in life and consciousness. Semiotica **123**, 231–246 (1999)
28. Igamberdiev, A.U.: Semiokinesis – semiotic autopoiesis of the Universe. Semiotica **133**, 1–26 (2001)
29. Igamberdiev, A.U.: Quantum computation, non-demolition measurements, and reflexive control in living systems. BioSystems **77**, 47–56 (2004)
30. Igamberdiev, A.U.: Objective patterns in the evolving network of non-equivalent observers. BioSystems **92**, 120–131 (2008)
31. Igamberdiev, A.U.: Time rescaling and pattern formation in biological evolution. BioSystems **123**, 19–26 (2014)
32. Igamberdiev, A.U., Shklovskiy-Kordi, N.E.: Computational power and generative capacity of genetic systems. BioSystems **142–143**, 1–8 (2016)
33. Igamberdiev, A.U., Shklovskiy-Kordi, N.E.: The quantum basis of spatiotemporality in perception and consciousness. Prog. Biophys. Mol. Biol. **130**, 15–25 (2017)
34. Igamberdiev, A.U.: Evolutionary transition from biological to social systems via generation of reflexive models of externality. Prog. Biophys. Mol. Biol. **131**, 336–347 (2017)
35. Igamberdiev, A.U., Brenner, J.E.: The evolutionary dynamics of social systems *via* reflexive transformation of external reality. BioSystems 104219 (2020)
36. Jaspers, K.: Existenzphilosophie (Philosophy of Existence). University of Pennsylvania Press, Philadelphia, PA (Originally published 1938) (1971)
37. Jaspers, K.: The Origin and Goal of History. Translated by Michael Bullock. Yale University Press, New Haven, CT (Originally published 1949) (1953)
38. Kant, I.: Critique of Pure Reason, translated by P. Guyer and A.W. Wood. Cambridge University Press, Cambridge UK (Originally published 1781) (1998)
39. Kastner, R.E.: The Transactional Interpretation of Quantum Mechanics: The Reality of Possibility. Cambridge University Press, Cambridge UK (2012)
40. Kondratieff, N.D.: Long Wave Cycle, p. 1984. E.P. Dutton, Boston, MA (1925)
41. Korotayev, A.N.: World Religions and Social Evolution of the Old World Oikumene Civilizations: a Cross-cultural Perspective. Edwin Mellen Press, New York, NY (2004)
42. Lacan, J.: Écrits. W.W. Norton and Co, New York NY (Originally published 1970) (2006)
43. Lefebvre, V.A.: Algebra of Conscience. A Comparative Analysis of Western and Soviet Ethical Systems. Dordrecht, Springer, 2001 (1982)
44. Lefebvre, V.A.: The principle of complementarity as the basis for the model of ethical cognition. J. Soc. Biological Struct. **7**, 243–258 (1984)
45. Lefebvre, V.A.: The fundamental structures of human reflection. J. Soc. Biol. Struct. **10**, 129–175 (1987)
46. Lefebvre, V.A.: A rational equation for attractive proportions. J. Math. Psychol. **36**, 100–128 (1992)
47. Lefebvre, V.A.: The Cosmic Subject. Russian Academy of Sciences Institute of Psychology Press, Moscow (1997)
48. Lefebvre, V.A.: Theoretical modeling of the subject: western and eastern types of human reflexion. Prog. Biophys. Mol. Biol. **131**, 325–335 (2017)
49. Lotman, Yu.M.: Universe of the Mind: a Semiotic Theory of Culture. I. B. Tauris & Co Ltd., London UK (1990)
50. Lotman, Yu. M.: Culture and Explosion. Berlin: Mouton de Gruyter (Originally published in Russia 1992) (2009)
51. Lupasco, S.: Logique et contradiction. Presses Universitaires de France, Paris (1947)

52. Malkov, SYu.: Social Self-organization and the Historical Process: the Possibilities of Mathematical Modelling. Librocom, Moscow (2009)
53. Mamardashvili, M.K.: Cartesian Meditations. Progress, Moscow (1993).((in Russian))
54. Matsuno, K.: Semantic commitments as a mode of non-programmable computation in the brain. BioSystems **27**, 235–239 (1992)
55. Matsuno, K.: Self-identities and durability of biosystems via their abstracting capacity. BioSystems **120**, 31–34 (2014)
56. Matsuno, K.: From quantum measurement to biology via retrocausality. Prog. Biophys. Mol. Biol. **131**, 131–140 (2017)
57. Nalimov, V.V.: In the Labyrinth of Language: a Mathematician's Journey. ISI Press, Philadelphia, PA, USA (1981)
58. Nalimov, V.V.: *Space, Time, and Life. The Probabilistic Pathways of Evolution.* Philadelphia, PA: ISI Press (1985)
59. Nefedov, S.A.: Theory of demographic cycles and social evolution of ancient and medieval societies of the East. Oriens **3**, 5–20 (2003)
60. Pomerants, G.S.: The steps of globalization. *Vestnik Evropy* 6. In Russian (2002)
61. Rosen, R.: Anticipatory Systems: Philosophical, Mathematical and Methodological Foundations. Pergamon Press, Oxford UK (1985)
62. Rosen, R.: Life Itself: a Comprehensive Inquiry into the Nature, Origin, and Fabrication of Life. Columbia University Press, New York, NY (1991)
63. Rosen, S.M.: Quantum gravity and Taoist cosmology: exploring the ancient origins of phenomenological string theory. Prog. Biophys. Mol. Biol. **131**, 34–60 (2017)
64. Sawa, K., Gunji, Y.P.: Dialogue and causality: global description from local observations and vague communications. BioSystems **90**, 783–791 (2007)
65. Sawa, K., Igamberdiev, A.U.: The Double Homunculus model of self-reflexive systems. BioSystems **144**, 1–7 (2016)
66. Sawa, K., Igamberdiev, A.U.: The dynamics of interaction of reflexive subjects operating with the two-valued versus many-valued logic. Prog. Biophys. Mol. Biol. **131**, 348–353 (2017)
67. Searle, J.: Consciousness and Language. Cambridge University Press, Cambridge UK (2002)
68. Searle, J.: Making the Social World: the Structure of Human Civilization. Oxford University Press, Oxford UK (2010)
69. Toynbee, A.J.: *The Disintegration of Civilizations* (Part One, Vol. 5 of "A Study of History"). Oxford UK: Oxford University Press (1939)
70. Trifonov, V.: GR-friendly description of quantum systems. Int. J. Theor. Phys. **47**, 492–510 (2008)
71. Turchin, P.: A theory of formation of large empires. Journal of Global History **4**, 191–217 (2009)
72. Turchin, V.: The Phenomenon of Science. A cybernetic approach to human evolution. Columbia University Press, New York NY (1977)
73. Vasiliev, N.A.: In: Vasyukov, V.L. (Ed.), Logic and Metalogic, trans. by Vasyukov, V.L. Axiomathes **4,** 329–351 (Originally published 1913) (1993)
74. Vavilov, N.I.: Studies on the origin of cultivated plants. Bulletin of Applied Botany, of Genetics and Plant Breeding **16**, 196–217 (1926)
75. Vygotsky, L.S.: The genesis of higher mental functions. In *The History of the Development of Higher Mental Functions*, vol. 4, Reiber, R. (Ed.), Plenum Press, New York, NY, pp. 97–120 (Originally published 1931) (1987)
76. Wallerstein, I.: World-systems Analysis: an Introduction. Duke University Press, Durham, NC (2004)

Chapter 19
Social Systems: The Global Sustainable Information Society. Ecology

19.1 Introduction. What is a Society?

The previous Chapter presented the dynamics of the human individual with his environment from a historical-philosophical perspective. We now discuss some of the same issues from the perspective of systems theory and its expression in the global dynamics of current social structures.

19.1.1 Evolutionary Systems Theory

From the vast domain of the philosophy and science of society, we have picked out a few subjects in which the operation of our dialectical logic can be observed. A society could be defined as a group of at least two human beings or animals, linked by communication. At a higher level, the unit of discussion is a group for which the minimum number of individuals is variable, but large enough for collective phenomena to appear, that is as a system. The difference between a social system and a society is then a matter of taste and emphasis, but the basic philosophical structures and properties are the same: on needs to take into consideration (a) the relations of individuals to their participation in the group with other individuals; (b) their individual relations to the group perceived as the set of individuals constituting the group; and (c) the relations of the group to its members perceived by those members or others.

We also must include, for a society, the term of the commons [6] since it gives a more dynamic description of the value-laden interaction between the human members of the society. Of most interest to us are *forms* as well as the relations that can be said to characterize a society and the *changes* or evolutions in some of these forms and relations. Regarding form, we note the recent work of Margaret Archer on the 'morphogenic' society. Her associates include Wolfgang Hofkirchner whose transdisciplinary approach to the information society was first mentioned in Chap. 13

J. E. Brenner and A. U. Igamberdiev, *Philosophy in Reality*, Studies in Applied Philosophy, Epistemology and Rational Ethics 60, https://doi.org/10.1007/978-3-030-62757-7_19

[20, 23]. Both authors use the term 'social formation' to describe an historical type of society that differs from others due to the "shape of its systemic make-up". We agree but suggest that the more dynamic term of 'forming' better conveys the changing aspects of this social 'ontolon'.

Like Lupasco (see Chap. 16), Hofkirchner also asks the necessary basic questions of "What is a System? and "What is a Social System?" Hofkirchner's principles for what he calls Evolutionary Systems Theory (EST) [21] can be usefully compared with those of LIR: both involve elements E and emergent relations R in interaction. In LIR, the actuality and potentiality is only partial, and new elements can emerge from the interaction. Hofkirchner refers to 'synergistic effects' which is very similar. However, in LIR, the Principle of Dynamic Opposition, operating *as* the evolution is sufficient, while EST requires a dynamics of self-organization. This is defined as the co-operation of a feed-forward and a feed-back loop between distinct macro- and micro- levels in the system. It is not clear to us, however, that same-level interactions are also significant for further behavior and the emergence of a 'next set' of interactions.

As far as the second question is concerned, it is clear that social systems have the capacities to *change* their form, transcend and re-create themselves. We insist, however, that these capacities are not totally actual but exist in part as potentialities. This concept applies to Hofkirchner's description of the commons—the common good—as an emergent quality, not fully determined by initial (actualized) conditions.

19.1.2 The Commons from a Critical Social Systems Perspective

Hofkirchner answers his question as to the shape of the systemic make-up of the society in dynamic terms that correspond to our vision of process in reality. Lower level societal relations shape the emerging edifice of society, but "every social formation is a formation in process: the new system is permanently on the point of being formed". We would only add, also of being 'de-formed' or 'un-formed' [20]. Changes are large or incremental, but the emergence is not univocal. The commons are defined as the emergent systemic effect of synergy between the actors and forces in social systems. It is quality, more than any quantitative measure of the well-being or contentedness of the actors. It is a term, therefore, that always has a positive valence "as a step towards the implementation of the good society."

As a system, the society is the locus of several obvious and well-studied subsystems, cultural, political and economic. However, Hofkirchner focuses on them in a new light as the contemporary 'battlefields' in the struggle for what we would call a more generous compassionate society. Hofkirchner states that such a society would be something new, in which "another mode of reflexivity establishes a meta-level thinking by human minds" ... which "mediates as a *third* between two other things on the human mind's object level, relates them in a new way and gives them a new meaning" [23]. We know of no better expression of the nature and need for our logic of an emergent, included middle, based on the principles established by Lupasco, than this one [27].[1]

19.1.3 The Dialectics of the Nature-Society System

Some of the key principles on which a satisfactory theory of our current society and of the way in which it should develop were laid out two articles by the sociologist Christian Fuchs. In the first, "The Dialectics of the Nature-Society System" [14], Fuchs defines nature as a self-organizing system that produces an evolutionary hierarchy of interconnected systems with specific qualities. Society is a product of nature where humans produce and reproduce structures that enable and constrain human practices in dynamic processes. Parts of nature are observed and appropriated by humans from within society, and these parts are socially constructed and form a subsystem of society. Conceiving the relationship of nature and society as a nature-society dialectic is one of several possibilities. After rejecting several reductionist alternatives, Fuchs concludes that a dialectical position sees nature and society both as identical and non-identical.

This statement appears so 'naturally' in the text to us that we almost forget that Logic in Reality is one of the few if not the only doctrine that grounds this statement in science. We have commented elsewhere (Chap. 18) why we think that a doctrine of self-organization, found also in the work of Fuchs' collaborator Wolfgang Hofkirchner, mentioned above is somewhat misleading. Self-organization theory can be considered as a reformulation of dialectical thinking; it conceives nature as complex, active, self-producing, and dynamic. However, it is exactly hetero-organization, which is a consequence of the dialectic between potentialized and actualized elements of a process and present throughout it that should be considered ontologically prior. Energy flows into the developing system in this physical and logical 'form'.

In defense of self-organization as such, all of the parameters cited by Fuchs [12] have a role to play (control parameters, critical values, bifurcation points, phase transitions, non-linearity). The methods of self-organization theory—selection, fluctuation and intensification—correspond to the dialectical principle of transition from quantity to quality, etc.) [13]. What is called emergence of order, production of information or symmetry breaking in self-organization theory corresponds to Hegel's notions of sublation (*Aufhebung*) and negation of negation (*ibid.*). The picture given by the term "self-organization of matter" to a certain extent occults the actual physical interactions that lead to the emergent hierarchical outcomes as a process. The

[1] In fairness, it should be stated that Hofkirchner finds support for his perspective in the semiotic categories of Peirce (see Chap. 9). In our view, however, Logic in Reality provides a better model for the dynamic relations involved in interaction of social agents, and the basis for developing and extending them.

operation of this process results in different organizational levels of matter (such as physical matter, biological matter, and society). They are organized in a systemic hierarchy in which upper levels have new emergent qualities that distinguish them from lower levels.

19.1.4 Nature and Society/Culture

The relationship of nature and society/culture is neither exclusive nor inclusive in character, i.e. nature and society are neither fully different nor fully identical. Nature is the totality of systems in the universe and their interactions, it is material and organizes itself on various levels, i.e. it consists of various evolving, interconnected system types. Systems of one type are interconnected and connected to systems of other types; hence nature is relational and dynamic in character. Society is the realm of human activity and interaction; it forms one specific, small part of nature. But for human beings this small part of the universe forms their overall context of activity. All human activity and observation takes place within society, there is no position of humans external to society. Hence nature as physical realm of activity of human labour, production, and communication is itself a part of society, in transforming and observing nature in economic, technological, cultural, and scientific processes, the human being integrates nature into society. A dialectical view of nature/society assumes that nature is the foundation of society, that there is a continuous metabolism between nature and society, but society has also emergent qualities that distinguish it from nature. Fuchs refers to Marx who pointed out that man like animals lives from inorganic nature, but he must remain in a continuing physical dialogue with nature in order to survive.

19.1.5 Political Implications

Our intention here is not to present all of the implications of Fuchs' critique of capitalism but mention a few which are logically and philosophically consistent with the LIR approach. For example, the still current capitalistic economic rationality is in direct antagonistic, contradiction with an ecological sustainability based on a more equitable distribution of value. The only true alternative seems to be the introduction of new technological systems based on renewable resources and energy forms and a solar revolution that requires a fundamental institutional change and an alternative economic and political model not oriented on exchange value and profit, i.e. a new society. More than for simple sustainability alone [14, 15], efforts toward *regeneration* of what has been and it being lost are becoming more and more necessary.

Some actors suggest that if natural resources were commodified and treated as economic resources the ecological crisis could be solved. Individual property would be more carefully treated than a global natural commons that is considered by corporations as a way of externalizing economic costs to society. Such arguments overlook that the root of the ecological crisis is the logic of accumulation that results in the depletion and pollution of natural resources and that the extension of property rights to nature does not challenge this root cause. Models that assume new technologies for elimination of pollution to solve the ecological crisis are short-sighted: they again ignore the root socio-economic causes of ecological degradation. Clean-up technologies require themselves high-cost energetic inputs.

A socially and ecologically sustainable society would have to be a self-managed, communalistic society based on co-operating decentralized direct democratic communes—a "commune of the communes" [2]. As cited by Fuchs, Bookchin's form of dialectic thinking argues that there is directionality toward ever-greater differentiation in the world that opens up ever new potentialities and that all Being is Becoming, as in the LIR approach: "Dialectical naturalism [...] conceives finiteness and contradiction as distinctly natural in the sense that things and phenomena are incomplete and un-actualized in their development—not 'imperfect' in any idealistic or supernatural sense. Until they have been constituted to become, they exist in a dynamic tension" There is a certain danger of anthropomorphism in saying that co-operation and participation are inherent in nature, but Bookchin [2] is well aware that social participation and co-operation are based on self-conscious activity that can make choices between different alternatives and can't be found in nonhuman nature. I would exercise more carefulness in employing the notions of co-operation and participation and would avoid speaking of an unfolding of potentialities in order to stress that social potentialities are not pre-existing in nature, but are produced by human beings in society.

The LIR physical-philosophical system replaces deterministic interpretations of the Marxist dialectic by emphasizing human praxis. A similar interpretation has been provided by Herbert Marcuse [30] who has argued that Marx's dialectic capitalism is an antagonistic system that negates human interests. A truly human society can only be established by a negation of the negation that is the historical practice of the human being. "It would be a distortion of the entire significance of Marxian theory to argue from the inexorable necessity that governs the development of capitalism to a similar necessity in the matter of transformation to socialism. ...Avoiding (simplistic) deterministic assumptions and linearism means not only to question the assumption that there is an automatic historical progress in the development of the productive forces, it also means criticizing the hypothesis that modern history must mean automatic historical regression." This means that both the simple 'assumption' and the simple 'hypothesis' are wrong. If so, then, more realistically, people do not see: "the dialectical character of modern technologies that puts forward (pits one against the other) antagonistic social tendencies, opportunities and risks at the same time. Human practice can appropriate nature in both sustainable and unsustainable ways; there is no pre-determined development of the nature-society-relationship."

This is exactly what Lupasco envisaged as a "para-deductive process". Ortho-deductive processes go in one of two specific directions, (1) toward limits of non-contradiction in which the two opposing elements are maximally actual or potential and vice versa and (2) toward maximum contradiction or opposition (T-state) from which new, third elements may emerge, say creative compromise. In addition, however, there are *para*-deductive processes characterized by not going in *any* particular direction, in part sideways or backwards [36, 37]).

The crucial concept here is that a dynamic principle, derived from physics, the Principle of Dynamic Opposition applies to the logical operators themselves. Thus disjunction itself also cannot be absolute and rigorous. Para-deductions, will always accompany, to some extent, ortho-deductions and vice versa. As long as the logical world exists, there will always be chains of deductions and causalities that have a sense, an orientation, and those that will have, rather, a non-sense, a negative sense of divergence. One can assign, in fact, greater existentiality to some para-deductions in that they do *not* go in the direction of an abstract absolute limit.

To us, this is a unique way of providing a rigorous and, finally, anti-ideological basis for describing existential situations. Although Marcuse has argued that dialectical reasoning doesn't apply to nature as such, he sees nature as a producing subject. But if nature is active and changes itself, this implies a dynamic and hence dialectical character. Nature and society are mutually connected. Man changes and socially constructs nature and parts of nature are incorporated into society as the latter's material foundation. Although there are some elements in the works of Marx and Engels where nature is considered as an endlessly exploitable resource, they can be seen as precursors of ecological thinking and sustainable development. In Orthodox Marxism, as noted, dialectical thinking has been interpreted as deterministic and its mechanical laws have been misused to argue that the Soviet system was a free society. An alternative is a dialectic that stresses human practice and that structures condition alternative possibilities for action. Dialectic thinkers like Marcuse and Ernst Bloch have argued that nature is a producing subject, a non-teleological subject [18]. Describing nature as a subject implies that if man destroys nature the latter as a producing subject will probably produce uncontrollable negative effects on society and that hence nature should be appropriated in sustainable ways. Matter is a natural subject that acts upon itself, whereas man is a human self-conscious subject that acts upon nature and society.

The important idea in Bookchin's thinking is that all world processes are based on dynamic interaction processes that act synergetically and produce new qualities of Being. The so-called self-organization of matter is an active, networked, synergetic, productive process. Following Wu Kun, we refer to such processes as hetero-organized.

19.2 A Logic of the Third in Society

Chapter 13 above included as discussion of the changes consequent on the major current role of information in society, in based in part on work by the information and systems scientist Wolfgang Hofkirchner. To further describe changes in society, Hofkirchner further introduces the concept of a Logic of the "Third" [19]. This is an informal logic that is unlike both standard propositional logic and our Logic in Reality of processes. The Logic of the Third is not and is not intended to be more than recognition of certain regularities in complex phenomena, of rigor in thinking about them and the value of a diagrammatic representation of key properties. The use of logic is metaphoric. Unlike Logic in Reality, or semantic and mathematical logics, it is not a logical *system*. LIR substitutes, for the syllogisms of propositional logics, a concept of dynamisms, descriptions of the changing energetic relations in phenomena for which we have coined the term 'ontolons'. We consider it permissible to say that the evolution of phenomena follows related logical laws. The logic of the "Third" for us is thus an epistemological categorization of aspects of phenomena, equated in part by Hofkirchner to the logics and categorizations of Peirce.

In his approach to a Unified Theory of Information (UTI),Hofkirchner does not refer to the fields of philosophy or Natural Philosophy of information as such. However, many of his concepts and formulations illuminate core philosophical, logical and ethical problems, in the context of the 'informational turn' and with a clearly stated objective of furthering a Global Sustainable Information Society (GSIS) [24, 25].

For the last twenty years, Hofkirchner's work has been directed toward the achievement of such a GSIS Raising the problem-solving capacity of the society, creation of a more efficient collective intelligence in a process of 'Informationalization' are processes necessary for society to cope with the problems resulting from its own development. For Hofkirchner, sustainability, denoting a society's ability to perpetuate its own development in an ethical manner is the most universal value for governing moral behavior. Measures of progress toward the goal proposed by Hofkirchner of a GSIS cannot be made in quantitative terms, for example of the extent of digitalization of the economy, material and intellectual. But how then can one avoid a foundational pessimism?

In our current view, the concepts offered to date for understanding, describing and changing behavior, for example, to achieve a GSIS, have been too diffuse to be effective. The countervailing forces at work operate with a logic of exclusiveness and disinformation, with socio-economic relations mediated by essentially criminal castes, gangs, and unscrupulous individual economic and political actors. Such negative forces are capable of impeding its reworking to put it on a path towards sustainable development. We see again here the catastrophic result of the underlying use of standard propositional logics or their mathematical equivalents, incapable of describing complex real processes and thus of providing the necessary theoretical support for them.

Let us now see how some of the above ideas might apply to the informational-scientific framework that Hofkirchner has developed for the discussion of the operation of information and information processes in the society and its functioning. To repeat, we feel this framework represents a significant advance over both simplistic semiotic and strictly computational approaches. Our objective is to place Hofkirchner's most recent case for the utility of a Universal Theory of Information (UTI) on a further sound physical and logical basis. This work is further supported by our view that our non-propositional logicss of processes, Logic in Reality (LIR), as also a logic of ecology—an eco-logic—that defines a basis for moral responsibility. Current work by Zhong [41], who has proposed the fields of Informational Ecology and the Ecology of Information Studies, suggests that both the phenomena and the studies of these phenomena respectively, follow this logic.

Sustainabilization has been defined by Hofkirchner in as "the process of putting society in a position to avoid anthropogenic breakdown and safeguard a stable path of development below the threshold of endangering the maintenance of society." Hofkirchner sees the information revolution in general as contributing to sustainabilism—the result of the process of sustainabilization—needed for the advent of a GSIS, of ecological and sustainable practices in society They receive justification and support from the principles of LIR as well as science. As Hofkirchner writes [22] a UTI should be a logical as well as an historical thesis, explaining not only the historical appearance (emergence) of new information processes and structures but how these processes and structures are logically linked. Hofkirchner explicitly excludes standard deduction as incapable of accomplishing this task, but LIR would appear to be an acceptable candidate for doing so. His combination of critical thinking and systems thinking—of Critical Theory and Systems Theory—includes the Logic of the Third [23]. The (informal) Logic of the Third is the foundation of a critical social systems theory, in which "criticism is a method oriented toward recognizing and sublating contradictions". Hofkirchner is an adept of Hegel; but we remind readers that real contradictions have a positive valence in Lupasco and LIR. 'Management' rather than elimination of them is our preferred strategy.

Hofkirchner states that given a society characterized by agonisms, legitimate agonistic differences can and do degenerate into antagonistic denials of the 'other' when social actors try to impose their own interests exclusively. This phenomenon can be considered as one principle of this Logic of the Third: "Only the social relations that reinforce one's own position enter into consideration as a 'mechanism' that operates for the transformation of agonisms into antagonisms. The most important question is if and how a needed 'mechanism' of turning antagonisms into agonisms can work, how it can take the edge off them, how it can sustainably de-escalate them—a question to be answered according to the same Logic of the Third". A conjunction between Hofkirchner's approach and Logic in Reality is easily found: Lupasco's central 1951 book is entitled, in English, "The Principle ofAntagonism and the Logic of Energy" [36]. LIR is, also, a logic of an emergent 'third' element in real processes and included without difficulty the real antagonism between human beings as logical in this sense.

Another classical philosophical problem, with which both LIR and Hofkirchner are concerned, is the relation between the 'one' and the 'many'; identity and diversity; and identity and difference. This is a key concept in Lupasco's dynamic logic of/in reality in which movement from primarily diversity or heterogeneity to unity orhomogeneity is as fundamental as that between actuality and potentiality in the basic structure of the universe. Hofkirchner proposes the relation between identity and difference as describing four ways of thinking: reductionism, projectivism, disjunctivism, and integrativism. The first two yield unity without diversity, the third diversity without unity. The fourth, as also expressed by the contemporary philosopher Edgar Morin, yields the necessary unity-in-diversity and diversity-in-unity. Hofkirchner applies these concepts to information in a doctrine of emergent materialism (EM) that goes beyond materialistic and idealistic monism and (even) standard interactive (mind–body) dualism. The dialectics of EM recognizes, like the informational metaphilosophy of Wu Kun (Chap. 12), the identity and difference ofmatter and information. EM is a philosophy of mind, "overarching all manifestations of information and not only mind."

Logic in Reality is compatible with part but not all of this approach. LIR provides a realistic interpretation of the physical, dialectical relation, a grounded 'interactive dualism' between identity and diversity, a unity-in-diversity and a diversity-in-unity (see Chap. 18) as well as between the terms of other critical physical and philosophical dualities. As stated above, LIR, in contrast to standard bivalent or multivalent propositional logics, provides the basis for an ethics as the finality for the intellectual process, a principle which also pervades the work of Hofkirchner. It thus satisfies both our formal and moral criteria for a Natural Philosophy (Chap. 14).

Hofkirchner's work provides valuable insights into the structuration and 'mechanisms' that operate in society and its evolution. The antagonisms in societal relations, for example the different views of the role and value of the common good are for him "the engine of *change*". They are seen as the manifestations of a deeper mechanism of social morphogenesis, to repeat, change of shape, embedded in systems theory. "Any system shows a dialectic if integration and differentiation". This is better restated by Hofkirchner in terms of an alternating and reciprocal formation of differentiated agents and integrated structures.

19.2.1 Society as a Process of Systemic Emergence

Hofkirchner's picture of the emergence of complex meta-reflexive relations in society leading eventually to a commons is a Lupascian one, more than neo-Marxist. The contradictions and antagonisms in social relations are real ones, also in the optimistic sense that there are present humanistic factors that may counterbalance and eventually outbalance the others.

Hofkirchner's picture of *evolution* is as the cumulative aspect of change in the sequence of historical (social) formations, whereas *revolution* signifies disruptive social change. However, the two are processes that, as "evo-revo", are inextricably intertwined, or as we would say, in a relation of dynamic opposition. Revolution is indeed the marker of *qualitative* change, independently of whether it is violent or not, or involves periods of stasis, morpho-static processes in Hofkirchner's apt term [23]. In fact, systems involved in social morphogenesis show dialectics of integration and differentiation: "there is divergence and convergence in one" and progress or, let us not forget, regress can be the result. In favorable cases, a desired 'unity through diversity' can be achieved.

We believe that in addition, attention should be called to the meaning of diversity through unity. If the thesis in this book represents a desirable reflexive "mode of thinking in a human mind", a facility of critique, we must call attention to the 'opposite' feature in reality of diversity through unity. Unity is thus not viewed only from the point of view of an absolute goal, an identity, but as a tool. When there is unity as a temporary interruption of change, it can have the desirable again consequence of giving the energy for emergence of diversity as variety.

19.3 The Logic of the Third and Lupasco

The logical system of Lupasco is described by both Lupasco and Nicolescu as a logic of the included third, or included third term (*tiers inclus*). Brenner has referred to it as a triadic logic or logic of the included middle, differentiating it from the standard logic of the excluded middle. The three 'terms' are the partially actual state of a phenomenon, the accompanying partially potential state and the emergent or T-state, as noted in Chap. 3.

As discussed in the previous Section, Hofkirchner defines a logic of the "third", using his notation as a common logic underpinning both information and self-organization. Given a complex system, any information is a "third" since it mediates between the system—the "first"—and its environment—the "second". It is a system that cannot be reduced to the direct, dyadic relation between the system and its environment. It relates the system and the environment by an emergent entity. The problem with this description is that the assignment of cardinality is arbitrary since the operational terms—"mediates" and "relates" are not given other than their usual epistemological definitions. What is the origin of the capacity for emergence of the emergent entity? Similarly, social information is generated by social systems, defined correctly as including human social actors. It is thus possible to say, and to subsume under a logic of the third, (or three terms), that the behavior of the social system comes "first", the environmental perturbation "second" and the postulated self-organization of the social information "third".

The problem we have with this formulation is that it is an epistemological classification without direct reference to the real forces in play. The temporal sequence of the phenomena is correct; any emergent phenomenon comes into existence 'after' a perturbation. But social information is not, in our view, constructed by the system by means of self-organization. The system is organized as a consequence of residual potentialities or process capacities which result in its emergence in and as a new form.

This formulation does not assign causal properties to a static epistemic principle. Nevertheless, again, we feel that the emergent third, amplified by the ontological properties it carries, is a proper characterization of the commons—the establishment of the common good. We note the pertinent description [11] by the French sociologist François Flahault of the antagonism between the rights of the individual (The Rights of Man), currently actualized at the expense of the common good, due in part to the 'communist' flavor of the term.

Social information and social organization are not identical—Hofkirchner considers them different in nature despite their underlying common logic. As he correctly notes, however, information can extend to organizational relations. In our preferred dynamic terms social information describes the processes involved in the evolution of a social system, while its organization is an epistemological 'slice' giving a picture of a system at a moment in time.

In the context of the 'mechanisms' at work in the emerging information society, Hofkirchner defines a functional role for additional 'thirds" which exemplify the logic of the "third". On the other hand, as stated above, we do not follow Hofkirchner is equating his conceptions of a Third with those of Peirce. As we have argued, the way in which they have been defined excludes them as natural elements, that is, within the purview of ontological reality and natural philosophy. In fact, we consider 3rd—ness, 2nd—ness 1st—ness as models of reality as weak and superfluous, and indeed Hofkirchner does not use them explicitly.

19.3.1 Stages in Society

As stages in the organizational development of a society, Hofkirchner lists Tribalism, Individualism and Cosmopolitanism. At the first stage of face-to-face" social system, communication and myths serve to internalize commonly accepted rules, and the groups psychological insights and traditions are conveyed by myths. It is perhaps still an open question if at this stage humanity consisted of a collective monad, in which Ego, Other and We were not differentiated. We believe it is preferable to consider that none of these terms were ever 'pure', and that a basis always existed for viewing the Other as partly Ego.

Clearly, Ego would be emphasized at the stage of Individualism, which might not be solely temporal. We can, based on our dialectics, reason that there always were individuals who placed Ego above We. The emergence of Cosmopolitanism, Hofkirchner proposes, is one of "unfettered system reflexivity", without restrictive barriers between them, enabling a consensus to be reached that can safeguard human values and enhance communications at a global level as the necessary basis for an emerging Global Sustainable Information Society (GSIS). According to Archer, a 'meta-reflexivity' is necessary to understand one's own position and that of others. Morphogenic society could be the term for this new social formation characterized by a new mode of social self-organization enabled by improved social information and exchanges:—"a mode that is characterized by system reflexivity such that unity and diversity is an achievable state of society in which humanity shapes its future".

For accuracy to the text, we have quoted the passage from Hofkirchner as it appears. However, we wish to state (restate) our concerns with the use of self-organization in this context also without qualification. If new social self-organization is 'brought about' by social information or anything else, it is *ipso facto* not independent of that information, its acquisition and its internalization, and the use of 'self-' becomes questionable. But this may be considered a minor criticism of Hofkirchner's vision. What is essential is the clear joint operation of communication and human social interactions and their logical evolution. It was again part of the genius of Lupasco, not yet recognized that his logic provided descriptions of *change* that were non-linear, not only reflexive but recursive or else moving in *no* clear overall direction. Until further notice, this description resembles best what we see happening in our day-to-day reality. Even the current catastrophic fixation on growth at all costs is in principle capable of being reversed.[2] However, energy considerations, following a LIR logic again suggest themselves as relevant. In a world with access to unlimited energy, capitalist accumulation may continue to exist locally, but its nefarious aspects are globally mitigated.

In a later paper (2015) by Hofkirchner on the "Mechanisms" at Work in the Information Society, the relation of his approach to the logic and dialectics in this book appear even more clearly. In particular, he goes more deeply into the nature of the 'Third' referred to above, seeing History and social totality as playing the role of a Third as an instance of downward causation of relations on the macro-level, and connects individual aspects of them. Hofkirchner's Evolutionary Systems Theory (EST) treats both entities and events as *processes*, as the consequences of other processes that support the emergence of systemic interconnectedness. This is the kind of philosophic concept to which LIR provides the further essential grounding in science.

In our view, Hofkirchner's emergent effects are the result of causality in operation between techno-social and social systems best described as 'spiral' rather than circular to avoid the implication that any process returns to exactly the same point from which it started. Hofkirchner further provides a detailed analysis of the structure of conflict which is compatible with our principle of dynamic opposition (PDO). In simple latent and manifest conflict, or antagonism, identity and difference are not reconciled, that is, misrelated or not integrated into something more 'noble'. Hofkirchner suggests this can come about by taking three 'measures': (1) the extension of cognition, breaking out of simple negation of an opponent's position and the accompanying loss of any dialectics; (2) the extension or transcending of the limits of restricted communication, communication that has been restricted to self-serving narrow goals; and, in the same spirit (3) extension of cooperation; *opponents* become *components*. As Hofkirchner succinctly states, the three measures need

---

[2]As this is being written, in the midst of the 2020 coronavirus catastrophe, the possibility of a far reaching change in the socio-economic paradigm is becoming greater.

not be conducted in any particular sequence, As LIR-type entities—ontolons—they mutually necessitate each other, forming a more complex dynamic ontolon.

19.4 Structures of Society: Modernism and Post-Modernism

The characterization of the nature and structure of society and is its functioning is in our view part of natural philosophy. The modes of communication between individuals and groups have always been a part of the operation of a society. Attitudes and doctrines can be causally effective, even if they themselves are nothing more than dogma and prejudice. As a test case for the applicability of our Philosophy in Reality, we may take the triad of modernism, post-modernism and post-post-modernism in which, regardless of the associations and interpretations one would like to make, a movement toward science, away from science and at least in part back science or to some of its characteristics can be discerned.

We are not totally convinced that the three members of the 'modernism-triad' are valid philosophical categories. The phenomena occurring in the society which they pick out are certainly dynamic trends, but how real are they? The Italian-French sociologist Michel Maffesoli [28] correctly places emphasis on an identifiable 'Real' that is fundamental and measures the fundamental limitations of human beings as well as their richnesses. But the term modernity, having evolved from an original acceptation in relation to art, has lacked a proper theoretical grounding.

19.4.1 Being Postmodern: "The Step Backwards"

If post-postmodernism is postmodernism with its drawbacks of a certain overdone anti-scientific bias eliminated, then the postmodernism of Maffesoli might as well be designated as the former rather than the latter. Maffesoli's thesis, in our view, is naturalized in its emphasis on nature as the origin and becoming of the world, natural intelligence, human-scale structures in society and the 'detranscendentalization' of thought. Despite the well-known problems of the digital society, also described by Capurro, new forms of "being-together" in a responsible way do exist, and they should be welcomed. In his most recent work [29], Maffesoli supports a view of postmodernism with concepts that are not present explicitly in standard philosophical discussions of the 'workings' of society, but some might be. For example, Maffesoli refers to the phenomenological 'step backwards (*Schritt zurück*) proposed by Heidegger in order to recover valid and necessary insights that have been occulted in the artificial 'flight forward' (*fuite en avant*) of Promethean modernism. Indeed, Maffesoli calls attention to the values of an *Epimethean* stance, and this is to his credit.

Maffesoli thus agrees with a need to, within limits, return to earlier attitudes to, for the example, the unity of philosophy and science described by Brenner [3, 4] as a 'metalogical rejunction'. Maffesoli suggests that 'dialogic' is the specific logic of postmodernity, a logic that can handle conflict and ambivalence, pluralism and holism. The formulation of return to tradition is acceptable, of course, but in our view all such formulations should be explicitly qualified so that they do not imply an equally unacceptable *fuite en arrière*.

The difficulty with Maffesoli's stance, and indeed all postmodernism is the lack of any specific rigorous philosophical research of the kind required, for example, by transdisciplinarity. As discussed elsewhere, dialogic is also not a satisfactory logic, either for postmodernity or anything else, as it lacks the grounding in science of Logic in Reality. Finally, as Wu Kun has also suggested [39], postmodern theory amounts to little more than the deconstruction of some traditional ideas, plus non-systematized concepts of future possibilities. Just as Peirce, in our view, makes no *ontological* commitment to the nature of the underlying units of existence (*phanerons*), Maffesoli, like many right-wing thinkers, makes no explicit *socio-political* commitment to the common good, and to the changes that are required to achieve it of the kind we find in Lupasco [4]. We will return to the discussion of the common good at the end of this book.

19.4.2 Rorty and Irony

All the arguments that can be made about modernity, postmodernity and post-postmodernity can, however, be distinguished by the parameter of absolutism. One need go no farther than the work of Richard Rorty to find the demonstration that a doctrine of absolute relativism is untenable. This is a vicious contradiction in the sense of lacking any new dialectical outcome, as in the case of other purely binary oppositions. Absolute cultural relativism, the idea that all cultural practices have equal rights to existence and execution, is no longer seriously entertained or defended as a doctrine, given its obvious self-contradiction. However, I believe that discussions of individual and group moral responsibility are still influenced by the only apparent human focus of ironists such as Richard Rorty and his followers, and a tendency remains to give some cultural practices more credit than they deserve.

We agree without reservation when Rorty [35] says that moral progress means the ability to see, "more and more, traditional differences (of tribe, religion, race customs, and the like), as unimportant when compared with similarities with respect to pain and humiliation." *Of course*, one wants to avoid perpetuating cruelty within social institutions by maximizing the quality of education, freedom of the press, educational opportunity and so on. But the kind of a-historical grounding, based on the contingency of language, on which Rorty bases the rationality of communication, is suspect as a basis for plurality if it eschews any universal validity. It is an illusion to believe that agreement on common purposes can be freely arrived at in reality as it would be in Rorty's "poeticized" culture (sic). He goes on to say, what he himself thinks might be considered "indecent", that responsibilities to others constitute only the public side of our lives, which do not have *automatic* priority over our private motives. The exercise of moral obligation will depend on the circumstances, but should not be the subject of a public–private dichotomy that does not exist absolutely. Habermas saw as objectionable Rorty's division of private self-development and a sense of obligation to other human beings, as it undermines its democratic intent, leaving the door open to anti-social policies and behavior. From the LIR standpoint, one of these antagonistic processes will predominate at any one time, but they cannot be considered as independent. The seemingly "intellectual" approach of Habermas, which Rorty criticized, is more pertinent to today's social dialogue than his irony.

In Rorty's account, an ironist culture is one in which the opposites of self-creation and social responsibility or politics can be combined in life but not in a theory, one which would be a "successor to Marxism". LIR provides exactly that, a basis for "combining" these opposites in theory, without adding arbitrary theoretical constructs or presuppositions. For me, liberal[3] political discourse does not have to and should not remain "as untheoretical and simpleminded as it looks" for fear of losing both credibility and influence. In fact, I will make the stronger claim that *unless* morality and individual and collective responsibility toward the environment and other beings, human and others, is grounded in basic physics, all arguments for any intrinsic moral values fail.

It is noteworthy that Maffesoli does not make any direct reference to Rorty or irony in his *Être Postmoderne*, although many of the desirable features of society that he seeks are to be found also in the latter. The conclusion to which we tentatively come is that the systems language of Minati may be a superior way in which to characterize the evolution of society in stricter philosophical terms.

The reason is that the 'trends' in the society are highly disparate entities, all constituted by information and carrying meaning, but all different with regard to their kind, involving groups of people, ideas, forms of debate and interaction, digital and non-digital, democratic and totalitarian. The new systems approach of Gianfranco Minati, which we discussed in detail in Chap. 18, is a framework for showing how such entities coexist, transform and are related to one another and to the 'whole' of society.

19.5 Public Philosophy

In considering both the sources and possible reception of this book, a brief reference is in order to Public Philosophy (PP) [34] as a current development in philosophy as a whole. The phrase first appeared in a 1955 book by the American writer Walter Lippmann On the Decline and Revival of the (sic) Western Society.

[3]Unless otherwise noted, we use the term "liberal" in the United States English acceptation, that is, predominantly democratic-socialist in opposition to conservative.

The term has two definitions, one of addressing issues of importance to the public—society—such as policy, morality and politics and economics from a philosophical stance. In addition to this concept of PP as content, there is also that of PP as style, engaging in discussions in public venues, including especially, now social media and one-line conferencing systems. It has been suggested that these two usages of PP are not totally separated but inform one another, very much in the spirit of this book. Although our arguments concern primarily content, we welcome any initiatives in which artificial barriers, real or implied between academic study and praxis in the commons are broken down. We suggest that application of the Principle of Dynamic Opposition field of PP would help to improve its quality as *science* and hence the acceptance of at least part of the debate as having another level or validity. The developing literature on PP [34] suggests that it is a response to a need in the commons for an intellectual dimension that has been lacking in the past, which is possibly contributing the continued strength of anti-social political and economic structures. This hypothesis deserves further elaboration.

19.6 Creativity, Love and Freedom

19.6.1 Critical Realism

An obvious requirement of any analysis relating the development of the society to human values is not only a restatement of those values, of the kind that is made by Maffesoli, but the grounding of those values in some acceptable notion of reality, such as that with which we have begun this book. The normative aspects of any theory should both of these in the foreground and not as a by-product of the analysis. We have accordingly used the sub-title of the book by Roy Bhaskar, *meta-Reality* [1], as the title of this Section, since these are certainly values everyone should be able to agree upon.

The work of Bhaskar and others can be subsumed under the term of critical realism, seen as a non-Marxist continuation of the Frankfurt school of philosophy. While the objectives of this movement are laudable, we see a structural weakness in Bhaskar's particular extension of it, namely, a form of transcendentalism that, to a non-reductionist realist seems naïve. According to LIR, if dualism is "hard-wired" into reality, then the solution to the problems of society cannot lie in transcending *its* duality but to live it. The former is what has been preached for centuries and is still being preached in a similar form, with the results that one observes. Reasoned and earnest restatements of ethical goals should always be welcomed. However, the value of such restatements should not be diluted by reintroduction of precisely those abstractions that are one cause of misunderstanding of the dynamics of society-in-conflict. The appeal of such positions makes it only more necessary to provide counterarguments in an appropriate dialectical manner.

In *meta-Reality*, Bhaskar makes several statements, especially the first about change, with which one can whole-heartedly agree (they have been edited slightly for clarity):

(1) There is no conflict between self-change and social change. There is no way to save society except through my actions in the here and now of my being, of course with others …
(2) There is no de-totalization, the characteristic form of dualistic and binary theories of perception. Thus, as in LIR, there is no total separation of perceiver and perceived, internality and externality, actuality and potentiality, etc.
(3) … we are inextricably connected with our fellow human beings, the social totality and the natural cosmos in which we are embedded and of which we are a part. … To be is in fact to be related.
(4) Unless we tackle questions at that deep, broad and categorial level, we will be reduced to such strategies as piecemeal social engineering; we will never be able to find systematic malaises or problems in our social order; and, incidentally, there will be no real use for philosophical or any other kind of macroscopic analysis.

On the other hand, Bhaskar's transcendental anti-realism results in apodictic statements that are unacceptable. Some examples follow:

(1) Concepts of identity, unity, etc. are ontologically, epistemologically and logically *prior* to concepts of non-identity, etc.
(2) To incorporate the transcendental or supramental, the intuitive, the mental, the emotional, vastly expands our ontology. *None of these can be derived in any way from physical reality* (emphasis ours).
(3) Truth is in fact a more basic conception than reality, for truth, unlike reality, is not necessarily (NB: there are many hedges of this kind) implicated in a dualistic mood or structure of thought.
(4) At a more prosaic level, the aletheic fact that metals conduct electricity may consist (another hedge) in their possession of a free electron, … (NB: this is a scientific tautology. What is 'prosaic' about our reality?)
(5) The primary (essential) level, which is good, autonomous, etc. could exist without the secondary level (i.e., the good or true or free could exist without the evil or false or oppressive), but not the other way round.
(6) (When we perceive) the world, we perceive it non-dualistically or we are not perceiving it at all. It is also the case that we understand it immediately, without mediation of physical signals or thought.
(7) It should be obvious that our ontology will need to be expanded to include …the emotional or astral realm—routinely accessed in sleep or after death.

In our dualistic philosophy, the difficult but necessary exercise is, the "trick", if you like, is to see that all the wonderful things in human life, imagination and spirituality as well as understanding are possible *solely* within a natural, energetic world, in which the laws of physics hold (Principle of Naturalistic Closure). This world has, however, as part of its reality, the *appearance* of another world—call it noumenal or some equivalent term. These phenomena, as well as the relation between them and the changes they undergo, are what we have tried to explicate in this book.

19.7 The Philosophy of Ecology

Both sustainability and ecology are current narratives about the behavior of humans toward themselves and the environment. For some, sustainability is an accepted, important moral goal. Unfortunately, the economic, ideological and psychological barriers to ethical behavior and corresponding social action remain great as well as obvious. We propose that our logic and philosophy in reality apply to the science and philosophy of these narratives and helps to explicate them. Recognition of the suggested new, transdisciplinary logical dimension of ecological processes, especially, of their informational characteristics, may provide credibility and support to new initiatives in ecology.

Ecology and ecological science are approaches to a better understanding of the complex interrelationships between humans and their environment. These are necessary for trying to achieve the goal of a sustainable world, in view of the resistance and barriers faced in the process. Concomitantly the necessity of managing the explosive expansion of scientific knowledge and the 'infosphere' of communication is apparent. The concept proposed by Capurro and others, of an Informational Ecology [12] as a global ecology of *intellectual* resources, now would include 'natural' ecology and sustainability themselves as domains of knowledge. We suggest that the dualism between altruism and selfishness, which is reflected in attitudes toward the environment, can be described by our logic and that this description, as a 'positivity factor' may contribute to predominance of the former.

19.7.1 The New Central Role of Ecology

As well expressed by Arran Gare [17], ecology is becoming central to the evolution of forms of thought—the science, philosophy and sociology—required to provide the basis for an ecological civilization. Recognition of the reality of the natural-logical principles underlying our Logic and Philosophy in Reality might help to focus debate not only on necessary changes in policy and behavior, but on the barriers to their implementation. We believe that strategies for the managed, sustainable use of resources are supported by the ethical implications of a logic, like LIR, that has a degree of credibility and rigor comparable to that of fundamental physics.

Gare says that "Ecological science is a fundamental challenge not only to the core assumptions of modern science, but of industrial civilization. Acceptance of ecology will involve not merely a transformation of science, but a transformation of the relationship between science and other domains of culture, impacting on people's lives, their institutions, and their organizations, and more fundamentally, on their image of the future and of the ultimate ideals and goals worth struggling for. By 'modern' science I mean the form science took with the emergence of industrial civilization, that is, a form of reductionist materialism that denied any purpose to human existence, rendered sentient life unintelligible, created a dualism between body and mind, divided science from and opposed it to the arts and humanities, and yet sustained a vision of the future as the total conquest of nature to serve human purposes, or at least the purposes of the winners in the struggle for survival and dominance."

Gare argues [16] that "ecology is developing the forms of thinking required to rethink the relationship between humanity and nature and between individuals and their communities, the nature of culture and civilization, and thereby to transform the way people live and organize themselves. It provides the basis for a different vision of the future, with a different kind of ethics and political philosophy than those which have dominated modernity. That is, ecology is developing the forms of thinking required to create an ecological civilization.[4]" He calls for new processual forms of ecology, philosophy and complexity for understanding life as ecosystems, that is, as communities of dynamical processes in symbiotic relation, constraining themselves and able to constrain each other directly or indirectly, thereby coordinating their activities to contribute to their common good. "Ecology is thus the science that thus integrates factors in the global ecosystem and thereby transcends the opposition between science and the humanities, giving a place to the narratives and more abstract thought through which humans create themselves and redefine their relation to each other and to the rest of nature." In my view, these positions, fully correct and desirable as goals, still require further scientific underpinning for maximum acceptability that a relevant logical system could provide.

This is a metaphilosophical principle, since it has to do with the content of philosophy itself. We recall from Chap. 13 that the justification for Wu's calling his Philosophy of Information a Metaphilosophy, "a highest philosophy", to be distinguished from all others, is its unique and universal character, its new worldview, as an informational conception of history, society, values, knowledge, science and technology. In the framework of this analysis of the domains of sustainability and ecology, one can show that the overlapping of scientific, ethical and social considerations follow the same logic of processes.

19.7.2 *The Information Ecology of Zhong*

Yixin Zhong, a major figure in Artificial Intelligence, derives his concept of an Information Ecology (2017) from the information involved in the survival conditions for human subjects (and by implication other animals). From the viewpoint

[4]Gare uses the term 'eco-poiesis' to describe a desired process of 'making homes', the creation of ecological niches in which the inhabitants can exist and develop. The usual meaning is the (hypothetical) artificial creation of a sustainable ecosystem on a lifeless planet.

of the subject-object interaction, information alone is not a complete information process including higher level products, knowledge and intelligence, constituting an information system. All information systems together with their environments constitute in turn an information ecosystem, whose finality is the optimization of the processes and the interactive relationships with other subjects, in other words, an ecology. We believe that the principles of Logic in Reality outlined above are pertinent to Information Ecology, as defined by Zhong, in several ways, but the most important is that there need be no separation between this concept of IE and the more familiar one of Capurro [9]. Both depend on the underlying evolution of science and philosophy under the influence of information.

The emphasis in the Information Ecology of Zhong, however, is on its function as a holistic *methodology* that can be applied for example to Intelligence Science but also to ecology. In Zhong's words: "The common feature of ecology and information ecology, the core of ecological methodology, is the concern with the interrelations among all the elements of an ecosystem and the ecosystem as a whole [5]." As Brenner showed elsewhere, Logic in Reality extends the categorial feature of non-separability to mereology, a new view of part-whole interactions of which Zhong gives a clear example in ecology.

The additional dimension which Zhong and his colleagues have given to their Information Ecology is its application to Information Studies, such studies themselves being considered as ecological systems. Zhong's Information Ecology is thus closely related and applies to the advances in the Philosophy and Metaphilosophy of Information of Wu Kun mentioned above. In addition to Information Philosophy, Information Science, Technology, Management, the Information Economy and so on can be viewed as together forming a large scale information ecosystem.

According to Burgin [6, 7], a major figure in the theory of information, the Information Ecology of Zhong is part of general ecology standpoint which studies information ecosystems either as physical, mental or structural ecosystems with natural, technological and social components. It is a holistic approach to the existence and functioning of information processing systems, as well as for better understanding of information processes in all spheres of reality. If ecology of plants studies structures and processes in systems of plants, information ecology studies structures and processes in organizations of information processing systems and formations.

Zhong's position has been echoed by his colleague Ouyang Kang [26]. From a perspective of science, information ecology involves the expansion and application of the principles and methods of ecology in information science research. From the perspective of practice, the real problems that information ecology encounters are the irrationality and injustice in information processing and information sharing, information overload, information pollution, information harassment and information crime. He calls for fairness but also cognitive justice in operating in an informational world. In addition to systematicity and objectivity, standard principles in all research, the methodological principles of subjectivity and justice should be developed to insure respect for all private and public actors.

19.7.3 Two Forms of Information Ecology

In the work of Capurro and Zhong [54], two forms of Informational Ecology can be discerned, characterized by their primary focus on the epistemological or ontological domain. Two different non-standard logics apply, that of the former, focused on Information Studies, is the recently introduced Transconsistent Logic of Zhang Jincheng [40]. This is a new logical approach, within the conception of Universal Logic of He Huacan (2005), who has described it as "a revolution in mathematics, logic and foundations of computer science". It which points to the errors and contradictions inherent in standard linguistic logics but remains a logic of propositions or their mathematical equivalents. In this work, The logic of the ontological domain is the non-linguistic, non-truth-functional logic of real processes, Logic in Reality. Both logical-philosophical approaches provide conceptual bridges to related developments in standard ecology, ethics and social action.

19.8 The Philosophy of Sustainability and Ecology

The concepts of ecology and sustainability help to naturalize the transdisciplinary attitudes of rigor, tolerance, openness and authenticity. As one moves into a field such as ecology in which ethics has an important role, the situation becomes similar to that described by the twentieth—Century Swiss philosopher and mathematician Ferdinand Gonseth [33] as requiring a new dialectical synthesis. The subject—we—construct a reality in which intuition and experience, theory and experiment have a new and dynamic, changing significance. Such a reality should have some 'logic' to it, and I have suggested Logic in Reality as a framework for discussing these complex links between science and philosophy.

As Gaspar has pointed out [18], most environmental education and behavior change projects and models regarding environmental behavior are still biased by a positivity fallacy, i.e., the belief that as long as people have the right attitudes, intentions, skills, information, etc., the right pro-ecological behavior should follow. However, the social sciences literature shows that there is an inconsistency between attitudes and behaviors in this regard and that the difficulty in changing behaviors is being underestimated. One reason for this might be that the role of dispositional and situational characteristics in inducing an inhibitive/constraining effect over pro-ecological behavior is being underestimated. Consequently, the barriers and constraints are not being analyzed and dealt with.

19.8.1 Strategies for Avoiding Tragedies of the Commons

As a way of looking at the dynamics of social, economic and political processes, the major work of Castells referred to above [8] has proven extremely prescient. He sees society as a complex system of networks that are a consequence of the new information and communication technologies. The LIR logical approach, applied to an analysis of the properties of the networks and their nodes, as well as to the segments of the society that are disfavored or excluded completely, complements Castells' informal reference to a "logic" of the network society and its dynamics. In his Chapter on the environmental movement, he distinguishes between environmentalism as ecology in practice, and ecology as environmentalism in theory and writes:

> By environmentalism I refer to all forms of collective behavior that, in their discourse and their practice, aim at correcting destructive forms of relationship between human action and its natural environment, in opposition to the prevailing structural and institutional *logic* (emphasis ours).

However, if environmentalism involves individual or collective moral responsibility to the environment and to other human beings, the question of the origin of this responsibility cannot be avoided. The operation of the principles of dynamic opposition in LIR, namely, that any complex process is accompanied by its more or less potentialized contradiction or "counteraction", is exemplified in the tragedy of the commons. Socially valid institutions put into place to solve instances of the tragedy of the commons often cause tragedy of another sort. In LIR terms, a tendency toward rigidity in an institution, viewed as a process, not a static entity, is a tendency toward an identity of non-contradiction. Rigid institutions, which Daniels called "tragic" [10], have an excess of stability (*vs.* what would be equally bad, absence of sufficient stability or anti-stability) involving both benefits and costs. My argument is that if this tendency is *anticipated,* programs to build responsiveness as an emergent property can be planned for. Delays that might otherwise significantly increase the costs of reform might accordingly be avoided.

19.8.2 Sustainability and Transdisciplinarity

Swilling and Annecke [36] have specifically proposed sustainability as an alternative to an on-going tragedy of the commons. They propose a "just transition" that reconciles the sustainable use of natural resources with a pervasive commitment to sufficiency, the (hypothetical?) situation where over-consumers are satisfied with less so that under-consumers can secure a minimum. Reference is made to the "perplexing logics" of the different literatures on the subject and the "compelling logics" of climate science. We are afraid, however, that the 'implacable logic of the market' may still be the one dominates. One can at least hope that new communities in South Africa and elsewhere may try to instantiate sustainability on a small scale.

John van Breda has placed sustainability and ecology in a transdisciplinary, hermeneutic framework in a major monograph, *Exploring Sustainability Science: A Southern African Perspective* [37]. Van Breda's key point in the context of this study is that we are dealing with complex social-ecological systems problems that simply cannot be approached from *within* the confines of the single disciplines or sub-disciplines only. Given the fragmentation of both the world and our knowledge-systems, van Breda's central argument is that the overcoming of disciplinary boundaries is a necessary prerequisite for finding sustainable solutions to complex planetary problems. However, we cannot merely assume that a trans-disciplinary dialogue will emerge when the different disciplines come together to look for solutions. What is essential is to see *how* such a trans-disciplinary dialogue between the different disciplines can be achieved. "It is then in this sense of a dynamic interplay between theory and praxis that the hermeneutics of a trans-disciplinary dialogue between the different disciplines should be seen as a *sine qua non* for the conceptualization of a 'sustainability science'. Without understanding *how* the different disciplines are to communicate and develop a shared understanding of the complex world, it remains highly improbable, if not impossible, to imagine the meaning and establishment of a 'sustainability science'. This means that overcoming the disciplinary divide is not only of theoretical interest, but is a necessary prerequisite in our search for sustainable solutions to today's problems. As required by van Breda, transdisciplinarity in this interpretation provides is a new way of understanding the non-separability of subject—object relation as a dynamic unity, an emergent property of the recursive, dynamic interplay between the subject and object. Sustainability thus becomes the new field in which the age-old philosophical questions receive new functional answers directly relevant to the ethical functioning of the society.

19.9 The Philosophies of Society and Their Study. The Sociotype

The lines of study to which this book refers have so far not included the formal analysis of the categories, entities and building blocks of society nor of the philosophies which are applicable to them. The philosophy of sociology names perhaps the most general applicable discipline since it spans the domain of social knowledge from theories of society to the closely related fields of anthropology, economics, political theory to military strategy and history. The classic Chinese *Art of War*, mentions or implies all of these subjects, and so the practical reason for not discussing the social sciences is clear; it would require another book. More to the point, it is generally accepted that the social sciences are sciences of 'Man', and thus overlap broadly with all knowledge that deals with human psychology and behavior as well as social structures and practices, culture, modernity and politics. In these areas, physical science and natural philosophy have less of a direct role to play, while the notion of complexity of situations and actors is taken for granted.

For the understanding of social processes, however, we would like to mention one concept, that of the sociotype introduced recently by Marijuán et al. [31, 32]. A social system is constituted by reflexive, i.e., communicating, interacting individuals in the sense of Chap. 18. In the social 'organism', the phenotype and genotype correspond to the biological realm, while the sociotype appears as the constituent imposed by the social structure. The sociotype represents an additional component to genotype and phenotype, forming the triad: genotype—phenotype—sociotype. A sociotype is an ontolon, a complex dynamic but relatively stable entity identifiable in the fabric, of the social world in which each individual life is developed [31]. It includes both the bonding structures of the social environment and the dynamics of social interactions to which the individuals of our species are adapted by evolution [32]. It can be considered as a pattern in the sense of Ladyman and Ross of the subjects forming social entity through their recursive mutual processing of information from the environment and themselves. The western and eastern types in the model of Lefebvre (see Sect. 18.2.2) are the most polarized formal representations of the sociotype for which a framework of Boolean lattices is sufficient. The real dynamics of sociotype formation and evolution is much richer, and, according to our approach, would require a *non*-Boolean logic for their description. The sociotype is in other words the kind of complex system comprising functionalities of different levels of complex and different logics and modes of evolution at the same time as envisioned by Minati (cf. Chap. 16).

19.10 Conclusion and Outlook. Sustainabilization and Regeneration

New insights from the science, philosophy and logic of information can thus be applied to the theoretical foundations and part of the practice of ecology and sustainability. We proposes that LIR has a role here as a logic of ecology—an *eco*-logic—that also defines a basis for moral responsibility. Both Informational Ecology and the Ecology of Information Studies, which emphasize the phenomena and the studies of these phenomena respectively, follow this logic. Sustainabilization was defined by Hofkirchner [24, 25] as "the process of putting society in a position to avoid anthropogenic breakdown and safeguard a stable path of development below the threshold of endangering the maintenance of society." As noted in Chap. 13, Wu Kun sees the philosophy of information as a the cause of a revolution in philosophy [38]; Hofkirchner sees the information revolution as contributing to the sustainabilism—the result of the process of sustainabilization—needed for the advent of a Global Sustainable Information Society (GSIS), mentioned in Sect. 3.1 above. The goals of Hofkirchner of a GSIS, of ecological and sustainable practices in society receive further justification and support from grounding in LIR as well as science. The same can said for the Ecological Civilization of Gare [16].

By seeing the logical relationships—between the human and non-human domains of nature—better strategies for influencing the society to move in an environmentally responsible manner may be possible. We realize that we are adding yet another domain of discourse to the vast array in place. However, our hope is that recognition of the validity of this logic grounded in physics, and exemplified in information science and systems science may suggest new strategies for to the transdisciplinary problems of society, facilitating the application of ecological and ethical principles. It is perhaps fitting to close the body of this book with this brief discussion of *non-change*, a domain in which stability is the greater good. Sustainability and the avoidance of further damaging change have become ethical and scientific imperatives. The term of regeneration, returning to a preferred state *before change*, is now justly receiving greater attention than simple sustainability. On this positive note, we will now go to the concluding Chapter of this book.

References

1. Bhaskar, R.: Meta-Reality. The Philosophy of meta-Reality, vol. 1, Creativity Love and Freedom. Sage Publications, London UK (2002)
2. Bookchin, M.: The Philosophy of Social Ecology. Essays on Dialectical Naturalism. Black Rose Books, Montréal/New York/London (1995)
3. Brenner, J. E.: Stéphane Lupasco et la rejonction métalogique. In: A *la Confluence de Deux Cultures—Lupasco Aujourd'hui*, Paris, France, 24 Mar 2010, ed. B. Nicolescu (2010)
4. Brenner, J.E.: Stéphane Lupasco: Vers la fin d'une noble marginalité. Sociétés **1**, 71–75 (2013)
5. Brenner, J.E.: The philosophy of ecology and sustainability: new logical and informational dimensions. Philosophies **3**, 16 (2018)
6. Burgin, M.: Theory of Information. Fundamentality, Diversity and Unification. World Scientific, Singapore (2010)
7. Burgin, M.: Principles of General Ecology. Proceedings **1**, 148 (2017)
8. Castells, M.: The Information Age: Economy, Society and Culture. Volume II the Power of Identity; Blackwell Publishing: Malden/Oxford, UK (2009)
9. Capurro, R.: Towards an Information Ecology. In: Worrell, I. (ed.) Information Quality. Definitions and Dimensions, London UK, Taylor Graham, pp. 120–139 (1990)
10. Daniels, B.: Emerging commons and tragic institutions. Environmental Law **37**, 515 (2007)
11. Flahault, F.: Où est passé le bien commun? Mille et une nuits (Fayard), Paris (2011)
12. Fuchs, C.: Dialectical Philosophy and Self-Organisation. In: Arshinov, V., Fuchs, C. (eds.) Causality, Emergence, Self-Organisation. Moscow, NIA- Priroda, 195–244 (2003).
13. Fuchs, C.: The antagonistic self-organization of modern society. Studies in Political Economy **73**, 183–209 (2004)
14. Fuchs, C.: The Dialectic of the nature-society system. Triple C **4**(1), 1–39 (2005)
15. Fuchs, C.: Towards a global sustainable information society (GSIS)? Triple C **4**(1), 40–99 (2005)
16. Gare, A.: From Kant to Schelling to process metaphysics: on the way to ecological civilization. Cosmos and History **7**, 6–69 (2011)
17. Gare, A.: Towards an ecological civilization: the science, ethics, and politics of eco-poiesis. Process Stud. **39**, 5–38 (2009)
18. Gaspar, R.: Understanding the reasons for behavioral failure: a process view of psychosocial barriers and constraints to pro-ecological behavior. Sustainability **5**, 296 (2013)
19. Hofkirchner, W.: Self-organisation as the mechanism of development and evolution in social systems. In: Archer, M.S. (ed.) Social morphogenesis, pp. 125–143. Springer, Dordrecht (2013)

20. Hofkirchner, W.: On the validity of describing 'morphogenic society' as a system and justifiability of thinking about it as a social formation, Chapter 6. In: Archer, M.S. (ed.) Late Modernity, pp. 119–141. Trajectories toward Morphogenic Society, Dordrecht, Springer (2014a)
21. Hofkirchner, W.: Emergent Information. A Unified Theory of Information Framework: World Scientific Series in Information Studies, vol. 3. World Scientific, Singapore (2013)
22. Hofkirchner, W.: How to achieve a unified theory of Information. triple-C: **7**, 357–358 (2009)
23. Hofkirchner, W.: The commons from a critical social systems perspective. Recerca, Revista De Pensamenti I Analisi **14**, 73–91 (2014b)
24. Hofkirchner, W.: Mechanisms at Work in Information Society. Chapter 5. In: Archer, M. (ed.) Late Modernity. Trajectories toward Morphogenic Society, Dordrecht, Springer, pp. 95–112 (2015)
25. Hofkirchner, W.: Potentials and risks for creating a global sustainable information society. In: Fuchs, C., Sandoval, M. (eds.) Critique, social media and the information society, London/New York, Routledge (2015)
26. Kang, O.: Information ecology and cognitive justice: core value and methodological principles of information ecology. Proceedings 1, 148–152 (2017)
27. Lupasco, S.: Le principe d'antagonisme et la logique de l'énergie. Paris: Editions du Rocher. (Originally published in Paris: Éditions Hermann, 1951) (1987)
28. Maffesoli, M.: Matrimonium. Petit traité d'écosophie. CNRS Éditions, Paris (2010)
29. Maffesoli, M.: Être postmoderne. Éditions du Cerf, Paris (2018)
30. Marcuse, H.: Reason and revolution. Hegel and the rise of social theory, Humanity Books, New York, NY (2020)
31. Marijuán, P.C., Del Moral, R., Ji, S., Gil Lacruz, M., Gómez-Quintero, J.D., Navarro, J.: Fundamental, quantitative traits of the "sociotype". BioSystems **180**, 79–87 (2019)
32. Marijuán, P.C., Navarro, J.: Sociotype and cultural evolution: The acceleration of cultural change alongside industrial revolutions. BioSystems 19, 104170 (2020)
33. Pouget, P.: Le rôle de la contradiction dans l'œuvre de Ferdinand Gonseth. Bulletin De l'Association F. Gonseth **166**, 15–28 (2017)
34. Public Philosophy.: *Wikipedia.* Retrieved 19 December 2019 from https://en.wikipedia.org/index.php?title=Public-philosophy&oldid=931473736(2019)
35. Rorty, R.: Contingency, Irony Solidarity. Cambridge University Press, Cambridge, U.K (1989)
36. Swilling, M., Annecke, E.: Just Transitions: Explorations of Sustainability in an Unfair World. UCT, Cape Town, South Africa (2012)
37. Van Breda, J.: Overcoming the Disciplinary Divide; Towards the Possibility of a Transdisciplinary Hermeneutics. In: Burns, M., Weaver, A. (eds.) Exploring Sustainability Science—A Southern Africa Perspective, pp. 1–46. SUN MeDIA, Stellenbosch, South Africa (2008)
38. Wu, K., Brenner, J. E.: Philosophy of information: revolution in philosophy. towards an informational metaphilosophy of science. Philosophies, **2**(4), 20 (2017)
39. Wu, K.: Personal communication to J.E. Brenner 20.09.2019 (2019)
40. Zhang, J.: Transconsistent Logic Principle (1). Beijing Book Publishing Company, Beijing (2017)
41. Zhong, Y.X.: Information Ecology and Information Studies. Proceedings **1**, 200 (2017)

Chapter 20
Summary and Conclusions: Is This a New Book of Changes?

20.1 The Path Towards a New Synthesis

We have described a new kind of logic, logic in reality (LIR) and a new dialectics. They offer a novel perspective on philosophy and its major fields: metaphysics, phenomenology, metaphilosophy and its recursive relation with philosophy. LIR naturalizes this relation, eliminating any implied circularity, since the total independence of premises and conclusion of standard logics is not required. Logic in Reality is the thread running through both knowledge and existence. Its explanatory power is not deep for specific scientific questions, but it is very broad in regard to philosophical questions that about existence in a rigorous scientific perspective. LIR offers a new 'Popperian' kind of criterion for judging not an absolute truth or falsity of a given propositional theory, a way of estimating the probability of the correctness of a theory. That criterion, as stated already in [1], is the extent to which a theory or, more broadly, a philosophical position, takes into account its opposite or contradiction.

We claim that this is a necessary part of a hermeneutic process of decision, a reflection of the basic asymmetry that accounts for existence and life in the first place. Above all, with inevitable exceptions, our approach insures the valuing of the 'other', a quasi-scientific basis for morality and the common good. In this, we have not discussed ethics as a 'field' with its literature and extremely complex dynamics since antiquity. What we have done is perhaps more useful: we have provided a framework for grounding morality and ethical behavior in the same physical theory as the rest of science and philosophy, and of the change they undergo.

Starting from the fundamental properties of matter-energy, LIR permits a new approach to critical issues in both philosophy and science. LIR incorporates and shows the underlying unity of concepts from fundamental physics and mathematics, in particular Gödel's theorems relating consistency and completeness. Its dynamic description of inter-theoretic relations is relevant to the understanding of metaphilosophy, in that it accepts the co-existence of conflicting or contradictory theories. The LIR philosophical view of nature is a form of scientific realism in which relations are as fundamental as entities, excluding the abstract entities of pure classical

J. E. Brenner and A. U. Igamberdiev, *Philosophy in Reality*, Studies in Applied Philosophy, Epistemology and Rational Ethics 60, https://doi.org/10.1007/978-3-030-62757-7_20

propositional logic. The most fundamental metalogical principle of LIR is that of opposition or antagonism, without which, in this view, nothing could exist. This is, therefore, at the same time the most fundamental metaphysical principle of LIR. LIR is therefore a system embodying aspects of logic, philosophy and metaphysics. It is philosophy and metaphilosophy *en soi*. We call attention to the metaphilosophy of information of Wu Kun and its implications for the convergence today of the science and philosophy of information and accordingly of science and philosophy in general.

We have thus established new relations between 'science' and philosophy, where science is not addressed as some featureless abstraction. Science is a set of forms of knowledge, linked not only by their methods but by their logic, and which in turn is linked to philosophy. We provide examples from the scientific disciplines of mathematics, physics, biology, information and communication and social systems.

In our introductory Chapter, we promised that the resultant of our analyses of the scientific and logical aspects of philosophy would lead to a new synthesis, a new Natural Philosophy. We have asked again the question of Deleuze and Guattari "What is philosophy?" [3]. The synthesis we have proposed is not dependent on any absolute criteria of truth or completeness, but incorporates and/or refers as far as possible to contradictory or opposing points of view. The second is that, as a metaprinciple, we facilitate the possibility of talking about principles at all, which some have contested (cf. 20.4). One is left with a criterion of utility like that of J. S. Mill which has been considered 'weak' and non-scientific. But this is exactly our thesis, even though by our own criteria we cannot prove or justify it, but support it on a methodological basis. This is that that dialectics and the Lupascian logic give the least abstract possible picture of the world. Our dialectical approach leads to recovery of both dialectics and semiotics from reductionist interpretations and to their reunification in a new synthetic paradigm centered on meaning and its communication. Our concept unites science, logic and philosophy in a common meta-thesis and provides the real contours of a basic understanding of nature and civilization. Formalization of any dialectical logic cannot be complete and that is why it tends, like other 'diversities', to be ignored in a scientific discourse which values, or rather overvalues, completeness and consistency. Defining a logic of real processes that establishes the ontological relations between meaning, communication and language will always remain a fundamental task; it forms the background of any description of nature that can accompany the new functional convergence of science and philosophy in progress. We consider the development of this 'dialectical realism' as a basis of the ethical development of knowledge for the common good.

20.1.1 The Philosophy of Science

We have not discussed in detail the Philosophy of Science as a discipline, with a substantial literature of its own. Rather we have been concerned primarily with a philosophy, a Philosophy in Reality, which is a Philosophy *of* the Reality that science is about. Our philosophy has, consequently the form of a metaphilosophy of science as discussed in Chap. 10, and in reference to the work of Wu Kun. Our intent throughout has been to focus on the non-separability and interactions of such concepts, including that of the 'Science of Philosophy'.

20.1.2 Science and Philosophy in Husserl. A Discontinuity

At several points in this book, we have referred to the philosophy of Edmund Husserl as being in opposition to that of Lupasco. We, like many others, have considered that the conceptual 'bracketing' by Husserl of the ontology of physical processes and the emphasis on appearance to human cognition have made his phenomenology unscientific. This was the basis for the critique by Sparrow, and, to a certain extent, that in some of Brenner's prior work. Wu's criticism of the phenomenological method of Husserl and others is that, due to the classical phenomenological suspension of natural objects and the human body, its description of the mechanism of human understanding is one-sided and incomplete.

However, to us as 'students' of the dialectical logic of Lupasco, the philosophy of Husserl represents a discontinuity for which another interpretation is possible. At the beginning of his major paper, "Philosophy as Rigorous Science" (1910), Husserl [5] implies that philosophy *should* have such a science as a goal, but that all the prior claims that have been made are not justified. There is of course a relation that philosophy as science should have to the natural and human sciences, but for Husserl philosophy is not an imperfect science; it is not a science at all. The emergence of a scientifically rigorous philosophy would be a "revolution in philosophy". A century later, Wu Kun referred to the impact of the science and philosophy of information as having a revolution in philosophy as a consequence [8]. For Husserl then, philosophy in the sense of rigorous science was a "pre-science", as in the new systems theory of Minati and Pessa [6].

In developing our position, because it is as important as the literal theses, we wish to call attention explicitly in ours to Husserl's creativity and humanity. They suggest to us a potentialized naturalization that we have tried to bring out. A parallel to Lupasco is also to be found in Husserl's negative view of Hegel's romanticism and its morphing into an historical "worldview philosophy", lacking adequate intellectual credentials. Lupasco showed that Hegel's 'drive' to idealist 'ultimate solutions' neglected the realties of a complex world.

We conclude that Husserl's main target was not empirical science per se, but its misuse by philosophers, as well as the lack of philosophical vision of scientists. Husserl believed that phenomenological eidetic inquiry can yield, and allow to be grasped, essences of experience—perceptions—without constant reference to the existence in reality of the physical world. To return from the noumenal world to the reality of phenomena, it is possible, necessary and sufficient to naturalize knowledge by accepting the combined operation of the actuality of a phenomenon and its potentiality. Existence and essence are not separated, and nature and Being are reconciled, which would mean than one can 'grasp/perceive' the potentiality as well as the actuality of a 'physical phenomenon'. Husserl's view on natural science is in fact not negative. However, his notion of essence without reference to existence means, to repeat, that it is taken without reference to actual existence. If we also consider potential existence, the outcome will be quite different but the question is how to properly grasp this through a comprehensible concept. This book has thus proposed Logic in Reality (LIR) and the dialectics and science related to it, in which we find its principles expressed, as a candidate for such a concept. We thus claim to have achieved, in some small but new measure, the goal that Husserl set himself of establishing what is specifically philosophical about the natural and human sciences. In Husserl's expression, LIR permits a rigorous "reminting of the conjectures of profundity" into new rigorous, rational formulations, whose targets are the sciences themselves, as well as knowledge in general. In this book, we have presented a text that, in its relatively detailed treatment of issues in some sciences, not only differentiates it from standard philosophy but also from the "worldview" philosophy attacked by Husserl for its having eliminated all that was correct in the scientific approach.

We thus can endorse, in part, Husserl's synthesis and its reference to direct intuition. As we have seen, Lupasco gave to intuition its deserved ontological, that is scientific, value without neglecting its obvious limitations. Husserl wrote: "Yet it lies precisely in the essence of philosophy, insofar as it returns to the ultimate origins, that its scientific work moves in spheres of direct intuition, and it is the greatest step our age has to make to see that with philosophical intuition in the right sense, the *phenomenological seizing upon essences*, an endless field of opens up and a science that, without any indirectly symbolizing and mathematizing methods, without the apparatus of inferences and proofs, nevertheless obtains an abundance of the most rigorous cognitions, which are decisive for *all* further philosophy". And, we would add, for all knowledge. We take this statement as the one of the most serious possible justifications for our work.

20.2 New Directions in Communication and Information Theory. Meaning

We have called attention to a convergence of science and philosophy, proposed most explicitly by the Chinese philosopher of information Wu Kun. This convergence is based on a new recognition/cognition of the overlooked ubiquity of information in cognitive processes and as a process in and of itself. The difficulty of defining information 'exactly' and its obvious dialectical properties, *e.g.*, of being partly present (actual) and absent (potential). These characterize information as a process following the rules of Logic in Reality in its evolution.

Based on our new conception of the energetic properties of information, we have revisited recent conceptions of communications such as those of Capurro, Hofkirchner and Leydesdorff. We have found a rational (logical) basis for distinguishing between the different forms of meaning present and carried in and by communications as another instance of the dialectics of reality. We have compared the notion of a meaning inherent in reality with *its* opposite, non-meaning or the absurd, and used the opposition to clarify the nature of both.

Our approach supports the extension to the content of communications theory proposed by Capurro which he and his associates designate as *Angeletics* of Messaging Theory [2]. Angeletics captures the active intentionality involved in generating a message, transmitting, receiving and processing. It thus naturalizes and extends Peircean views of communication based on his abstract categorizations of natural phenomena to the phenomena and agents in reality. These theories, including ours, are ways of talking about real, ontological changes with moral and social implications.

20.3 The Philosophization of Systems Theory

Appropriate concepts of systems are as fundamental and necessary for philosophy as they are for science and knowledge and their meaning. We undertook a program of rethinking systems science and systems theory which begins with a critique of some their assumptions about the structure and evolutionary dynamics of real systems, that is, their changes. We have accomplished two objectives in this direction: (1) we have established a grounding for the existence and properties of systems which is the same as that of our other groundings of logic and dialectics and (2) we have positioned our LIR approach with relation to the post-Bertalanffy systemics under current development by Minati and Pessa [6]. We have also included the authoritative formulation of the philosophy of complex systems by Hooker and his associates [4]. We suggested the 'recovery' of a philosophy of complex systems may be a valid topic of research.

Minati's universe of complex systems is larger than that to which the principles of LIR apply. For example, LIR and Philosophy in Reality apply to complex systems whose complexity does not depend on a multiplicity of elements. The 'three-body' problem of a system of husband, wife and child is well-known to be non-computable, and PIR and LIR are not mathematically capable of changing this fact. The Minati view of society describes it as complex system in which sub-systems of different forms and having different dynamics co-exist. Thus, the properties and behavior of some complicated systems like crowd and flocks can be understood using "old-fashioned" notions of systems, while those involving fewer members but more complex cognitive relations require something like our logic and dialectics. It is our hope that by seeing the relations in systems and *their* dynamics as the expression of underlying conflictual forces that are *logical* in nature and logical *in* nature, further tools, which complement existing systems approaches, can be designed, and some are suggested for further work.

20.4 Principles and the Common Good

This book has been about the logical and philosophical principles underlying science, here, the study of chemical, biological, cognitive and social systems. Its potential utility for contributing to the common good has been in the background of the authors' minds throughout. We thus consider it as a statement *of* principle: we do not believe that any philosophic 'work' of the kind we are engaged in can be justified, as can pure science, if it is 'pure' philosophy, that is, non-natural philosophy. The use of human and natural resources can be justified for science, as such and as a part of natural philosophy, *because* is directed at increases in understanding of the real world. An example at the other extreme is the discussion, following Kripke and others, of possible worlds that have no existence other than that of fictional objects.

There are two other general operative principles that we can state as a conclusion of this study: any synthesis that may emerge in further work will not be dependent on any absolute criteria of truth or completeness, but will seek to incorporate or in any event refer as far as possible to contradictory or opposing points of view. The second is that, as a metaprinciple, one can talk about principles at all, which some have contested. We are left with a criterion of utility like that of J. S. Mill which has been considered 'weak' and non-scientific. But this is exactly our thesis, even though by our own criteria we cannot prove or justify it, but support it on a methodological basis, as being closest to reality, and the best expression of a Philosophy in Reality.

In thinking and discussion of philosophical issues, we have been able to establish a link to Russian philosophers and summaries of some of their work that bears a close relation to ours but is not readily accessible. It is thus not only dialectics in abstract sense that we have recovered but valuable expressions of it in human thought. These have tended to be lost in a mass of 'negative thinking' about Russia, but they add further conceptual dimensions and further universality to our approach.

20.5 Is This Book a 'New Book of Changes'?

All of the considerations in this book are directed toward the development of new and better tools for the understanding of man and the world for the common good. This implies the understanding, adaptation to and management of change that is no different in form from that undertaken by the Chinese in the second and first centuries BCE. We do not aspire to a listing of this book by UNESCO. There are, nevertheless, several tasks that are similar to those recorded in the *I Ching* that we believe we have accomplished. This book has discussed the phenomenon of change from many different points of view, epistemological and ontological. We have sought in each phenomenon the components of the change in progress as constituting the essence of it as philosophy, its philosophical essences—ontolons. We should emphasize again that by their nature, ontolons are not ends-in-themselves of thought; they are cognitive entities, constituted by energy in the same way that information is, and are both meaning and carriers of meaning. They are tools for thought, better thinking, to be used by people, in and by the society, to try to make it more just and sustainable.

Among the most significant ontolons we can imagine are stances or attitudes, for example the Informational Stance proposed by Wu Kun and Brenner (2013) and the Transdisciplinary Attitude of Basarab Nicolescu [7]. Such process objects *are* change, distant cousins, so to speak, of the phanerons of Peirce, but whose evolution and changing characteristics are in stark contrast to the inert signs with which he constructs them. As authors of this book, we hope we have demonstrated the value, for debate if nothing else, of the some new considerations of change in logical and dialectical terms. These considerations constitute the core thesis of a Philosophy in and of Reality.

The question can still be asked whether our thesis is sufficient for this book to claim to be modern version of a classic and universal Chinese text which is part of the heritage of humanity. We therefore ask ourselves and the reader again whether this book can be considered as a new *Book of Changes*. Even a superficial reading of the *I Ching* and other Chinese classics is sufficient to show the presence throughout them of a concern for society, and we have directed Chap. 19 toward this objective. Direct application of some classical Chinese concepts referring only to a monarchical system can be translated to refer to modern democracies, and modern authoritarian governments and their leaders. These are (unfortunately) part of our everyday reality.

We have thus attempted to draw clear lines of correspondence between the philosophical logic of Stéphane Lupasco, in its elaboration by Brenner, and central ideas of Chinese philosophy. While they accept the transcendental nature of questions posed by existence, both doctrines are realist, philosophies focussed on immanence and real change. The essence of Chinese and Japanese thought in this sense are their 'down-to-earth' realism and practicality, to be developed in the interest of the common good.

The prevailing fragmentation of knowledge and of the structures of current society, much greater today than in antiquity, renders problematic any similar direct application by its leaders. It would be naïve and misleading to expect any direct impact on those structures. Nevertheless, for individuals, even the existence of our *New Book of Changes* could be, like that of the Chinese one, a source of hope for a better, more rational world. The society is an organism, and even if is not conscious in the sense of an individual human being or animal. It has a distributed systemic sensitivity that in principle is open to influence, as Minati suggests. In some favorable circumstances, a knowledge structure like this book could function as an operator, changing the way in which even a small number of people reason. This would be for us ample justification for *its* existence.

In this book, we have argued in favor of some novel concepts, but our intention has not been to persuade or convince, let alone prove. Rather, we have to tried open up new areas and ways of discussion and debate. In this, we have been guided by another Fragment of Heraclitus, N° 93, which makes a statement that we believe Chuang Tzü would readily endorse: "The Master, whose oracle is at Delphi, neither affirms, nor denies, but suggests..."

References

1. Brenner, J.E.: Logic in Reality. Springer, Dordrecht (2008)
2. Capurro, R., Holgate, J. (eds.): Messages and Messengers. Angeletics as an Approach to the Phenomenon of Communication, ICIE Series. Wilhelm Fink, Munich (2011)
3. Deleuze, G., Guattari, F.: Qu'est-ce que la Philosophie? Les Éditions de Minuit, Paris (1991)
4. Hooker, C. (ed.).: Philosophy of complex systems, vol.10, Handbook of the Philosophy of Science. Elsevier, Oxford/Amsterdam (2011)
5. Husserl, E.: Philosophy as Rigorous Science, M. Brainerd (Trans.). *The New Yearbook for Phenomenology and Phenomenological Philosophy II*: 249–295 (2002). (Originally published in *Logos. Intenationale Zeitschrift fur Philosophie der Kultur 1* (1910–11)
6. Minati, G., Pessa, E.: From Collective Beings to Quasi-Systems. Springer Science+Business Media, New York NY (2018)
7. Nicolescu, B.: Manifesto of Transdisciplinarity. State University of New York Press, Albany NY (2002)
8. Wu, K., Brenner, J.E.: The informational stance: philosophy and logic. Part I: The basic theories. Logic and Logical Philosophy **20**, 453–493 (2013)

Index

A

J. E. Brenner and A. U. Igamberdiev, *Philosophy in Reality*, Studies in Applied Philosophy, Epistemology and Rational Ethics 60,
https://doi.org/10.1007/978-3-030-62757-7

N

The manufacturer's authorised representative in the EU is Springer Nature Customer Service Centre GmbH, Europaplatz 3, 69115 Heidelberg, Germany. If you have any concerns regarding our products, please contact ProductSafety@springernature.com

Printed and bound by CPI Group (UK) Ltd, Croydon, CR0 4YY
30/09/2026
02225453-0020